Hadron Spectroscopy and the Confinement Problem

NATO ASI Series

Advanced Science Institutes Series

A series presenting the results of activities sponsored by the NATO Science Committee, which aims at the dissemination of advanced scientific and technological knowledge, with a view to strengthening links between scientific communities.

The series is published by an international board of publishers in conjunction with the NATO Scientific Affairs Division

A	**Life Sciences**	Plenum Publishing Corporation
B	**Physics**	New York and London
C	**Mathematical and Physical Sciences**	Kluwer Academic Publishers Dordrecht, Boston, and London
D	**Behavioral and Social Sciences**	
E	**Applied Sciences**	
F	**Computer and Systems Sciences**	Springer-Verlag
G	**Ecological Sciences**	Berlin, Heidelberg, New York, London,
H	**Cell Biology**	Paris, Tokyo, Hong Kong, and Barcelona
I	**Global Environmental Change**	

PARTNERSHIP SUB-SERIES

1. **Disarmament Technologies**	Kluwer Academic Publishers
2. **Environment**	Springer-Verlag
3. **High Technology**	Kluwer Academic Publishers
4. **Science and Technology Policy**	Kluwer Academic Publishers
5. **Computer Networking**	Kluwer Academic Publishers

The Partnership Sub-Series incorporates activities undertaken in collaboration with NATO's Cooperation Partners, the countries of the CIS and Central and Eastern Europe, in Priority Areas of concern to those countries.

Recent Volumes in this Series:

Volume 352—History of Original Ideas and Basic Discoveries in Particle Physics
edited by Harvey B. Newman and Thomas Ypsilantis

Volume 353—Hadron Spectroscopy and the Confinement Problem
edited by D. V. Bugg

Volume 354—Physics and Chemistry of Low-Dimensional Inorganic Conductors
edited by Claire Schlenker, Jean Dumas, Martha Greenblatt, and Sander van Smaalen

Volume 355—Stability of Materials
edited by A. Gonis, P. E. A. Turchi, and Josef Kudrnovsky

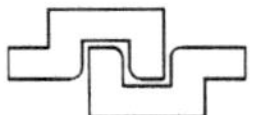

Series B: Physics

Hadron Spectroscopy and the Confinement Problem

Edited by

D. V. Bugg

Queen Mary and Westfield College
London, England

Plenum Press
New York and London
Published in cooperation with NATO Scientific Affairs Division

Proceedings of a NATO Advanced Study Institute on
Hadron Spectroscopy and the Confinement Problem,
held June 27–July 8, 1995,
in London, England, and Swansea, Wales

NATO-PCO-DATA BASE

The electronic index to the NATO ASI Series provides full bibliographical references (with keywords and/or abstracts) to about 50,000 contributions from international scientists published in all sections of the NATO ASI Series. Access to the NATO-PCO-DATA BASE is possible in two ways:

—via online FILE 128 (NATO-PCO-DATA BASE) hosted by ESRIN, Via Galileo Galilei, I-00044 Frascati, Italy

—via CD-ROM "NATO Science and Technology Disk" with user-friendly retrieval software in English, French, and German (©WTV GmbH and DATAWARE Technologies, Inc. 1989). The CD-ROM contains the AGARD Aerospace Database.

The CD-ROM can be ordered through any member of the Board of Publishers or through NATO-PCO, Overijse, Belgium.

Library of Congress Cataloging-in-Publication Data

Hadron spectroscopy and the confinement problem / edited by D.V. Bugg.
p. cm. -- (NATO ASI series. B, Physics ; v. 353)
"Proceedings of a NATO Advanced Study Institute on Hadron Spectroscopy and the Confinement Problem, held June 27-July 8,1995, in London, England, and Swansea, Wales"--T.p. verso.
"Published in cooperation with NATO Scientific Affairs Division."
Includes bibliographical references and index.
ISBN 0-306-45303-7
1. Hadron spectroscopy--Congresses. 2. Meson spectroscopy--Congresses. I. Bugg, D. V. (David Vernon) II. North Atlantic Treaty Organization. Scientific Affairs Division. III. NATO Advanced Study Institute on Hadron Spectroscopy and the Confinement Problem (1995 : London, England, and Swansea, Wales) IV. Series.
QC793.5.H327H35 1996
539.7'44--dc20 96-17036
CIP

ISBN 0-306-45303-7

A Division of Plenum Publishing Corporation
233 Spring Street, New York, N. Y. 10013

10 9 8 7 6 5 4 3 2 1

Printed in the United States of America

PREFACE

This meeting on Hadron Spectroscopy and the Confinement Problem took place from June 27 to July 8, 1995. The first four days were at Queen Mary and Westfield College, London and the last six at the University College of Swansea, Wales. The reason for the split sites was a question of accomodating the 54 students and 12 lecturers. However, it was enjoyed by all concerned, providing the opportunity to sample the nightlife of London one week and enjoy the delightful coastal area around Swansea the following week. The meeting immediately preceded the Hadron'95 conference in Manchester.

The contents of this volume run roughly parallel to the order of the lectures. Martin Faessler's brief and that of Madjid Boutemeur was to review experimental techniques as well as physics results. We have tried to review all the areas which are currently active experimentally. This means mostly Light Meson Spectroscopy. However, Simon Capstick gave us a welcome review of Baryon Spectroscopy and the coming CEBAF program.

This is an opportunity to thank NATO for their generous financial support, and also the two host institutions for excellent coordination and an enjoyable stay.

CONTENTS

LIGHT MESONS WITH OPEN AND HIDDEN STRANGENESS

Martin Faessler
Sektion Physik, Ludwig-Maximilians-Universität,
D-80799 Munich, Germany

Abstract Mesons are the most fundamental objects allowing a study of quark and gluon confinement and, more generally, the strong interaction at low energy. The spectroscopy of light mesons is a challenge: the glueballs predicted on the basis of Quantum Chromodynamics are expected amidst the light mesons. However, identifying light mesons and determining their quantum numbers is particularly difficult due to their extremely short lifetimes. The strangeness quantum number is easy to detect and indicates the presence of a strange quark. Thus the field of (openly) strange mesons is relatively well explored. But that of mesons with hidden strangeness, more directly relevant to the glueball search, still needs more experimental input, in particular from studies of final states with kaon-antikaon pairs.

Introduction

February 7, 1913 was the day when Niels Bohr first heard of the Balmer formula, known since 1885. By March 6, he had completed a paper containing its interpretation. This event marks the beginning of quantum mechanics [1]. The spectral lines of hydrogen, the simplest atom, bound by the electromagnetic force, played a leading role at several stages in the development of the present theory of the electromagnetic force, Quantum ElectroDynamics (QED).

Now, if you learn during this school, perhaps for the first time, about the spectra of mesons - the simplest systems bound by the strong force - it is not very likely (although a great hope of the teachers) that in a month from now you will come up with an elegant theory explaining all their properties.

Firstly, all experts are convinced that we already have the best theory of strong interactions, Quantum ChromoDynamics (QCD). The groundstates of mesons and baryons have played a key role for the first big step on the way to QCD, the conception of the quark model. But the proper domain of QCD nowadays concerns rare processes, where quarks are asymptotically free. "Asymptotically" implies that really free quarks have never been observed. They are confined to hadrons. And our best theorists have failed so far to apply their pet theory QCD to the simplest bound states. In this latter domain of "nonperturbative" QCD, the calculations become so complex that nobody has succeeded yet in deriving a truly quantitative description of meson spectra from QCD. Clearly, one may comfortably live with such a situation and look for other interesting objects like the Higgs particles. However, there were quite a few similar situations in history which may serve as a warning. Think of Ptolemaios who had a simple and elegant theory postulating just epicycles and epicenters as basic elements of the orbits of stars and planets. Yet, as more and more orbits were observed the application of that theory turned into an increasingly complex exercise. (And at the end the theory turned out to be a rather bad approximation of nature.)

The second reason which could prevent you from finding an elegant description of the spectra of mesons may be that the experimental input is too fragmentary. Ultimately, the strength of the interaction is to blame for this, as it is for the difficulty in applying QCD to nonperturbative processes. Since the interaction is so strong, the meson states, whenever their decay is induced by the strong force, are extremely short-lived. Bound by the strongest of all interactions, their constituents stay together for a shorter moment than any other bound state of nature. Their extremely short lifetime renders the identification particularly difficult. Detected through their decay, these bound states do not show up as narrow, separate lines on the scale of invariant energy or mass of the decay products but as broad diffuse enhancements of intensities, called resonances. (The linewidth Γ is inversely proportional to the lifetime, $\Gamma = \hbar/\tau$). In general, various resonances with different quantum numbers, produced in a given process, overlap in energy. This renders the determination of the quantum numbers cumbersome. Thus, even after several decades of research the measurements are still not as complete as one may wish.

The subject of the three lectures are light strange mesons, i.e. those containing a strange quark (s) or antiquark ($\bar{s}$), and mesons with hidden strangeness, i.e. those containing an ($\bar{s}s$)-pair. The first two chapters are rather fundamental. They were meant as an introduction for an audience of students or physicists with a general physics education but not yet experts in particle physics and hadron spectroscopy. The first chapter ends with a survey of known quantum states of light mesons which can be considered as a starting point for experts in the field. The second chapter outlines the experimental methods of meson spectroscopy, in particular detection of mesons with strange quarks. The last chapter is a review of some of the open problems in the special field of strange and hidden strange light mesons. The lectures, as they are given by an experimentalist, are oriented towards the experimental rather than the theoretical work in this field of meson spectroscopy,

Quantum states of light mesons

The basic constituents

The systematics of hadrons (mesons and baryons) first observed in the 50s and 60s, have led to the **quark model**. It describes hadrons as composite objects made of hypothetical constituents, the quarks. Those quarks had to carry the quantum numbers or fractions thereof which had been observed to be conserved in strong interactions, like charge, spin, baryon number (also conserved in other interactions), isospin, strangeness etc. Since these are lectures on strange mesons it seems appropriate to recall briefly the discovery of **strangeness**. Soon after the existence of hyperons (unstable particles heavier than a nucleon) and K mesons (particles with masses between the pion and nucleon mass) had been established through various decay modes, and the rate of production had been studied in a pion beam at the Brookhaven Cosmotron, it was noted that there was a problem. These particles were copiously produced by the strong interaction, at a rate of 10% of that for pions; but their decay was very slow (mediated

Table 1: Quarks, their quantum numbers and masses. Flavour quantum numbers: I_z = isospin-z-component, S = strangeness, C=charm, B = bottomness, T = topness, Q = charge in units of the proton charge, m = current quark mass, M = constituent quark mass. The corresponding antiquarks have equal masses but opposite signs for all other quantum numbers.

name	symbol	I_z	S	C	B	T	Q	m[MeV/c^2]	M [MeV/c^2
down	d	-1/2	0	0	0	0	-1/3	10 ± 5	300 ± 100
up	u	+1/2	0	0	0	0	+2/3	5 ± 3	300 ± 100
strange	s	0	-1	0	0	0	-1/3	200 ± 100	450 ± 100
charm	c	0	0	1	0	0	+2/3	1300 ± 300	≈ 1500
bottom	b	0	0	0	-1	0	-1/3	4300 ± 200	≈ 5000
top	t	0	0	0	0	1	+2/3	174000 ± 1700	

by the weak interaction). As an explanation, Pais(1952) and Nambu introduced the rule of "associated production", requiring hyperon and kaon to be produced in pairs by the strong interaction, e.g.

$$\pi^+ + n \rightarrow \Lambda + K^+.$$

Inverting this process, the strong decay of Λ and K would either be

$$\Lambda \rightarrow \pi^+ + n + K^- \text{ or } K \rightarrow \pi^+ + n + \bar{\Lambda}$$

Both decays are forbidden by energy conservation. An explanation of the rule was given by Gell-Mann(1953) and Nishijima(1955) by inventing a new additive quantum number, called strangeness (S) conserved in strong interactions. The Λ and K^- were assigned the same S, and K^+ was given the opposite S. Strangeness nonconservation in the decays rules out the action of the strong force and thus the long time scale of the decay was explained.

The **pointlike nature** and some other properties of the quark constituents were inferred from many subsequent experiments, which studied hard interactions involving hadrons. ("Hard" means that high momentum transfers are involved; to probe short distances, large momentum transfers are needed according to the Heisenberg uncertainty relations). As the best description of all these experimental data, the present theory of strong interactions, QCD, has emerged. Ironically, the "stable" hadrons and their lowest excitations which started this chain of progressive understanding, are the objects where the theory fails most strikingly to produce a quantitative description. Nonetheless, the qualitative description provided by QCD for the laws governing how hadrons are built from basic constituents is a good starting point and guideline which we will use in the following. Of course, doing so we start from the back end of research, i.e. from theory instead of the empirical basis which led to the theory. However, there is a definite advantage for this access to the subject. The quantum numbers and names of mesons are much more easily memorized if one already has an idea of how mesons are composed.

QCD is a gauge theory modeled after QED. It describes a world where the elementary fermions are the quarks and antiquarks, with spin quantum number 1/2, baryon number +1/3 for quarks and -1/3 for antiquarks , charge $\pm 1/3$ or $\pm 2/3$ of the proton charge and a quantum number called flavour which is characteristic of each quark species (Table 1). An additional property of the quarks, called colour charge, plays an analogous role for the strong interaction, as does electric charge for the electromagnetic interaction and mass for gravitation. Colour charge can have 3 different nonzero values. The sum of 3 different colours or of a colour and the corresponding anticolour yields zero, analogous to the additive mixing of three primary colours, hence the name "colour charge". The quanta of the strong force field, analogous to the photon (the quantum of the electromagnetic field), are the **gluons**. There are 8 = 3x3-1 gluons, because there are three colours and three anticolours. Gluons have spin quantum number 1; their electric charge, baryon number, flavour quantum numbers are all zero. But they carry colour charges, and as a consequence gluons interact directly not only with quarks and antiquarks but also with other gluons . (The photon can interact with another photon only via charged particles.)

Neither quarks nor gluons have ever been observed as free particles. Therefore, they are believed to be confined within hadrons. **Confinement** is built into the theory via the assumption that free particles have to be colourless or, more precisely, singlets with respect to colour-SU(3) transformations.

Hadrons

A hadron is, according to QCD, a very complex object, consisting of an indefinite number of quarks and/or gluons such that together they form a colour singlet. Two distinctions are necessary in talking about the constituents of hadrons. i) Between **"valence"** quarks and **"sea"** quarks and ii) between **"current"** quarks and **"constituent"** quarks. Valence quarks are those responsible for the quantum numbers like charge, baryon number, isospin and strangeness. Sea-quarks are present as pairs of a quark and the corresponding antiquark, so that they do not contribute to these quantum numbers. They may be considered as vacuum fluctuations. The strong field, whose quanta are the gluons, creates virtual quark-antiquark pairs as the electromagnetic field creates electron positron pairs (the latter effect is known from the Lamb shift in the hydrogen atom). The number of sea-quarks is indefinite. Current quark is synonymous with elementary quark. The usual description of a constituent quark is that it contains an elementary valence quark surrounded by a "polarization cloud" of sea-quark pairs and gluons. A similar distinction is made between elementary gluon and **constituent gluon**.

The known hadrons are divided into three classes: mesons, baryons and antibaryons, corresponding to their baryon number, which has been observed to take the values 0, +1 and -1, respectively. The baryon number equals the sum of the baryon numbers of the valence quarks contained in the hadron. The simplest hadrons are the **mesons**, which contain a valence quark q_i and a valence antiquark $\bar{q}_j$, and an indefinite number of sea-quark pairs and gluons. The indices i and j stand for the five quark flavours - the sixth, recently discovered top quark is too short lived (although elementary!) to build a hadron. Using the notion of constituent quarks, these mesons, called **"normal"** or **"quark-model" mesons** consist of a constituent quark and a constituent antiquark.

Baryons like the proton or neutron contain 3 valence quarks, antibaryons 3 valence antiquarks. The proton consists of (uud), the neutron of (udd) and the antiproton of $(\bar{u}\bar{u}\bar{d})$. According to the prescription mentioned above, the colour combination has to be singlet. This would not be possible for one or two single quarks, but it is for the combination to a normal meson, baryon or antibaryon. In addition to these three classes, there ought to be other possible combinations to obtain a color singlet, for instance two or three constituent gluons ("gluonium" or "glueball") or a quark, antiquark and a constituent gluon ("hybrid") or two quarks and two antiquarks (4-quark meson). These other combinations, predicted in principle by QCD, are so far assigned to the category **"exotic" mesons**.

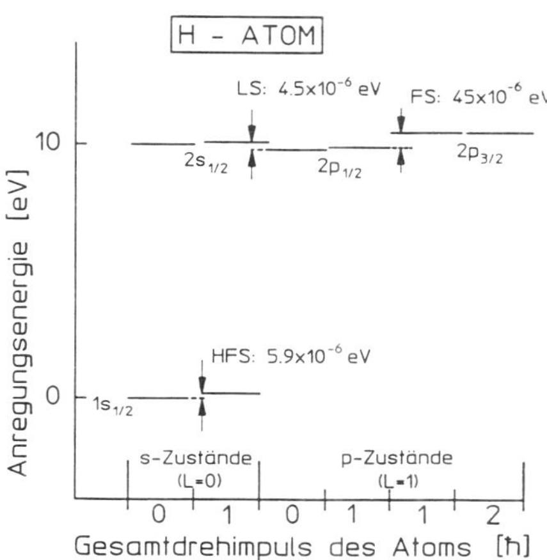

Figure 1: The 8 lowest energy levels of the hydrogen atom. Horizontal axis: total angular momentum (sum of electron and proton spin and orbital angular momentum). Vertical axis: excitation energy relative to the n=1 ground states. HFS = hyperfine splitting, LS = Lamb shift, FS = fine structure splitting

Constructing mesons: Spatial quantum numbers

The normal mesons are the simplest systems bound by the strong interaction. Given the fact that single free quarks or gluons have not been observed, one should expect that these objects are particularly important for an understanding of both the strong force and the nature of the constituents. One should think that they play a similar role for the development of the theory of strong interactions as did the **hydrogen atom** for the development of QED. However, this has not been the case (so far, at least). The mesons, in particular the light mesons built from the light quarks (u, d, and s) and their antiquarks, have scarcely penetrated the text books of particle physics. The reasons have already been indicated above: the lack of true understanding and the incompleteness of the data.

Not only the key role of the H-atom for QED can serve as a reference but also its level scheme (Fig. 1). It is useful for a comparison because it is the best known scheme of quantum states. Moreover, the spin and orbital angular momentum structure of mesons is very similar to that of the H-atom. Both bound systems consist of two fermions (spin 1/2). For an orbital angular momentum L =0 between the two fermions there are two states, the "para" configuration with antiparallel spins, total spin S=0 and total angular momentum J=0, and the "ortho" configuration with S=J=1. For a relative orbital angular momentum L=1, there are 4 states with total angular momenta J=0,1,1, and 2. Using a notation which is customary for atoms with several electrons $^{2S+1}L_J$ where L=0,1,2,.., expressed by the letters S,P,D,.., the corresponding six meson states are:

$$^1S_0,^3S_1,^3P_0,^3P_1,^1P_1,^3P_2.$$

A different notation, also frequently applied for mesons, uses the total angular momentum J, the parity P and the C-parity C, in the case that the state is an eigenstate of the charge conjugation operator **C**. The strong interaction is symmetric for an inversion of the coordinates (reflection at the origin), the parity operation **P**. The eigenvalue P can be ±1. For a normal $(q_i\bar{q}_j)$ meson it is given by the expression

$$P = (-1)^{L+1}.$$

Table 2: Names and quantum numbers of normal mesons for L=0 and L=1

name	J^P	$^{2S+1}L_J$
pseudoscalar	0^-	$^1S_0$
vector	1^-	$^3S_1$
scalar	0^+	$^3P_0$
axial	1^+	$^3P_1$ or $^1P_1$
tensor	2^+	$^3P_2$

The factor $(-1)^L$ is due to the action of **P** on the spatial wave function and the additional factor -1 is due to the opposite intrinsic parities of fermions and antifermions (derived from the Dirac equation). The charge conjugation **C** transforms a particle into its antiparticle, $\mathbf{C}u = \bar{u}$, $\mathbf{C}d = \bar{d}$, $\mathbf{C}s = \bar{s}$. The C-parity is another conserved quantum number. Mesons which consist of a quark q_i and the corresponding antiquark $\bar{q}_i$ are eigenstates of **C** and of the total spin S. The C -parity is given by

$$C = (-1)^{L+S}.$$

This relation is derived by writing the wave function as a product of a spin part, a spatial part and a flavour part and applying the generalized Pauli principle, which requires the complete wave function to be antisymmetric with respect to interchange of the two fermions. With these relations we can now make the transformation from one notation to the other (Table 2).

The two axial-vector mesons are distinguished by their C-parity in case of eigenstates of **C**: $J^{PC} = 1^{++}$ for $^3P_1$ and $J^{PC} = 1^{+-}$ for $^1P_1$. For mesons which are not eigenstates of **C**, like the K-mesons, the two components $^3P_1$ and $^1P_1$ can mix. These mesons are eigenstates of L and J but not of S.

Only the eight lowest-energy spatial configurations of mesons were discussed above; they correspond to those shown for the hydrogen atom in Fig. 1, which included the first radial excitations of the S-states and the lowest orbital excitations with L=1. Of course, as for the H-atom, higher orbital and radial excitations ($L \geq 2$ and $n \geq 2$) are possible.

All the listed quantum number combinations are characteristic for mesons. You find that half the J^{PC}-combinations are excluded even if higher orbital angular momenta are considered, e.g. $J^{PC} = 0^{--}$ or 1^{-+}. Those "**J^{PC}-exotic**" combinations are, however, possible for gluonia or hybrids since the gluon constituent has integer spin ($J^{PC} = 1^{--}$). If found, they would constitute a very direct proof of the existence of exotic mesons.

Constructing mesons: Flavour quantum numbers and the nonet

Combining the first three quarks with the first three antiquarks yields a nonet of normal quark model mesons for each spatial configuration (Fig. 2a). They are called the **light mesons**, whereas those containing at least one charm or bottom quark are called heavy. There is, for instance, a nonet of vector mesons, as shown in Fig. 2b. All the (additive) flavour quantum numbers and the charge can be obtained by summing up the corresponding quantum numbers of the quark constituents. The 6 states at the corners of the hexagon are easily distinguishable by their external quantum numbers, strangeness and charge. The three states in the center of the hexagon , with $S = I_z = 0$ are less easy to identify. Their wave functions are superpositions of the 3 combinations

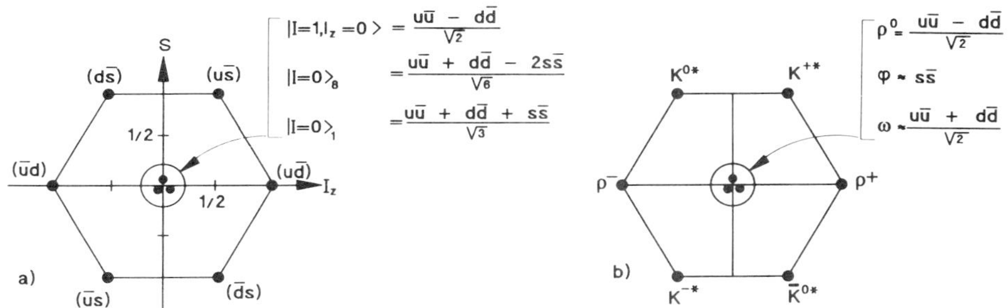

Figure 2: The flavour nonet of mesons and the wave functions of the three states in the center. Horizontal axis: Isospin z-component I_z. Vertical axis: strangeness S. a) The case of ideal SU(2) and SU(3) symmetry and b) the vector mesons, where the two isosinglet states are "ideally" mixed, i.e. almost pure $s\bar{s}$ and $\bar{u}u + \bar{d}d$.

$\bar{u}u, \bar{d}d, \bar{s}s$. These wave functions are shown, for the ideal case of perfect SU(2) and SU(3) symmetry, in Fig. 2a and for a real case, the vector mesons, in Fig. 2b.

Without going into details of SU(2) and SU(3) which can be found in text books, the following explanation is meant as a mnemonic for the signs and weights contained in these wave functions. I assume that the isospin formalism is known, e.g. from nuclear physics courses. It is an empirical fact that the strong interaction does not distinguish between up and down quarks, to a very good approximation. (QCD incorporates this fact by the assumption that the strong interaction only couples to the colour charge of quarks and by having nearly equal masses of up and down quarks.) We may then consider the up and down quark as the two states of an isospin doublet (u,d) = (+1/2, -1/2), as (following Heisenberg's suggestion) is done for p and n in nuclear physics, in analogy to the two orientations up and down for a spin 1/2 particle. The doublet of the antiquarks is (+1/2, -1/2) = $(\bar{d}, -\bar{u})$ with a minus sign in front of $\bar{u}$, such that this doublet transforms properly under isorotations [2]. Combining the doublet with the antidoublet, we obtain 4 states (as in atomic physics combining two spin doublets, i.e. combining two particles with spin $s = 1/2$ and $s_z = \pm 1/2$). Remember the four states obtained when the spins of two electrons are summed up. Using the notation $\uparrow, \downarrow$ for spin up and down, s_z = +1/2, -1/2), the four spin wave functions are

$$\begin{aligned}
|S=0, S_z=0> &= (\uparrow\downarrow - \downarrow\uparrow)/\sqrt{2} \\
|S=1, S_z=+1> &= \uparrow\uparrow \\
|S=1, S_z=-1> &= \downarrow\downarrow \\
|S=1, S_z=0> &= (\uparrow\downarrow + \downarrow\uparrow)/\sqrt{2}.
\end{aligned}$$

For the combination of our two isospin doublets, or more precisely, of an isodoublet with the isodoublet of its antiparticles, the only difference is that we have just the opposite signs in the wave functions with $I_z = 0$ because of the minus sign in front of the $\bar{u}$ quark:

$$|I=0, I_z=0>=|\bar{d}d + \bar{u}u> /\sqrt{2}$$

$$|I=1, I_z=0>=|\bar{d}d - \bar{u}u> /\sqrt{2}.$$

The first state is the isosinglet, and behaves like a scalar under SU(2) transformations, i.e. rotations in isospace. The second state is the neutral partner of the three isovectors (the isotriplet); for the case of pseudoscalar ($J^P = 0^-$) mesons this triplet consists of the three pions, π^+, π^0, π^-.

Now let us add a third quark, the strange quark. Four additional particles are obtained at the corners, the K-mesons, with **"open strangeness"** and two more particles in the center, ($S = I_z = 0$), with **"hidden strangeness"**, Fig.2. These 6 meson species are the subjects of the present lectures.

The nonet of mesons is divided into a singlet and an octet. The combination

$$|I = 0 >_1 = |\bar{u}u + \bar{d}d + \bar{s}s > /\sqrt{3}$$

behaves like a scalar under flavour-SU(3) transformations; it is the SU(3) singlet, and the subscript "1" refers to this property. For the colour-SU(3) of QCD, it is the corresponding colour singlet which is realised in free hadrons; it is found by combining either three different quark colours or colour and anticolour. Likewise, there are only 8 coloured gluons, since one SU(3) combination is colourless.

The third state in the center can now be found by requiring orthogonality to the first two states $|I = 1, I_z = 0 >$ and $|I = 0 >_1$:

$$|I = 0 >_8 = |\bar{u}u + \bar{d}d - 2\bar{s}s > /\sqrt{6}.$$

The index 8 refers to the fact that it is a member of the octet. Thus it is seen that hidden strangeness is mixed with $\bar{u}u + \bar{d}d$ in a certain ratio for this ideal SU(3) case. If nature were flavour-SU(3) symmetric, all the members of the octet would be mass degenerate. It is not - the presumed reason being the significantly heavier mass of the s quark compared to the masses of the u and d quarks, see Table 1. Nature also mixes the two states $|I = 0 >_1$ and $|I = 0 >_8$ the way it likes. We will come back to the octet-singlet mixing angle below. For the moment, let us assume that flavour-SU(3) is not a good symmetry, in contrast to SU(2) for up and down quarks. This also explains why in Table 1 the up and down quarks appear to play a special role. Instead of assigning them the quantum numbers "upness" and "downness", they were labelled $I_z = +1/2, -1/2$. As a consequence of SU(2) symmetry, total isospin I and I_z are good quantum numbers, i.e. conserved in strong interactions, and hadrons with the same I but different I_z are approximately mass degenerate.

Thus the mesons of every nonet can be grouped into four multiplets of particles with (almost) degenerate masses:

- the 3 isovectors (an isotriplet, I = 1);
- the two (I=1/2) doublets of K-mesons, $K^+, K^0, \bar{K}^0, K^-$;
- the first isosinglet (I=0) and
- the second isosinglet (I=0).

The mesons "constructed" in the last two sections were all **"normal" or "quark-model" mesons**. So far no mesons have been proven to exist which have **flavour-exotic** quantum numbers like strangeness S =2 or charge = 2. They would indicate that bound states of more than one quark and antiquark exist. In addition to the notion flavour-exotic and the notion J^{PC}- exotic explained above, there is also the notion **crypto-exotic** for conceivable meson states which have the same quantum number combinations as normal mesons but a different composition, for instance, gluonia with $J^{PC} = 0^{++}, 2^{++}$ or 0^{-+}, and with S = I = 0.

A survey of all mesons

You are now ready for a general survey of all mesons observed so far, after about four decades of research. After that survey the lectures will specialize to the strange sector.

A complete up-to-date survey of mesons with references to every single mesurement can be found in the Review of Particle Properties by the Particle Data Group (PDG) – the bible of the particle physicists [3], pages 1443 to 1672. There is also a handy, recommendable resumeé of the quark model in this review, page 1319ff. This book is indispensible for every meson or hadron expert. Fig. 3 compares the level schemes of the majority of mesons as compiled in the meson summary table of the PDG [3]. This figure also serves to introduce all the symbols for the names of the mesons. (You have to be a botanist, according to E.Fermi, to remember these names.) Only states with $L \leq 1$ and $n \leq 2$ are displayed in Fig. 3, as in Fig. 1 for the H-atom. Not too many known mesons have been omitted from these plots. Omitted: 11 higher radial and angular excitations of non-strange light mesons, 5 higher radial excitations of strange mesons, 4 higher J/ψ states and 4 higher Υ states. Moreover, all the 9 known "D" mesons with open charm and the 4 "B" mesons with open bottomness were left out.

Charmonium and bottonium (Fig. 3a,b) are eigenstates of the charge conjugation operator **C**, hence can be characterized by the three quantum numbers J^{PC} (horizontal coordinate). The vertical coordinate is the energy of the state relative to the lowest state with $J^{PC} = 1^{--}$. All the displayed states are very narrow, i.e. have a large lifetime in comparison with most of the light mesons. This is one of the reasons why these two level schemes have found their way into high energy physics text books more rapidly than the light mesons.

The cause for the long life time is decribed by the so called OZI rule [4]. According to this rule, decays where $\bar{c}c$ and $\bar{b}b$ annihilate are suppressed compared to those decays where these quarks are tranferred to the decay products. An example for an OZI forbidden decay is $\bar{c}c \rightarrow \bar{u}d + \bar{d}u$ (Fig. 4a). An example of an OZI allowed decay is $\bar{c}c \rightarrow \bar{c}d + c\bar{d}$ (Fig. 4b). The OZI rule also applies to decays of $\bar{s}s$ pairs and to the inverse processes when heavy quark pairs $(\bar{s}s, \bar{c}c, \bar{b}b)$ are produced, see (Fig. 4c,d). The thresholds for the OZI allowed decays to $D\bar{D}$ and $B\bar{B}$ are shown in Fig. 3.

Comparing with the H-atom (Fig. 1), one notices - apart from the gigantic difference of the energy scales (a factor of 10^8)- that the splitting between S (L=0) and P (L=1) states is significant in both cases and the hyperfine splitting between η_c and J/ψ is large. The comparison between $\bar{c}c$ and $\bar{b}b$ shows that, in spite of the hugely different rest energies, 3097 MeV for J/ψ and 9460 MeV for $\Upsilon(1S)$, the energy separations between the levels are very similar, a remarkable fact.

The most prominent difference between the preceding level schemes and the next one of K-mesons, Fig. 3c, is that all but one states are presented by boxes in the latter figure. The height of the boxes is equal to the width Γ of the resonance. Only the pseudoscalar ground state of the kaons cannot decay by the strong interaction; it decays weakly. All others decay strongly and rapidly; there is no OZI suppression of the decay. The first radially exited state with $J^P = 0^-$ has not yet been established with certainty. The hyperfine splitting between 0^- and 1^- ground states is 400 MeV, much larger than the corresponding splitting of the $\bar{c}c$ mesons. The first radial exitation of the 1^- and the P-states appear as almost degenerate, at least relative to the widths of these states. The gap between the 1^- ground state and its first radial exitation is 500 MeV, which seems to be a canonical number for **all** mesons. The K-mesons are not eigenstates of **C**, therefore the states with $J^P = 1$ ($^1P_1$ and $^3P_1$) mix.

The isospin triplet states $\pi, \rho, a_0, ..$ are shown in Fig. 3d. Only the neutral member of each triplet is an eigenstate of **C**. The similarity of the energy gaps to the previous schemes is striking, apart from the hyperfine splitting between π and ρ, which is further increased compared to the previous cases. The $a_0(980)$ looks like an outsider because of its mass, width and decay branching ratios. It is discussed as a candidate for a

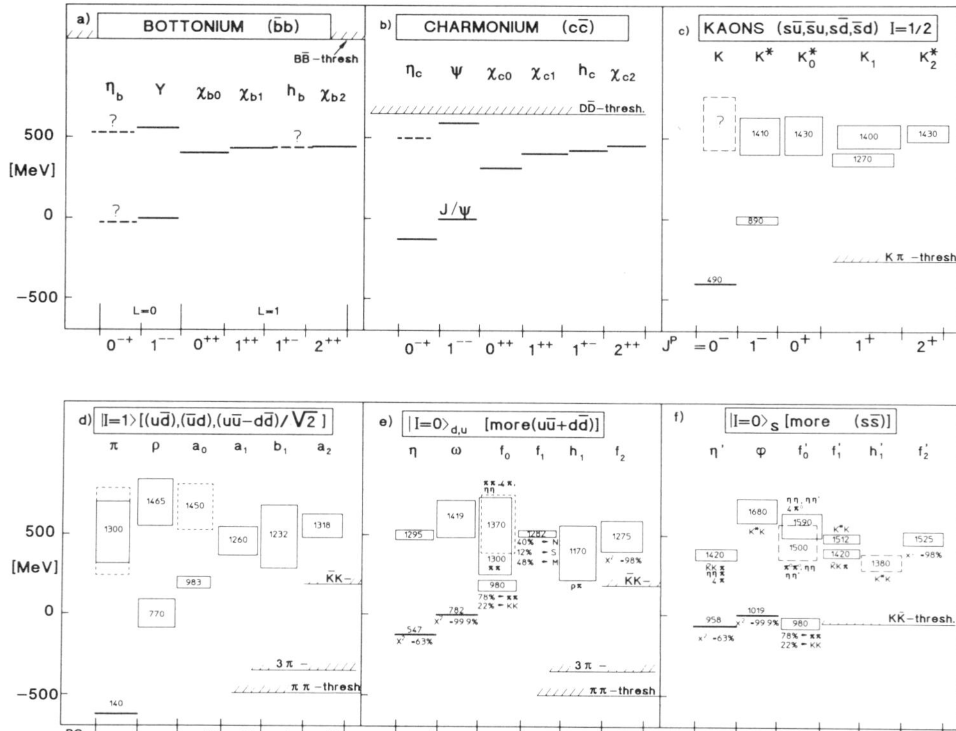

Figure 3: Level schemes of mesons, only $L \leq 1$ corresponding to the levels shown in Fig. 1. Vertical axis: energy difference to the 1^- ground state. This scale was chosen to facilitate a comparison of the level schemes of mesons with different total mass. Horizontal axis: J^P or J^{PC}. The symbols for the names of the mesons are given under the figure titles, for each J^{PC}. Each state is represented by a box or line, the height of which corresponds to the width Γ. The number in the box or above the line indicates the rest energy in MeV. a) Bottonium; b) charmonium; c) kaons (two isodoublets per box); d) isovectors (I=1), a triplet of states per box. The isoscalars have been separated according to their dominant quark content. Those with more $\bar{u}u + \bar{d}d$ are put in e), those with more $\bar{s}s$ in f). The fraction of the dominant component is indicated by $X^2 = ..\%$ if the nonet mixing angle is known. Otherwise the dominant decay modes are indicated. There are three exceptions to this rule of dividing: The $f_0(980)$ has been put into both e) and f), and $f_0(1500)$ and $f_0(1590)$ have been put into f), because the corresponding place in e) was already too crowded. Based on the present knowledge of their decay modes, at least $f_0(1500)$ should rather be in e), but anyhow this is perhaps not a normal meson.

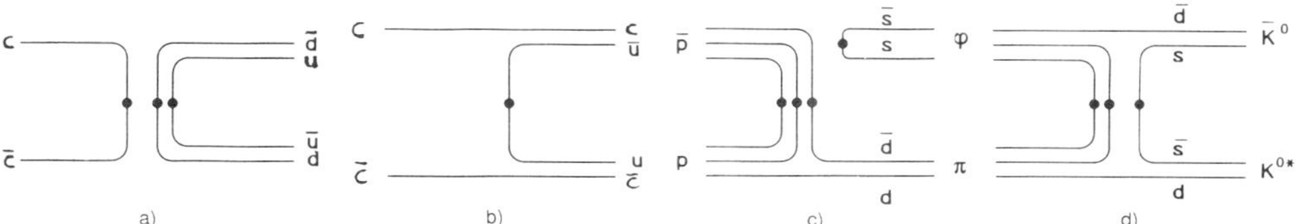

Figure 4: Examples for OZI forbidden and allowed processes a) forbidden decay of charmonium into non-charmed mesons; b) allowed decay of charmonium into mesons with open charm; c) forbidden ϕ meson production in $\bar{p}p$ annihilation; d) allowed production of mesons with open strangeness.

$\bar{K}K$ molecule, together with the $f_0(980)$ which appears to be the I=0 partner of this state. The scalar state at 1450 MeV has been discovered only recently [5] and needs confirmation. It could be another candidate for the quark model ground state.

For the isoscalars shown in Fig. 3e) and f), the mixing between the SU(3) singlet and octet states makes the identification of the two physical states even more difficult. (Moreover, if glueballs exist, additional states come in and mix.) If we take as a new basis for the two quark model states $|\bar{u}u + \bar{d}d > /\sqrt{2}$ and $|\bar{s}s >$, since we are interested in the $\bar{s}s$ component of mesons, the physical states are given by

$$|I = 0 >_{d,u} = X|\bar{u}u + \bar{d}d > /\sqrt{2} + Y|\bar{s}s >$$

$$|I = 0 >_s = -Y|\bar{u}u + \bar{d}d > /\sqrt{2} + X|\bar{s}s >$$

with $X^2 + Y^2 = 1$. Here X= $\cos(\theta - \theta_{ideal})$, where θ is the conventional nonet mixing angle and θ_{ideal} is the ideal conventional mixing angle of 35.3^o. For $X > 0.5$ the state $|I = 0 >_{d,u}$ consists dominantly of $\bar{u}u + \bar{d}d$, and $|I = 0 >_s$ consists dominantly of $\bar{s}s$.

According to their dominant $\bar{q}q$ content, the mesons have been put either into e) or f) if the mixing angle is known (taking the average of the mixing angles obtained with the linear and quadratic mass formula). If the mixing angle is not known, the meson is put into e) if the decay is dominantly into non-strange mesons and into f) in the opposite case. The three exceptions are $f_0(980)$ which is diplayed in e) and f), and $f_0(1500)$ and $f_0(1590)$, which are put into f) just for the sake of clarity.

Detecting strange mesons and hidden strangeness

How are mesons detected and how does one learn they contain $s, \bar{s}$ or $\bar{s}s$? This will be our next subject. I shall first describe the basic features of a modern detector for meson spectroscopy. (Of course, it will be the one the author is involved with). And then how with this detector "stable" strange mesons, unstable strange mesons and mesons with $\bar{s}s$ content are identified. First of all it is necessary that you understand the time and space evolution of events in which mesons are produced and detected.

Space-time evolution of events

As an example, let us consider $\bar{p}p$ annihilation. This is a process which is especially well suited for meson spectroscopy. When a $\bar{p}$ annihilates at rest, practically only mesons are produced. Among them, exotic mesons (gluonia, hybrids etc.) are also expected.

An interesting sequence of events takes place even before the annihilation. The antiproton, slowed down in the target, is caught by an H_2 molecule in some high atomic orbit, ejecting an electron if the binding energy of the $\bar{p}$ is larger than that of

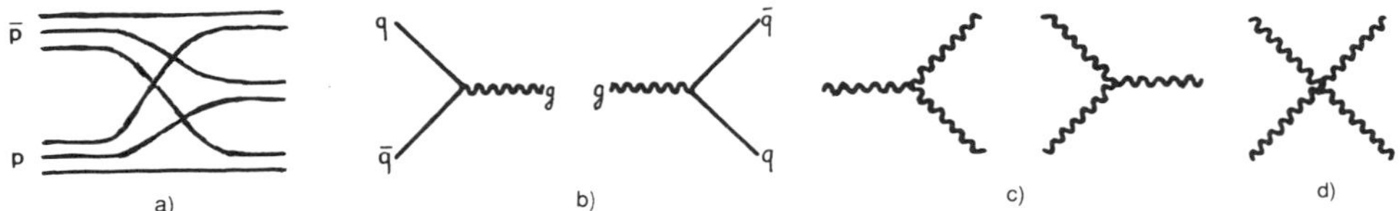

Figure 5: QCD processes taking place in $\bar{p}p$ annihilation a) rearrangement, b) fusion and pair creation, c) gluon dissociation, d) 4-gluon vertex.

the electron. From the ratio of reduced masses of the $\bar{p}p$ atom and electron, the initial main quantum number can be estimated to be around 30. After formation, the $\bar{p}p$ atom is de-exited by radiation and the Auger effect until $\bar{p}$ and p have approached each other closely enough that the strong force (range of about a femtometer) starts acting.

What are the **elementary QCD processes** that take place at annihilation? The incoming $\bar{p}$ consists of three valence antiquarks and of sea quarks and gluons, the p consists of 3 valence quarks, sea quarks and gluons. The elementary processes are (Fig. 5):

1. rearrangement of quarks and antiquarks into pairs (mesons),

2. fusion of a quark with the corresponding antiquark into a gluon, and the inverse process, dissociation of a gluon into a $\bar{q}q$ pair,

3. dissociation of a gluon into two gluons, and the inverse process, the fusion of two gluons into one,

4. scattering of two gluons (4-gluon vertex).

It is obvious that by appropriate combination of these processes, normal meson states, gluonia, 4-quark states or hybrids can be formed. These processes are confined in hadrons and take very short times, of the order of $10^{-24}s$. The lifetimes of the hadrons emerging from the interaction vary from 10^{-24} to $5 \cdot 10^{-8}$ s. Consider a specific example, the annihilation

$$\bar{p}p \rightarrow K^{+*}K^{-},$$

which implies the fusion of an incoming $\bar{u}u$ and a $\bar{d}d$ pair and the creation of an $\bar{s}s$ pair:

$$\bar{u}\bar{u}\bar{d} + uud \rightarrow u\bar{s} + \bar{u}s.$$

The K^{+*} decays with a lifetime $\tau = \hbar/\Gamma$ of about 10^{-23} s into $K\pi$, i.e. either $K^{+}\pi^{0}$ or $K^{0}\pi^{+}$. In the laboratory, due to time dilation, the distance the K^{+*} traverses before it decays, its decay path, is

$$\tau \cdot \gamma \cdot v = c\tau \cdot pc/(mc^{2}),$$

where $\gamma = (1-\beta^{2})^{-1/2}$, with $\beta = v/c$, $v =$ the velocity, $p =$ the momentum, $mc^{2} =$ the rest energy of K^{+*}. For a typical momentum of around 500 MeV/c the decay path is 3fm, truly short! Due to its short lifetime the state $K^{+*}K^{-}$ is called an **intermediate state**.

Let us assume we have after the decay of the K^{+*} the three particles $K^{+}\pi^{0}K^{-}$. This state is called the **"final" (hadron) state.** It is by no means final: only with

Table 3: Lifetimes and products $c\tau$

	K^*	π^0	K_S	K_L	$K^\pm$	$\pi^\pm$	$\mu^\pm$
$\tau[s]$	10^{-23}	$8.4 \cdot 10^{-17}$	$8.9 \cdot 10^{-11}$	$5.2 \cdot 10^{-8}$	$1.2 \cdot 10^{-8}$	$2.6 \cdot 10^{-8}$	$2.2 \cdot 10^{-6}$
$c\tau$ [m]	$3 \cdot 10^{-15}$	$2.5 \cdot 10^{-8}$	$2.7 \cdot 10^{-2}$	15.5	3.7	7.8	659

respect to the appearance of hadrons is it final. They decay due to the action of weak and electromagnetic forces. The π^0 decays electromagnetically in $8.4 \cdot 10^{-17}$s, mainly (99%) into two photons, 2γ. The corresponding decay path is still very short, some tens of nm. The K^+ and K^- decay weakly, most frequently into $\mu\nu$ or $\pi\pi$. The charged pions decay into $\mu\nu$; finally, the μ's decay into $e^-(e^+)$ and a pair of ν's. The relevant lifetimes τ and the products $c\tau$ are listed in Table 3.

The decay paths of the weakly decaying particles are several meters or longer. The really final, really stable particles are the photons, electrons or positrons and neutrinos. In vacuo they would live forever. In the real world they interact; this is how they are detected. (It is difficult to detect the interaction of neutrinos but possible).

Our presumed detector, about 2 m in diameter, will intercept the sequence of events before 10^{-8} seconds have passed, i.e. the π^0 has decayed but K^+ and K^- are still alive. Hence these are the detected particles in this special event. In the general case of $\bar{p}p$ interaction at rest, the detector has to detect π^+ and π^- in addition to γ, K^+, K^- .

Detecting strange mesons and $\bar{s}s$ components

Now it is straightforward to understand how (openly) strange mesons, i.e. those containing one s or $\bar{s}$ quark, are detected. Since strangeness is conserved in strong interactions, the strange quark has to be found in the decay products. There has to be a K^+, K^-, K^0 or $\bar{K}^0$ in the final hadron state. If a meson resonance is found to decay into an odd number of K-mesons, it must be strange.

Because strangeness is conserved in strong interactions and the initial state has S=0, an even number of K's, i.e. $\bar{K}K$ pairs, is always observed in $\bar{p}p$ annihilation. They can come from three different sources:

1) A pair of strange mesons in the intermediate state;

2) a meson with hidden strangeness;

3) a meson without any strange quark constituent where $\bar{s}s$ is created in the decay.

An example for a prediction from flavour-SU(3), relevant to the cases 2) and 3), is

If $f_0 = (\bar{u}u + \bar{d}d)/\sqrt{2}$ then $BR(f_0 \rightarrow \bar{K}K)/BR(f_0 \rightarrow \pi\pi) = 1/6$;

if $f_0' = (\bar{s}s)$ then $BR(f_0' \rightarrow \bar{K}K)/BR(f_0' \rightarrow \pi\pi) = 1/0$;

if f_0 = glueball then $BR(f_0 \rightarrow \bar{K}K)/BR(f_0 \rightarrow \pi\pi) = 4/3$;

if f_0 = meson-antimeson ($M\bar{M}$) molecule, then $BR(f_0 \rightarrow M\bar{M})$ is dominant.

A modern detector–the Crystal Barrel spectrometer

The detector is required to detect photons, charged kaons and charged pions, and to measure their momenta and angles for momenta up to about 2GeV/c. In order to detect all particles produced in an event, it has to cover the complete solid angle 4π sr, or at least a big fraction of it. Fig. 6 shows how this is realized with the Crystal Barrel detector at LEAR. The $\bar{p}$ beam from LEAR (the Low Energy Antiproton Ring at CERN) enters from the left, along the z-axis. This beam is the best antiproton beam in the world, with the following characteristics (in brackets the characteristics of a conventional $\bar{p}$ beam, like those at BNL(USA) or KEK(Japan)): Intensity: 10^4 to 10^7

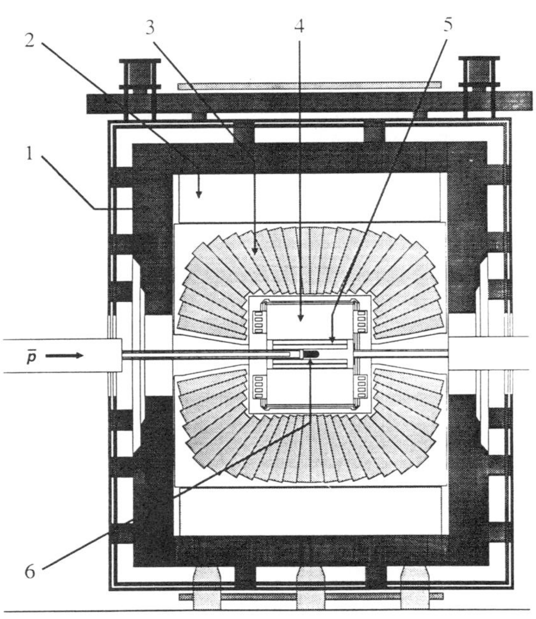

Figure 6: The Crystal Barrel spectrometer at LEAR cut along the cylinder symmetry axis (z-axis). Components: (1) magnet yoke, (2) magnet coil, (3) electromagnetic calorimeter, (4) jet drift chamber, (5) proportional chamber (6) target.

$\bar{p}/s$ ($< 10^5\bar{p}/s$); momentum resolution: $\Delta p/p = 10^{-3}(3 \cdot 10^{-2})$; purity : $100\%\bar{p}(1\%\bar{p})$; beam cross section: 1 mm^2 (5 cm^2).

The beam is stopped or interacts in flight in a tiny target containing liquid or gaseous H_2 or D_2. The target can be so tiny because of the beam quality. A tiny target is a big advantage: precise definition of the primary interaction vertex, few secondary interactions in the target, close approach of detectors to the point of interaction. The target is surrounded by a 2-layer multiwire proportinal chamber (PWC) or, more recently, by a silicon microstrip vertex detector (SVX). The PWC or SVX are surrounded by a cylindrical Jet Drift Chamber (JDC) with 23 layers of wires and 30 azimuthal sectors. The JDC is located inside a cylindrical photon calorimeter, consisting of 1380 Cs(Tl) crystals, each 30 cm long. This crystal barrel gave the name to the whole spectrometer. All the active detectors are embedded in a solenoidal magnet which produces a homogeneous axial field (parallel to the cylinder, beam or z-axis) of strength 1.5 T. The detector is described in detail in a long paper [6]. Here we only want to sketch the particle detection methods with this apparatus.

With the JDC, charged tracks are measured at up to 23 space points. The wires are parallel to the z axis. Thus each drift time provides precise information on r (distance of the wire to the beam axis) and ϕ (azimuthal angle). The left-right ambiguity is solved by staggering the wires. The spatial resolution (σ) in the r-ϕ plane is about 100 μ for each point. A measurement of the z-coordinate is provided by charge division: the sense wires have a suitable resistance and the signal is read out on both sides of the wire. From the charge ratio, the z location of the track can be determined to 5 mm precision (σ). The total charge is proportional to the energy loss of the particle in the chamber (dE/dx). This energy loss is useful information to distinguish between $K^{\pm}$ and $\pi^{\pm}$. Since the tracks are curved in the magnetic field [$pc/MeV = 300 \cdot (B/Tesla)(\rho/m)$, where ρ/m is the radius of the orbit in units of a meter for $\vec{p}$ orthogonal $\vec{B}$], the momentum can be measured, with a precision (σ) between 2% and 6% for momenta p_T between 0.1 and 1 GeV/c. (p_T is the momentum component orthogonal to z).

The main role of the PWC or SVX is to provide a fast trigger on the multiplicity of charged particles in the final state. For instance, if one wants to record only events with no charged particle emerging from the target, PWC or SVX are used as veto counters and events are rejected if a charged particle is detected. In combination with the requirement of observing charged tracks, this can be used to set up a powerful trigger for events with strange particles, see below.

The CsI(Tl) crystals of the calorimeter are read out by photodiodes via a wave length shifter. Inorganic crystals are the best but also most expensive material to detect high energy photons. The granularity of the calorimeter has been chosen such that for the given distance to the target an optimal angular resolution and shower separation is achieved. Each high energy photon induces an electromagnetic shower (a cascade driven by pair production and bremsstrahlung, which involves thousands of photons and electrons). Each shower is distributed in general over several crystals. From the center of gravity, the impact point of the primary photon is determined. The energy resolution of the calorimeter for photons is

$$\sigma_E/E = 2.5\%(E/GeV)^{-1/4}.$$

Great care has been taken to minimize the noise in the read-out of the crystals in order to be able to extend the measurement to the lowest possible energy. The noise of the crystals corresponds to less than 300 keV. However, the lowest reasonable energy threshold is 5 MeV. At these low energies "physical noise" becomes dominant: statistical shower fluctuations which seem to come from separated showers. These fluctuations are called "split offs". The collaboration has learnt progressively how to recognize split-offs, employing neural networks among other techniques, in order to improve the calorimetry in the low energy region.

Detecting "stable" mesons

The weakly decaying charged kaons and pions live long enough that they decay outside our detector, which occupies a cylinder of 1.5 m diameter and 1.5 m length. Together with the long-lived neutral K meson, the K_L to be discussed below, they are the only mesons with such long lifetimes.

How are the charged kaons detected? First, their momentum is measured in the JDC. In many cases, this is already sufficient to identify them, i.e. to distinguish them from the other charged mesons, the pions. Take for example the annihilation into two charged stable mesons,

$$\bar{p}p \rightarrow K^+K^-, \quad \bar{p}p \rightarrow \pi^+\pi^-.$$

The particles are monoenergetic, their momentum is determined by energy and momentum conservation and related to the mass of the particles produced (Fig. 7a). If there are more than two particles in the final state, and if one wants to distinguish for instance between the two reactions $\bar{p}p \rightarrow K^+K^-\pi^0$ and $\bar{p}p \rightarrow \pi^+\pi^-\pi^0$, energy and momentum conservation in general allows this to be achieved, although the particle momenta vary. This is done in the kinematic fit. (The higher the precision of energy and momentum measurement, the cleaner the distinction.)

The additional tools to distinguish between kaons and pions are in general: measurement of time of flight, Cherenkov radiation and energy loss dE/dx. All these measurements provide information on the velocity v of the particle which, combined with the momentum p = mv, fixes m. Fig. 7b shows a dE/dx measurement in the JDC of

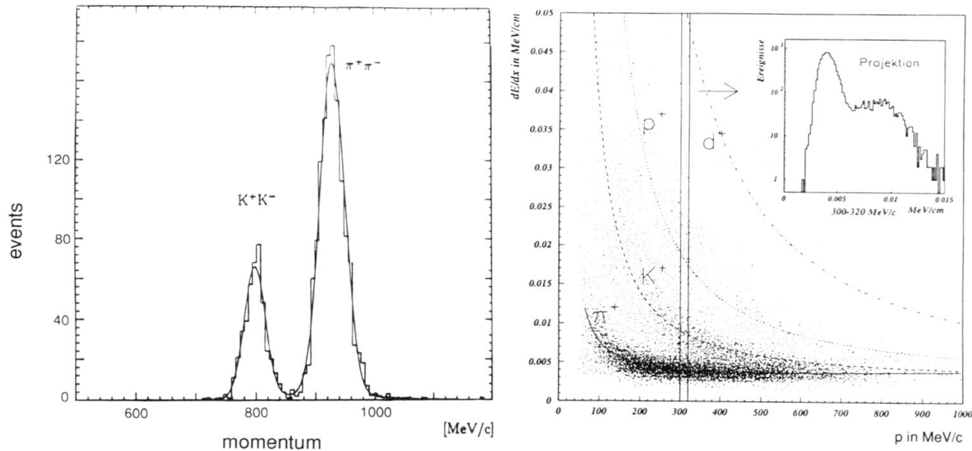

Figure 7: a) Momentum distribution of particles in collinear two-prong events due to the reactions $\bar{p}p \rightarrow K^+K^-, \bar{p}p \rightarrow \pi^+\pi^-$. b) Energy loss versus momentum for charged tracks, measured in two-track events with two to four additional showers in the calorimeter. (From a study of $\bar{p}p \rightarrow K_L K^{\pm}\pi^{\mp}\pi^0$, C.Kolo, thesis, Munich 1995).

the Crystal Barrel detector as a function of momentum. The bands which are seen can be associated with π's and K's. The separation is good for momenta up to 400 MeV/c.

Less stable mesons

Neutral π^0 , η and η' mesons have very long lifetimes on the scale of strong interactions but decay practically at the primary interaction vertex. These particles are identified through their decay products. All of them have a decay mode into two photons, with the following branching ratios: the π^0 with 99%, the η with 39%, the η' with 2%.

The technique to identify them is to calculate the invariant mass $M_{\gamma\gamma}$ of any pair of γ's produced in the event,

$$M_{\gamma\gamma} = \sqrt{(p_1 + p_2)^2},$$

where p_1 and p_2 are the 4-momenta of the two photons. This invariant mass equals the mass of the mesons within the measurement errors if the two γ's were indeed the decay products (Fig. 8)[6].

In the sort of event used for Fig. 8 there were 6 photons and hence 15 possible pairs. Supposing that all photons come from either one of the possible 3-meson final states: $\pi^0\pi^0\pi^0, \pi^0\pi^0\eta, \pi^0\eta\eta, \pi^0\pi^0\eta'$, ... there are only three good combinations for each event. The remaining 12 wrong combinations are called combinatorial background. This usually flat background sits under the signal peaks. Since a photon pair is associated to a meson if the invariant mass falls within an appropriate window containing the signal peak, it is obvious that a smaller window implies a smaller contribution from the combinatorial background. This is why good energy and momentum resolution of the detector is a definite advantage. The better the mass resolution, the cleaner the selected sample of events.

The figures of merit of the Crystal Barrel detector for the mass resolution of π^0, η, η' in 6-photon events are

$$\sigma_{\pi^0} = 10MeV, \quad \sigma_\eta = 16MeV, \quad \sigma_{\eta'} = 17MeV.$$

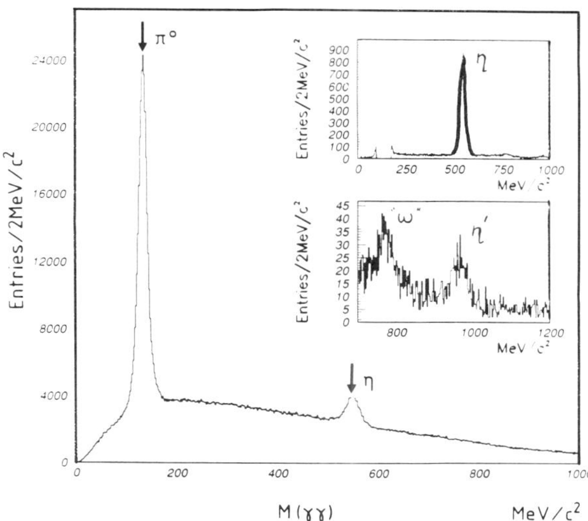

Figure 8: Invariant mass of two photons, $M_{\gamma\gamma}$, for 6 photon final states in $\bar{p}p$ annihilation at rest ($\bar{p}p \rightarrow 6\gamma$).

Other decay modes can also be used to identify these particles, as for instance $\eta \rightarrow 3\pi^0$, by calculating the invariant mass of all the particles produced in the decay. In the latter case

$$M = \sqrt{(p_1 + p_2 + p_3)^2}.$$

In Fig. 8 a signal called "ω" is seen. The ω decays 8% into $\pi^0\gamma$ and is frequently produced in the reaction

$$\bar{p}p \rightarrow \pi^0\pi^0\omega.$$

It is background from this channel which is seen in Fig. 8. Suppose a very low energy γ is lost from the decay of the ω, such that the loss is not recognized by the requirement of energy and momentum conservation. Then the event will only have 6 photons and the invariant mass of the remaing two photons from the ω decay will almost be correct. This is the origin of the signal in Fig. 8 and it shows again how important are good momentum and energy resolution and a low energy threshold for detecting photons in this kind of physics.

Neutral K mesons

The neutral K mesons are very special in many respects. They are described in all particle physics books. A beautiful chapter of the Feynman lectures (Vol. 3) is dedicated to them; it was even written before the discovery of CP-violation, which added another beautiful chapter to neutral K physics.

For the present context of detecting kaons, it is important to realise that the particles observed are not the K^0 or $\bar{K}^0$, the eigenstates of the strong interaction, but K_L and K_S, which are introduced in text books as the two eigenstates of **CP** with the eigenvalues CP = + or - 1, plus some small CP-violating admixtures of the eigenstate of opposite CP. Neglecting these small admixtures, which barely lead to any observable effects in meson spectroscopy, we may consider them as CP-eigenstates and as eigenstates of C separately , since they are eigenstates of P, separately. With $\mathbf{C}K^0 = -\bar{K}^0$, $\mathbf{C}\bar{K}^0 = -K^0$ (see appendix), $\mathbf{P}K^0 = -K^0$ and $\mathbf{P}\bar{K}^0 = -\bar{K}^0$ (pseudoscalars, i.e.

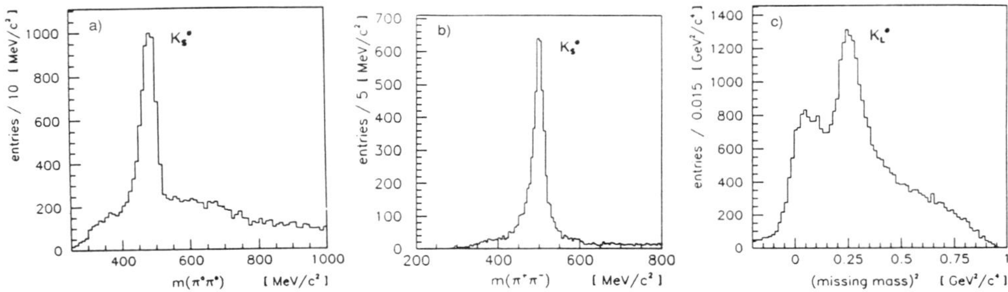

Figure 9: Identification of K_S via its decay into two charged or neutral pions. Invariant mass distributions of a) $\pi^0\pi^0$ from an analysis of the reaction $\bar{p}p \rightarrow K_S K_L \pi^0$; b) of $\pi^+\pi^-$ from an analysis of the reaction $\bar{p}p \rightarrow K_S K_S \pi^0$ with one K_S decaying into $2\pi^0$, the other one into $\pi^+\pi^-$; c) Identification of K_L's by missing mass technique. (Thesis of H.P.Dietz, University Munich 1994.)

P= -1) the eigenstates are

$$K_S \approx K_1 = (K^0 + \bar{K}^0)/\sqrt{2}; \qquad CP = +1, \;\; C = -1$$

$$K_L \approx K_2 = (K^0 - \bar{K}^0)/\sqrt{2}; \qquad CP = -1, \;\; C = +1.$$

The lifetime of the K_S is such that it decays a few cm from the vertex, see Table 3. This is normally within the detector but well separated from the primary interaction vertex. The secondary interaction vertex provided a powerful tag to identify K_S in bubble chambers many years ago. It can also be used in electronic counter experiments to provide a trigger on events with a K_S. This is an interesting possibility since these events are relatively rare, and enhancement by a specific trigger is in general required to obtain high statistics samples for final states which are rare. The dominant decays of the K_S are

$$K_S \rightarrow \pi^+\pi^-, \qquad K_S \rightarrow \pi^0\pi^0.$$

The K_S is identified by calculating the invariant mass of the two pions, see Fig. 9.

The charged decay mode is used to trigger on events with a K_S. The trigger of the Crystal Barrel spectrometer requires simply that in the JDC two additional tracks are seen compared to those detected in the SVX (or PWC before it was replaced by the SVX). The PWC had a radius of 2.6cm, the SVX has a radius of 1.2cm. The smaller radius of the SVX significantly improves the trigger efficiency at lower K_S momenta (100 - 200 MeV/c).

The lifetime of the K_L and its decay path is so long (Table 3) that it usually decays outside the detector. Since it is neutral it is not seen in the tracking chambers, but there is a 50% chance that it interacts in the crystal calorimeter. These properties of the K_L can be used to identify it, as long as all the other particles in the final state are detected. (For this purpose, the detector has to be hermetic, i.e. cover 4π sr.) If it has not interacted, the missing mass

$$M_{miss} = \sqrt{(p_{miss})^2} = \sqrt{(E_{miss}^2 - \vec{p}_{miss}^{\,2})}$$

can be calculated from missing energy and momentum, $(E_{miss}, \vec{p}_{miss})$, i.e. the differences between the sum of detected particle energies and momenta and the incoming energy and momentum (Fig. 9b). If it has interacted, the energy deposit in the crystals does not provide any good information on the energy of the K_L, since the fluctuations of

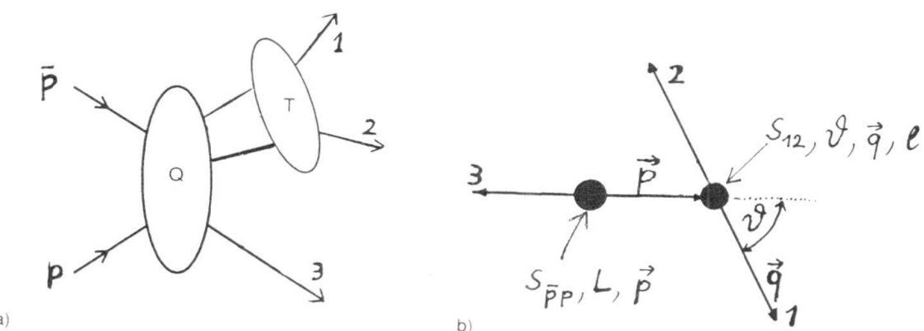

Figure 10: a) Production of a 3-body final state in $\bar{p}p$ annihilation with a resonance between particles 1 and 2. b) Definition of the corresponding kinematic variables s_{12}, s_{13} (invariants), $\vec{p}$ and L (defined in the $\bar{p}p$ cm system), and $\vec{q}, \theta$ (defined in the rest system of the resonance, the Gottfried Jackson system, see text).

such hadronic showers are very large. However, the information on the angle is still useful. This technique allows identification of events of the type

$$\bar{p}p \rightarrow K_L K_L \pi^0$$

and determination of their complete kinematics.

Short lived states

Short lived states, also called resonances (between the hadrons into which they decay), are in principle detected, like "stable" hadrons, by calculating the invariant mass of the decay products. Whereas the width of a stable or very narrow resonance is usually determined by the detector resolution, broad resonances have a natural width which is much larger than the detector resolution.

Detecting and identifying resonances is, in general, much more difficult if they are broad than if they are narrow. In many cases, it is even justified to say that for the hunters of short lived hadrons the real data analysis work begins where it has ended for the other particle physicists. After the selection of the final state and the subtraction of the background (this is the point where the normal particle physicist is ready to publish the observation of a signal or an upper limit) an **amplitude analysis** has to be performed which can become very tedious. The main reason for the difficulty is that short lived resonances, since they are broad, very rarely "stand alone" but rather overlap and interfere with each other.

Let us first consider the hypothetical production of an isolated resonance in $\bar{p}p$ annihilation Fig. 10. In order to form a resonance in production, the final state has to contain at least three particles and the total energy $\sqrt{s_{\bar{p}p}} = M_{\bar{p}p} = \sqrt{(p_{\bar{p}} + p_p)^2}$, i.e. the available phase space, must be sufficiently large. The case of 3 particles is of course the simplest. Such a final state can be described by two kinematic variables if we assume that there is no preferred direction, as is the case for the annihilation of unpolarized $\bar{p}p$ at rest. The usual variables are the invariant masses squared between pairs of particles, for instance

$$m_{12}^2 \equiv s_{12} = (p_1 + p_2)^2$$

$$m_{13}^2 \equiv s_{13} = (p_1 + p_3)^2$$

where the p_i are the 4-vectors. The third invariant mass is then given by the relation

$$s_{12} + s_{23} + s_{13} = s_{\bar{p}p} + s_1 + s_2 + s_3,$$

where $s_i \equiv m_i^2$. Each event can be represented by a point in the plane s_{13} versus s_{12} and the collection of points (scattergram) corresponding to a sample of many events is called a **Dalitz plot.** It can be shown that a homogeneous distribution of points corresponds to a homogeneous phase space distribution of the final state particles.

Suppose now only one initial $\bar{p}p$ state with quantum numbers J^{PC} contributes and only particles 1 and 2 resonate with angular momentum l. Assuming for simplicity that all three final state particles are (pseudo)scalars, l equals the spin of the resonance. Then the observed intensity (normalized density of points in the Dalitz plot) is

$$I(s_{12}, s_{13}) = \quad ds_{12}ds_{13} \mid \sum A_{JPC,L,l}(s_{12}, s_{13}) \mid^2,$$

where the sum $\sum$ runs over L, the possible angular momenta between particle 3 and the resonance. The total amplitude is a (coherent) sum over L since, in general, different L may combine with a fixed l to the same J. The partial amplitude $A_{JPC,L,l}$ is given by the following product:

$$A_{JPC,L,l}(s_{12}, s_{13}) = Z_{JPC,L,l}(\vec{p}/p, \vec{q}/q) D_L(p) F_l(q),$$

where $\vec{p}/p$ is the direction of the resonance in the lab and $\vec{q}/q$ is the direction of particle 1 relative to the $\vec{p}$ after a Lorentz boost into the rest sytem of the resonance (Gottfried-Jackson system), see Fig. 10.

The momenta p and q are given in terms of s_{12}, s_{13} by

$$p^2 = [s_{\bar{p}p} - (m_{12} + m_3)^2][s_{\bar{p}p} - (m_{12} - m_3)^2]/4s_{\bar{p}p},$$

$$q^2 = [s_{12} - (m_1 + m_2)^2][s_{12} - (m_1 - m_2)^2]/4s_{12}.$$

The angle θ (Fig. 10) between $\vec{p}$ and $\vec{q}$ is given by

$$2s_{13} = [(s_{13})_{max} + (s_{13})_{min}] + [(s_{13})_{max} - (s_{13})_{min}] cos\theta,$$

where $(s_{13})_{min,max}$ depend on s_{12}. The function $Z_{JPC,L,l}(\vec{p}/p, \vec{q}/q)$ is the spin-parity amplitude describing the angular correlations between the final state particles due to angular momentum conservation. We have used notations which are customary when the non-relativistic Zemach formalism [8] is applied. Other formalisms are in use: the helicity formalism [9] and the Rarita-Schwinger formalism [10]. Recently, covariant formulations of the two former methods have been advocated [11, 12]. The spin parity function Z seems to depend on more variables than s_{12}, s_{13}; however, when the sum of partial amplitudes is squared, the squares or mixed products of the Z-functions (which are in reality tensors) are reduced to scalar functions of the angle θ. The function $D_L(p)$ is the "angular momentum barrier" or "penetration factor, taken to be proportional to p^L or a more complicated expression acccording to Blatt-Weisskopf [7] and the factor $F_l(q)$, the "dynamical function", describes the dynamics. For the case of a single isolated resonance it is given by the relativistic Breit-Wigner amplitude [17]

$$F_l = g^2/(m_0^2 - s_{12} - i\rho g^2)$$

with $\rho = 2q/\sqrt{s_{12}}$, m_0 the nominal resonance mass and $g^2 = m_0\Gamma_0 D_l(q)^2/\rho$.

The process leading to the production of the resonance may be conceptually factorized into two: first, the production of the three-body final state - described by the blob called Q in Fig. 10a. And second, by the resonance, or, more generally, final state interaction between particle 1 and 2, described by the blob called T in Fig. 10a. Thus this production process can be related to a different kind of process, the scattering of particle 1 on particle 2, described by the scattering amplitude T. If the scattering can be inelastic, i.e. in addition to the elastic scattering (e.g. $\pi\pi \rightarrow \pi\pi$), other final states are possible at the energy $\sqrt{s_{12}}$ (e.g. $\pi\pi \rightarrow K\bar{K}$); then T has to be extended to be a matrix, the "T-matrix". For our example with two possible channels, T_{11} and T_{22} describe the elastic scattering $\pi\pi \rightarrow \pi\pi, K\bar{K} \rightarrow K\bar{K}$, resepectively, and T_{12}, T_{21} the transitions $\pi\pi \rightarrow K\bar{K}, K\bar{K} \rightarrow \pi\pi$. The factor Q turns into a vector, the "Q-vector" where Q_1 describes the production $\bar{p}p \rightarrow \pi\pi + spectator$ and Q_2 the production $\bar{p}p \rightarrow K\bar{K} + spectator$. Two diagrams contribute in our example to the final state, with the intermediate particles 1 and 2 between the blob Q and T being $\pi\pi$ or $K\bar{K}$.

We have sketched here what is called the "Q-vector" approach [13] for the production. Other descriptions: the P-vector approach [14] (see for example the publication [15]) or the N/D method (see for example the publication [16]) are being used to describe the production of resonances which couple to more than one continuum state. So far, we have made the simplifying assumption that only one resonance is being produced, between 1 and 2. However, there may be more than one resonance between 1 and 2, with different l, and there may be also resonances, or more generally, final state interactions between 1 and 3 and 2 and 3 and even final state interactions involving all three particles (like the rescattering of one of the decay products of the resonance on the spectator, which of course cannot then be called spectator anymore). Usually, the interactions involving three particles are ignored and it is assumed that the intermediate state always consists of a resonance and a stable hadron, the "spectator" or "bachelor", which do not interact (**isobar model**). (Extending this model to final states with more than three particles it is assumed that they are reached by sequences of this kind where eg. first two resonances are produced which then in turn decay into two resonances and so on.) All these amplitudes interfere with each other as long as they involve the same initial state and overlap in phase space, which is more likely the broader the resonances are. (If different initial states are involved, the corresponding intensities are summed incoherently.) For instance, in one of the examples given above, the annihilation

$$\bar{p}p \rightarrow K^{+*}K^- \rightarrow (K^+\pi^0)K^-,$$

the same final state can also be reached by annihilation via different intermediate states

$$\bar{p}p \rightarrow K^+K^{-*} \rightarrow K^+(\pi^0 K^-) \qquad \text{or} \qquad \bar{p}p \rightarrow \phi\pi^0 \rightarrow (K^+K^-)\pi^0$$

from the same initial state. The bracket signifies a resonance between the final state particles in the bracket. As in the "classic" 2-slit experiment of quantum mechanics, the corresponding three amplitudes have to be summed coherently.

Thus the sum over L in the expression above for the intensity has to be extended to a sum over various l, L and over all permutations of particles to include also resonances of other pairs of particles. Taking this extended sum, phase correlations following from symmetries should be respected. If, for instance, 1 and 2 are identical bosons, the sum of amplitudes has to be symmetrical with respect to the interchange of particles 1 and 2. (Automatically, this eliminates all odd angular momenta l in our example). Or if the final state is an eigenstate of C or G, various terms contributing to the total amplitude are connected by plus or minus signs (relative phase of 0 or 180°). For the case of states with $\bar{K}K$ pairs these eigenfunctions are derived and explicitely listed in the appendix, since they are relevant for the search for mesons with open or hidden strangeness.

In Fig. 11 a series of four Dalitz plots for $\bar{K}K\pi$ final states with different charge combinations are shown as examples. The possible quantum numbers of the $\bar{K}K$ resonances are listed in Table 4.

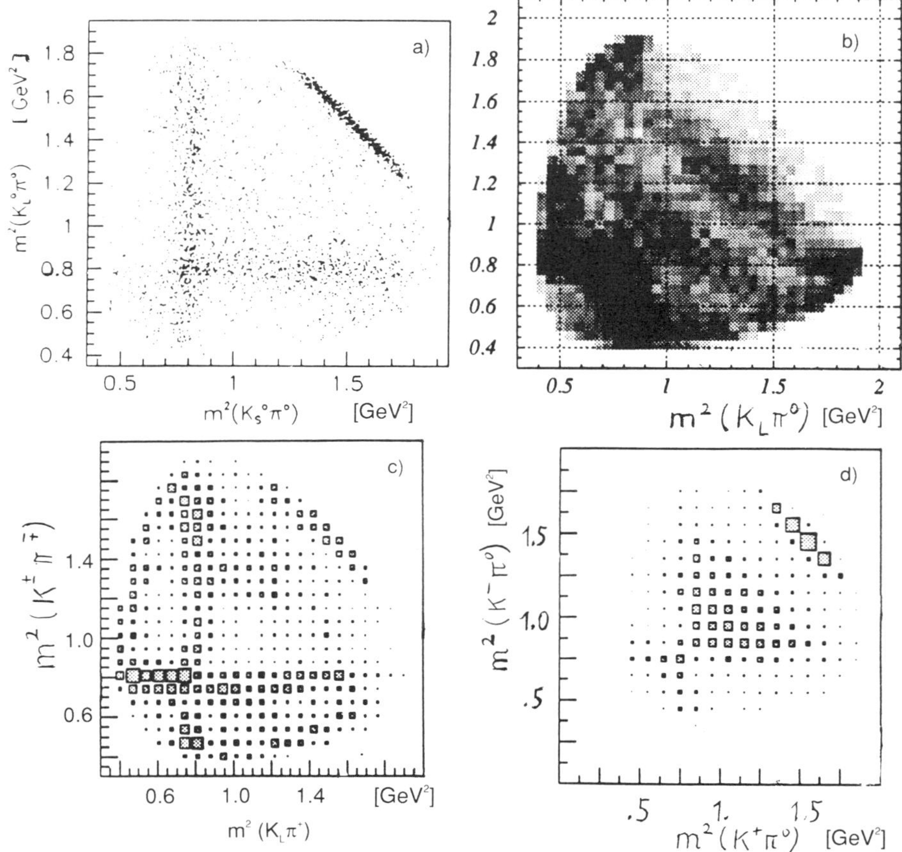

Figure 11: Dalitz plots for four different $\bar{K}K\pi$ final states produced in $\bar{p}p$ annihilation at rest: a) $\bar{p}p \to K_L K_S \pi^0$; b) $\bar{p}p \to K_L K_L \pi^0$; c) $\bar{p}p \to K_L K^\pm \pi^\mp$; d) $\bar{p}p \to K^+ K^- \pi^0$. All the data are from the Crystal Barrel experiment.

Table 4: Quantum numbers of $\bar{K}K$ resonances

$\bar{K}K$	C	Isospin	J^{PC}	possible resonances
$K_L K_S$	-1	0,1	1^{--}	ϕ
$K_L K_L, K_S K_S$	1	0,1	$0^{++}, 2^{++}$	a_0, f_0, a_2, f_2
$K_L K^\pm, K_S K^\pm$		1	$0^+, 1^-, 2^+$	$a_0, a_2, \rho, ..$
$K^+ K^-$	± 1	0,1	$0^{++}, 1^{--}, ..$	$\phi, a_0, f_0, a_2, f_2, ..$

Our first example, (Fig. 11a), from the reaction

$$\bar{p}p \rightarrow K_L K_S \pi^0$$

is a particularly simple case since only few resonances occur. Remember that if no resonances at all would occur and the events were homogeneously distributed over the available phase space, the density of points in this Dalitz plot would be homogeneous within the kinematic borderlines. Resonances show up as vertical and horizontal bands for $K_L\pi^0$ and $K_S\pi^0$ and as antidiagonal bands for the third pair of particles, K_LK_S. Fig. 11a shows exactly one resonance for each case: a vertical and a horizontal band due to $\bar{p}p \rightarrow K^*(890)\bar{K} + cc$, forming a "$K^*$ cross, and an antidiagonal band due to $\bar{p}p \rightarrow \phi(1020)\pi^0$; (cc means charge conjugate, here $\bar{K}^*(890)K$.) The combination K_LK_S with C parity = -1 acts like a filter on possible C-eigenstates leaving in this case only a $J^{PC} = 1^{--}$ resonance, the ϕ. Since the whole final state is a C-parity eigenstate, each individual particle K_L, K_S and π^0 being a C-parity eigenstate, there is a phase relation between the two amplitudes $K_L^*K_S$ and $K_S^*K_L$ where $K_{L,S}^*$ means that this K^* decays to $K_{S,L}\pi^0$ (see appendix). There is no overlap of the three bands, except in the crossing of the vertical and horizontal band. This is a classic example of a clean Dalitz plot with nothing but three resonances and little interference.

The next example, the Dalitz plot of Fig. 11b for the reaction

$$\bar{p}p \rightarrow K_L K_L \pi^0$$

still shows, as dominant neutral $K\pi$ resonance, the $K^*(890)$ but different $\bar{K}K$ resonances are now filtered out, having positive C parity, i.e. f_0's and a_0's. Here interferences play a much stronger role. Through interference with the diagonal $\bar{K}K$ bands the upper parts of the K^* cross are almost wiped out. Next, let us see in Fig. 11c what happens in the final state with a charged pair of K's, where ϕ's and f_0's cannot be produced, see Table 4. This Dalitz plot is ideally suited to extract the strength of a_0 and a_2 production, without the problem of overlap with nearby f_0 and f_2. Some indications for diagonal activity can be seen which interferes with the K* cross.

Now the last Dalitz plot, Fig. 11d, shows a fascinating alternative of interferences although the plot is a bit crude. Interference between the K* cross and the diagonal a's and f's is constructive in the upper right and lower left quarter and destructive elsewhere. It changes sign on the middle lines of the cross, e.g. the left side of the vertical K_S^0 band is eaten up in the upper part and the right side in the lower part.

These Dalitz plots, not completely analysed yet, show in a qualitative way the power of interference. It was said above that interference brings in another complication. True, but it sounds too negative. Once interferences are understood and the right formalism is found to describe them, they are a powerful tool to determine amplitudes. Interferometry in general is one of the finest and strongest methods in physics and not only there. It was by measuring interference of X-rays that the double helical structure of the DNA was found, and looking at the X-ray picture and the corresponding structure of DNA (Fig. 12) it can be seen that some experience is needed to interpret these patterns.

Biographies of strange characters

The sector of strange mesons is particularly rich in experimental information, mainly owing to many productive years of the LASS experiment at SLAC. Among the lowest lying kaons, only the first radially excited pseudoscalar, $K(1460)$, needs confirmation.

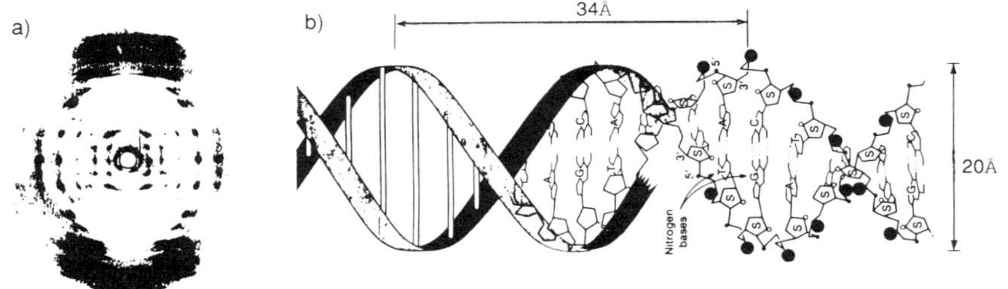

Figure 12: a) X ray diffraction photograph of crystalline DNA by Rosalind Franklin and b) the structure of DNA.

An extraordinarily large number of higher kaon excitations has been found, too. I will discuss in the following sections only some of the open problems for mesons with hidden strangeness. Let us first discard those mesons in this sector which do apparently not pose any problem: These are the ϕ mesons -groundstate and first radial excitation are confirmed- and the tensor meson $f_2'(1525)$. For the vector ground state and first excited state and for the lowest tensor state, the whole nonets are in ideal shape. Ideal in a general and a specific sense: complete information is available and the two isoscalars are ideally mixed, i.e. almost pure $\bar{s}s$ and $(\bar{u}u + \bar{d}d)$. The nonet mixing angles can be found on page 1320 of the Review of Particle Properties [3], called "bible" in the following.

Who or where is f_0'?

The scalar nonet is a mystery and, in particular, the lack of a confirmed f_0' seems scandalous. Since this nonet is the subject of lectures by T.Burnett (0^+ and 2^+ mesons) and also by M.Boutemeur (Glueballs and Hybrids), I restrict the discussion to the question where or what is the isoscalar scalar meson consisting dominantly of $\bar{s}s$, which should be seen via the preferred decay mode into $\bar{K}K$ (for instance K_SK_S, K_LK_L).

According to the "bible" there were three confirmed f_0 mesons, in the year 1994: $f_0(980), f_0(1300)$ and $f_0(1590)$. The $f_0(1500)$ [15], not in the 1994-list of confirmed mesons yet, will certainly make it for the next edition of the "bible" and $f_0(1370)$ [18] will probably be merged with the $f_0(1300)$. Reading the "Note on S-wave $\pi\pi, \bar{K}K$ and $\eta\eta$ interactions in the "bible" [3], page 1478, you get the impression how complicated and interesting the real situation is and how different its interpretation may be from what we think now. The dominant prejudices are that $f_0(980)$ is a $\bar{K}K$ molecule, $f_0(1300)$ is a good, normal meson, mainly $\bar{u}u + \bar{d}d$, $f_0(1500)$ is marketed as a glueball and $f_0(1590)$, which has been one of the best glueball candidates before $f_0(1500)$ had arrived, has been pushed into a corner. In a recent letter by Amsler and Close [20], the $f_0(1590)$ is not even mentioned a single time.

According to prejudice, the f_0' should be about 200 to 300 MeV (twice the quark mass difference $m_s - m_u$) heavier than the $f_0(1300)$ and thus be situated around 1500 MeV. There is a candidate for an f_0' seen only in two experiments and therefore omitted from the "bible"s summary table. Fig.13 shows the modest evidence presented by the LASS collaboration [21] for a K_SK_S S-wave resonance f_0'. The perfect degeneracy with the strong peak $f_2'(1525)$ seen in D-wave is suspicious, although there is no strong reason to exclude that f_0' and f_2' are indeed degenerate, see Fig.3. It is also not excluded that this resonance has the same origin as the $f_0(1500)$ of the Crystal Barrel.

From a bubble chamber experiment ($\bar{p}d$) an upper limit has been published for the production of an $f_0(1500)$ decaying to K_SK_S [19]. (They have, by the way, observed the

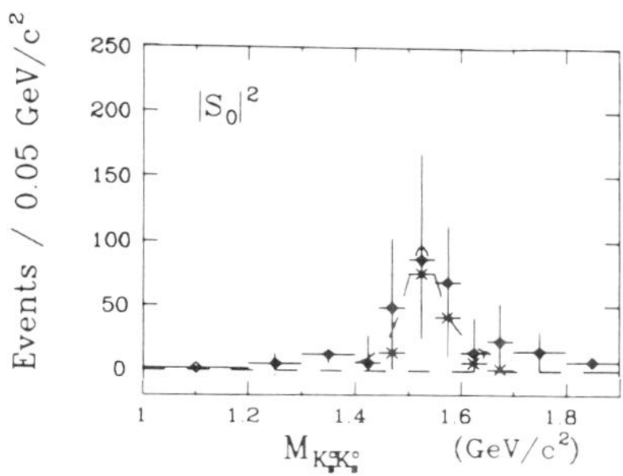

Figure 13: Invariant mass distribution for S-wave $K_S K_S$ produced in $K^- p \to K_S K_S \Lambda$ (LASS experiment).

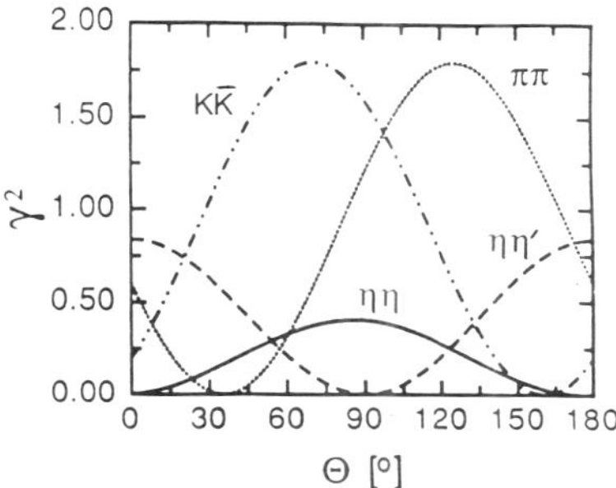

Figure 14: Relative couplings (arbitrary units) to $\pi\pi, \eta\eta, \eta\eta', \bar{K}K$ as a function of the nonet mixing angle from reference Amsler and Close.

first indication of this resonance which they called f_2' in $\pi^+\pi^-$ and gave it a tentative - and correct- assignment J=0!) Their limit on the $K_S K_S$ decay can be translated into a conservative upper limit of 0.1 for the ratio of phase space corrected branching ratios $f_0(1500) \to \bar{K}K$ to $f_0(1500) \to \pi\pi$. This value has been used in the letter [20] as the key result to prove that the $f_0(1500)$ found by Crystal Barrel is a glueball. We discuss their argument in the following for the sake of its didactic value.

The evidence for $f_0(1500)$ being a scalar glueball is based on the ratios of decay branching ratios for $f_0(1500)$:

$$\begin{array}{ccccccc} \pi\pi & : & \eta\eta & : & \eta\eta' & : & \bar{K}K \\ 1.00 & : & 0.27 \pm 0.11 & : & 0.19 \pm 0.08 & : & < 0.1 (95\% CL) \end{array}$$

corrected for phase space. The first three branching ratios were measured by the Crystal Barrel experiment, the latter is the limit from the bubble chamber experiment [19]. The key point is that the branching ratio into $\bar{K}K$ is too low in comparison with the other numbers to be accommodated by any quark composition (mixing angle) of normal mesons. This can be seen in Fig.14 which shows the relative invariant couplings (phase space corrected decay branching ratios, predicted from flavour-SU(3) symmetry) as a function of the octet-singlet mixing angle for the decays of an f_0 into various pairs of pseudoscalars. The $\eta\eta/\pi\pi$ ratio puts us in the range $90^0 \leq \theta \leq 140^0$ (the small range near $\theta = 0$ allowed by this ratio is excluded since it demands a dominant $\bar{K}K$ decay.) The ratio $\eta\eta/\eta\eta'$ which can be anything between 0.6 and 3 favours the region $110^0 \leq \theta \leq 135^0$. As a rough guess, the mixing angle should therefore be around $120^0 - 130^0$. But then the $\bar{K}K/\pi\pi$ ratio ought to be between 0.2 to 0.5, clearly larger than the limit of reference [19].

It remains to be seen what the analyses of the Dalitz plots Fig.11b-d will teach us about contributions of $f_0(f_0')$. A preliminary analysis of the $K_LK_L\pi^0$ and $K_SK_S\pi^0$ Dalitz plots has found an f_0 resonance at 1600 MeV with mass and width compatible with the GAMS $f_0(1590)$ [22]. Its contribution to the intensity of the final state was 30% which combined with the branching ratio

$$BR(\bar{p}p \to K_SK_S\pi^0) = (7.8 \pm 0.6)10^{-4}$$

yields a branching ratio

$$BR(\bar{p}p \to f_0(1600)\pi^0 \to K_SK_S\pi^0) = (2.3 \pm 0.3)10^{-4}.$$

Now let us make the bold assumption that $f_0(1500)$ and $f_0(1600)$ are the same particle –since representatives of GAMS and Crystal Barrel at a recent meeting on a remote island have agreed that the GAMS $f_0(1590)$ and the Crystal $f_0(1500)$ can well be identical; our assumption may not be completely absurd, but should not be taken too seriously at present. Comparing the last branching ratio with

$$BR(\bar{p}p \to f_0(1500)\pi^0 \to 3\pi^0) = (8 \pm 3)10^{-3},$$

we obtain a ratio of branching ratios

$$BR(f_0(1600) \to K_SK_S)/BR(f_0(1500) \to \pi\pi) \approx 0.3.$$

After correcting for phase space ($p_K/p_\pi = 580/750$) and applying a weight of 4 to the K's and of 3 to the π's we get

$$BR(f_0(1500/1600) \to \bar{K}K)/BR(f_0(1500/1600) \to \pi\pi) \approx 0.5.$$

Thus this object $f_0(1500)/(1600)$ could indeed be a $\bar{q}q$ meson with a mixing angle of 120^0.

However, we may have been presumptuous since analysis of all data is not yet complete. Determining the nature of the $f_0(1500)$ and searching for the $\bar{s}s$ partner f_0' of $f_0(1300)$ remains a great challenge.

The main content of this section was based on a common prejudice: The $f_0(980)$ although it prefers to decay into $\bar{K}K$ cannot be the $\bar{s}s$ meson f_0' because its mass is too low relative to that of the $\bar{u}u + \bar{d}d$ resonance $f_0(1300)$. In a recent, highly sophisticated paper (discussed in detail by Burnett in his lectures) Törnqvist [23] presents a coherent interpretation of the lowest scalar mesons $f_0(980), a_0(980), K_0^*(1430)$ and $f_0(1300)$ in the framework of the unitarized quark model. The $f_0(980)$ emerges as a dominantly $\bar{s}s$ resonance with a large $\bar{K}K$ component. If all this is true, the $f_0(1500/1600)$ have to be considered as candidates for first radial excitations!

The radially excited pseudoscalar η'

The lower one of the excited η mesons, $\eta(1285)$, has been detected in four experiments. It has been seen to decay into $\eta\pi^+\pi^-$ and $a_0\pi$. It is natural to identify this meson with a dominantly $(\bar{u}u + \bar{d}d)$ radial excitation of the $\eta(547)$.

Under the entry $\eta(1440)$ in the "bible" you find a plethora of experimental references and a special "Note on the $\eta(1440)$". This material is too rich to be covered in a lecture like this.

In short, an "E" meson with $J^{PC}(I^G) = 0^{-+}(0^+)$ has been first seen in a bubble chamber experiment $\bar{p}p \to (\bar{K}K\pi)\pi\pi$ at a mass of 1425 MeV with a width of 80 MeV by

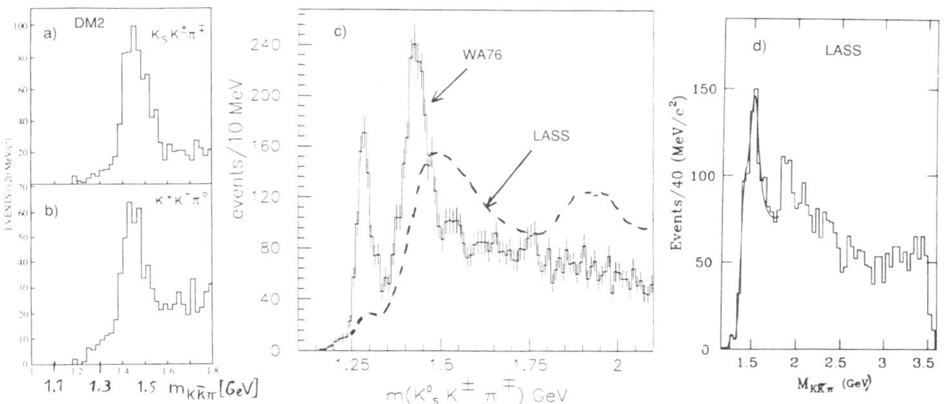

Figure 15: Measured invariant $\bar{K}K\pi$ mass spectra from three experiments: a) and b) DM2 (J/ψ radiative decay); c) WA76 ($\pi^- p \rightarrow K_S K^{\pm}\pi^{\mp} n$) compared with LASS (dashed line) and d) LASS ($K^- p \rightarrow K_S K^{\pm}\pi^{\mp}\Lambda$).

Baillon et al. [24], decaying via $a_0(980)\pi$ and KK^* to $\bar{K}K\pi$. Some time later, confusion was created when in the reaction $\pi^- p \rightarrow (\bar{K}K\pi)n$ a 1^{++} resonance was observed at almost exactly the same mass, also decaying to into $\bar{K}K\pi$; people decided the E meson was an f_1 rather than an η. However, years later a 0^{-+} resonance was rediscovered in radiative J/ψ decays: $J/\psi \rightarrow (\bar{K}K)\gamma$ and was called "iota (ι)". Two groups (Mark III [25] and DM2 [26]) studying radiative J/ψ decays concluded that there are two overlapping 0^{-+} resonances, one at 1420 MeV, decaying into $\bar{K}K\pi$ and $a_0\pi$ and the other one at 1490 MeV mainly decaying into $\bar{K}K\pi$, see Fig.15a,b and Fig.16b-d. Today, the most likely interpretation of all the results is that there are indeed, in addition to two f_1 mesons at 1420 and 1510 MeV, two 0^{-+} (η or η') mesons in this narrow mass range, and perhaps, an h_1' meson as well, see below. (In Fig.3e only one η' state has been plotted.)

Recently, even more experimental information has been obtained and has to be inserted into the frame called the "E/iota" puzzle. One comes from the Crystal Barrel [29]: a 0^{-+} resonance is observed in $\bar{p}p \rightarrow \eta\pi^0\pi^0\pi^+\pi^-$ at rest, decaying into $\eta\pi\pi$ via $a_0\pi$ and $\eta(\pi\pi)_S$ and having a mass of 1409 ± 3 MeV and a width of 86 ± 10 MeV; it is perfectly consistent with the lower iota. The other piece of information comes from the OBELIX experiment, which studied $\bar{p}p \rightarrow \bar{K}K\pi\pi\pi$ with larger statistics than the earlier bubble chamber experiment [30]. OBELIX finds evidence for two pseudoscalar states, at 1406 and 1499 MeV with widths 75 and 240 MeV, respectively; the first one decays mainly via $a_0\pi$, the second one via K^*K to $\bar{K}K\pi$, in agreement with the results from radiative J/ψ decays. Also hints of a 1^{++} resonance are seen, mainly in K^*K around a mass of 1470 MeV and with a width of 80 MeV.

The axial vector meson ground state h_1'

The experimental information is rather restricted for h_1 and h_1'. A very wide $\rho\pi$ resonance with $J^{PC} = 1^{+-}(I = 0)$ has been seen in three experiments at a mass of 1170 MeV. The width is $\approx$ 340 MeV! This resonance is identified with the dominantly $(\bar{u}u + \bar{d}d)$ meson h_1. The $\bar{s}s$ partner h_1' is expected to decay dominantly to $\bar{K}K\pi$. So far, in a single experiment, by the LASS collaboration, a candidate has been observed in

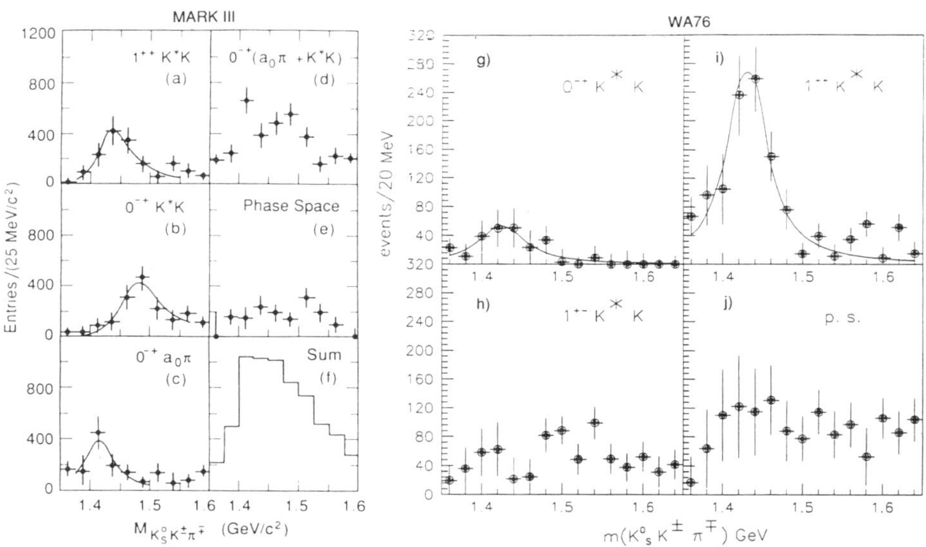

Figure 16: Results of partial wave analyses a)-f) from Mark III ($J/\psi \to \gamma K_S K^{\pm}\pi^{\mp} n$) and g)-j) from WA76 ($\pi^- p \to K_S K^{\pm}\pi^{\mp} n$).

$K^- p \to K_S K^{\pm}\pi^{\mp}\Lambda$ at a mass ≈ 1380 MeV, with a width ≈ 80 MeV, decaying into $K^*\bar{K}$ (+cc) [31]. The invariant mass distribution of $K_S K^{\pm}\pi^{\mp}$ shows a prominent structure around 1.5 GeV (Fig.15d) –and more structure at higher masses, not to be discussed here. The partial wave analysis finds that the 1.5 GeV mass region is dominated by $K^*\bar{K} + cc$ with $J^P = 1^+$. However, the structure cannot be due to a single resonance. The sign of the interference of the two K^* bands changes between the two $\bar{K}K\pi$ mass windows 1.34-1.46 GeV and 1.46 - 1.58 GeV (Fig. 1b,c of ref.[31]): The interference appears to be destructive at the lower mass indicating a G = -1 object there, whilst for the higher mass it is instead constructive as expected for G=+1, see Fig.18e,f. From the measured $K^0\bar{K}^*$ and $\bar{K}^0 K^*$ intensities and the interference between these two amplitudes the intensity of the G parity eigenstates has been extracted and is shown in Fig.17a,b; the authors conclude that two 1^+ resonances are observed in this mass region, one with G=-1 and the other one with G=+1. Arguing that isospin = 0 is more likely in the peripheral hypercharge exchange reaction $K^- p \to K_S K^{\pm}\pi^{\mp}\Lambda$ and hence G=C and that the production of $\bar{s}s$ dominates in this reaction, it is then claimed that the h_1' ground state (Fig.17b) and an $f_1(1520)$ meson (Fig.17a) have been observed.

More recently, the Crystal Barrel collaboration has studied the reaction $\bar{p}p \to K_L K_S \pi^0\pi^0$ ($\to 8\gamma+$ missing energy and momentum)[32]. The advantage of this reaction is that all particle combinations immediately tell the C parity if the parent state was a C eigenstate. The final state is dominated by the strange resonances $K_1(1400)$, $K_1(1270)$, $K_0^*(1430)$ and $K^*(892)$. With a small intensity, a 1^+ state is also needed at 1385 MeV mass with width 200 MeV, decaying to K^*K. The intensity is 3% if it is assumed that $\bar{p}p$ annihilates exclusively from the $^3S_1$ state and 10% if also the $^3P_1$ initial state is allowed in the amplitude analysis. The $K_S K_L \pi^0$ Dalitz plot for a mass window of ± 50 MeV around 1380 MeV shows a clearly destructive interference between the two K^* bands which is consistent with a $J^{PC} = 1^{+-}$ state (G= C). However, background from other amplitudes and the 1^{+-} signal superimpose in this window. The resonant

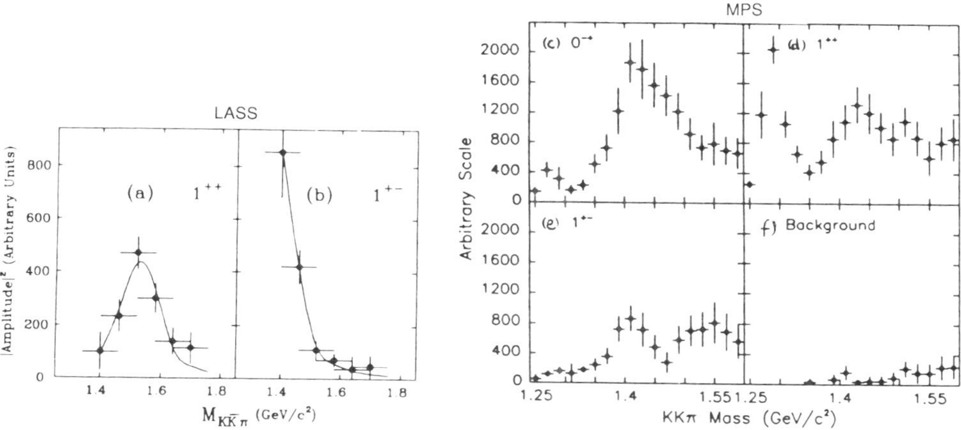

Figure 17: Results from partial wave analyses a,b) from LASS ($K^-p \to K_S K^\pm \pi^\mp \Lambda$); c)-f) from MPS ($\pi^-p \to K_S K^+ \pi^- n$).

1^{+-} amplitude has the weakness that, although being clearly needed in the partial wave analysis, it cannot be made clearly visible to the eye in the $\bar{K}K\pi$ mass projection.

Axial vector mesons f_1'

According to the PDG [3], there are three narrow f_1 resonances in the mass range from 1250 to 1550 MeV where only two quark model mesons are expected: the $f_1(1285)$, $f_1(1420)$, and $f_1(1510)$, with widths of 24, 52 and 35 MeV, respectively. The first one decays with a probability of 54% to $\eta\pi\pi$ (mainly $a_0(980)\pi$), 30% to 4π and 10% to $\bar{K}K\pi$. It appears to be a reliable $\bar{u}u + \bar{d}d$ quark meson. The other two mesons have both been observed exclusively through their decay into $\bar{K}K\pi$; thus they are a priori candidates for $\bar{s}s$ mesons. The $f_1(1510)$ was seen in the LASS experiment, which also saw the h_1' meson [31] and in two more experiments; the $f_1(1420)$ has been identified in many different experiments, in $\bar{p}p$ annihilation, J/ψ decays, $\gamma\gamma$ collisions, and in πp interactions, see for instance Fig.15a-c [28], Fig.16a,i [28] and Fig.17d [27].

On the one hand, it is completely excluded in the quark model to have two narrow states with identical quantum numbers at a distance of only 90 MeV. Moreover (still on the same hand), the situation reminds us of that for the $\eta(1410)/\eta(1490)$ and makes one wonder what the presence of these doublets might signify. The "bible" quotes as preferred interpretation that one of them, the $f_1(1420)$ is not a normal meson, whilst the $f_1(1510)$ is the dominantly $\bar{s}s$ f_1' meson.

On the other hand, a close inspection and comparison on the same mass scale of, the LASS [31] and the WA76 [28] data (Fig.15c) makes one wonder whether these two f_1 mesons are really distinguished from each other. Why is the $f_1(1420)$ meson only seen in one experiment and not in the other one and vice versa for the $f_1(1510)$? The questions raised here at the end of the lectures do not really concern the $f_1(1420)/f_1(1510)$ problem. (I am not in the position, at present, to judge the quality and consistency of various experiments.) But they have a rather general didactic purpose. After having referred to the Review of Particle Properties [3] of the Particle Data Group (PDG) throughout these lectures as something like the bible for the meson and baryon spectroscopist, it is a moral duty to say at least once that even this bible should sometimes be questioned. This is **our** bible. Therefore, whilst we try to forward as many of the miracles that we, the experimentalist, accomplish, into that book, we should not believe that every miracle that has been adopted into the meson (or baryon) summary table, exists beyond doubt.

Appendix: C and G parity eigenstates of $\bar{K}K$ pairs

We adopt the following sign convention for the charge conjugation C and the G parity operation [2]:

$Cu = +\bar{u}$	$Cd = +\bar{d}$	$Cs = +\bar{s}$	$C\bar{u} = +u$	$C\bar{d} = +d$	$C\bar{s} = +s$
$Gu = +\bar{d}$	$Gd = -\bar{u}$	$Gs = +\bar{s}$	$G\bar{d} = -u$	$G\bar{u} = +d$	$G\bar{s} = +s$

where C is the charge conjugation operator and $G = e^{i\pi I_y}C$.
The flavour wave functions of Table 4.2 in [2] are adopted. Thus, for instance:

$$
\begin{aligned}
\pi^+ &= +[u(1)\bar{d}(2) + \bar{d}(1)u(2)]/\sqrt{2} \\
\pi^0 &= +[d(1)\bar{d}(2) - u(1)\bar{u}(2) + \bar{d}(1)d(2) - \bar{u}(1)u(2)]/2 \\
K^+ &= +[u(1)\bar{s}(2) + \bar{s}(1)u(2)]/\sqrt{2} \\
K^- &= -[s(1)\bar{u}(2) + \bar{u}(1)s(2)]/\sqrt{2} \\
K^0 &= +[d(1)\bar{s}(2) + \bar{s}(1)d(2)]/\sqrt{2} \\
\bar{K}^0 &= -[s(1)\bar{d}(2) + \bar{d}(1)s(2)]/\sqrt{2} \\
\rho^+ &= +[u(1)\bar{d}(2) - \bar{d}(1)u(2)]/\sqrt{2} \\
K^{+*} &= +[u(1)\bar{s}(2) - \bar{s}(1)u(2)]/\sqrt{2} \\
K^{-*} &= -[s(1)\bar{u}(2) - \bar{u}(1)s(2)]/\sqrt{2} \\
K^{0*} &= +[d(1)\bar{s}(2) - \bar{s}(1)d(2)]/\sqrt{2} \\
\bar{K}^{0*} &= -[s(1)\bar{d}(2) - \bar{d}(1)s(2)]/\sqrt{2}.
\end{aligned}
$$

The following signs for the action of C, G and G^2 on pseudoscalar mesons are obtained:

$C\pi^0 = +\pi^0$	$G\pi^0 = -\pi^0$	$G^2\pi^0 = +\pi^0$
$C\pi^+ = -\pi^-$	$G\pi^+ = -\pi^+$	$G^2\pi^+ = +\pi^+$
$C\pi^- = -\pi^+$	$G\pi^- = -\pi^-$	$G^2\pi^- = +\pi^-$
$CK^+ = -K^-$	$GK^+ = -\bar{K}^0$	$G^2K^+ = -K^+$
$CK^- = -K^+$	$GK^- = -K^0$	$G^2K^- = -K^-$
$CK^0 = -\bar{K}^0$	$GK^0 = +K^-$	$G^2K^0 = -K^0$
$C\bar{K}^0 = -K^0$	$G\bar{K}^0 = +K^+$	$G^2\bar{K}^0 = -\bar{K}^0$

For the strange **vector** mesons one obtains just the opposite signs for C and G:

$CK^{+*} = +K^{-*}$	$GK^{+*} = +\bar{K}^{0*}$	$G^2K^{+*} = -K^{+*}$
$CK^{-*} = +K^{+*}$	$GK^{-*} = +K^{0*}$	$G^2K^{-*} = -K^{-*}$
$CK^{0*} = +\bar{K}^{0*}$	$GK^{0*} = -K^{-*}$	$G^2K^{0*} = -K^{0*}$
$C\bar{K}^{0*} = +K^{0*}$	$G\bar{K}^{0*} = -K^{+*}$	$G^2\bar{K}^{0*} = -\bar{K}^{0*}$

The action of G^2 causes a minus sign for all kaons. A 2π rotation in isospace for an isospin 1/2 particle has the same effect as the corresponding rotation in space on the wavefunction of a spin 1/2 particle.

Pairs of $K\bar{K}$.

Here we are mainly interested in the C and isospin eigenstates or alternatively, G eigenstates. The $|I, I_z = 0, 0>$ and $|I, I_z = 1, 0>$ wave functions of $\bar{K}K$, combinations of an I- doublet with an I-antidoublet, like ρ^0 and ω, or π^0 and η, are

$$|I, I_z = 0, 0> \quad = |K^0(1)\bar{K}^0(2) + K^+(1)K^-(2) > /\sqrt{2}$$
$$|I, I_z = 1, 0> \quad = |K^0(1)\bar{K}^0(2) - K^+(1)K^-(2) > /\sqrt{2}.$$

(Notice that the signs are opposite to those of the standard Clebsch-Gordan coefficients, see chapter 1 for an explanation). These are not yet eigenfunctions of C, but from the action of C on the K's the eigenstates can be constructed. (In the following, the particle indices in brackets which refer to a given spatial state of one particle will be omitted).

I	I_z	C	$\bar{K}K$ *flavour wavefunctions*	G	L
0	0	+1	$[K^0\bar{K}^0 + K^+K^- + \bar{K}^0K^0 + K^-K^+]/2$	+1	$0, 2, \ldots$
0	0	−1	$[K^0\bar{K}^0 + K^+K^- - (\bar{K}^0K^0 + K^-K^+)]/2$	−1	$1, 3\ldots$
1	0	+1	$[K^0\bar{K}^0 - K^+K^- + (\bar{K}^0K^0 - K^-K^+)]/2$	−1	$0, 2, ..$
1	0	−1	$[K^0\bar{K}^0 - K^+K^- - (\bar{K}^0K^0 - K^-K^+)]/2$	+1	$1, 3, ..$

What do these eigenfunctions tell us? For meson antimeson pairs, parity P, C-parity and G-parity are related to L,S and I by similar relations as for fermion-antifermion pairs (we use capital L here for the orbital angular momentum, S for the total spin of the pair):

$$P = (-1)^L$$
$$C = (-1)^{L+S}$$
$$G = (-1)^I C = (-1)^{I+L+S}.$$

Thus it is seen from the expression for P (or more directly, from the symmetries of the wave function with respect to particle interchange) that the first and third of the 4 eigenfunctions correspond to a wave function with L = 0, 2 .. and the remaining two have odd L, see the last column. It can be verified with the relation for G, or directly from applying the rules listed above for the action of G on the K's and $\bar{K}$'s, that G takes the values given in the next to last column.

Similarly, the wave functions for other pairs, like KK^*, K^*K^* can be obtained. The C, I and G eigenstates for neutral ($I_z = 0$) pairs of $K^*\bar{K}^* + cc$ have the same sequence of signs:

I	I_z	C	$K^*\bar{K}^*$ *flavour wavefunctions*	G	$S+L$
0	0	+1	$[K^{0*}\bar{K}^{0*} + K^{+*}K^{-*} + \bar{K}^{0*}K^{0*} + K^{-*}K^{+*}]/2$	+1	$0, 2, \ldots$
0	0	−1	$[K^{0*}\bar{K}^{0*} + K^{+*}K^{-*} - (\bar{K}^{0*}K^{0*} + K^{-*}K^{+*})]/2$	−1	$1, 3\ldots$
1	0	+1	$[K^{0*}\bar{K}^{0*} - K^{+*}K^{-*} + (\bar{K}^{0*}K^{0*} - K^{-*}K^{+*})]/2$	−1	$0, 2, ..$
1	0	−1	$[K^{0*}\bar{K}^{0*} - K^{+*}K^{-*} - (\bar{K}^{0*}K^{0*} - K^{-*}K^{+*})]/2$	+1	$1, 3, ..$

For pairs of $K\bar{K}^*, K^*\bar{K}$, the signs in the middle are reversed because the C parity of K^* is opposite to that of K:

I	I_z	C	$K\bar{K}^*(\bar{K}^0K^*)$*flavour wavefunctions*	G
0	0	+1	$[K^0\bar{K}^{0*} + K^+K^{-*} - (\bar{K}^0K^{0*} + K^-K^{+*})]/2$	+1
0	0	−1	$[K^0\bar{K}^{0*} + K^+K^{-*} + (\bar{K}^0K^{0*} + K^-K^{+*})]/2$	−1
1	0	+1	$[K^0\bar{K}^{0*} - K^+K^{-*} - (\bar{K}^0K^{0*} - K^-K^{+*})]/2$	−1
1	0	−1	$[K^0\bar{K}^{0*} - K^+K^{-*} + (\bar{K}^0K^{0*} - K^-K^{+*})]/2$	+1

In the last case the four terms in the sum seem to be physically distinguished, not just permutations of particle indices as in the case of $K\bar{K}$ and $K^*\bar{K}^*$. However, the final states resulting from the decay of the K^*s overlap in some regions of phase space and therefore interference between the various terms takes place. We need additional relations to obtain the correct signs of interference in the final state, namely those which connect the various K^*s with their decay products:

$$\begin{aligned}
K^{+*} &= -\sqrt{1/3}(K^+\pi^0) + \sqrt{2/3}(K^0\pi^+) \\
K^{-*} &= +\sqrt{1/3}(K^-\pi^0) + \sqrt{2/3}(\bar{K}^0\pi^-) \\
K^{0*} &= +\sqrt{1/3}(K^0\pi^0) - \sqrt{2/3}(K^+\pi^-) \\
\bar{K}^{0*} &= -\sqrt{1/3}(\bar{K}^0\pi^0) - \sqrt{2/3}(K^-\pi^+).
\end{aligned}$$

Here, we have used standard Clebsch Gordan coefficients for the coupling of I = 1 with I = 1/2 particles to I = 1/2 particles, **with two exceptions** where the signs are opposite [33]: the K^- and K^{-*}, which contain the $\bar{u}$ quark have to be provided with a minus sign, consistent with the change of sign in the isospin wave functions of $\bar{K}K$ pairs, see above. In the old literature, see for instance [34, 35], which may lead to some confusion today, standard Clebsch Gordan coefficients were used throughout. However, their final result for the sign of interference in the final state is consistent with ours.

Since the experiments identify and distinguish certain charge combinations in the final state, for instance, in the case of KK^*, the combination $K^+K^0\pi^-$ (with $I = 1$ or $0, I_z = 0$) or $K^0K^-\pi^0$ (with $I = 1, I_z = -1$), only some of the terms have to be considered at the end. With the formulae given above one can derive the expression for the G parity eigenstates, valid for arbitrary isospin I and I_z:

$$(\bar{K}K\pi)_G = (K(1)\bar{K}(2)\pi(3) + G\bar{K}(1)K(2)\pi(3))/\sqrt{1/6},$$

where G refers to the state $\bar{K}K\pi$. For specific charge states one obtains, for instance

$$(\bar{K}K\pi, I = 0 \;\; or \;\; 1, I_z = 0)_G = (K^+\bar{K}^0\pi^- + G\bar{K}^0K^+\pi^-)/\sqrt{1/6}$$

or

$$(\bar{K}K\pi, I = 1, I_z = -1)_G = (K^0K^-\pi^0 + GK^-K^0\pi^0)/\sqrt{1/6}.$$

A positive G parity does not necessarily imply that the interference in the region of overlap is constructive. This depends, in addition, on the spatial, or more specifically on the angular wave function, i.e. its symmetry for interchange of the particle indices. Fig.18 shows some interesting cases of J^{PC}, G eigenstates decaying into KK^*. These patterns can be used in order to deduce the G parity from the observed interference pattern if J^{PC} is known or assumed to be known.

If the final state under consideration is not a C eigenstate, the G eigenstates are still useful since G is conserved in the strong interaction. We consider as an example pairs with negative charge only, having isospin $I, I_z = 1, -1$, i.e. $K^0K^-, K^{0*}K^{-*}$ and $K^0K^{-*}, K^{0*}K^-$. The G eigenstates can be constructed as above.

$\bar{K}K$ *flavour wavefunctions*	G	L
$(K^0K^- + K^-K^0)/\sqrt{2}$	-1	$0, 2, \ldots$
$(K^0K^- - K^-K^0)/\sqrt{2}$	$+1$	$1, 3, \ldots$

$K^*\bar{K}^*$ *flavour wavefunctions*	G	$S + L$
$(K^{0*}K^{-*} + K^{-*}K^{0*})/\sqrt{2}$	-1	$0, 2, \ldots$
$(K^{0*}K^{-*} - K^{-*}K^{0*})/\sqrt{2}$	$+1$	$1, 3, \ldots$

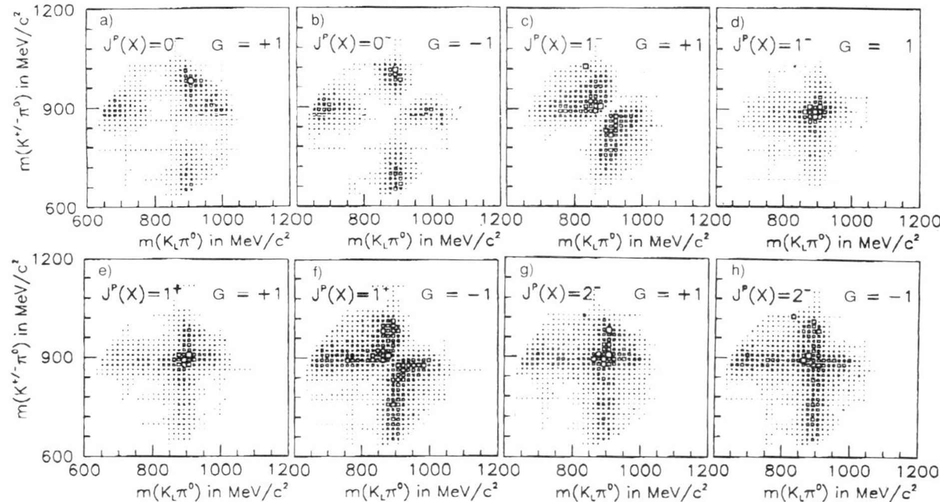

Figure 18: Dalitz plots (Monte Carlo generated by C.Kolo) showing the interference patterns for various J^P and G combinations of a meson with mass 1530 MeV decaying into $\bar{K}K\pi$ via $K^*\bar{K} \pm cc$. J^{PG} = a) 0^{-+}, b) 0^{--}, c) 1^{-+}, d) 1^{--}, e) 1^{++}, f) 1^{+-}, g) 2^{-+}, h) 2^{--}.

KK^* *flavour wavefunctions*	G
$(K^0K^{-*} + K^-K^{0*})/\sqrt{2}$	$+1$
$(K^0K^{-*} - K^-K^{0*})/\sqrt{2}$	-1

Now there is another set of expressions which are useful when one deals with neutral K's. The eigenstates above were expressed in terms of $K^0, \bar{K}^0$ which are eigenstates of the strong interaction, but not the mesons seen by the detector. The observed particles are the K_L and K_S, which are to a good approximation C eigenstates as explained in the text and to a better approximation CP eigenstates. For our purpose we can use

	C	P	CP
$K_S \approx K_1 = (K^0 + \bar{K}^0)/\sqrt{2}$	-1	-1	$+1$
$K_L \approx K_2 = (K^0 - \bar{K}^0)/\sqrt{2}$	$+1$	-1	-1

The K_1 has exotic quantum numbers $J^{PC} = 0^{--}$! (but no defined isospin). For pairs of neutral K's one finds the following identities by substitution:

$C = +1$	$K^0\bar{K}^0 + \bar{K}^0K^0$	$K_SK_S - K_LK_L$
$C = -1$	$K^0\bar{K}^0 - \bar{K}^0K^0$	$K_LK_S - K_SK_L$

It seems as if we have too many terms here since each pair (like K_SK_S) is in itself already an eigenstate of C, each particle being a C-eigenstate. However, substituting the the inverse relations

$$K^0 = (K_S + K_L)\sqrt{2}$$

$$\bar{K}^0 = (K_S - K_L)\sqrt{2}$$

it is seen that the differences of the two terms are necessary to cancel pairs with double strangeness, like K^0K^0.

As an example, for a neutral $K\bar{K}^*$ pair decaying to neutral K's, consider the final state $K_L K_S \pi^0$. The C eigenstates are:

$C = +1$	$K^0\bar{K}^{0*} - \bar{K}^0 K^{0*}$	$K_L K_L{}^* - K_S K_S{}^*$
$C = -1$	$K^0\bar{K}^{0*} + \bar{K}^0 K^{0*}$	$K_S K_L{}^* - K_L K_S{}^*$

where "$K^*_{L,S}$" is a short-hand for $K^0(\bar{K}^0)$ decaying to $K_{L,S}\pi^0$ with angular momentum L = 1. Similarly, for $K^*\bar{K}^*$

$C = +1$	$K^{0*}\bar{K}^{0*} + \bar{K}^{0*}K^{0*}$	$K_L{}^* K_L{}^* - K_S{}^* K_S{}^*$
$C = -1$	$K^{0*}\bar{K}^{0*} - \bar{K}^{0*}K^{0*}$	$K_L{}^* K_S{}^* - K_S{}^* K_L{}^*$

Again, as above, each term in the right column is a C eigenstate and the differences are necessary in order to cancel double strangeness terms.

Acknowledgements: I wish to thank the students and my coteachers and above all, the organizer, David Bugg for a very stimulating and enjoyable school.

References

[1] A.Pais, Niels Bohr's Times, Clarendon Press Oxford 1991.
[2] F.E.Close, An Introduction to quarks and Leptons, Academic Press, London 1979.
[3] L.Montanet et al., Review of Particle Properties, Particle Data Group, Phys. Rev. D 50 (1994) 1173.
[4] S. Okubo, Phys. Lett. 5 (1963) 165; G. Zweig, CERN Repor TH412 (1964); I. Iizuka, Prog. Theor. Phys. Suppl. 37 (1966) 21.
[5] C.Amsler et al., Phys. Lett. B333 (1994) 277.
[6] E. Aker et al., Nucl. Instrum. & Methods A 321 (1992) 69.
[7] J.M.Blatt and V. Weisskopf, Theoretical Nuclear Physics, Wiley, New York (1952)
[8] C.Zemach, Phys.Rev. B133 (1964) 1201; B140 (1965) 97, 109.
[9] M.Jacob and G.C.Wick, Ann. Phys. (USA) 7 (1959) 404.
[10] W.Rarita and J.Schwinger, Phys.Rev. 60 (1941) 61.
[11] V.Filippini, A.Fontana, and A. Rotondi, Phys.Rev. D51 (1995) 2247.
[12] S.U.Chung, Preprint BNL-QGS94-22, submitted to Phyds.Rev. D.
[13] R.N.Cahn and P.V.Landshoff, Nucl. Phys. B266 (1986) 451.
[14] I.J.R.Aitchison, Nucl.Phys. A189 (1972) 417.
[15] C.Amsler et al., Phys. Lett. B355 (1995) 425.
[16] V.V.Anisovich et al., Phys.Rev. D50 (1994) 1972; D.V.Bugg et al., Phys.Rev. D50 (1994) 4412.
[17] J.D.Jackson, Nuovo Cimento, 34 (1964) 1644.
[18] C.Amsler et al., Phys. Lett. B322 (1994) 431.
[19] L.Gray et al., Phys.Rev.D 27 (1983) 307.
[20] C.Amsler, F.E.Close, Phys.Lett. B 353 (1995) 385.
[21] D.Aston et al., Nucl. Phys. B301 (1988) 525.
[22] H.P.Dietz, PhD thesis, University Munich 1994.
[23] N.A.Törnqvist, Understanding the scalar mesons $\bar{q}q$ nonet, submitted to Z. Phys. C (1995).
[24] P. Baillon et al., Nuov. Cim. 50 A (1967) 393.
[25] Z.Bai et al. (MARKIII),, Phys. Rev. Lett. 65 (1990) 2507.
[26] J.-E.Augustin et al. (DM2), Phys. Rev. D 46 (1992) 1951.
[27] A.Birman et al.(MPS), Phys. Rev. Lett. 61 (1988) 1557.
[28] T.A.Armstrong et al.(WA76), Z.Phys. C 56 (1992) 29.

[29] C.Amsler et al., (Crystal Barrel Coll.) " E decays to $\eta\pi\pi$ in $\bar{p}p$ annihilation at rest", Phys. Lett B (1995) accepted for publication.
[30] V.G.Ableev et al.(OBELIX Coll.) " Study of the E/ι decay to $\bar{K}K\pi$ in $\bar{p}p$ annihilation at rest with the OBELIX spectrometer", presented by A.Masoni at LEAP94, Bled.
[31] D.Aston et al., Phys. Lett B 201 (1988) 573.
[32] C.Felix (Crystal Barrel Coll.), presented at the High Energy Physics conference, Marseille 1993.
[33] I thank C.Zupancic for helping me at this point after a long battle with signs.
[34] N.Barash et al., Phys. Rev. B139 (1965) 1659.
[35] R.Armenteros and B.French in High Energy Physics ed. E.H.S.Burhop, Academic Press New York, London 1969.

THEORETICAL ASPECTS OF LIGHT MESON SPECTROSCOPY

T. Barnes

Computational and Theoretical Physics Group
Oak Ridge National Laboratory
Oak Ridge, TN 37831-6373
and Department of Physics and Astronomy
University of Tennessee, Knoxville, TN 37996-1200

Abstract

In this pedagogical review we discuss the theoretical understanding of light hadron spectroscopy in terms of QCD and the quark model. We begin with a summary of the known and surmised properties of QCD and confinement. Following this we review the nonrelativistic quark potential model for $q\bar{q}$ mesons and discuss the quarkonium spectrum and methods for identifying $q\bar{q}$ states. Finally we review theoretical expectations for non-$q\bar{q}$ states (glueballs, hybrids and multiquark systems) and the status of experimental candidates for these states.

1 INTRODUCTION

1.1 The QCD and QED Lagrangians

Quantum chromodynamics (QCD) describes the strong interaction as a quantum field theory of interacting spin-1/2 Dirac fermions (the quarks) and eight vector gauge fields (the gluons). (For a review of early QCD references see Marciano and Pagels [1].) There are six different types "flavors" of quark, which have different masses, electric charges and weak interactions but identical strong interactions. For each of these flavors of quarks there is an internal label called "color", $i = 1, 2, 3$, and the three colors of quarks are coupled to the $a = 1..8$ colors of gluons in a manner described by the QCD Lagrangian. The QCD Lagrangian density for the gluons and a *single* quark flavor is

$$\mathcal{L}_{QCD} = \bar{\psi}_i \Big(\delta_{ij}(i\not\partial - m_q) - g\frac{\lambda^a_{ij}}{2}\not A^a \Big)\psi_j - \frac{1}{4}F^a_{\mu\nu}F^a_{\mu\nu} \tag{1}$$

where the gluon field strength tensor $F^a_{\mu\nu}$ is defined in terms of the gluon field by

$$F^a_{\mu\nu} = \partial_\nu A^a_\mu - \partial_\mu A^a_\nu + gf^{abc}A^b_\mu A^c_\nu \ . \tag{2}$$

The $\{\lambda^a_{ij}\}$ are the 3×3 Gell-Mann SU(3) matrices which couple a quark, antiquark and gluon together in a "color-singlet" combination that is invariant under SU(3) gauge

Hadron Spectroscopy and the Confinement Problem
Edited by D.V. Bugg, Plenum Press, New York, 1996

transformations. (These are mathematically analogous to the Pauli matrices in the isospin-conserving pion-nucleon interaction $\bar{\Psi}\vec{\tau}\Psi \cdot \vec{\pi}$. This coupling implies that the quark-gluon interaction has an internal SU(3) gauge symmetry.)

The QCD Lagrangian is quite similar to the QED form

$$\mathcal{L}_{QED} = \bar{\psi}\left(i\partial\!\!\!/ - m - eA\!\!\!/\right)\psi - \frac{1}{4}F_{\mu\nu}F_{\mu\nu} \tag{3}$$

$$F_{\mu\nu} = \partial_\nu A_\mu - \partial_\mu A_\nu \tag{4}$$

except that in QED there is only a single "color" degree of freedom. These two Lagrangians are invariant under the gauge transformations

$$\psi \to e^{ie\Lambda(x)}\psi \quad \text{(QED)}$$

$$A_\mu \to A_\mu - \partial_\mu\Lambda(x) \quad \text{(QED);} \tag{5}$$

$$\psi_i \to e^{ig\frac{\lambda^a_{ij}}{2}\Lambda^a(x)}\psi_j \quad \text{(QCD)}$$

$$A^a_\mu \to A^a_\mu - \partial_\mu\Lambda^a(x) + gf^{abc}\Lambda^b(x)A^c_\mu \quad \text{(QCD).} \tag{6}$$

The invariance of the Lagrangians under these gauge transformations is crucial, since without this symmetry, interacting theories involving vector bosons in three space dimensions are nonrenormalizable. With this gauge invariance only a finite number (three) of independent infinities arise in perturbative QED and QCD calculations of scattering amplitudes, so once these infinities are absorbed in the definitions of three quantities taken from experiment, the remaining physical quantities calculated are finite and can be compared with experiment.

In both theories the $\bar{\psi}\psi A$ interaction implies an elementary fermion-antifermion-vector vertex, the difference in QCD being the presence of the λ^a_{ij} color matrix that gives the couplings of the different quark and gluon colors. The most important difference between QED and QCD is the presence in QCD of gluon self-couplings. Since the gluon field strength tensor $F^a_{\mu\nu}$ contains both linear and quadratic A terms, its square in the QCD Lagrangian gives three-gluon and four-gluon interactions. These gluon self-interactions qualitatively change the nature of the spectrum of QCD relative to QED.

The lowest-order interaction between an electron-positron pair in QED is due to one photon exchange, which for slowly moving fermions gives the familiar $V(r) = -\alpha/r$ Coulomb potential. At next order in an expansion in v^2/c^2 this one photon exchange matrix element gives the spin-dependent terms of the Breit-Fermi Hamiltonian, spin-spin, spin-orbit and tensor. Since the strength of the electron-photon interaction at laboratory scales is quite weak, one photon exchange is a very good approximation. One may of course calculate the electron-positron interaction due to more complicated processes such as vacuum polarization loops and vertex corrections, which at the one-loop level in perturbation theory give $O(\alpha^2)$ contributions to the effective e^+e^- interaction. One finds that these one-loop corrections have a momentum dependence that make the electron-positron interaction appear slightly weaker with increasing distances than predicted by one photon exchange alone; this effect is known as screening, and gives logarithmic corrections to the $-\alpha/r$ potential.

In QCD there are additional one-loop corrections due to the gluon self-coupling, and these are large enough to change the sign of the screening effect relative to QED. This results in an effective quark "color charge" that increases with increasing separation between sources. This is evident in the increased α_s required in potential models to describe the spectroscopy of lighter quarks, which have wavefunctions with larger length scales; typically, fits to spectroscopy require $\alpha_s(u\bar{u}, d\bar{d}, s\bar{s}) \approx 0.6, \alpha_s(c\bar{c}) \approx 0.4$ and $\alpha_s(b\bar{b}) \approx 0.15$. This "antiscreening" is a perturbative indication of the presence of the remarkable effect known as confinement.

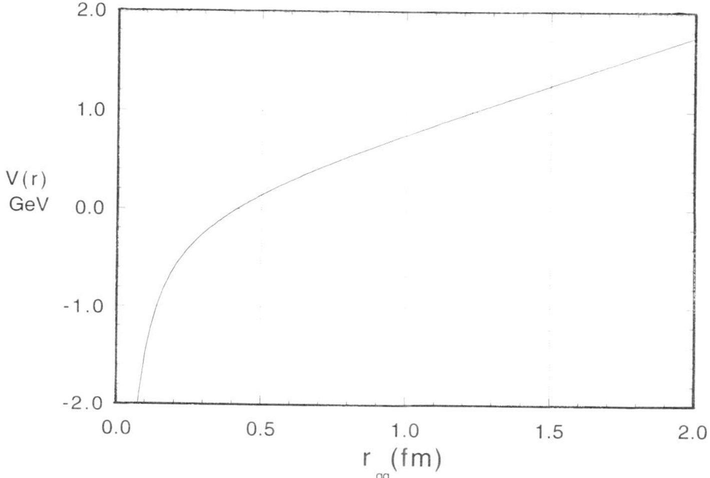

Figure 1. Light quarkonium Coulomb plus linear potential; alpha$_s$=0.6, b=0.18 GeV$_2$.

If we attempt to separate a quark-antiquark pair in a meson, we find that at large separations this antiscreening effect dominates the physics, and replaces the $1/r$ term in the OGE color Coulomb interaction by what appears to be a linear potential,

$$\lim_{r\to\infty} V_{q\bar{q}}(r) = b\, r \ , \tag{7}$$

where the constant r is known as the "string tension", and is numerically found to be

$$b \approx 0.18 \text{ GeV}^2 \approx 0.9 \text{ GeV/fm} \ . \tag{8}$$

The origins of confinement are poorly understood; it is widely believed to be a nonperturbative effect, and in lattice gauge theory simulations is seen to involve the formation of a "flux tube" of chaotic color fields between color charges. In potential models of mesons such as we will discuss, the confining interaction is treated phenomenologically as a linear potential which has a Lorentz scalar coupling to quarks. (This is not obvious, and the earliest models assumed that the coupling was between timelike vector currents, as for the Coulomb interaction.) The combination of OGE color Coulomb and scalar confining interactions gives an effective potential for a slowly moving color-singlet $q\bar{q}$ pair of

$$V_{q\bar{q}}(r_{q\bar{q}}) = -\frac{4}{3}\frac{\alpha_s}{r} + b\, r \ . \tag{9}$$

This potential is shown in Figure 1 with parameters typical of those found in fits to light $q\bar{q}$ systems, $\alpha_s = 0.6$ and $b = 0.18$ GeV^2. The crossover from the perturbative OGE regime to separations at which confinement is dominant evidently takes place at $r_{q\bar{q}} \approx 0.4$ fm.

An additional constant V_0 is usually added to this potential, and to fit light $q\bar{q}$ spectroscopy a large negative value of $V_0 \approx -0.8$ GeV is required. This V_0 is largely needed to cancel the fictitious $2m_q$ present when "consitituent" light quark masses of ≈ 330 MeV are used in the Schrödinger equation, so we do not include a V_0 in the $V(r)$ of Figure 1.

1.2 Color Singlets in QCD

Since the confining interaction operates to confine any colored objects, the physically observed hadrons must be color singlets. Since quarks transform as a fundamental representation 3 of the color SU(3) gauge group, and antiquarks as the conjugate representation $\bar{3}$, we may use group theory to determine the color representations of product states and search for the physically allowed color singlets. First, for two quarks,

$$q \otimes q = 3 \otimes 3 = 6 \oplus \bar{3} \tag{10}$$

so "diquarks" cannot exist as separate hadrons. At the three quark level however we find

$$q \otimes q \otimes q = 10 \oplus 8 \oplus 8 \oplus 1 \ , \tag{11}$$

so three quark systems can exist as physical states; these are the conventional baryons, and the color-singlet wavefunction is the obvious contraction of three vectors to a singlet,

$$|qqq(1)\rangle = \frac{1}{\sqrt{6}} \sum_{i,j,k=1}^{3} |i,j,k\rangle \ . \tag{12}$$

Similarly any system of q^{3N} quarks has color singlet combinations, and can exist as physical states; the question is then the dynamical one of whether such multiquark systems are realized in nature as single multiquark clusters, or (as appears more likely) as loosely associated systems of q^3 color singlet baryons, such as the atomic nuclei.

The same arguments applied to quark and antiquark systems tell us that $q\bar{q}$ can exist as a physical state,

$$q \otimes \bar{q} = 3 \otimes \bar{3} = 8 \oplus 1 \ , \tag{13}$$

which are the conventional $q\bar{q}$ quark model mesons. Again higher states are possible, and the first new color singlet which appears beyond $q\bar{q}$ in the quark-antiquark Hilbert space is in the combination $q^2\bar{q}^2$,

$$q^2 \otimes \bar{q}^2 \subset 1 \ , \tag{14}$$

and similarly for any combination of the form $(q^3)^m(q\bar{q})^n$. Although much effort in the early days of the quark model was expended on calculations of the properties of hypothetical multiquark hadrons such as $q^2\bar{q}^2$ "baryonia", q^6 "dibaryons" and so forth, it now appears that these may not exist as resonances (with rare exceptions) because they can be rearranged into separate $q\bar{q}$ mesons and q^3 baryons *without interaction.*

We can also look for color singlet basis states in combinations involving gluons, which obviously cannot be arranged into separate $q\bar{q}$ and q^3 systems. Since the gluon transforms as a color octet, the first physically allowed combination appears at the gg level,

$$g \otimes g = 8 \otimes 8 = 27 \oplus 10 \oplus \overline{10} \oplus 8 \oplus 8 \oplus 1 \ . \tag{15}$$

Since combining another gluon will repeat the $8 \otimes 8 \subset 1$ product, we can make physically allowed hadron basis states from any number of gluons greater than 1.

The final type of color-singlet combination involves quarks, antiquarks and gluons. Since $q\bar{q}$ contains a color octet, if we form the combination $q\bar{q}g$ we find a color singlet basis state,

$$q \otimes \bar{q} \otimes g = 3 \otimes \bar{3} \otimes 8 = (8 \oplus 1) \otimes 8 \subset 1. \tag{16}$$

This $q\bar{q}g$ is the simplest possible color-singlet combination of quarks, antiquarks and gluons, and is known as a hybrid. We shall see that physical states made from this basis state and others can have quantum numbers forbidden to $q\bar{q}$ states, which makes the search for hybrids one of the most interesting topics in hadron spectroscopy today.

Similarly one may form color singlet "hybrid baryon" basis states, since

$$q \otimes q \otimes q \otimes g = 3 \otimes 3 \otimes 3 \otimes 8 = (10 \oplus 8 \oplus 8 \oplus 1) \otimes 8 \subset 1 \ , \tag{17}$$

so hybrid baryon resonances are expected in addition to the conventional q^3 quark model baryons. Although the combinations $(q\bar{q})^{\ell}(q^3)^m g^n$ all contain color singlets for $\ell + m + n \geq 2$, they need not lead to additional hybrid resonances beyond $q\bar{q}g$ and q^3g because of the fall-apart effect discussed above for $q^2\bar{q}^2$.

2 $Q\bar{Q}$ MESONS

2.1 $q\bar{q}$ States and Quantum Numbers

We will begin our discussion of spectroscopy with a summary of the properties of the conventional quark model "$q\bar{q}$" mesons, and we will specialize to mesons with identical quark and antiquark. The quarks are spin-1/2 Dirac fermions and have charges of $+2e/3$ (for u,c,t quarks) and $-e/3$ (for d,s,b quarks). Our discussion will deal with the $(u,d)(\bar{u},\bar{d})$, $d\bar{d}$, $s\bar{s}$ and $c\bar{c}$ systems. All these except $c\bar{c}$ are highly relativistic bound states and are rather similar physically; $c\bar{c}$ is only quasirelativistic and hence has a relatively simple level structure, which makes it a useful model system for our discussion of quark-antiquark forces. Since the u and d quarks are quite similar in mass, $\approx 5 - 10$ MeV and this is much smaller than the intrinsic mass scale of QCD itself, it is convenient to treat them as partners in an "isospin" doublet,

$$[u, d] = [(1/2, 1/2), (1/2, -1/2)]_{I,I_z} \ , \tag{18}$$

and an antiquark isodoublet, with a (-1) phase required to write the antiquark basis states (which transform according to the conjugate representation of SU(2)) as members of the fundamental representation;

$$[-\bar{d}, \bar{u}] = [(1/2, 1/2), (1/2, -1/2)]_{I,I_z} \ . \tag{19}$$

Hadrons containing s quarks have rather similar properties to (u,d) systems, although the heavier strange quark mass of ≈ 150 MeV means that the $s\bar{s}$ states are well displaced in mass from the (u,d) $q\bar{q}$ states, so with the exception of the $\eta - \eta'$ system there appears to be little mixing between $s\bar{s}$ and light $q\bar{q}$ states.

When we form the physically allowed $q\bar{q}$ combinations, for light quarks we can make isospins $I = 0, 1$;

$$q \otimes \bar{q} = 2 \otimes 2 = 3 \oplus 1 \tag{20}$$

In SU(2) these representations are $2I+1$ dimensional, and are $I=1$ and $I=0$ respectively. With the antiquark basis phases given above and conventional SU(2) Clebsch-Gordon coefficients these states are

$$|I=1,I_z\rangle = \begin{cases} -|u\bar{d}\rangle & I_z=1 \\ \frac{1}{\sqrt{2}}(|u\bar{u}\rangle - |d\bar{d}\rangle), & I_z=0 \\ |d\bar{u}\rangle, & I_z=-1 \end{cases} \tag{21}$$

and

$$|I=0,I_z=0\rangle = \frac{1}{\sqrt{2}}(|u\bar{u}\rangle + |d\bar{d}\rangle) \ . \tag{22}$$

These isospin eigenstates are physically relevant because $q\bar{q}$ annihilation processes in QCD such as $u\bar{u} \leftrightarrow d\bar{d}$ are large relative to $m_u - m_d$ and so drive the physical states into the linear combinations $u\bar{u} \pm d\bar{d}$. In contrast, mixing with $s\bar{s}$ states is much less important, and where this mixing can be studied (as in the tensors) it is usually found to be quite small. The single well established exception is in the $\eta - \eta'$ system, which is driven into approximately equal flavor mixtures of $s\bar{s}$ and nonstrange systems,

$$|\eta\rangle \approx \frac{1}{\sqrt{2}}(|n\bar{n}\rangle - |s\bar{s}\rangle) \tag{23}$$

and

$$|\eta'\rangle \approx \frac{1}{\sqrt{2}}(|n\bar{n}\rangle + |s\bar{s}\rangle) \ , \tag{24}$$

where $|n\bar{n}\rangle \equiv (|u\bar{u}\rangle + |d\bar{d}\rangle/\sqrt{2}$. These can be written in terms of a flavor mixing angle ϕ as

$$|\eta\rangle = \cos(\phi)|n\bar{n}\rangle - \sin(\phi)|s\bar{s}\rangle \ , \tag{25}$$

$$|\eta'\rangle = \sin(\phi)|n\bar{n}\rangle + \cos(\phi)|s\bar{s}\rangle \ . \tag{26}$$

One often sees these states expressed as a linear combination of flavor SU(3) octet and singlet states, which was especially common in the early days of the quark model;

$$|\eta\rangle = \cos(\theta)|8\rangle + \sin(\theta)|1\rangle \ , \tag{27}$$

$$|\eta'\rangle = \sin(\theta)|8\rangle - \sin(\theta)|1\rangle \ . \tag{28}$$

This basis now appears less physically relevant than the flavor mixing angle ϕ, which should probably be used instead. In any case they are trivially related, $\phi = \theta + \tan^{-1}(\sqrt{2})$.

Now we will consider the quantum numbers of $q\bar{q}$ mesons in the nonrelativistic quark model, assuming that they are states of definite total quark spin $S_{q\bar{q}}$ and orbital angular momentum $L_{q\bar{q}}$. Combining two spin-1/2 fermions gives total spin 0 and 1;

$$2 \otimes 2 = \begin{cases} 3 \quad (S_{q\bar{q}}=1), \text{ "spin triplets"}, \\ \oplus \\ 1 \quad (S_{q\bar{q}}=0), \text{ "spin singlets"}. \end{cases} \tag{29}$$

These are combined with orbital wavefunctions to makes states of total J, J_z;

$$|q\bar{q}(J,J_z)\rangle = \sum_{L_z,S_z} < L, L_z; S, S_z | J, J_z > \sum_{s_z,\bar{s}_z} < 1/2, s_z; 1/2, \bar{s}_z | S, S_z >$$

$$\cdot \int d\Omega \, Y_{\ell m}(\Omega) \cdot \int_0^\infty r^2 dr \, \psi(r) \, |q_{s_z}(r/2,\Omega), \ \bar{q}_{\bar{s}_z}(r/2,-\Omega)\rangle \ . \tag{30}$$

In the nonrelativistic quark model this is the state vector of a $q\bar{q}$ meson with a radial wavefunction $\psi(r)$, which is determined by the $q\bar{q}$ potential. This form assumes that the spin-dependent OGE tensor interaction is not present, since it mixes spin-triplet states such as ${}^3S_1$ and ${}^3D_1$ that have $S=1, J=L\pm 1$. Given the quark-model meson state (30) we can easily derive quantum numbers. First, under parity $\mathcal{P}$ we transform $\Omega \to -\Omega$, which with the parity of $(-1)^L$ of the spherical harmonics and the intrinsic negative parity of a fermion-antifermion pair gives

$$\mathcal{P}|q\bar{q}(J,L,S)\rangle = (-1)^{L+1}|q\bar{q}(J,L,S)\rangle \ . \tag{31}$$

Another useful quantum number is C-parity; the $\mathcal{C}$ operator changes particles into antiparticles and vice versa, without changing any other state labels. Thus for example

$$\mathcal{C}|u_\uparrow \bar{d}_\downarrow\rangle = |\bar{u}_\uparrow d_\downarrow\rangle \ . \tag{32}$$

If this operator is diagonal, the phase is called C-parity. Since $\mathcal{C}$ switches quarks and antiquarks, it is only diagonal on states with q and $\bar{q}$ of the same flavor. Carrying out $\mathcal{C}$ on the $|q\bar{q}(J,L,S)\rangle$ states defined above, we find a phase of

$$\mathcal{C}|q\bar{q}(J,L,S)\rangle = (-1)^{L+S}\ |q\bar{q}(J,L,S)\rangle \ . \tag{33}$$

C-parity is useful primarily because it is conserved in both QCD and QED, provided that we define the photon to have negative C-parity. Thus for example for the π^o, with $S=L=0$, $C=(+1)$ and therefore annihilation of a π^o to only even numbers of photons is allowed. In contrast, for neutral vectors we have $C=(-1)^{L+S}=(-1)$, and for this reason the electromagnetic interaction of the ρ^o is dominated by virtual transitions through a single photon. The strong reactions of single photons with light hadrons are often dominated by the reverse transition to a vector $q\bar{q}$ state, notably the ρ^o, which is known as "vector dominance".

A final useful multiplicative quantum number is G-parity. This is a combination of C-parity and isospin symmetry which is diagonal on charged states within an isomultiplet, and is also conserved in QCD in the limit of equal u, d quark masses. G-parity combines C with a rotation by π in isospin,

$$\mathcal{G} = \mathcal{C}\ e^{i\pi I_y} \ . \tag{34}$$

C-parity on a $|\pi^+\rangle = -|u\bar{d}\rangle$, for example, gives a $(-1)^{L+S}|\bar{u}d\rangle = |d\bar{u}\rangle$ state, which is a $|\pi^-\rangle$; thus if applied to S-matrix elements, C-parity typically relates processes involving different asymptotic meson states. To obtain instead a diagonal operator $\mathcal{G}$ we follow this with a rotation about I_y, which from the theory of angular momentum gives a factor of $(-1)^I$. Thus we find for u, d $q\bar{q}$ isospin states in general

$$\mathcal{G}|q\bar{q}(J,L,S,I)\rangle = C_n\,(-1)^I\,|q\bar{q}(J,L,S,I)\rangle = (-1)^{L+S+I}|q\bar{q}(J,L,S,I)\rangle \ , \tag{35}$$

where C_n is the C-parity of the neutral member of the multiplet. Since G-parity is diagonal on isospin states it can be used to exclude many *a priori* possible reactions. For example,

$$\mathcal{G}|\pi\rangle = C_n\,(-1)^I\,|\pi\rangle = \ (-1)\ |\pi\rangle \ , \tag{36}$$

so any reaction which would scatter an even number of pions into an odd number is forbidden. Other examples are $\rho \not\to \pi\pi\pi$ (since ρ has $(+1)$ G-parity), and $\omega \not\to \pi\pi$.

Finally we will discuss angular quantum numbers and state names. For spin-singlet $|q\bar{q}\rangle$ states we trivially have $J=L$ only, which in spectroscopic notation are ${}^1L_{J=L}$. For spin-triplet states we have $J = L+1, L, L-1$, and the states are called ^3L_J. In 1992 the Particle Data Group [2] standardized the names for states based on their quantum numbers, which for the flavor states we consider here, $q\bar{q} = u\bar{d}, u\bar{u}-d\bar{d}, d\bar{u}; u\bar{u}+d\bar{d}, s\bar{s}; c\bar{c}$, are as follows:

$$\begin{cases} \pi; \eta, \eta'; \eta_c & J^{PC} = \text{even}^{-+} \\ b; h, h'; h_c & J^{PC} = \text{odd}^{+-} \\ \rho; \omega, \phi; \psi & J^{PC} = (\text{even or odd}, \geq 1)^{--} \\ a; f, f'; \chi_c & J^{PC} = (\text{even or odd})^{++} \end{cases} ,$$

and a subscript J is added when necessary to avoid ambiguity. The J^{PC} cases not listed are not allowed for $q\bar{q}$ states, and are known as "spin-parity exotics". These exotics are given the same name as a $q\bar{q}$ state which differs only in C-parity, thus for example an $I=1$, $J^{PC_n} = 1^{-+}$ exotic at 1900 MeV would be named for the 1^{--} ρ as $\hat{\rho}(1900)$.

2.2 $q\bar{q}$ Mesons: Spectrum of States

Having discussed the quantum numbers expected for $q\bar{q}$ states, we can now proceed to dynamical models of the spectrum of $q\bar{q}$ resonances. The most successful models for the description of spectroscopy are quark potential models, and in particular we shall consider the nonrelativistic quark potential model. The most widely used potential model for $q\bar{q}$ mesons is the relativized model of Godfrey and Isgur [3], and for qqq baryons is the essentially equivalent model of Capstick and Isgur [4]. (Relativized versions differ from the nonrelativistic quark potential model only in relatively minor ways, such as the use of $H_{KE} = 2\sqrt{\vec{p}^{\,2} + m_q^2}$ as the kinetic term in place of $2m_q + \vec{p}^{\,2}/m_q$, the retention of m/E factors in matrix elements and the introduction of coordinate smearing in singular terms such as $\delta(\vec{r})$.)

In the nonrelativistic quark potential model one typically assumes that the $q\bar{q}$ wavefunction in (30) is a solution of a Schrödinger equation with a potential that combines the color Coulomb term from one gluon exchange with a phenomenological linear confining term,

$$\Big(H_{KE} + V(r_{q\bar{q}})\Big)\psi_n(r) = E_n \psi_n(r) , \tag{37}$$

$$V(r) = -\frac{4}{3}\frac{\alpha_s}{r} + br , \tag{38}$$

which is known as a "Coulomb plus linear" potential model. In addition at $O(v^2/c^2)$ in the quark motion one finds a series of spin-dependent terms from the nonrelativistic reduction of the Dirac spinor matrix elements. For the OGE interaction these are the Breit-Fermi Hamiltonian, familiar from atomic physics. The $O(v^2/c^2)$ spin-dependent interaction from the linear confining interaction, assuming it to be pure Lorentz scalar, is an inverted spin-orbit term. In summary, the $O(v^2/c^2)$ spin-dependent $q\bar{q}$ Hamiltonian is

$$H_{spin-dep.} = \\ + \underbrace{\frac{32\pi\alpha_s}{9m_q^2}\vec{S}_q\cdot\vec{S}_{\bar{q}}\,\delta(\vec{x})}_{OGE\ spin-spin} + \underbrace{\frac{4\alpha_s}{m_q^2 r^3}\Big(\vec{S}_q\cdot\hat{r}\vec{S}_{\bar{q}}\cdot\hat{r} - \frac{1}{3}\vec{S}_q\cdot\vec{S}_{\bar{q}}\Big)}_{OGE\ tensor} + \underbrace{\frac{2\alpha_s}{m_q^2 r^3}\vec{L}\cdot\vec{S}}_{OGE\ spin-orbit} - \underbrace{\frac{b}{2m_q^2 r}\vec{L}\cdot\vec{S}}_{scalar\ spin-orbit} . \tag{39}$$

Since these include singular terms such as $-1/r^3$, they are usually treated perturbatively in the nonrelativistic Schrödinger formalism. As an example of a quarkonium spectrum predicted by this interaction we show the relatively straightforward $c\bar{c}$ system in Figure 2. We shall use $c\bar{c}$ for our discussion of the Hamiltonian, because this spectrum is well established up to the second radial excitation.

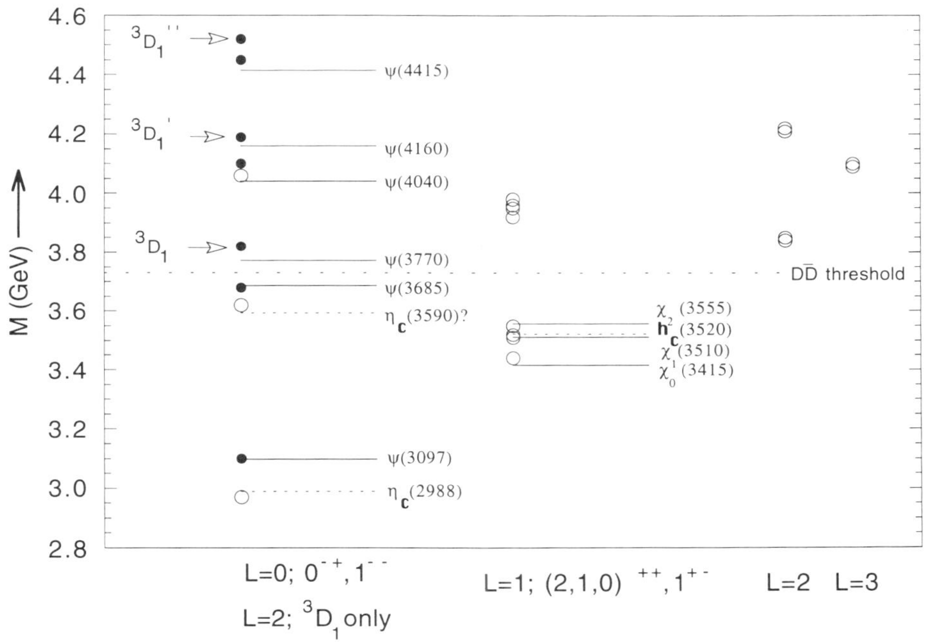

Figure 2. The charmonium spectrum; experiment (lines) versus a relativized potential model.

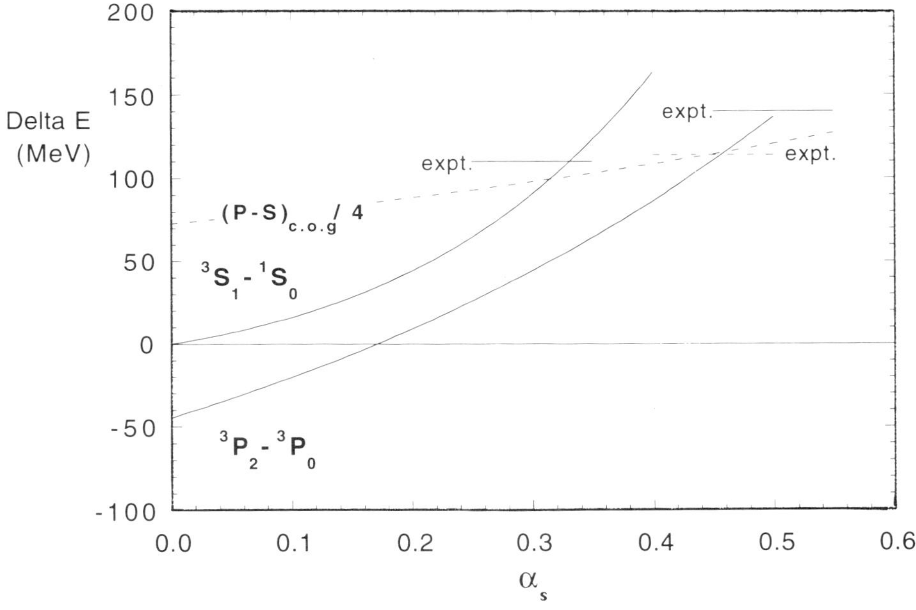

Figure 3. Multiplet splittings $^3S_1$-$^1S_0$ and $^3P_2$-$^3P_0$ and mean (P-S) splitting in charmonium. (For b=0.18 GeV^2 and m_c=1.4 GeV.)

The qualitative effect of the various terms is as follows. The OGE spin-spin force is a contact interaction (in the nonrelativistic limit *only*), so it causes a large positive splitting between the $L = 0$ vector $^3S_1$ and pseudoscalar $^1S_0$ states, such as $\rho - \pi$ and $K^* - K$. Since the strength falls with m_q as $1/m_q^2$, it is much weaker at the $c\bar{c}$ mass scale, but still gives a $\psi - \eta_c$ splitting of about 110 MeV. In the Cornell potential model of *ca.* 1980 [6] it was assumed that the confining interaction acted as the time-time components of Lorentz vectors ($\gamma_0 \otimes \gamma_0$) in Dirac spinor matrix elements, rather than as a scalar, $I \otimes I$. Such a vector-vector interaction at $O(v^2/c^2)$ leads to a spin-spin interaction of

$$H_{vector\,spin-spin} = +\frac{2}{3m_q^2}\,\vec{S}_q \cdot \vec{S}_{\bar{q}}\,\nabla^2 V(r) \;=\; \frac{4b}{3m_q^2 r}\;. \tag{40}$$

This vector confinement is now strongly ruled out by the discovery of the $^1P_1$ h_c spin-singlet $c\bar{c}$ charmonium state by the E760 Collaboration [7]; with typical $c\bar{c}$ potential model parameters, the vector confinement spin-spin term predicts a splitting between the c.o.g. of the $\{\chi_j\}$ ^3P_j triplet states and the h_c $^1P_1$ singlet of ≈ 30 MeV, whereas experimentally the splitting is about 1 MeV. Scalar confinement expects a very small $h_c - P_{c.o.g}$ splitting, because there is no scalar spin-spin interaction, and the (nonrelativistic) OGE spin-spin interaction is a contact term.

The spin-orbit interaction is quite interesting because the OGE and scalar confining terms have opposite signs. The OGE spin-orbit is shorter ranged, so we expect small-L quarkonium states with sufficiently large masses to have an ordering of $M(J = L+1 > L > L-1)$. For charmonium this is observed in the χ_j states. At sufficiently large L the scalar spin-orbit should be largest, so multiplet inversion should take place. Models usually predict that this should happen by $L \sim 3$ for light $q\bar{q}$ systems [8], although it may be masked by mass shifts due to other interactions such as couplings to virtual two-meson channels [5].

Finally, there is a tensor OGE interaction, which modifies the characteristic $\vec{L} \cdot \vec{S}$ splitting pattern and mixes spin-triplet basis states related by $\Delta L = 2$, for example $^3S_1 \leftrightarrow {}^3D_1$. The effect of the tensor term can be seen in the $c\bar{c}$ system in the χ_j states: In the ^3P_j and $^1P_1$ states the spin-orbit and tensor matrix elements are

$$\langle P|\vec{L} \cdot \vec{S}|P\rangle = \begin{cases} +1, & ^3P_2 \\ 0, & ^1P_1 \\ -1, & ^3P_1 \\ -2, & ^3P_0 \end{cases} \tag{41}$$

and

$$\langle P|T|P\rangle = \begin{cases} -1/30, & ^3P_2 \\ 0, & ^1P_1 \\ +1/6, & ^3P_1 \\ -1/3, & ^3P_0\;. \end{cases} \tag{42}$$

The masses reported for the corresponding $c\bar{c}$ states by the 1994 PDG are

$$\langle P|T|P\rangle = \begin{cases} 3556.17(13)\ MeV & \chi_2 \\ 3526.14(24)\ MeV & h_c \\ 3510.53(12)\ MeV & \chi_1 \\ 3415.1(1.0)\ MeV & \chi_0\;. \end{cases} \tag{43}$$

Clearly the experimental masses are not well described by the equally-spaced spin-orbit pattern, whereas P-wave mass splittings not far from experiment are observed when we include both spin-orbit *and* tensor terms with conventional parameter values (see

Figure 2). We illustrate this in more detail in Figure 3, which shows the spin-spin splitting of the $\psi - \eta_c$ and the splitting of the $\chi_2 - \chi_0$ as we increase α_s. (This is with a plausible charmed quark mass of $m_c = 1.4$ GeV and a string tension of $b = 0.18$ GeV^2; this test is rather insensitive to m_c and b, and these values are typical for $c\bar{c}$ spectroscopy.) The experimental S- and P-multiplet splittings evidently require a value of $\alpha_s(c\bar{c}) \approx 0.4$, which is consistent with the value of $\alpha_s = 0.45$ required to fit the *spin-independent* $P_{c.o.g} - S_{c.o.g}$ of ≈ 456 MeV. This is a very nontrivial test of the validity of the OGE interaction in the $c\bar{c}$ spectrum, since it compares the leading order color Coulomb potential and the linear confining potential with the $O(v^2/c^2)$ spin-dependent interactions implied by OGE and scalar confinement.

Note that the value $\alpha_s \approx 0.4$ we need to describe $c\bar{c}$ spectroscopy is rather larger than the $\alpha_s \approx 0.2$ which was advocated long ago from fits to strong decay widths. For example, the Cornell collaboration [6] note that the ψ total width would imply $\alpha_s = 0.19(3)$ if we use the orthopositronium decay formula for ${}^3S_1(e^+e^-) \to \gamma\gamma\gamma$ to model $\psi \to$ hadrons by $\psi \to ggg$. These tree-level nonrelativistic wavefunction-at-contact formulas are now known to have very important corrections from $O(\alpha_s)$ QCD radiative diagrams and especially from finite-range corrections to the contact approximation, so the small earlier value is presumably less reliable.

Now we shall begin our discussion of the light $q\bar{q}$ spectrum, as a preparation for our discussion of non-$q\bar{q}$ states. In Figures 4 and 5 we show the experimental $I = 1$ and $I = 0$ spectrum (shaded) and the predictions of Godfrey and Isgur [3] (lines). (The data represents the experimental situation in 1985.) Note that especially in $I = 1$ there is quite good agreement between theory and experiment, and the missing levels are typically the lower-spin radial excitations or lower-J members of multiplets, which are more difficult to extract from angular distributions. The largest discrepancy found since 1985 is the $a_0({}^3P_0)$ state, which was predicted to lie near 1.09 GeV, whereas the new Crystal Barrel candidate [9] is a much more massive $a_0(1450)$. Although the discrepancy is disturbing, it may not be a serious problem, because Godfrey and Isgur chose their parameters to give a rather light ${}^3P_0$ $f_0(q\bar{q})$ state. Certainly the question of whether or not an $a_0(1450)$ state can be accommodated in the $q\bar{q}$ quark model should be reinvestigated, given what is now known about the importance of open-channel couplings for scalar $q\bar{q}$ states [5].

In the $I = 0$ sector (Figure 5) the situation is much more complicated. Recent experimental work, carried out largely at LEAR, has added several new f_0 levels, and we now have *six* experimental f_0 resonances in the mass range between 980 and 1710 MeV. (These will be discussed in the following sections.) The quark model can perhaps accommodate three of these states as quarkonia, a ${}^3P_0$ $q\bar{q}$, a ${}^3P_0$ $s\bar{s}$ and a first radial excitation ${}^3P_0'$ $q\bar{q}$. Evidently there is much more to the resonance spectrum than just quarkonia, at least in the f_0 channel. In the following sections we review theoretical suggestions for what states other than $q\bar{q}$ might appear in the meson spectrum as resonances, and how these states might be distinguished from $q\bar{q}$.

2.3 Identification of $q\bar{q}$ States

Traditionally meson resonances have been given $q\bar{q}$ quark model assignments based on their flavor and J^{PC} quantum numbers, masses, and to a lesser extent their decay modes and production characteristics. This procedure works quite well in the mass region below about 1.5 GeV, where relatively few $q\bar{q}$ resonances are expected. As classic examples we might cite the tensor mesons $f_2(1270)$, $a_2(1320)$ and $f_2'(1525)$; the near degeneracy of the lighter $I = 0$ and $I = 1$ states suggests that they are nonstrange

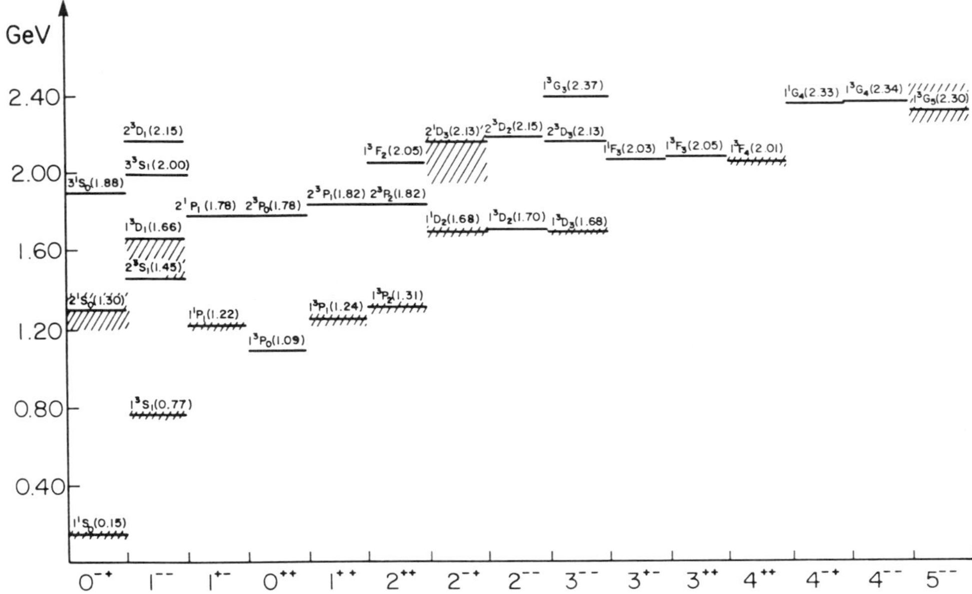

Figure 4. I =1 meson resonances, as of 1985; experimental (shaded) and theoretical qqbar levels (lines) (Godfrey and Isgur [3]).

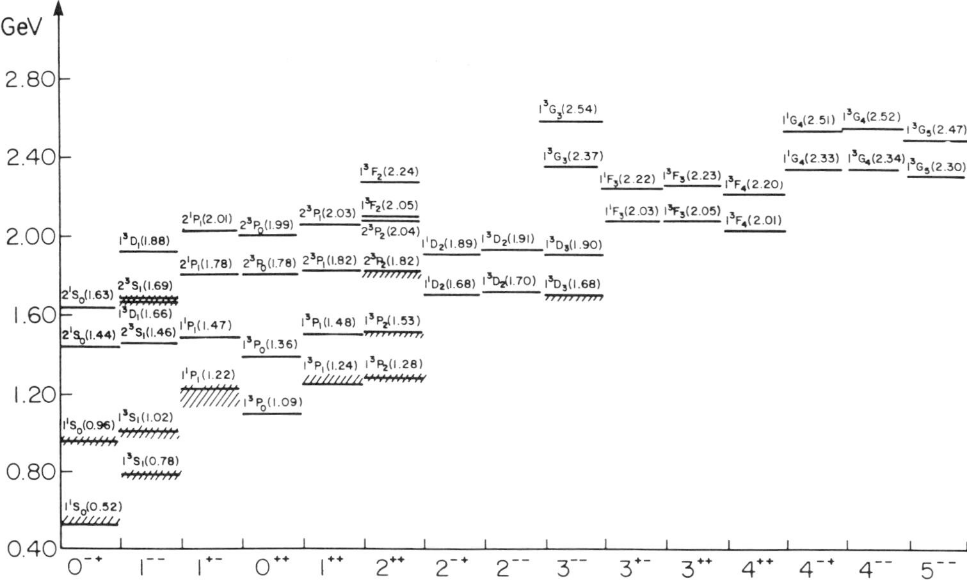

Figure 5. As in Figure 4 but for I =0 states; uubar, ddbar and ssbar.

u, d isospin partners, and the heavier $f_2'(1525)$ might be an $s\bar{s}$ partner of these states. (Note the mass separation of ≈ 230 MeV, quite similar to $\phi - \omega \approx \phi - \rho \approx 240$ MeV.) This is supported by the strong preference of the $f_2'(1525)$ for $K\bar{K}$ final states over $\pi\pi$, as expected from the Zweig rule, which states that the initial $q\bar{q}$ pair in a meson (here $s\bar{s}$) cannot easily annihilate during a strong decay. In contrast, the lighter $f_2(1270)$ has comparable couplings to $\pi\pi$ and $K\bar{K}$. (The observed $f_2(1270) \to \pi\pi : K\bar{K}$ branching fraction ratio of $\approx 20 : 1$ merely reflects the P^5 suppression expected for a D-wave decay.)

Unfortunately these simple ideas are not generally applicable to mesons in the mass region above 1.5 GeV. The complication is that many of the decay amplitudes of orbitally and (especially) radially excited states have nodes as functions of the outgoing hadron momentum, and if the physical momentum for a given mode is accidentally near a node, one could erroneously conclude that the coupling was weak for more fundamental reasons, such as flavor content. A simple example is the decay of the $f_0(q\bar{q})$ state to two pseudoscalars; in the "$^3P_0$" decay model of LeYaouanc *et al.* [10], with SHO wavefunctions the decay rate to $\pi\pi$ is given by [11]

$$\Gamma(f_0 \to \pi\pi) = \frac{2^6}{3^3}\gamma_0^2\, M_{f_0}\, \frac{P}{\beta}\left(1 - \frac{2}{9}\frac{P^2}{\beta^2}\right)e^{-P^2/6\beta^2} \tag{44}$$

Evidently this rate has a node at an outgoing pion momentum of $P = 3\beta/\sqrt{2}$, which is not far above the physical point; typically, decay models use $\beta \approx 0.4$ GeV, which would give a vanishing width to $\pi\pi$ for an $f_0(q\bar{q})$ state with $M_{f_0} = 1.6$ GeV. In the $\approx 1.5 - 2.5$ GeV mass region which is of greatest interest in the search for non-$q\bar{q}$ mesons, the conventional $q\bar{q}$ states which we must identify as a background are orbital and radial excitations. Many of these will have complicated and counterintuitive decay amplitudes due to nodes. Resolution of this problem will require a careful investigation of radially and orbitally excited $q\bar{q}$ states, in particular their strong decay modes and branching fractions, so that we can test and improve the decay models. Given an accurate description of strong decays, it may be possible to identify all the $q\bar{q}$ levels in this mass region reliably.

Another approach, which has seen little application thus far, is to make use of electromagnetic couplings of meson resonances. One example is the two-photon couplings, which are measured in the reaction $e^+e^- \to e^+e^- +$ hadrons; from a measurement of this cross section to a particular hadronic final state such as $\pi^o\pi^o$ one can see for example the $f_2(1270)$ as a clear bump, and the cross section implicitly determines the two-photon width times the branching fraction, here $\Gamma_{\gamma\gamma}(f_2) \cdot B(f_2 \to \pi^o\pi^o)$. In principle even the absolute two-photon widths can be calculated from quark model wavefunctions, but for light $q\bar{q}$ mesons the results are rather sensitive to relativistic effects. (For $\Gamma_{\gamma\gamma}(c\bar{c})$ as an example see [12].)

The most convincing tests for $q\bar{q}$ candidates using electromagnetic couplings come from comparisons of the relative rates of a set of resonances with the same J^{PC} which are possible members of a u, d, s flavor multiplet. In $\gamma\gamma$ couplings, the decay amplitude involves the charge matrix element of two electromagnetic vertices,

$$A(q\bar{q} \to \gamma\gamma) \propto \langle q\bar{q}|e_q^2|0\rangle \tag{45}$$

in a notation in which we suppress the photon state and current matrix elements. This gives relative amplitudes of

$$\langle f : a : f'|e_q^2|0\rangle = \frac{(2/3)^2 + (-1/3)^2}{\sqrt{2}} : \frac{(2/3)^2 - (-1/3)^2}{\sqrt{2}} : (-1/3)^2\,. \tag{46}$$

Given these relative amplitudes, the relative $\gamma\gamma$ decay rates for $I = 0 : I = 1 : s\bar{s}$ mesons of the same spatial state, neglecting phase space differences, are

$$\Gamma_{\gamma\gamma}(f : a : f') = 25 : 9 : 2 \ . \tag{47}$$

In comparison, the measured two-photon widths of the corresponding tensor states are

$$\Gamma_{\gamma\gamma}(f_2(1270) : a_2(1320) : f'(1525)) = 2.8(4)\ KeV \ \ : 1.04(9)\ KeV \ \ : 0.105(17)\ KeV \ , \tag{48}$$

which clearly supports the $(u\bar{u} \pm d\bar{d})/\sqrt{2}$ and $s\bar{s}$ assignments. (There is some additional suppression of the $s\bar{s}$ coupling to $\gamma\gamma$, due to the strange quark mass.) To date no radial excitations have been reported in $\gamma\gamma$, but once this data becomes available it should be very useful for identifying the members of $q\bar{q}$ flavor multiplets. Higher orbital excitations have already been reported, notably the $L = 2$ $\pi_2(1670)$ and probably an η_2 state [13].

Finally, we note that single photon transitions, $(q\bar{q})_i \to \gamma(q\bar{q})_f$, would also be extremely useful for identifying $q\bar{q}$ states; these are more readily calculable theoretically, and also have a characteristic pattern of rates based on flavor,

$$\Gamma((q\bar{q})_i \to \gamma(q\bar{q})_f) = 9 : 4 : 1 \tag{49}$$

for

$$\Delta I = 1 : s\bar{s} : \Delta I = 0 \ .$$

The $\Delta I = 0, 1$ ratios for transitions of (u, d) meson pairs can be seen for example in decays of the light vectors,

$$\Gamma(V \to \gamma\pi) = \begin{cases} 720(40) & \text{KeV } \omega \to \pi^o\gamma \ , \\ 68(7) & \text{KeV } \rho \to \pi\gamma \ . \end{cases} \tag{50}$$

Although mesaurements of these radiative transitions would be very useful for the classification of higher $q\bar{q}$ and non-$q\bar{q}$ states, little work has been done along these lines. For example, of the many f_J and a_J states, only two transition rates to (u, d) mesons have been measured, $a_2 \to \pi\gamma = 295(60)$ KeV and $a_1 \to \pi\gamma = 640(246)$ KeV. Since measurements of radiative transitions could determine the nature of controversial states such as the $f_1(1420)$ (in $\gamma K\bar{K}$, to be discussed), they should certainly be carried out if possible. For $\pi\gamma$ modes this could use the inverse reaction (electroproduction) rather than photon emission, which has a large hadronic background.

3 NON-$Q\bar{Q}$ MESONS: INTRODUCTION

The most exciting developments in QCD spectroscopy involve searches for resonances which are external to the conventional $q\bar{q}$ quark model of mesons. There are two general classes of such states, which are those with dominant gluonic excitations "gluonic hadrons" and states with more quarks and antiquarks than the familiar $q\bar{q}$ states.

Since QCD is a theory which contains both quarks *and* gluons as dynamical degrees of freedom, we would expect to see evidence of both these building blocks in the spectrum of physical color-singlet hadrons. It is remarkable, however, that of the hundreds of hadronic states now known, most can be described as states made only of quarks and antiquarks in the nonrelativistic quark model, and none of the remaining problematic resonances have been established as having dominant gluonic valence components. The best evidence for the presence of gluons at low energies is indirect, for example in the Breit-Fermi one-gluon-exchange Hamiltonian used in potential models and in the $q\bar{q} \leftrightarrow s\bar{s}$ configuration mixing evident in the η and η'.

In addition to these gluonic states, one may also form color singlet combinations from multiquark systems of quarks and antiquarks, beginning with $q^2\bar{q}^2$. Although such resonances have been quite controversial, it now appears that such resonances do exist in nature, although they are dominantly "molecular" associations of meson pairs rather than single four-quark clusters.

Experimental studies now in progress may radically alter this situation, since there are now several resonances that, if confirmed, appear to be likely candidates for glueballs, hybrids and molecules. As we shall see, these states share several common features with theoretical expectations for these unusual hadronic states.

In this contribution we review current theoretical expectations for these gluonic hadrons and molecules, and briefly discuss some of the more recent and interesting experimental candidates for these states.

4 GLUEBALLS

4.1 Glueball masses

A priori one would expect glueballs to be the most attractive gluonic hadrons experimentally, since they might be expected to differ most noticibly from $q\bar{q}$. In practice this naive expectation may not be realized; studies of the light glueball spectrum using lattice gauge theory have found that the lowest-lying glueball is a scalar, and its coupling to two-pseudoscalar final states suggests a typical hadronic width. The next glueballs encountered at higher masses are predicted to be 0^{-+} and 2^{++}, and states which couple to two transverse gluons (presumably the lightest glueballs) do not contain exotic J^{PC}.

Although there have been many studies of the spectrum and quantum numbers expected for glueballs [14], the results of lattice gauge theory should be treated as the most relevant to experiment, since they bear the closest resemblance to full QCD. The assumptions of quenched lattice gauge theory are that decay channels do not modify glueball masses significantly (since the neglect of quarks implies stable light glueballs) and that the extrapolations to small lattice spacing and large lattive volume do not introduce important biases. If glueballs are not very broad objects, the assumption of stable glueballs should not introduce large errors in masses.

There are lattice predictions for the masses of glueballs with various J^{PC} [15]; the most reliable is presumably for the scalar glueball ground state, which is predicted to have a mass of

$$(0^{++}) = \begin{cases} 1.550(50) & \text{GeV [16]} \\ 1.740(71) & \text{GeV [17]} \end{cases} . \tag{51}$$

The corresponding mass estimate for the tensor glueballs is in the 2.2-2.4 GeV range,

$$(2^{++}) = \begin{cases} 2.270(100) & \text{GeV [16]} \\ 2.359(128) & \text{GeV [17]} \end{cases} ; \tag{52}$$

with the pseudoscalar glueball at a similar mass.

There are obvious problems associated with the identification of a scalar state near 1.5 GeV. The f_0 sector is the most complicated of all meson sectors, with at least six problematical states, $f_0(980)$, $f_0(1300)$, $f_0(1365)$, $f_0(1500)$ $f_0(1590)$ and $f_0(1710)$. Since these are broad (except the $f_0(980)$) and overlapping resonances, the problem of identifying unusual states against the $q\bar{q}$ and $s\bar{s}$ background, and the related problems of separating individual resonances from interference and threshold effects are daunting ones. If the scalar glueball does have a typical hadronic width, as suggested by the work of Sexton *et al.* [18], it may be quite difficult to identify the scalar glueball convincingly. Amsler and Close [19] note that the near degeneracy of the pure LGT glueball and the $L = 1$ $q\bar{q}$ and $s\bar{s}$ multiplets may lead to complicated mixing effects, so the physical states may be nontrivial combinations in flavor space, as in the η and η' sector.

The tensor glueball may be an easier experimental target, since the expected mass is far above the lowest-lying 2^{++} quarkonium states. Here the problem is that the mass region above 2 GeV is poorly explored, so the problem of distinguishing a tensor glueball from the background of radial-$^3P_2$ and $^3F_2$ $q\bar{q}$ and $s\bar{s}$ states cannot yet be addressed. This lack of adequate information regarding the higher mass quarkonium spectrum is even more of a problem in the 0^{-+} sector.

4.2 Expectations for Glueball Properties

Since we have no confirmed glueballs and the states predicted are in channels with a complicated or poorly explored resonance spectrum, it would be useful to have reliable theoretical predictions of glueball properties as a guide. The data we are likely to have on gluonic candidates in the near future are their strong branching fractions. Here a very characteristic naive glueball signature can be given, although it is easy to imagine ways in which this signature might be violated.

As gluons at the bare lagrangian level have equal strength couplings to quarks of all flavors, one can make the assumption that flavor-symmetric couplings to hadron final states are approximately valid for physical glueballs. This gives a characteristic flavor-singlet branching fraction to pseudoscalar pairs, which is (neglecting phase space differences)

$$\Gamma(G \to \pi\pi : K\bar{K} : \eta\eta : \eta\eta' : \eta'\eta')/(phase\ space) = 3 : 4 : 1 : 0 : 1\,. \tag{53}$$

Of course this simple pattern should at least incorporate the $|\vec{P}\,|$ from phase space for an S-wave decay, and there is in addition a decay form factor which depends on the unknown scalar glueball wavefunction and decay mechanism. Experience with the $^3P_0$ f_0 $q\bar{q}$ decay amplitude to $\pi\pi$, which has a node near the physical point [11], suggests that the naive pattern of flavor-singlet decay amplitudes may indeed be far from the physical couplings.

The accuracy of naive flavor-singlet couplings can be tested for a pure (quenched) scalar glueball in lattice gauge theory through a determination of the glueball-Ps-Ps three point function. Preliminary results for this coupling [18] indicate that flavor-singlet symmetry may indeed be badly violated at the amplitude level, and higher-mass Ps pairs are preferred in the decay. In view of the relatively large errors it will be important in future work to improve the statistics of this important lattice gauge theory measurement, and to extend these decay amplitude studies to the tensor and pseudoscalar glueball sectors.

In future experimental work it may be possible to determine or limit electromagnetic couplings of glueball candidates. Measurements of electromagnetic couplings of these resonances in one-photon ($R \to \gamma q\bar{q}$) and two-photon ($R \to \gamma\gamma$) transitions would be extremely important for the clarification of the nature of these states; since theorists can calculate radiative transitions of $q\bar{q}$ states with reasonably accuracy, the radiative transition rates of a relatively pure glueball would clearly be anomalous relative to expectations for an $f_0(q\bar{q})$ state. If glueballs are indeed strongly mixed with $q\bar{q}$ and $s\bar{s}$ states, a convincing way to identify the flavor components of these mixed states would be through a comparison of the relative rates

$$\Gamma(R \to \gamma\rho^0 : \gamma\omega : \gamma\phi)$$

since these act as flavor tags. Similarly, $\gamma\gamma$ couplings can be used to locate the scalar nonstrange f_0 $q\bar{q}$ signal, since this state should have a strong coupling to $\gamma\gamma$. Results on this reaction have already been obtained by the Crystal Ball in the reaction $\gamma\gamma \to \pi^o\pi^o$.

Since a glueball should have suppressed couplings to $\gamma\gamma$, measurements of the $\gamma\gamma$ couplings of the various f_0 states and other light resonances would be very important contributions to light meson spectroscopy. This would be possible at e^+e^- facilities such as B-factories or a TCF.

4.3 Summary of Glueball Candidates

At present there are two prominent experimental candidates for glueballs, a scalar $f_0(1500)$ and the $\xi(2230)$, which is probably a tensor. The scalar candidate as reported by the Crystal Barrel [20] has a mass and width of

$$M(f_0) = 1520 \begin{array}{c} +20 \\ -55 \end{array} MeV \tag{54}$$

and

$$\Gamma(f_0) = 148 \begin{array}{c} +20 \\ -25 \end{array} MeV \ . \tag{55}$$

At this mass the $f_0(1500)$ cannot easily be assigned to a $^3P_0$ nonstrange $q\bar{q}$ state, but is consistent with the lower mass estimates from LGT for a scalar glueball. The width is also quite narrow for a $^3P_0$ $q\bar{q}$ state at this mass. The decay pattern to pseudoscalar pairs is however inconsistent with flavor symmetry; the squared invariant couplings cited by Crystal Barrel [20] are

$$\Gamma(f_0(1500) \to \pi\pi : K\bar{K} : \eta\eta : \eta\eta')/(p.s.) =$$

$$1 \ : < 1/8.6 \ (95\% c.l.) \ : \ 0.24 \pm 0.12 \ : \ 0.35 \pm 0.15 \ . \tag{56}$$

A priori this argues against a pure glueball interpretation, and subsequent work by Amsler and Close [20] has investigated the possibility that these decays may be consistent with a scalar glueball that has important $q\bar{q}$ and $s\bar{s}$ components, leading to an $\eta\eta'$ mode and suppressing the $K\bar{K}$ mode. The limit on the coupling to $K\bar{K}$ is actually inferred from another experiment, and a more careful study of this coupling including interferences at the Crystal Barrel appears to find a much larger $K\bar{K}$ coupling [21]. This state may also have been seen in a recent reanalysis of the MarkIII data on $\psi \to \gamma\pi^+\pi^-\pi^+\pi^-$ by Bugg *et al.* [22]; in this channel the $f_0(1500)$ appears dominantly in the "$\sigma\sigma$" mode of two S-wave $\pi\pi$ pairs.

The second glueball candidate, which might be a narrow part of the $\xi(2230)$ previously reported by MarkIII in ψ radiative decays, is reported by BES [23] to have very anomalous properties for a tensor above 2 GeV. The mass and width BES cite for this state in K_SK_S are

$$M(\xi) = 2232 \begin{array}{c} +8 \\ -7 \end{array} \pm 15 \ MeV \tag{57}$$

and

$$\Gamma(\xi) = 20 \begin{array}{c} +25 \\ -16 \end{array} \pm 10 \ MeV \ , \tag{58}$$

with similar results in $P\bar{P}$, K^+K^- and $\pi^+\pi^-$. If this state is confirmed it is a remarkable discovery indeed. The mass is consistent with LGT expectations for the lightest tensor glueball (2), and the narrow width implies that this is certainly not a tensor quarkonium state. Since the couplings to $\pi\pi$ and $K\bar{K}$ appear to be approximately flavor symmetric, this does appear to be a natural glueball candidate.

Although Godfrey, Kokoski and Isgur [24] noted that the $^3F_2$ and $^3F_4$ $s\bar{s}$ states expected near this mass could be relatively narrow, subsequent work by Blundell and Godfrey [25] has shown that other modes such as $K_1(1270)K$ are large, so $\Gamma(f_2(s\bar{s})) \geq$ 400 MeV. Similarly for the $^3F_4$ Blundell and Godfrey now find a broader state given these additional modes, $\Gamma(f_4(s\bar{s})) \geq 130$ MeV. Thus the $s\bar{s}$ assignments now appear implausible for an experimental width of ≤ 50 MeV.

There are several disturbing aspects of the experimental properties reported for this narrow effect. It has surprisingly small branching fractions to pseudoscalar pairs in view of the available phase space; branching fractions of a few percent are implied by the PS185 limit on $P\bar{P} \to \xi \to K\bar{K}$. A more important concern is that the reported statistical significance in each of the four channels studied by BES is rather small, $\approx 3\sigma$. A caution is appropriate because many previously reported narrow states have been found to be artifacts of limited statistics (for example the $\zeta(8.3)$). In view of the remarkable properties reported for this state, measurements of these channels with higher statistics are extremely important tasks for any e^+e^- facility operating at the ψ mass.

5 HYBRIDS

5.1 Hybrid Quantum Numbers

Hybrid mesons may be defined as resonances in which the dominant valence basis state is $q\bar{q}$ combined with a gluonic excitation. Hybrids are attractive experimentally because, unlike glueballs, they span complete flavor nonets and hence provide many possibilities for experimental detection. In addition, the lightest hybrid multiplet is expected to include at least one J^{PC}-exotic (forbidden to $q\bar{q}$). For hybrids in the MIT bag model [26], for example, the lightest gluon mode has $J^P = 1^+$, so the lowest-lying $q\bar{q}g$ multiplet contains the quantum numbers

$$J^{PC_n}(q\bar{q}g) = \begin{cases} 0^{-+}, 1^{-+}, 2^{-+} & (S_{q\bar{q}} = 1), \\ 1^{--} & (S_{q\bar{q}} = 0) \end{cases} . \qquad (59)$$

The flux tube model extends this bag model list by adding a degenerate set with reversed $\{P, C\}$ to the lowest hybrid multiplet. Constituent gluon models of hybrids differ in that their lowest hybrid multiplet has P-wave $q\bar{q}$ quantum numbers [27] and so is nonexotic, although exotics appear in excited hybrid multiplets. An investigation of $q\bar{q}g$ interpolating fields [28] shows that hybrids can have *any* J^{PC}.

5.2 Hybrid Masses

Hybrids have been studied in the literature using a wide range of models and techniques. These are the MIT bag model [26], constituent gluon models [27, 29, 30], the flux tube model [31, 32, 33, 34, 35, 36, 37, 38, 39], an adiabatic heavy-quark bag model [40] and heavy-quark lattice gauge theory [41]. Of the techniques now available to theorists for determining the properties of resonances in QCD, only Monte Carlo lattice gauge theory has not yet been applied. In all the approaches employed to date the lightest hybrids (H_q, involving u, d, s flavors) are predicted to have masses in the $\approx 1\frac{1}{2}$-2 GeV

Table 1: Predicted 1^{-+} Hybrid Masses.

state	mass (GeV)	model	Ref.
$H_{u,d}$	1.3-1.8	bag model	[26]
	1.8-2.0	flux tube model	[31, 32, 33, 34, 37]
	2.1-2.5	QCD sum rules (most after 1984)	[42, 43, 44]
	2.1	constituent gluon model	[30]
H_c	≈ 3.9	adiabatic bag model	[40]
	4.1-4.5	flux tube model	[32, 33, 34, 37]
	4.1-5.3	QCD sum rules (most after 1984)	[42, 43, 44]
	4.19(3) $\pm$ sys.	HQLGT	[41]
H_b	10.49(20)	adiabatic bag model	[40]
	10.8-11.1	flux tube model	[32, 33, 34]
	10.6-11.2	QCD sum rules (most after 1984)	[42, 43, 44]
	10.81(3) $\pm$ sys.	HQLGT	[41]

region. A summary of hybrid mass predictions for the especially interesting 1^{-+} exotic is given in the table below, taken from [37].

A more detailed discussion of these predictions and the literature on hybrids is given by Barnes, Close and Swanson [37]; for other recent reviews of hybrids see [45].

Much of the recent interest in hybrids has derived from the predictions of the flux tube model, which gives rather precise predictions for masses and decay modes of hybrids. The original flux tube references [32, 33, 34] cited masses of $\approx$ 1.9 GeV for the lightest ($q = u, d$) hybrid multiplet, $\approx$ 4.3 GeV for $c\bar{c}$ hybrids and $\approx$ 10.8 GeV for $b\bar{b}$ hybrids. There is an overall variation of about 0.2-0.3 GeV in these predictions, as indicated in Table I. Multiplet splittings are usually neglected in the flux tube model. This approximation may not be justified; a large inverted spin-orbit Thomas term for hybrids was found by Merlin and Paton [34].

A recent Hamiltonian Monte Carlo study [37] of the flux tube model determined hybrid masses without using the various questionable approximations of the early flux tube model studies, such as an adiabatic separation of quark and flux-tube motion and a small oscillation approximation for the flux tube. This Monte Carlo study generally confirmed the accuracy of the earlier flux-tube model mass estimates, for both $q\bar{q}$ and $c\bar{c}$ mesons (compared to experiment) and for hybrids (compared to the earlier approximate analytical calculations). These Monte Carlo predictions for light and charmonium levels are shown in Figs.1 and 2. By varying the model parameters over a plausible range, this study concluded that the lightest hybrid masses in the flux tube model were

$$(H_{u,d}) = 1.8 - 1.9\ GeV \tag{60}$$

for light quark hybrids and

$$(H_c) = 4.1 - 4.2\ GeV \tag{61}$$

for charmonium hybrids. Excited hybrids were also considered, and the first hybrid orbital excitation (${}_\Lambda L = {}_1D$) was found to appear at 2.30 GeV, 400 MeV above the lightest (${}_1P$) hybrids. The same numerical result was found earlier by Merlin [35] using the adiabatic approximation. This ${}_1D$ multiplet contains the J^{PC} states $(1,2,3)^{\pm\mp}$ and $2^{\pm\pm}$, which includes the exotics $1^{-+}, 2^{+-}$ and 3^{-+}. If the experimental candidates for ground-state hybrids near 1.8 GeV [46] and 1.6-2.2 GeV [47] are confirmed, it may be useful to search for members of the excited ${}_1D$ hybrid multiplet about 0.4 GeV higher in mass.

5.3 Light Hybrid Decay Modes

Theoretical models predict rather characteristic two-body decay modes for hybrids. Both flux tube [36] and constituent gluon [29] models find that the lightest hybrids decay preferentially to pairs of one $L_{q\bar{q}}=0$ and one $L_{q\bar{q}}=1$ meson, for example πf_1 and πb_1. These unusual modes have received little experimental attention because they involve complicated final states, which may explain why hybrids were not been discovered previously. The original predictions for hybrid two-body modes given by Isgur, Kokoski and Paton [36] are given below in modern notation, together with information on secondary decay branches.

These decay predictions are quite interesting because they suggest that many hybrids are so broad that they will be effectively invisible, whereas a few hybrids should be narrow enough to be easily observable in certain channels. The $I = 1$ $J^{PC} = 1^{-+}$ exotic had already been cited as an attractive experimental candidate, and this work suggested that this state should be relatively narrow for a hybrid $\Gamma_{tot} \approx 200$ MeV, and that the $S + P$ modes πb_1 and πf_1 should be the dominant final states populated by decays of this state. This work has motivated several studies of πb_1 and πf_1 final states, which show possible indications of resonant amplitudes in 1^{-+}.

These original flux tube decay calculations were for the three exotic J^{PC} quantum numbers in the lowest flux-tube multiplet. Since this multiplet contains a total of eight J^{PC} assignments, $1^{\pm\pm}$ (for $S_{q\bar{q}} = 0$) and $2^{\pm\mp}$; $1^{\pm\mp}$; $0^{\pm\mp}$; (for $S_{q\bar{q}} = 1$), one might wonder whether the nonexotic hybrids might be narrow enough to be observable. The decay amplitudes of these nonexotic hybrids were recently calculated by Close and Page [38], who also checked the exotic decay amplitudes and found reasonable agreement with Isgur, Kokoski and Paton. The Close-Page results for decays of $(u\bar{u} \pm d\bar{d})/\sqrt{2}$ hybrids are reproduced below, since they suggest some very important experimental tests of the flux tube model.

Close and Page predict that many of the *nonexotic* hybrids are also so broad as to be effectively unobservable. There are two striking exceptions. One is a 1^{--} ω-hybrid with a total width of only ≈ 100 MeV, which decays to $K_1(1270)K$ and $K_1(1400)K$; this should be searched for in K_1K final states, perhaps in photoproduction. A second interesting nonexotic hybrid is a π_2, with $\Gamma_{tot} \approx 170$ MeV. This may be the high-mass π_2 state which has been reported repeatedly in photoproduction reactions at a mass near 1775 MeV [46]. Other notable conclusions are that 1) several other hybrids, including exotics, have total widths near 300 MeV and so should be observable, and 2) the $I = 0$ 0^{+-} exotic found by Isgur *et al.* to have $\Gamma_{b_1\pi} = 250$ MeV actually has very large K_1K modes and so should be unobservable.

In addition Close and Page investigate the "forbidden" decay modes such as $\hat{\rho}(1900) \to \rho\pi$, and find that, due to differences in the ρ and π spatial wavefunctions, these S+S modes are present with partial widths of typically ~ 10 MeV. Thus it is interesting to search relatively straightforward modes such as $\rho\pi$ for hybrids, in addition to the favored but more difficult S+P modes such as $b_1\pi$, $f_1\pi$ and K_1K.

5.4 Prospects for Charmonium Hybrids

The predictions of the flux tube model and heavy-quark LGT (Table I) that hybrid charmonium states should exist at 4.1-4.2 GeV are expecially interesting. Charmonium spectroscopy is rather well understood up to about 3.8 GeV, so searches for unusual states should be straightforward near this mass. Since only a few open charm channels occur below 4.3 GeV, for a considerable range of hybrid masses one might anticipate rather narrow hybrid resonances. This possibility receives additional support from the

Table 2: J^{PC_n}-Exotic Hybrid Meson Decays in the Flux Tube Model

Hybrid Meson					Secondary Decays		
State	IKP label	$J^{PC_n}(I^G)$	Decay	$\Gamma_{H\to AB}$ (MeV)	Γ_B (MeV)	Decay	$b.f.$
$\hat{a}_2(1900)$	x_2^{+-}	$2^{+-}(1^+)$	$[\pi a_2(1320)]_P$	450	103	$a_2 \to \rho\pi$	70%
						$\to \eta\pi$	15%
						$\to \omega\pi\pi$	11%
						$\to K\bar{K}$	5%
			$[\pi a_1(1260)]_P$	100	~400	$a_1 \to \rho\pi$	most
			$[\pi h_1(1170)]_P$	150	360	$h_1 \to \rho\pi$	seen
$\hat{f}_2(1900)$	y_2^{+-}	$2^{+-}(0^-)$	$[\pi b_1(1235)]_P$	500	155	$b_1 \to (\omega\pi)_{S,D}$	most
						$\to \eta\rho$	seen
$\hat{f}_2'(2100)$	z_2^{+-}	$2^{+-}(0^-)$	$[KK_2^*(1430)]_P$	250	98	$K_2^* \to K\pi$	50%
						$\to K^*\pi$	25%
						$\to K^*\pi\pi$	13%
			$[KK_1(1400)]_P$	200	174	$K_1 \to K^*\pi$	94%
$\hat{\rho}(1900)$	x_1^{-+}	$1^{-+}(1^-)$	$[\pi b_1(1235)]_{S,D}$	100,30	155	$b_1 \to (\omega\pi)_{S,D}$	most
						$\to \eta\rho$	seen
			$[\pi f_1(1285)]_{S,D}$	30,20	24	$f_1 \to \eta\pi\pi$	50%
						$\to 4\pi(\rho\pi\pi)$	38%
						$\to a_0(980)\pi$	37%
$\hat{\omega}(1900)$	y_1^{-+}	$1^{-+}(0^+)$	$[\pi a_1(1260)]_{S,D}$	100,70	~400	$a_1 \to \rho\pi$	most
			$[\pi\pi(1300)]_P$	100	200-600	$\pi(1300) \to \rho\pi$	seen
			$[KK_1(1400)]_S$	100	174	$K_1 \to K^*\pi$	94%
$\hat{\phi}(2100)$	z_1^{-+}	$1^{-+}(0^+)$	$[KK_1(1270)]_D$	80	90	$K_1 \to K\rho$	42%
						$\to K_0^*\pi$	28%
						$\to K^*\pi$	16%
						$\to K\omega$	11%
			$[KK_1(1400)]_S$	250	174	$K_1 \to K^*\pi$	94%
			$[KK(1460)]_P$	30	250	$K \to K\pi\pi$	seen
$\hat{a}_0(1900)$	x_0^{+-}	$0^{+-}(1^+)$	$[\pi a_1(1260)]_P$	800	~400	$a_1 \to \rho\pi$	most
			$[\pi h_1(1170)]_P$	100	360	$h_1 \to \rho\pi$	seen
			$[\pi\pi(1300)]_S$	900	200-600	$\pi(1300) \to \rho\pi$	seen
$\hat{f}_0(1900)$	y_0^{+-}	$0^{+-}(0^-)$	$[\pi b_1(1235)]_P$	250	155	$b_1 \to (\omega\pi)_{S,D}$	most
						$\to \eta\rho$	seen
$\hat{f}_0'(2100)$	z_0^{+-}	$0^{+-}(0^-)$	$[KK_1(1270)]_P$	800	90	See $\hat{\phi}(2100)$	
			$[KK_1(1400)]_P$	50	174	$K_1 \to K^*\pi$	94%
			$[KK(1460)]_S$	800	250	$K \to K\pi\pi$	seen

Table 3: Flux tube results for the dominant decay modes (in MeV) of $I = 1$ hybrids, from Close and Page [38].

A	B, C	L	Γ	A	B, C	L	Γ	A	B, C	L	Γ
2^{-+}	$f_2(1270)\pi$	S	40	1^{+-}	$a_2(1320)\pi$	P	175	1^{-+}	$f_1(1285)\pi$	S	40
		D	20		$a_1(1260)/\pi$	P	90			D	20
	$b_1(1235)\pi$	D	40		$h_1(1170)\pi$	P	175		$b_1(1235)\pi$	S	150
	$a_2(1320)\eta$	S	~ 40		$b_1(1235)\eta$	P	150			D	20
	$K_2^*(1430)K$	S	~ 30		$K_2^*(1430)K$	P	60		$a_1(1260)\eta$	S	50
2^{+-}	$a_2(1320)\pi$	P	200		$K_1(1270)K$	P	250		$K_1(1270)K$	S	20
	$a_1(1260)\pi$	P	70		$K_0^*(1430)K$	P	70		$K_1(1400)K$	S	~ 125
	$h_1(1170)\pi$	P	90	1^{++}	$f_2(1270)\pi$	P	175	0^{-+}	$f_2(1270)\pi$	D	20
	$b_1(1235)\eta$	P	~ 15		$f_1(1285)\pi$	P	150		$f_0(1300)\pi$	S	~ 150
0^{+-}	$a_1(1260)\pi$	P	700		$f_0(1300)\pi$	P	~ 20		$K_0^*(1430)K$	S	~ 200
	$h_1(1170)\pi$	P	125		$a_2(1320)\eta$	P	50	1^{--}	$a_2(1320)\pi$	D	50
	$b_1(1235)\eta$	P	80		$a_1(1260)\eta$	P	90		$a_1(1260)\pi$	S	150
	$K_1(1270)K$	P	600		$K_2^*(1430)K$	P	~ 20			D	20
	$K_1(1400)K$	P	150		$K_1(1270)K$	P	40		$K_1(1270)K$	S	40
					$K_1(1400)K$	P	~ 20		$K_1(1400)K$	S	~ 60

Table 4: As in Table III, for $I = 0$ nonstrange hybrids.

A	B, C	L	Γ	A	B, C	L	Γ	A	B, C	L	Γ
2^{-+}	$a_2(1320)\pi$	S	125	2^{+-}	$b_1(1235)\pi$	P	250	1^{++}	$a_2(1320)\pi$	P	500
		D	60		$h_1(1170)\eta$	P	30		$a_1(1260)\pi$	P	450
	$f_2(1270)\eta$	S	~ 50	0^{+-}	$b_1(1235)\pi$	P	300		$f_2(1270)\eta$	P	70
	$K_2^*(1430)K$	S	~ 30		$h_1(1170)\eta$	P	90		$f_1(1285)\eta$	P	60
1^{+-}	$b_1(1235)\pi$	P	500		$K_1(1270)K$	P	600		$K_2^*(1430)K$	P	~ 20
	$h_1(1170)\eta$	P	175		$K_1(1400)K$	P	150		$K_1(1270)K$	P	40
	$K_2^*(1430)K$	P	60	1^{-+}	$a_1(1260)\pi$	S	100		$K_1(1400)K$	P	~ 20
	$K_1(1270)K$	P	250			D	70	0^{-+}	$a_2(1320)\pi$	D	60
	$K_0^*(1430)K$	P	70		$f_1(1285)\eta$	S	50		$f_0(1300)\eta$	S	~ 200
1^{--}	$K_1(1270)K$	S	40		$K_1(1270)K$	S	20		$K_0^*(1430)K$	S	~ 200
	$K_1(1400)K$	S	60		$K_1(1400)K$	S	~ 125				

flux tube model [36, 38], which predicts that the dominant two body decay modes of the lowest lying hybrids are an $L = 0$ and $L = 1$ $q\bar{q}$ meson pair. These $S + P$ thresholds are rather high in mass, about 4.3 GeV for $c\bar{c}$ hybrids and 11.0 GeV for $b\bar{b}$ hybrids. The possibility that relatively narrow hybrid charmonium states may exist within this 3.8-4.3 GeV window provides an exciting opportunity for e^+e^- facilities such as BEPC, KEK and a Tau-Charm Factory.

Calculations of the decay widths of charmonium hybrids have been carried out in the flux tube model by Close and Page [48], assuming masses of $\approx$ 4.1-4.2 GeV. These partial widths (to D^*D) are found to be quite small, typically only $\sim 1-10$ MeV. Thus if there are relatively unmixed charmonium hybrids, the 1^{--} vector hybrids should appear as narrow spikes in R in this mass range. For this reason a detailed scan of R starting near the open charm threshold would be a first priority at a Tau-Charm Factory. Close and Page speculate about a more complicated possibility, which is that the $\psi(4040)$ and $\psi(4160)$ may be equal-weight linear combinations of $3S$ $|c\bar{c}\rangle$ and the 1^{--} $c\bar{c}$-hybrid. (The usual assignment is that the $\psi(4040)$ is $3S$ and the $\psi(4160)$ is $2D$ [49].) The Close-Page linear combinations would explain why the e^+e^- widths are approximately equal and relatively large for both states, which is surprising if one is a D-wave $c\bar{c}$. The assignments for the ψ states above open-charm thresholds can be tested by measurements of their branching fractions to $DD, D^*D, ..., D_s^*D_s^*$. The branching fractions predicted by these models are very characteristic; unfortunately they have not yet been measured accurately. Determination of these branching fractions would be a high priority at an e^+e^- machine operating at the ψ mass.

Finally, we note that the non-vector hybrids can also be produced in e^+e^- through a "continuum cascade", as suggested by D.Bugg, and discussed in references [49, 50]. In this approach one produces a high-mass $c\bar{c}$ system in the continuum, perhaps at 5 GeV; this may then decay hadronically to hybrid charmonium levels of various J^{PC}, and the hybrid in turn decays hadronically to a characteristic state such as the ψ. Thus one can search for the decay chain

$$e^+e^- \to c\bar{c} \to H_c\,\eta; \quad H_c \to \eta\psi; \quad \psi \to e^+e^-$$

in the final state $\eta\eta e^+e^-$, triggering on a lepton pair from the ψ and $\gamma\gamma$ pairs from the two ηs. The $\eta\psi$ invariant mass distribution can then be studied for evidence of hybrids or $c\bar{c}$ states. Other quantum numbers can be investigated by replacing η by other hadrons, by $(\pi\pi)_S$ for example, in the hadronic cascades.

5.5 Hybrid Experimental Candidates

There are several experimental candidates for hybrids, but as in the case of glueballs there are no generally accepted states at present.

In the exotic channels (which would be most convincing evidence for a hybrid), previous claims by GAMS that a resonant signal had been detected in the 1^{-+} wave of $\eta\pi$ [51] have now been withdrawn. A KEK experiment [52] instead finds evidence for a resonant 1^{-+} $\eta\pi$ wave but with the mass and width of the $a_2(1320)$; this surprising result obviously must be checked carefully for "feedthrough" of the a_2 amplitude. VES [53] has studied $\pi\eta$ and $\pi\eta'$ and report a broad, higher-mass effect in $\pi\eta$ and especially in $\pi\eta'$, with a mass near 1.6 GeV. The phase motion of the 1^{-+} component has not yet been determined.

Studies of the final state $f_1\pi$ suggested by flux tube model decay calculations are underway [53, 47], and preliminary evidence for a possible 1^{-+} signal has been reported by E818 at BNL [47].

There have been several observations of a photoproduced $I = 1$ state at about 1775 MeV in $\rho\pi$ and $f_2\pi$ [46], which is too heavy to be the $\pi_2(1670)$ without complicated interference effects. Although the quantum numbers of this state have not been determined definitively, 1^{-+} is preferred over 2^{-+}. A possible narrow 1^{-+} state has been reported by GAMS in $\eta\eta'$ at a mass of 1910 MeV; here there are rather few events, so it will be important to improve the statistics. Several experiments plan future studies of these channels, including E818 (to study $\pi^- f_1$) [54] and E852 (to study πf_1 and $\pi\eta$) [55] at BNL.

In addition to exotic hybrid candidates there are several nonexotic candidates; recall for example the Close-Page result that a hybrid with π_2 quantum numbers is expected to be relatively narrow, and should be visible in $f_2\pi$. One way to distinguish hybrids from $q\bar{q}$ spin-singlet states is through their strong branching fractions; a selection rule in the $^3P_0$ decay model forbids the decay of a spin-singlet $q\bar{q}$ state to two final spin-singlet quarkonia,

$$(q\bar{q})_{S=0} \not\to (q\bar{q})_{S=0} + (q\bar{q})_{S=0} \ .$$

In the π_2 channel this selection rule forbids the decay of a $^1D_2$ $q\bar{q}$ π_2 to a $^1P_1$ b_1 plus a $^1S_0$ π,

$$\pi_2(q\bar{q}) \not\to b_1\pi$$

but allows it for a hybrid π_2 which does not have the $q\bar{q}$ pair in an $S = 0$ configuration. Close and Page find the $b_1\pi$ mode of a π_2 hybrid should be rather large, so it is especially important to search the $b_1\pi$ channel for evidence of a 2^{-+} hybrid signal. The Crystal Barrel collaboration has preliminary results for πb_1 but at present sees no evidence for unusual states [56].

Other nonexotic hybrid candidates which have been suggested recently are a $\pi(1800)$ reported by VES [57] and the nonstrange 1^{--} states near 1.4-1.7 GeV [58]. The $\pi(1800)$ is cited as a possible hybrid because it has unusual branching fractions, including a significant coupling to $\pi\eta\eta$, apparently through the glueball candidate $f_0(1500) \to \eta\eta$. This $\pi(1800)$ is also reported by VES in $\omega\rho$, $a_0(980)\eta$, $\pi f_0(980)$ and $\pi f_0(1300)$. The decay mode $\pi(1800) \to \pi\rho$ is notably absent, and $f_2\pi$ is also weak or absent.

Although the weakness of the $\rho\pi$ S+S mode is indeed suggestive of a hybrid, since a $\pi(1800)$ is expected in the quark potential model as a second radial excitation, one should be careful to exclude this possibility. Radial quarkonia can have unusual branching fractions due to the nodes in their decay amplitudes, and in the $^3P_0$ decay model with SHO wavefunctions the amplitude for $\pi(3S) \to \rho\pi$ actually has a node at $M = 1.88$ GeV for $\beta = 0.35$ GeV, so the weakness of the $\rho\pi$ mode is understandable for a 3S state. However the same model predicts a weak $f_0(1300)\pi$ mode which disagrees with experiment. The decay amplitude for $\pi(3S) \to \rho(2S)\pi$ is predicted to be quite large, so a search for a $\rho(1450)\pi$ final state would be useful.

Similarly the unusual properties of the nonstrange $I = 0$ and $I = 1$ vectors near 1.5 GeV have led to suggestions that hybrid vector states may be present near this mass [58, 59]. In $I = 1$, for example, the two states $\rho(1450)$ and $\rho(1700)$ are usually assigned to 2^3S_1 and $^3D_1$ respectively, but the very large $\rho(1450) \to 2(\pi^+\pi^-)$ mode [59] is in conflict with quark model expectations for a 2^3S_1 state [59, 60, 61]. A better understanding of these vector states would be possible with a detailed isobar analysis of their quasi two-body strong decay modes.

As a general conclusion regarding nonexotic hybrids, these comparisons of strong decay modes illustrate the importance of having an accurate understanding of the decays of radially excited $q\bar{q}$ states. Both experimental and theoretical studies of the strong decays of radially excited $q\bar{q}$ candidates such as the $\pi(1300)$, $\rho(1450)$, $\phi(1680)$ and so forth will be required if we are to distinguish $q\bar{q}$ from non-$q\bar{q}$ glueballs and hybrids with identical quantum numbers.

6 MOLECULES

6.1 Introduction

Here we consider the experimental and theoretical status of hadronic molecules, which are weakly-bound states of two or more hadrons. We begin with a brief history of the subject of multiquark systems, following which we suggest some signatures for molecules and review experimental candidates. A more general understanding of $2 \to 2$ hadron-hadron scattering amplitudes will be crucial for molecule searches, since bound states are a special feature of the more general problem of determining hadron-hadron scattering amplitudes. We conclude with a short summary of some recent theoretical work on determining scattering amplitudes from the quark model.

In the 1970s it was widely believed that there would be a very rich spectrum of discrete levels of multiquark resonances. The argument was that the many known $q\bar{q}$ and qqq resonances exist because they are color singlets, so we should expect other color-singlet sectors of Hilbert space to possess resonances as well. The "four-quark" $q^2\bar{q}^2$ system was the subject of many detailed studies because it contains the first color-singlet multiquark system beyond three quarks, and because this system could couple to baryon-antibaryon systems through a single $q\bar{q}$ annihilation. Partly for this reason $q^2\bar{q}^2$ states were referred to as "baryonia". Although there were many reports of possible experimental baryonium states, and many detailed spectra were published in various models, no such states have yet been established. In sectors which support $q\bar{q}$ states the spectrum is already very complicated, so the issue of multiquark states remains somewhat obscure. However when one specializes to "smoking gun" systems such as the exotic $I = 2$ channel, which is predicted to support a light 0^{++} $q^2\bar{q}^2$ level (at about 1.2 GeV in the MIT bag model) but cannot have a $q\bar{q}$ state, there is no resonance in evidence [62].

The problems with the various theoretical models that led to erroneous predictions of discrete multiquark levels have been discussed by Isgur [63]. The novel feature of multiquark systems which the models missed is that, unlike $q\bar{q}$ and qqq, they need not exist as single color-singlet hadronic clusters; a $q^2\bar{q}^2$ system in general has some projection onto two color-singlet $q\bar{q}$ mesons, and continuous deformation into two separate mesons appears to be energetically favored in most cases. This rearrangement into color singlets is called "fall-apart" [64], and apparently excludes most single-hadron $q^2\bar{q}^2$ clusters as resonances. Fall-apart would not be possible if the cluster had a mass lower than the threshold of the two-hadron system it can rearrange into, which is why the question of the existence of multiquark clusters such as the H $u^2d^2s^2$ system is so interesting. The bag model predicts this state 81 MeV below $\Lambda\Lambda$ threshold [64], but this prediction should be treated with caution because the bag model has previously given a misleading picture of multiquark states. The tentative evidence for dilambda hypernuclei [65] (if confirmed) makes the existence of an H six-quark resonance well below $\Lambda\Lambda$ threshold appear very unlikely. Whether single multiquark clusters exist as resonances under any conditions is a detailed dynamical question, which should be investigated using models that allow the system itself freedom to choose between a single cluster or separate color singlets. At present it appears that single $q^2\bar{q}^2$ hadronic clusters may only exist as resonances in heavy-light systems such as $c^2\bar{q}^2$ [66].

It was the lack of sufficient freedom in the wavefunctions that led to the spurious prediction of many discrete baryonium levels; the models *assumed* that such states existed, and then gave predictions for the spectrum of these discrete levels. The first detailed study that allowed the $q^2\bar{q}^2$ system to choose between clusters and separate mesons as ground states was the variational study of the 0^{++} sector by Weinstein and Isgur [67], which found that continuous deformation of a cluster into separate $q\bar{q}$ mesons was usually preferred energetically. The two exceptions found by Weinstein and Isgur will be discussed below.

6.2 Experimental Molecule Candidates

1) Nuclei

Lest one form the impression that hadronic molecules are controversial, note that the $\sim 10^5$ known nuclear levels are all hadronic molecules. Of course the term is usually applied to hadron pairs; even if we specialize to this restricted case, the deuteron can be cited as a noncontroversial example of a dominantly S-wave hadronic molecule. Its almost-bound $I = 1, S = 0$ partner is an example of another phenomenon which may appear elsewhere in the spectrum, a molecular resonance above threshold which is due to a strongly attractive final-state interaction. The existence of so many nuclear species is especially notable since the nucleon-nucleon system is rather unfavorable for the formation of bound states, due to the strong short-distance repulsive core. This suggests that many other families of bound hadrons may exist, although they may not be experimentally accessible except in special cases.

2) $f_0(975)$ and $a_0(980)$: The "$K\bar{K}$-molecules".

Weinstein and Isgur [67] found an exception to the fall-apart phenomenon in the scalar sector, with parameters corresponding to the $qs\bar{q}\bar{s}$ system. Here weakly-bound deuteronlike states of kaon and antikaon were found to be the ground states of the four-quark system; Weinstein and Isgur refer to these as "$K\bar{K}$ molecules". The scalars $f_0(975)$ and $a_0(980)$ were obvious candidates for these states, having masses just below $K\bar{K}$ threshold and strong couplings to strange final states. Subsequently the $\gamma\gamma$ couplings of the $f_0(975)$ and $a_0(980)$ were found to be anomalously small relative to expectations for light ${}^3P_0$ $q\bar{q}$ states ($q = u, d$), as discussed in Refs. ([68, 69]). The status of the $K\bar{K}$ molecule assignment and the many points of evidence in its favor have been discussed recently by Weinstein and Isgur [70, 71].

Although Morgan and Pennington have argued against a molecule interpretation of the $f_0(975)$ [72], their criticism applies to a $K\bar{K}$ potential model in which the $f_0(975)$ is a single pole in the scattering amplitude. The more recent work of Weinstein and Isgur [70, 71] incorporates couplings to open meson-meson channels and heavier ${}^3P_0$ $q\bar{q}$ states, so the physical resonances are not only $|K\bar{K}\rangle$. Since there has been much criticism of the idea of a pure $K\bar{K}$ bound state, a direct quote from Weinstein and Isgur [70] (regarding the $I = 0$ state) is appropriate:

"Despite its name and location, the "$K\bar{K}$ molecule" is not a simple $K\bar{K}$ bound state. Its stability is dependent on its couplings to the other $I = 0$ channels and at $E = M_{S^*}$ the coupled-channel wavefunction has substantial components of the other states."

Although the f_0 and a_0 states remain dominantly $K\bar{K}$, these modifications may answer the objections of Morgan and Pennington. Pennington suggests that the term "deuteronlike" may be a misnomer, if couplings to other states than $K\bar{K}$ play an important rôle in these states [68]. Thus it appears that the important question regarding the f_0 and a_0 may be one of detail, specifically how large the subdominant non-$K\bar{K}$ components are in these states and how they can be observed experimentally.

3) $f_1(1420)$

Since the $f_1(1420)$ is above the K^*K threshold of 1390 MeV it is a candidate for a nonresonant threshold enhancement ($K^*\bar{K}$ + $h.c.$) rather than a molecular bound state. This possibility was suggested by Caldwell [73], and satisfies the criteria of lying just above the K^*K threshold (antiparticle labels are implicit) and having quantum numbers allowed for that pair in S-wave. The apparent width of the enhancement should not be narrower than the intrinsic width of the K^*, and indeed the PDG values are similar, $\Gamma(f_1(1420)) = 56 \pm 3$ MeV and $\Gamma(K^*) = 50$ MeV. Longacre [74] found that a model with an S-wave nonresonant ($K^*\bar{K} + h.c.$) enhancement gives a good description of this state, and Isgur, Swanson and Weinstein [75] also favor this possibility. The (off-shell) $\gamma\gamma^*$ couplings of the $f_1(1420)$ relative to expectations for a 1^{++} $s\bar{s}$ state may provide a test of the hadron-pair model.

Another test of this K^*K-assignment is in radiative transitions; the dominant radiative mode of a K^*K system will arise from the radiative transition of the K^* constituent, $K^* \to K\gamma$, implying a partial width of

$$\Gamma(f_1(K^*K) \to \gamma K\bar{K}) \approx 80 \ KeV \ , \tag{62}$$

and a characteristic pattern of preference for $K^o\bar{K}^o$ over K^+K^- by about a factor of two. An $s\bar{s}$ state would give a similar radiative partial width, $\Gamma(f_1(1420)(s\bar{s}) \to \gamma\phi) \approx 50$ KeV if we scale from the $\Gamma(f_2' \to \gamma\phi) = 96$ KeV of Godfrey and Isgur [3]. Although the radiative rates are similar, there is a crucial difference in the intermediate states: The $s\bar{s}$ decay is to $\gamma\phi$, so the final $K\bar{K}$ pair will originate from a $\phi(1019)$ peak. The $K\bar{K}$ events from a K^*K system will instead peak at the invariant mass which results from the radiative decay $K^* \to K\gamma$ of an initial $K^*\bar{K}$ pair at rest, which is slightly *above* the ϕ mass. Due to a numerical coincidence these two masses are rather similar, so the most convincing tests would be the $\gamma K^o\bar{K}^o$ to γK^+K^- ratio and the γK invariant mass distribution (obviously this will have a K^* peak in the molecule assignment). Thus, the two $f_1(1420)$ assignments can be distinguished through a study of the charged and neutral final states and γK invariant mass distributions observed in $f_1(1420) \to \gamma K\bar{K}$.

4) $f_0(1710)$

Early references often considered this a glueball candidate, since it was discovered in a ψ radiative decay [76] and has no obvious assignment in the $q\bar{q}$ spectrum. It was also discussed as a possible $qs\bar{q}\bar{s}$ multiquark state (a single cluster rather than a molecule), although the presence of a fall-apart coupling to $K\bar{K}$ makes this appear untenable. The $f_0(1710)$ appears unlikely to be a radially excited nonstrange $^3P_0$ $q\bar{q}$ because it has a mass 110 MeV below the Godfrey-Isgur prediction [3] and it has a very weak coupling to $\pi\pi$ final states, which makes both nonstrange $q\bar{q}$ and glueball assignments rather implausible (assuming naive flavor-singlet glueball couplings).

The $f_0(1710)$ should also be considered a molecule candidate, of two vector mesons. There have been several suggestions in the literature regarding molecule assignment for this state, which differ primarily in the proposed binding mechanism; this leads to observable differences in the predicted decay modes.

Törnqvist [77] suggests a $K^*\bar{K}^*$ assignment for this state. If the K^*s decay as free hadrons this leads one to expect a large partial width of $\Gamma(f_0(1710) \to K^*\bar{K}^*) \approx 2\,\Gamma(K^*) \approx 100$ MeV, making it the dominant decay mode, since the 1992 PDG gives the $f_0(1710)$ a total width of 146 ± 12 MeV. This prediction appears to disagree with the branching fractions to other final states estimated by the PDG. Ericson and Karl [78] have studied the one-pion-exchange binding mechanism proposed by Törnqvist, and conclude that it may just provide sufficient attraction to bind $K^*\bar{K}^*$. They also conclude that this mechanism would predict several other vector-vector and meson-baryon molecules, such as B^*B^* and $N\Sigma$.

Dooley *et al.* [79] assumed that a different mechanism was responsible for the dominant vector-vector interactions, specifically a OGE constituent interchange model which they had previously applied to several other channels (see previous section). This model predicts a very strong coupling between $K^*\bar{K}^*$ and $\omega\phi$ channels, found by Swanson [80] and represented using off-diagonal hadron-hadron potentials. Dooley *et al.* used these potentials in a multichannel generalization of the Weinstein-Isgur work, and found weakly bound molecules in several vector-vector systems. In the $I = 0$ $qs\bar{q}\bar{s}$ sector they found a linear combination of two-meson basis states as a 0^{++} ground state,

$$|\Psi_0(qs\bar{q}\bar{s}\ 0^{++})\rangle = \frac{1}{\sqrt{2}}\Big(|K^*\bar{K}^*\rangle + |\omega\phi\rangle\Big)\ , \tag{63}$$

and a 2^{++} excited state somewhat higher in mass. This linear combination predicts $K\bar{K}$ and $\eta\eta$ branching fractions close to experimental values [81] and also gives reasonable results for the flavor-tagging $\psi \to VX$ hadronic decays, which suggest that the $f_0(1710)$ couples as if it were a strange-nonstrange mixed-flavor state. (A similar result is known for the $f_0(975)$ in these decays.) A $K^*\bar{K}^*$ decay mode is expected as in the pure $K^*\bar{K}^*$-molecule picture, but with a smaller branching fraction of about 35%. The branching fraction of a $(K^*\bar{K}^* + \omega\phi)/\sqrt{2}$ molecule to $\pi\pi$ is more problematical; Dooley found it to be about 5% [81], consistent with experiment. A very characteristic electromagnetic decay mode in this assignment is $f_0(1710) \to \phi\pi^0\gamma$, with a branching fraction of 0.3%, due to constituent-ω radiative decay. It should also have an anomalously small $\gamma\gamma$ coupling relative to a nonstrange $I = 0$ $q\bar{q}$ state such as the $f_2(1274)$ [82].

A search for a $K^*\bar{K}^*$ mode is clearly the most important test of the vector-molecule models of the $f_0(1710)$, since both models expect a large $K^*\bar{K}^*$ branching fraction. Of course one $K\pi$ combination will be skewed downwards from the free K^* mass by slightly more than E_B, assuming that this acts as a sequential decay, $f_0(1710) \to K^*(\bar{K}\pi)$ followed by decay of the free recoiling K^*.

5) $\Lambda(1405)$

In meson-baryon sectors, the $\frac{1}{2}^-$ $\Lambda(1405)$ has long been considered a candidate $\bar{K}N$ bound state [83], since it is just below the $\bar{K}N$ threshold, has S-wave $\bar{K}N$ quantum numbers, and is nearly 150 MeV below the mass expected for the lightest $\frac{1}{2}^-$ Λ^* baryon in the quark model [4]. (The decay amplitudes however are consistent with a uds assignment [4].) Theoretical study of this channel has been incomplete, however, because of the complications of open channels and $q\bar{q}$ annihilation; a model which includes mixing in the full $N\bar{K}$-$\Sigma\pi$-uds system is required for a complete study of this state. The radiative partial widths to $\gamma\Lambda$ and $\gamma\Sigma^0$ compared to quark model predictions for a uds baryon may allow a convincing test of the molecule assignment. A wide range of theoretical numbers has been reported for these radiative widths [84], so one should be careful to use techniques which give reliable results for well-established qqq quark model states.

6) Z^*s and dibaryons

We conclude this section with a reminder that there may be hadronic molecules in other sectors of Hilbert space, which have received little recent attention because they do not correspond to $q\bar{q}$ or qqq flavor states in the quark model or because of the lack of appropriate experimental facilities.

One especially interesting system, which should soon be accessible to experiments at DAΦNE and perhaps CEBAF, is the kaon-nucleon system. Possible resonances in the $q^4\bar{s}$ sector are known as Z^*s, and although there have been indications of such states for many years in elastic KN scattering [85], the lack of clear evidence for multiquark states and the uncertainties of partial wave analyses in the KN system have left the possibility of such states a controversial question. In our recent theoretical study of KN, K^*N, $K\Delta$ and $K^*\Delta$ systems [86] we found that several of these channels, notably the minimum total spin, minimum total isospin ones, have strongly attractive interactions. The experimental reports of Z^* resonances may represent observations of final state enhancements or even of meson-baryon bound states just below threshold. Clarification of this issue will require accurate partial wave analyses of KN scattering, and data on the inelastic channels $KN \to K^*N, K\Delta$ and $K^*\Delta$ would also be very valuable for the study of possible Z^* states. Here the theoretical calculations should be more reliable since these reactions are annihilation-free at valence quark level.

Finally, there are controversial reports of resonances in partial wave analyses of elastic NN scattering [87], and these "dibaryon" resonances may also include S-wave baryon-baryon molecule states. This system too is relatively straightforward theoretically because it is annihilation-free, and a few $\Delta\Delta$ channels (notably $I = 0, S = 1, I = 1, S = 0, I = 0, S = 3$ and $I = 3, S = 0$) have been cited by theorists as the most likely for the formation of nonstrange baryon-baryon resonances [88, 89].

7) Other possible molecules.

There are many other possible molecular states, which can only be mentioned briefly here. In the meson sector these include the $f_0(1500)$ glueball candidate, which might be a nonstrange vector-vector system [77, 90]. This would explain the weakness of the $K\bar{K}$ mode. The $f_0(1365)$ should also be considered a possible vector-vector molecule, in view of its very large coupling to $\rho\rho$ despite the near absence of phase space. The 2^{++} state reported by VES [57] in $\rho^o\rho^o$ is another possible vector-vector molecule, although its appearance in the $\rho\rho$ D-wave may be a problem. The $\psi(4040)$, which shows a strong preference for $D\bar{D}$ over $D^*\bar{D}^*$ (opposite to expectations from phase space) was one of the earliest molecule candidates [91]. As the $c\bar{c}$ assignment for this state is a $3S$ radial excitation, this anomalous branching fraction may be due to a node in the decay amplitude near the $D^*\bar{D}^*$ momentum [92].

6.3 Signatures for Molecules

Leaving aside questions of detailed dynamics, there are several obvious signatures for hadron-pair molecules that may be abstracted from the more convincing molecule candidates. These signatures are:

1) J^{PC} and flavor quantum numbers of an $L = 0$ hadron pair.

The residual "nuclear" strong forces that bind molecules are of such short range that $L > 0$ molecules appear unlikely in light hadronic systems. Actually there is a possible exception, the $\psi(4040)$, which couples so strongly to $D^* \bar{D}^*$ that it was suggested as a P-wave $D^* \bar{D}^*$ molecule (note $E_B \approx 0$) some time ago [91]. This exception may be possible because heavier mesons bind more easily, and the light quarks ensure relatively strong interactions between them. Should this state actually be a P-wave charmed-meson molecule, a spectrum of more deeply-bound S-wave charm molecules is anticipated [93].

2) A binding energy of at most about $50 - 100$ MeV.

From the uncertainty principle; a minimum separation of ≈ 1 fm is required for hadrons to maintain separate identities, which gives $E_B \sim 1/(2\mu R^2) \sim 1/(1 \text{ GeV}) \cdot (1 \text{ fm})^2 \approx 50$ MeV, and of course a factor-of-two uncertainty is plausible in this simple estimate. For comparison, Weinstein and Isgur find an rms $K\bar{K}$ separation of about 1.7 fm in their model of the $f_0(975)$ and $a_0(980)$, which have $E_B \approx 10 - 20$ MeV. Note also that an attractive interaction may lead to a final-state enhancement in S-wave just *above* threshold, which may or may not be resonant; the $I = 1, S = 0$ partner of the deuteron is an example of such a resonance, and the $f_1(1420)$ may be an example of a nonresonant final-state enhancement.

3) Strong couplings to constituent channels.

As an example, the anomalously large coupling of the $f_0(975)$ to $K\bar{K}$, as indicated by $B(K\bar{K})/B(\pi\pi) \approx 1/4$ despite the near absence of $K\bar{K}$ phase space, is an important clue that it is not a nonstrange $q\bar{q}$ state.

4) Anomalous EM couplings relative to expectations for conventional quark model states.

The $f_0(975)$ for example has a tiny $\gamma\gamma$ partial width of perhaps 0.2 Kev to 0.6 Kev (depending on the analysis) [68]. This small $\gamma\gamma$ width is expected for a $K\bar{K}$ molecule [69], but for a nonstrange $^3P_0$ $q\bar{q}$ state the quark model predicts about 3 Kev [94], as has been found for the broad $f_0(\approx 1300)$ [95]. Close, Isgur and Kumano [96] suggest a related test for the $f_0(975)$ and $a_0(980)$ involving the radiative decays $\phi \to \gamma(f_0, a_0)$, which may be possible at DAΦNE and CEBAF [97].

6.4 The Importance of Scattering Amplitudes

Since molecular bound states are a special aspect of $2 \to 2$ hadron scattering amplitudes, one might anticipate that an understanding of these scattering amplitudes will lead to reliable predictions of molecules. Theoretical work along these lines is in progress; at present there are different predictions for molecules depending on the scattering mechanism assumed.

In one pion exchange models [77, 78] many bound states are predicted which should be experimentally observable. Here I would like to discuss scattering calculations in terms of quark-gluon forces (see [79, 80, 88, 86, 98, 99] and references cited therein).

Hadron-hadron scattering amplitudes at low energies are generally thought to involve the quark-gluon interaction nonperturbatively, so although calculations of meson-meson and baryon-baryon interactions at the quark-gluon level have been rather successful, they have typically used complicated nonperturbative methods such as resonating group or variational techniques. Extension of this work to channels such as vector-vector has been slow largely because of the difficulty of applying these methods, although some variational and Monte Carlo results have been reported for special cases, including an extension of the Weinstein-Isgur work to I=2 $\rho\rho$ [100].

Recently our collaboration has found evidence that the Ps-Ps scattering amplitudes found by Weinstein and Isgur (in channels without $q\bar{q}$ annihilation) are actually dominated by perturbative diagrams, although "higher-twist" contributions in the form of external $q\bar{q}$ wavefunctions attached to the diagrams are an essential, nonperturbative aspect of the scattering amplitudes [80, 98]. We initially studied $I = 2$ $\pi\pi$ [98] and $I = 3/2$ $K\pi$ [99] and found that OGE followed by constituent interchange dominates these scattering amplitudes, and leads to results which are numerically very similar to the Ps-Ps potentials found variationally by Weinstein and Isgur. Recent lattice QCD results for the $I = 2$ $\pi\pi$ scattering length support our conclusion regarding the dominance of these diagrams [101]. We refer to these perturbative diagrams with external wavefunctions attached as "quark Born diagrams". With SHO quark model wavefunctions these lead to overlap integrals that can often be evaluated in closed form, and the results for $\pi\pi$ and $K\pi$ S-wave phase shifts are in excellent agreement with experiment over the entire range of energies studied given standard quark model parameters. We have similarly found good agreement in the $I = 0$ and $I = 1$ KN system [86] (*albeit* with some problems at higher energies which may be due to the assumption of single-Gaussian nucleon wavefunctions). In our study of NN, NΔ and $\Delta\Delta$ [88] we found the strongest diagonal attraction in the $I = 0, S = 1$ $\Delta\Delta$ channel, in agreement with the variational work of Maltman [89]. Very recently we have studied the $N_s = 2$ baryon-baryon channel [102], and we agree with Oka, Shimizu and Yazaki and Straub *et al.* [103] that quark model forces lead to a repulsive $\Lambda\Lambda$ core interaction; of the six $N_s = 2$ octet-octet channels we find that only $I = 0, S = 0$ $\Sigma\Sigma$ has an attractive core. It is reassuring that exactly the same conclusion was reached by Oka *et al.* [103] using nonperturbative resonating group methods.

Given this reasonably successful description of hadron-hadron scattering at low energies, can we proceed to study all experimentally accessible channels and see in which we expect hadronic molecules to form? Unfortunately this is not yet possible. The difficulty is that $q\bar{q}$ annihilation appears to be the dominant effect when allowed (as in I=0,1 $\pi\pi$ and I=1/2 $K\pi$), so a realistic description of scattering amplitudes in these channels requires accurate modelling of the couplings of different sectors of Hilbert space (as in $\pi\pi \to f_0(q\bar{q}) \to \pi\pi$). On closer examination of the Weinstein-Isgur results [71] it now appears that both level repulsion against higher-mass s-channel $q\bar{q}$ resonances and nonresonant scattering are needed for the simultaneous binding of $I = 0$ and $I = 1$ $K\bar{K}$ systems. (Without level repulsion against the $q\bar{q}\,{}^3P_0$ states at ≈ 1.3 GeV only the $I = 0$ molecule binds.) The existence of molecules in some channels may be due entirely to level repulsion against a more massive quark model state, and the hadronic couplings of most quark-model states are not well enough established to model this level repulsion accurately. Remarkably, the $q\bar{q}$ pair production process is still rather poorly understood (for a recent study see [104]), and is usually treated using phenomenological models such as ${}^3P_0$ that have no clear relation to QCD. Accurate modelling of hadronic forces in channels with annihilation, and hence reliable predictions of molecules, must await a better understanding of the $q\bar{q}$ annihilation mechanism.

7 ACKNOWLEDGEMENTS

It is a pleasure to thank D.Bugg for his kind invitation to present these lectures to the NATO Summer School on the Confinement Problem, held at Queen Mary and Westfield College and at the University of Swansea. I would also like to thank my colleagues and collaborators, notably E.S.Ackleh, C.Amsler, J.Beene, S.Capstick, S.U.Chung, F.E.Close, G.Condo, K.Danyo, A.Dzierba, P.Geiger, S.Godfrey, N.Isgur, P.R.Page and E.S.Swanson for their assistance in preparation of these notes. I sincerely hope that the attendees of the Summer School will find an exciting and rewarding career in the study of QCD and hadron physics.

This research was sponsored in part by the United States Department of Energy under contract DE-AC05-840R21400, managed by Lockheed Martin Energy Systems, Inc.

REFERENCES

[1] W.Marciano and H.Pagels, Phys. Rep. 36C, 137-276 (1978).

[2] Particle Data Group, Phys. Rev. D50, 1177 (1994).

[3] S.Godfrey and N.Isgur, Phys. Rev. D32, 189 (1985).

[4] S.Capstick and N.Isgur, Phys. Rev. D34, 2809 (1986).

[5] P.Geiger and N.Isgur, Phys. Rev. D47, 5050 (1993); Phys. Rev. Lett. 67, 1066 (1991); Phys. Rev. D44, 799 (1991); *ibid.*, D41, 1595 (1990).

[6] E.Eichten, K.Gottfried, T.Kinoshita, K.D.Lane and T.M.Yan, Phys. Rev. D17, 3090 (1978); Phys. Rev. D23, 203 (1980).

[7] T.A.Armstrong *et al.*, E760 Collaboration, Phys. Rev. Lett. 69, 2337 (1992).

[8] S.Godfrey, Phys. Rev. D31, 2375 (1985).

[9] The $a_0(1450)$ is reviewed by S.Spanier, contribution to HADRON95.

[10] A.LeYaouanc, L.Oliver, O.Pene and J.Raynal, Phys. Rev. D8, 2223 (1973); *ibid.*, D9, 1415 (1974); *ibid.*, D11, 1272 (1975); see also L.Micu, Nucl. Phys. B10, 521 (1969).

[11] E.S.Ackleh, T.Barnes and E.S.Swanson, Oak Ridge National Laboratory report ORNL-CTP-95-09.

[12] E.S.Ackleh and T.Barnes, Phys. Rev. D45, 232 (1992); T.Barnes, in Proceedings of the Ninth International Workshop on Photon-Photon Collisions (La Jolla, 22-26 March 1992), eds. D.O.Caldwell and H.P.Paar (World Scientific, 1992), pp.263-282.

[13] D. Antreasyan *et al.*, Z. Phys. C48, 561 (1990); H.J. Behrend *et al.*, Z. Phys. C46, 583 (1990); K.-H. Karch *et al.*, Z. Phys. C54, 33 (1992).

[14] F.E.Close, Rep. Prog. Phys. 51, 833 (1988).

[15] For the earlier LGT glueball literature see for example C.Michael and M.Teper, Nucl. Phys. B314, 347 (1989); P.deForcrand, G.Schierholz, H.Schneider and M.Teper, Phys. Lett. B152, 107 (1985); G.Berg, Nucl. Phys. B221, 109 (1983); and references cited therein.

[16] G.Bali *et al.*, Phys. Lett. B309, 378 (1993).

[17] H.Chen *et al.*, IBM report IBM-HET-94-1 (contribution of A. Vaccarino to Lattice 93).

[18] J. Sexton *et al.*, IBM report IBM-HET-94-5 (contribution to Lattice 94).

[19] C.Amsler and F.E.Close, "Evidence for Glueballs", Rutherford Laboratory and CERN report CCL-TR-95-003 (April 1995); *ibid.*, Phys. Lett. B353, 385 (1995) and Rutherford Laboratory report RAL-TR-95-003 (July 1995).

[20] V.V.Anisovich *et al.*, Phys. Lett. B323, 233 (1994); C.Amsler, in Proceedings of the XXVII Int. Conf. on High Energy Physics (Glasgow, 20-27 July 1994), Zürich report UZH-PH-50/94.
[21] C.Amsler, personal communication.
[22] D.Bugg *et al.*, Phys. Lett. B353, 378 (1995).
[23] T.Huang, contribution to the Argonne Workshop on a Tau-Charm Factory (June 1995); see also T.Huang *et al.*, CCAST report BIHEP-TH-95-11.
[24] S.Godfrey, R.Kokoski and N.Isgur, Phys. Lett. B141, 439 (1984).
[25] S.Godfrey, personal communication; H.G.Blundell and S.Godfrey, "The $\xi(2220)$ Revisited: Strong Decays of the 1^3F_2 and 1^3F_4 $s\bar{s}$ Mesons.", Carleton University report OCIP/C 95-11.
[26] T.Barnes, Caltech Ph.D. thesis (1977), unpublished; T.Barnes, Nucl. Phys. B158, 171 (1979); T.Barnes and F.E.Close, Phys. Lett. 116B, 365 (1982); M.Chanowitz and S.R.Sharpe, Nucl. Phys. B222, 211 (1983); T.Barnes, F.E.Close and F.deViron, Nucl. Phys. B224, 241 (1983); M.Flensburg, C.Peterson and L.Sköld, Z. Phys. C22, 293 (1984).
[27] D.Horn and J.Mandula, Phys. Rev. D17, 898 (1978).
[28] T.Barnes, "The Bag Model and Hybrid Mesons", in Proceedings of the SIN Spring School on Strong Interactions (Zuoz, Switzerland, April 9-17, 1985). Also distributed as University of Toronto report UTPT-85-21 (April 1985).
[29] M.Tanimoto, Phys. Lett. 116B, 198 (1982); Phys. Rev. D27, 2648 (1983); A.LeYaouanc, L.Oliver, O.Pène, J.-C.Raynal and S.Ono, Z. Phys. C28, 309 (1985); F.Iddir, A.LeYaouanc, L.Oliver, O.Pène, J.-C.Raynal and S.Ono, Phys. Lett. B205, 564 (1988); S.Ishida, H.Sawazaki, M.Oda and K.Yamada, Phys. Rev. D47, 179 (1992); Prog. Theor. Phys. 82, 119 (1989).
[30] J.M.Cornwall and S.F.Tuan, Phys. Lett. B136, 110 (1984).
[31] N.Isgur and J.Paton, Phys. Lett. 124B, 247 (1983).
[32] J.Merlin and J.Paton, J. Phys. G11, 439 (1985).
[33] N.Isgur and J.Paton, Phys. Rev. D31, 2910 (1985).
[34] J.Merlin and J.Paton, Phys. Rev. D35, 1668 (1987).
[35] J.Merlin, Oxford University Ph.D. thesis (unpublished); J.Paton, personal communication.
[36] N.Isgur, R.Kokoski and J.Paton, Phys. Rev. Lett. 54, 869 (1985).
[37] T.Barnes, F.E.Close and E.S.Swanson, Oak Ridge National Laboratory / Rutherford Laboratory Report ORNL-CTP-95-02 / RAL-94-106, hep-ph/9501405, Phys. Rev. D (to appear).
[38] F.E.Close and P.R.Page, Nucl. Phys. B443, 233 (1995).
[39] F.E.Close and P.R.Page, Rutherford Laboratory report RAL-94-122, hep-ph/9412301.
[40] P.Hasenfratz, R.R.Horgan, J.Kuti and J.-M.Richard, Phys. Lett. 95B, 299 (1980).
[41] S.Perantonis and C.Michael, Nucl. Phys. B347, 854 (1990), and references cited therein.
[42] J.I.Latorre, S.Narison, P.Pascual and R.Tarrach, Phys. Lett. 147B, 169 (1984); J.I.Latorre, P.Pascual and S.Narison, Z. Phys. C34, 347 (1987); S.Narison, "QCD Spectral Sum Rules", Lecture Notes in Physics Vol.26, p.375 (World Scientific, 1989).
[43] J.Govaerts, F.deViron, D.Gusbin and J.Weyers, Phys. Lett. 128B, 262 (1983); (E) Phys. Lett. 136B, 445 (1983); J.Govaerts L.J.Reinders, H.R.Rubinstein and J.Weyers, Nucl. Phys. B258, 215 (1985); J.Govaerts, L.J.Reinders and J.Weyers,

Nucl. Phys. B262, 575 (1985); J.Govaerts, L.J.Reinders, P.Francken, X.Gonze and J.Weyers, Nucl. Phys. B284, 674 (1987).

[44] J.Govaerts, F.deViron, D.Gusbin and J.Weyers, Nucl. Phys. B248, 1 (1984).

[45] See for example T.Barnes, ORNL-CCIP-93-11 / RAL-93-065 and F.E.Close, RAL-93-053, in Proceedings of the Third Workshop on the Tau-Charm Factory (Marbella, Spain, 1-6 June 1993); T.Barnes, ORNL-CCIP-93-14 / RAL-93-069 in Proceedings of the Conference on Exclusive Reactions at High Momentum Transfers (Marciana Marina, Elba, Italy, 24-26 June 1993); F.E.Close, Rep. Prog. Phys. 51, 833 (1988); C.Dover, in Proceedings of the Second Biennial Conference on Low Energy Antiproton Physics (Courmayeur, 14-19 Sept. 1992); A.Dzierba, Indiana University report IUHEE-93-2, in Proceedings of the BNL meeting on Future Directions in Particle and Nuclear Physics at Multi-GeV Hadron Facilities (Brookhaven, N.Y. 4-6 March 1993); S.Godfrey, in Proceedings of the BNL Workshop on Glueballs, Hybrids and Exotic Hadrons (AIP, 1989), ed. S.-U. Chung; D.Hertzog, Nucl. Phys. A558, 499c (1993); N.Isgur, CEBAF-TH-92-31, in Proceedings of the XXVI International Conference on High Energy Physics (Dallas, August 1992); G.Karl, Nucl. Phys. A558, 113c (1993).

[46] G.Condo *et al.*, Phys. Rev. D43, 2787 (1991); the same state may have been seen earlier by D.Aston *et al.*, Nucl. Phys. B189, 15 (1981).

[47] J.H.Lee *et al.*, Phys. Lett. B323, 227 (1994).

[48] F.E.Close and P.R.Page, Oxford University / Rutherford Laboratory report OUTP-95-13P / RAL-95-122, hep-ph/9507407.

[49] T.Barnes, in Proceedings of the 3rd Workshop on the Tau Charm Factory (Edition Frontieres 1994), eds. J.Kirkby and R.Kirkby, p.41.

[50] F.E.Close, in Proceedings of the 3rd Workshop on the Tau Charm Factory (Edition Frontieres 1994), eds. J.Kirkby and R.Kirkby, p.73.

[51] D.Alde *et al.*, Phys. Lett. B205, 397 (1988).

[52] H.Aoyagi *et al.*, Phys. Lett. B314, 246 (1993).

[53] G.M.Beladidze *et al.*, Phys. Lett. B313, 276 (1993).

[54] S.U.Chung, personal communication.

[55] A.Dzierba, personal communication.

[56] M.Doser, personal communication; see also J.LoSecco, E852 Collaboration, in preparation.

[57] D.Rybachikov, contribution to HADRON95.

[58] A.Donnachie and Yu. Kalashnikova, Z.Phys C59, 621 (1993); A.Donnachie, Yu. Kalashnikova and A.Clegg, Manchester University Report C-TH-93/10.

[59] A.B.Clegg and A.Donnachie, Z. Phys. C62, 455 (1994).

[60] G.Busetto and L.Oliver, Z. Phys.C 20, 247 (1983).

[61] R.Kokoski and N.Isgur, Phys. Rev. D35, 907 (1987).

[62] W.Hoogland *et al.*, Nucl. Phys. B126, 109 (1977).

[63] N.Isgur, Acta Physica Austraica, Suppl. XXVII, 177 (1985).

[64] R.L.Jaffe, Phys. Rev. Lett. 38, 195, 617E (1977).

[65] S.Aoki *et al.*, Prog. Theor. Phys. 85, 1287 (1991), and references cited therein.

[66] J.P.Ader, J.M.Richard and P.Taxil, Phys. Rev. D25, 2370 (1982); G.Grondin, unpublished.

[67] J.Weinstein and N.Isgur, Phys. Rev. Lett. 48, 659 (1982); Phys. Rev. D27, 588 (1983); see also A.Astier *et al.*, Phys. Lett. B25, 294 (1967); A.B.Wicklund *et al.*, Phys. Rev. Lett. 45, 1469 (1980).

[68] M.R.Pennington, University of Durham report DTP-94/26 (April 1994), to appear in Proceedings of the Meeting on Two-Photon Physics from DAΦNE to LEP200 and Beyond, eds. F.Kapusta and J.Parisi (Paris, February 1994).
[69] T.Barnes, Phys. Lett. 165B, 434 (1985); E.P.Shabalin, Yad. Fiz. 46, 852 (1987); T.N.Truong, in Proceedings of the HADRON '89 International Meeting on Hadron Spectroscopy (Ajaccio, 1989), pp.645; N.Brown and F.E.Close, Rutherford Laboratory report RAL-91-085.
[70] J.Weinstein and N.Isgur, Phys. Rev. D41, 2236 (1990).
[71] J.Weinstein, Phys. Rev. D47, 911 (1993).
[72] K.L.Au, D.Morgan and M.R.Pennington, Phys. Rev. D35, 1633 (1987); D.Morgan and M.R.Pennington, Phys. Lett. 258B, 444 (1991); *ibid.*, Rutherford Laboratory report RAL-92-070 (December 1992).
[73] D.O.Caldwell, Mod. Phys. Lett. A2, 771 (1987); Proceedings of the BNL Workshop on Glueballs, Hybrids and Exotic Mesons (Upton, N.Y. 29 August - 1 September 1988), ed. S.U.Chung (AIP 1989), pp.465-471.
[74] R.S.Longacre, Phys. Rev. D42, 874 (1990).
[75] N.Isgur, E.S.Swanson and J.Weinstein, work in progress.
[76] C.Edwards *et al.*, Phys. Rev. Lett. 48, 458 (1982).
[77] N.Törnqvist, in Proceedings of the International Conference on Hadron Spectroscopy "HADRON '91", (World Scientific, 1992; eds. S.Oneda and D.C.Peaslee), pp.795-798; Phys. Rev. Lett. 67, 556 (1991).
[78] T.E.O.Ericson and G.Karl, Phys. Lett. B309, 426 (1993); G.Karl, Nucl. Phys. A558, 113c (1993).
[79] K.Dooley, E.S.Swanson, and T.Barnes, Phys Lett. 275B, 478 (1992); K.Dooley, in Proceedings of the 4th International Conference on Hadron Spectroscopy "Hadron '91" (College Park, Md. 12-16 August 1991), (World Scientific, 1992), pp.789-794.
[80] E.S.Swanson, Ann. Phys. (NY) 220, 73 (1992).
[81] K.Dooley, Ph.D. thesis, University of Toronto (1993).
[82] E.S.Ackleh, K.Dooley and Z.P.Li, University of Toronto report UTPT-92-03.
[83] R.H.Dalitz and S.F.Tuan, Ann. Phys. (NY) 3, 307 (1960); J.J.Sakurai, Ann. Phys. (NY) 11, 1 (1960).
[84] H.Burkhardt and J.Lowe, Phys. Rev. C44, 607 (1991).
[85] J.S.Hyslop, R.A.Arndt, L.D.Roper and R.L.Workman, Phys. Rev. D46, 961 (1992).
[86] T.Barnes and E.S.Swanson, Phys. Rev. C49, 1166 (1994); T.Barnes, in Proceedings of the DAPHNE95 Workshop on Physics and Detectors for the Frascati Φ Factory (4-7 April 1995, Frascati, Italy). See also K.Maltman and S.Godfrey, Nucl. Phys. A452, 669 (1986), who reach rather different conclusions regarding attractive channels.
[87] R.A.Arndt, L.D.Roper, R.L.Workman and M.W.McNaughton, Phys. Rev. D45, 3995 (1992).
[88] T.Barnes, S.Capstick, M.D.Kovarik and E.S.Swanson, Phys. Rev. C48, 539 (1993).
[89] K.Maltman, Nucl. Phys. A438, 669 (1985);see also T.Goldman *et al.*, Phys. Rev. C39, 1889 (1989); M.Cvetic *et al.*, Phys. Lett. 93B, 489 (1980); T.Kamae and T.Fujita, Phys. Rev. Lett. 38, 471 (1977).
[90] Yu.S.Kalashnikova, in Proceedings of the International Conference on Hadron Spectroscopy "HADRON '91", (World Scientific, 1992; eds. S.Oneda and D.C.Peaslee), pp.777-782.
[91] V.A.Novikov *et al.*, Phys. Rep. C41, 1 (1978); M.B.Voloshin and L.B.Okun, JETP Lett. 23, 333 (1976).

[92] P.R.Page, Nucl. Phys. B446, 189 (1995); see also A.LeYaouanc *et al.*, Phys. Lett. B71, 397 (1977); *ibid.*, Phys. Lett. B72, 57 (1977).

[93] A.DeRújula, H.Georgi and S.L.Glashow, Phys. Rev. Lett. 38, 317 (1977); S.Iwao, Lett. Nuovo Cimento 28, 305 (1980).

[94] Z.P.Li, F.E.Close and T.Barnes, Phys. Rev. D43, 216 (1991); E.S.Ackleh, University of Tennessee Ph.D. thesis (1993); T.Barnes, in Proceedings of the Ninth International Workshop on Photon-Photon Collisions (La Jolla, 22-26 March 1992), eds. D.O.Caldwell and H.P.Paar (World Scientific, 1992), pp.263-282; S.Gerasimov, unpublished.

[95] J.K.Bienlein (Crystal Ball Collaboration), in Proceedings of the Ninth International Workshop on Photon-Photon Collisions (La Jolla, 22-26 March 1992), eds. D.O.Caldwell and H.P.Paar (World Scientific, 1992), pp.241-257.

[96] F.E.Close, N.Isgur and S.Kumano, Nucl. Phys. B389, 513 (1993).

[97] A.Dzierba *et al.*, CEBAF proposal E-94-016; A.Dzierba, "Measuring Rare Radiative Decays of the Φ Meson at CEBAF", in Proceedings of the Second Workshop on Physics and Detectors for DAPHNE (DAPHNE95).

[98] T.Barnes and E.S.Swanson, Phys. Rev. D46, 131 (1992); for closely related work on meson-meson scattering see B.Masud, J.Paton, A.M.Green and G.Q.Liu, Nucl. Phys. A528, 477 (1991); B.Masud, "A Model for $q^2\bar{q}^2$ Systems, Illustrated by an Application to $K\bar{K}$ Scattering", University of Helsinki report HU-TFT-93-30 (November 1993); D.Blaschke and G.Röpke, Phys. Lett. B299, 332 (1993); A.LeYaouanc, L.Oliver, O.Péne and J.-C.Raynal, Phys. Rev. D42, 3123 (1990).

[99] T.Barnes, E.S.Swanson and J.Weinstein, Phys. Rev. D46, 4868 (1992).

[100] G.Grondin, E.S.Swanson and J.Weinstein, unpublished.

[101] S.R.Sharpe, personal communication; R.Gupta, A.Patel and S.R.Sharpe, Nucl. Phys. B383, 309 (1992).

[102] T.Barnes and J.Weinstein, in preparation.

[103] M.Oka, K.Shimizu and K.Yazaki, Nucl. Phys. A464, 700 (1987). U.Straub, Z.-Y.Zhang, K.Bräuer, A.Fäβler and S.B.Khadkikar, Phys. Lett. 200B, 241 (1988).

[104] P.Geiger and E.S.Swanson, "Distinguishing Among Strong Decay Models", MIT report MIT-CTP-2295 (April 1994).

[105] C.Michael, in Proceedings of the Workshop on QCD: 20 Years Later (Aachen, 9-13 June 1992).

[106] C.Amsler *et al.*, Phys. Lett. B322, 431 (1994).

ELEMENTS OF SCATTERING THEORY

V.V. Anisovich

Petersburg Nuclear Physics Institute,
Gatchina, St. Petersburg 188350, Russia

These lectures don't give a full presentation of scattering theory. They are devoted to some key topics which are important for the analysis of the analytical structure of amplitudes. At the present time, the study of analytical properties of the amplitudes is dictated by the necessities of the experiments: the discovery and investigations of new particles are mainly based on the study of leading singularities of amplitudes. A systematic presentation of the scattering theory problems which are touched here can be found in various textbooks and monographs, see, for example, refs. [1-6]. The presentation of some special problems is based on original articles, see refs. [7-12].

1 Scattering in quantum mechanics

Let us start with the non-relativistic scattering theory discussed in the framework of quantum mechanics. The quantum mechanical approach is a good basis for defining notions and sketching problems of scattering theory.

1.1 Schrödinger equation and the wave function of two scattered particles

Let us consider elastic scattering of the spinless particle. It means that particles do not change their inner state and do not produce new particles. We suppose also that interaction forces are centrally symmetrical ones: they depend only on relative separation of the scattered particles.

The problem of two-particle elastic scattering in quantum mechanics can be reformulated as a problem of scattering of one particle in the field $V(r)$. It is done by considering two-particle scattering in the center mass system (center of inertia of two particles). In more detail, the Hamiltonian of two interacting particles is equal to

$$\widehat{H} = -\frac{1}{2m_1}\Delta_1 - \frac{1}{2m_2}\Delta_2 + V(r). \qquad (1.1)$$

Hadron Spectroscopy and the Confinement Problem
Edited by D.V. Bugg, Plenum Press, New York, 1996

Here Δ_1 and Δ_2 are Laplace operators for the coordinates of particles 1 and 2, $\Delta_i = \frac{\partial^2}{\partial x_i^2} + \frac{\partial^2}{\partial y_i^2} + \frac{\partial^2}{\partial z_i^2}$, m_i's are their masses and $V(r)$ is the interaction potential depending on the distance between particles:

$$\vec{r} = \vec{r}_1 - \vec{r}_2. \tag{1.2}$$

Let us introduce the coordinates of the center of inertia for these two particles:

$$\vec{R} = \frac{m_1\vec{r}_1 + m_2\vec{r}_2}{m_1 + m_2}. \tag{1.3}$$

The Hamiltonian written as a function of variables $\vec{r}$ and $\vec{R}$ is equal to

$$\widehat{H} = -\frac{1}{2(m_1 + m_2)}\Delta_R - \frac{1}{2m}\Delta + V(r), \tag{1.4}$$

where $m = m_1m_2/(m_1+m_2)$. So the Hamiltonian is reduced to a sum of two independent terms, one standing for a free movement of the center of mass, another for the interaction of the particles. This latter is equivalent to the Hamiltonian of a particle with reduced mass m moving in the potential $V(r)$. Correspondingly, the wave function written for the two particles, $\psi(\vec{r}_1, \vec{r}_2)$, can be presented in factorized form:

$$\psi(\vec{r}_1, \vec{r}_2) = \phi(\vec{R})\psi(\vec{r}). \tag{1.5}$$

The wave function $\phi(\vec{R})$ describes the center-of-mass movement (free movement of a particle with the mass $m_1 + m_2$), while $\psi(\vec{r})$ describes the relative movement of particles 1 and 2 (movement of a particle with mass m in the centrally symmetrical field $V(r)$).

The Schrödinger equation for $\psi(\vec{r})$ reads:

$$\left[-\frac{\Delta}{2m} + V(r)\right]\psi(\vec{r}) = E\psi(\vec{r}), \tag{1.6}$$

E is the energy of relative movement. The same equation written in spherical coordinates is

$$\frac{1}{r^2}\frac{\partial}{\partial r}\left(r^2\frac{\partial\psi}{\partial r}\right) + \frac{1}{r^2}\left[\frac{1}{\sin\theta}\frac{\partial}{\partial\theta}\left(\sin\theta\frac{\partial\psi}{\partial\theta}\right) + \frac{1}{\sin^2\theta}\frac{\partial^2\psi}{\partial\varphi^2}\right] + 2m[E - V(r)]\psi = 0. \tag{1.7}$$

After introducing the operators $\hat{\ell}^2$ and $\hat{p}_r$,

$$\frac{1}{2m}\left[-\frac{1}{r^2}\frac{\partial}{\partial r}\left(r^2\frac{\partial\psi}{\partial r}\right) + \frac{\hat{\ell}^2}{r^2}\psi\right] + V(r)\psi = E\psi, \tag{1.8}$$

$$\hat{p}_r\psi = -i\frac{1}{r}\frac{\partial}{\partial r}(r\psi) = -i\left(\frac{\partial}{\partial r} + \frac{1}{r}\right)\psi,$$

the Hamiltonian is written as

$$\widehat{H} = \frac{1}{2m}\left(\hat{p}_r^2 + \frac{\hat{\ell}^2}{r^2}\right) + V(r). \tag{1.9}$$

The wave function $\psi(\vec{r})$ is a product

$$\psi = R(r)Y_{\ell\mu}(\theta, \varphi), \tag{1.10}$$

where $Y_{\ell\mu}$ is an eigenfunction of $\hat{\ell}^2$ and $\hat{\ell}_z$:

$$\hat{\ell}^2 Y_{\ell\mu} = \ell(\ell+1) Y_{\ell\mu}. \tag{1.11}$$

The radial wave function $R(r)$ obeys the equation

$$\frac{1}{r^2}\frac{d}{dr}\left(r^2\frac{dR}{dr}\right) - \frac{\ell(\ell+1)}{r^2}R + 2m[E - V(r)]R = 0. \tag{1.12}$$

The wave function ψ is a function of the energy E, total angular momentum ℓ and its projection μ. The normalization condition for $R(r)$ is defined by the integral

$$\int_0^\infty |R(r)|^2 \, r^2 dr = 1. \tag{1.13}$$

At large distances, $r \to \infty$, we can neglect $V(r)$. Then Eq. (1.12) reads

$$\frac{1}{r}\frac{d^2(rR)}{dr^2} + k^2 R = 0, \tag{1.14}$$

where $k = \sqrt{2mE}$. At large r, the general solution of Eq. (1.14) can be written in the form:

$$R \approx \sqrt{\frac{2}{\pi}}\,\frac{\sin(kr - \ell\pi/2 + \delta_\ell)}{r}, \tag{1.15}$$

where δ_ℓ is a constant which is defined by the behaviour of the radial function $R(r)$ at comparatively small r, where $V(r)$ is not negligible; it is called a phase shift.

1.2 Scattering process

Now let us reformulate the scattering of two particles as a scattering of one particle on the stationary field $V(r)$. The reformulation is done in the c.m.s. of particles 1 and 2, where the zero total momentum, $\vec{P}(1+2) = 0$. Therefore the wave function of the center-of-mass movement can be chosen to be equal to unity

$$\phi(\vec{R}) = 1. \tag{1.16}$$

In the c.m.s. the two-particle wave function is determined by $\psi(\vec{r})$ only:

$$\psi(\vec{r}_1, \vec{r}_2) = \psi(\vec{r}). \tag{1.17}$$

Just this wave function, $\psi(\vec{r})$, gives us the necessary reformulation. Let a free particle

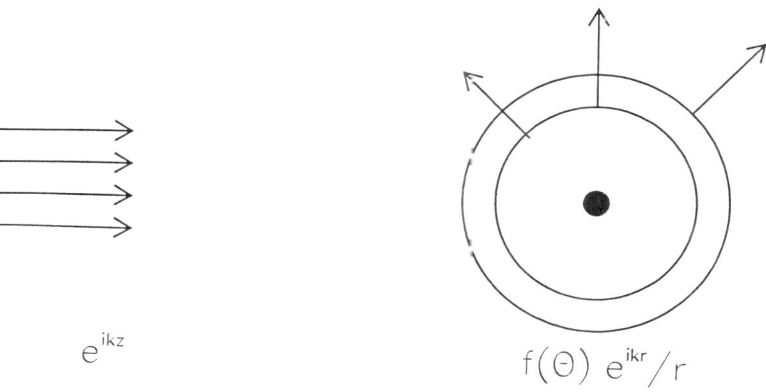

Fig. 1.1. Plane and outgoing waves.

before scattering be moving along the z-axis. It is described by a plane wave

$$e^{ikz}, \tag{1.18}$$

while an outgoing particle after scattering is described, at asymptotically large distances, by the spherical outgoing wave

$$f(\theta)\,\frac{e^{ikr}}{r}. \tag{1.19}$$

Here $f(\theta)$ is the scattering amplitude which depends on the polar scattering angle θ. So, the wave function, being a solution of Eq. (1.6) and describing a scattering process, has the following asymptotics at large r (see Fig. 1.1):

$$\psi(r) \simeq e^{ikz} + \frac{f(\theta)}{r}\,e^{ikr}. \tag{1.20}$$

The scattering amplitude $f(\theta)$ is completely determined by the phase shifts δ_ℓ.

1.3 Free movement: plane waves and spherical waves

The wave function

$$\psi_{\vec{k}}(r) = \text{ const } \cdot e^{i\vec{k}\vec{r}} \tag{1.21}$$

describes the state with momentum $\vec{k}$ (and energy $E = k^2/2m$).

A state with orbital momentum ℓ and its projection μ is characterized by

$$\psi_{k\ell\mu} = R_{k\ell}(r) Y_{\ell\mu}(\theta, \varphi). \tag{1.22}$$

The radial wave function is determined by the equation

$$R''_{k\ell} + \frac{2}{r}\,R'_{k\ell} + \left[k^2 - \frac{\ell(\ell+1)}{r^2}\right] R_{k\ell} = 0, \tag{1.23}$$

where $\psi_{k\ell\mu}$ and $R_{k\ell}$ obeying to normalization conditions:

$$\begin{aligned} \int \psi^*_{k'\ell'\mu'}\psi_{kl\mu} dV &= \delta_{\ell\ell'}\delta_{\mu\mu'}\delta(k'-k), \\ \int_0^\infty r^2 R_{k'\ell} R_{k\ell} dr &= \delta(k'-k). \end{aligned} \tag{1.24}$$

The solution of Eq. (1.23) finite at $r \to 0$ is

$$R_{k\ell}(r) = (-1)^\ell \sqrt{\frac{2}{\pi}}\,\frac{r^\ell}{k^\ell}\left(\frac{d}{rdr}\right)^\ell \frac{\sin kr}{r}. \tag{1.25}$$

The plane wave can be presented as an expansion in a series of functions $R_{k\ell}$'s:

$$e^{ikz} = \sqrt{\frac{\pi}{2}} \sum_{\ell=0}^{\infty} i^\ell (2\ell+1) P_\ell(\cos\theta) R_{k\ell}(r). \tag{1.26}$$

Here $kz = kr\cos\theta$. When $r \to \infty$

$$e^{ikz} \approx \frac{1}{kr} \sum_{\ell=0}^{\infty} i^\ell (2\ell+1) P_\ell(\cos\theta) \sin\left(kr - \frac{l\pi}{2}\right). \tag{1.27}$$

1.4 Scattering process: cross section, partial wave expansion, and phase shifts

The asymptotic expression for the wave function

$$\psi(\vec{r}) = e^{ikz} + \frac{f(\theta)}{r} e^{ikr} \tag{1.28}$$

describes the flux of incoming particles with density

$$v|\psi_{in}|^2 = v|e^{ikz}|^2 = v \tag{1.29}$$

and the flux of outgoing particles. The probability for the scattered particle to go through an element of the surface $dS = r^2 d\Omega$ in unit time is equal to

$$v|\psi_{out}|^2 d\Omega = v|f(\theta)|^2 d\Omega, \tag{1.30}$$

and its ratio to the flux of the incoming particles is the cross section:

$$d\sigma = |f(\theta)|^2 d\Omega. \tag{1.31}$$

If in Eq. (1.31) the integration over $d\varphi$ is performed, using azimuthal symmetry, then $d\Omega = 2\pi \sin\theta d\theta$. This is the cross section for scattering into angular interval $(\theta, \theta + d\theta)$:

$$d\sigma = 2\pi |f(\theta)|^2 \sin\theta \, d\theta. \tag{1.32}$$

Now let us express the scattering amplitude $f(\theta)$ in terms of the phase shift. The wave function $\psi(\vec{r})$ satisfies Eq. (1.8). At large r the solution of this equation is

$$R_\ell(r) \approx \frac{a_\ell}{r} \sin\left(kr - \frac{\ell\pi}{2} + \delta_\ell\right) \tag{1.33}$$

(see Eq. (1.15)). So the general form of the asymptotic wave function can be written as a series in R_ℓ defined by Eq. (1.33):

$$\psi \simeq \sum_{\ell=0}^{\infty} (2\ell+1) A_\ell P_\ell(\cos\theta) \frac{1}{kr} \sin\left(kr - \frac{\ell\pi}{2} + \delta_\ell\right) = \tag{1.34}$$

$$= \sum_{\ell=0}^{\infty} (2\ell+1) A_\ell P_\ell(\cos\theta) \frac{i}{2kr} \left\{ \exp\left[-i\left(kr - \frac{\ell\pi}{2} + \delta_\ell\right)\right] - \exp\left[i\left(kr - \frac{\ell\pi}{2} + \delta_\ell\right)\right]\right\}.$$

The coefficients should be chosen in such a form that $\psi(\vec{r})$ has an asymptotic form given by Eq. (1.28). In other words, at large r the expression $\psi(\vec{r}) - e^{ikz}$ should contain outgoing waves only. Comparing Eqs.(1.27) and (1.34) one gets the following values for A_ℓ:

$$A_\ell = i^\ell e^{i\delta_\ell}. \tag{1.35}$$

Thus, the wave function of Eq. (1.34) has the following asymptotic form:

$$\psi \simeq \frac{1}{2ikr} \sum_{\ell=0}^{\infty} (2\ell+1) P_\ell(\cos\theta) [(-1)^{\ell+1} e^{-ikr} + e^{2i\delta_\ell} e^{ikr}]. \tag{1.36}$$

For $\delta_\ell = 0$ there is no scattering, and the right-hand side of Eq. (1.36) turns into $\exp(ikz)$. Elastic scattering does not change the probability of outgoing particles, $|e^{2i\delta_\ell}| = 1$; the outgoing wave changes its phase only.

Eq. (1.36) gives the following expression for the scattering amplitude:

$$f(\theta) = \frac{1}{2ik} \sum (2\ell+1)(e^{2i\delta_\ell} - 1) P_\ell(\cos\theta). \tag{1.37}$$

Partial wave amplitudes are defined as

$$f_\ell = \frac{1}{2ik}(e^{2i\delta_\ell} - 1). \tag{1.38}$$

The value $e^{2i\delta_\ell}$ is the element of the S-matrix:

$$S_\ell = e^{2i\delta_\ell}. \tag{1.39}$$

The operator $\hat{S}$ is unitary

$$\hat{S}\hat{S}^+ = 1. \tag{1.40}$$

This unitarity condition reflects the fact that the number of particles in elastic scattering is conserved.

The unitarity condition for the partial amplitude reads as

$$\text{Im } f_\ell = k f_\ell^* f_\ell. \tag{1.41}$$

In field theory another normalization condition is used for the scattering amplitude

$$\begin{aligned} A_\ell &= \frac{1}{2i\rho}(e^{2i\delta_\ell} - 1), \qquad (1.42) \\ \rho &= \frac{k}{8\pi(\sqrt{m_1^2 + k^2} + \sqrt{m_2^2 + k^2})}; \end{aligned}$$

ρ is the invariant two-particle phase space factor. Then

$$\text{Im } A_\ell = \rho |A_\ell|^2. \tag{1.43}$$

1.5 K-matrix representation

The partial wave amplitude can be represented in the K-matrix form. This representation is rather useful because it correctly reproduces the singularities of the amplitude related to the two-particle rescatterings.

Let us introduce the partial wave amplitude in the following form

$$T_\ell = \frac{1}{2i}(e^{2i\delta_\ell} - 1) = e^{i\delta_\ell} \sin \delta_\ell. \tag{1.44}$$

The K-matrix form of the T_ℓ-amplitude is

$$T_\ell = \frac{K_\ell}{1 - iK_\ell}, \qquad K_\ell = \tan \delta_\ell. \tag{1.45}$$

Here K is a real function of k^2. The singularity for two-particle scattering is extracted by the substitution

$$K_\ell = k a_\ell(k^2). \tag{1.46}$$

Presentation of the T_ℓ in the form

$$T_\ell = \frac{k a_\ell(k^2)}{1 - ik a_\ell(k^2)} \tag{1.47}$$

is widely used in nuclear physics for the description of the nucleon-nucleon interaction at low energies.

The scattering-length approximation corresponds to the case

$$\ell = 0, \qquad a_0(k^2) \equiv a = \text{ Const.} \tag{1.48}$$

This case corresponds to a point-like interaction. Partial scattering-lengths behave near the elastic threshold like

$$a_\ell(k^2) \sim k^{2\ell}. \tag{1.49}$$

This can be easily seen by considering analytical properties of the amplitude; we shall do it below.

The resonance relates to a pole of the a_ℓ-amplitude:

$$a_\ell = \frac{g_\ell^2 k^{2\ell}}{k_0^2 - k^2} \tag{1.50}$$

Here g_ℓ^2 is a constant. Then

$$T_\ell = \frac{g_\ell^2 k^{2\ell+1}}{k_0^2 - k^2 - ig_\ell^2 k^{2\ell+1}} = \frac{\gamma_\ell k^{2\ell+1}}{E_0 - E - i\gamma_\ell k^{2\ell+1}}. \tag{1.51}$$

Here $E_0 = k_0^2/2m, E = k^2/2m$, and $\gamma = g_\ell^2/2m$. Such a form for the amplitude is frequently used for the description of the low-energy nucleon-nucleon interaction, namely, the S-wave amplitudes with $I = 0$ ($^3S_1$-state) and $I = 1$ ($^1S_0$-state). In both cases the coupling constant g^2 is rather large. If the coupling constant is small with E_0 being positive, then eq. (1.51) stands for the Breit-Wigner resonance.

1.6 Scattering with absorption

Scattering without absorption is described by the wave function of Eq. (1.36): at large r the intensities of incoming and outgoing waves are the same. Absorption means that the

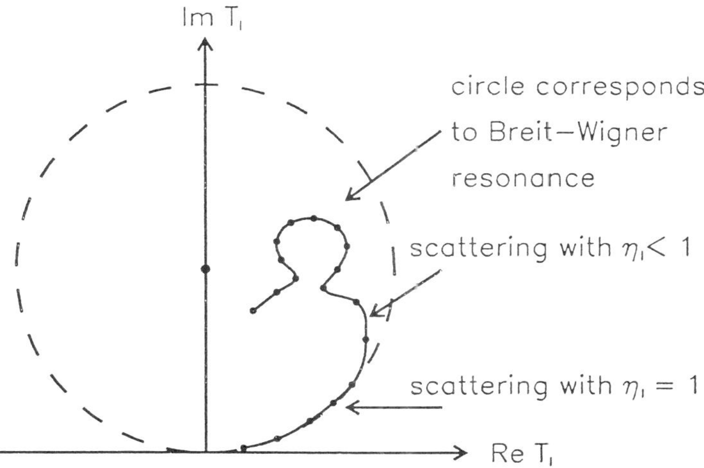

Fig. 1.2. Argand-diagram for T_ℓ: points give values of T_ℓ at different energies (or k^2).

intensity of the outgoing wave is decreased. Therefore, the scattering with absorption is described by the following wave function

$$\psi \simeq \frac{1}{2ikr}\sum_{\ell=0}^{\infty}(2\ell+1)P_\ell(\cos\theta)\left[(-1)^{\ell+1}e^{-ikr}+\eta_\ell e^{2i\delta_\ell}e^{ikr}\right], \tag{1.52}$$

where the inelasticity parameter η varies within the limits

$$0 \le \eta_\ell \le 1. \tag{1.53}$$

The case $\eta = 0$ corresponds to full absorption.

Partial amplitudes are equal to

$$T_\ell = \frac{\eta_\ell e^{2i\delta_\ell}-1}{2i}. \tag{1.54}$$

The complete absorption relates to $T_\ell = i/2$. The value of T_ℓ is imaginary and maximal in the case of the Breit-Wigner resonance at $k^2 = k_0^2$, $T_\ell = i$.

The k^2-dependence (or the energy dependence) of T_ℓ can be displayed on the Argand-diagram (see Fig. 1.2): the points on the Argand diagram correspond to T_ℓ at different energies.

The unitarity condition for the scattering amplitude (1.54) reads:

$$\operatorname{Im} T_\ell = T_\ell T_\ell^* + \frac{1}{4}(1-\eta_\ell^2). \tag{1.55}$$

In a graphical form this unitarity condition is shown in Fig. 1.3: the term $(1-\eta_\ell^2)/4$ in Eq. (1.55) corresponds to the contribution of inelastic processes to the imaginary part of the scattering amplitude. The first term on the r.h.s. of Eq. (1.55) describes elastic rescattering.

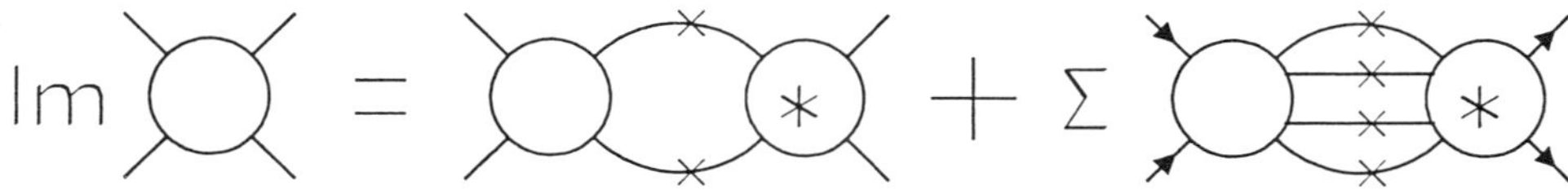

Fig. 1.3. Unitarity condition for the scattering amplitude.

2 Analytic properties of the amplitudes, Green functions, Feynman diagrams

This lecture is devoted to a discussion of the analytical properties of amplitudes. The extraction of leading singularities of the amplitudes is a standard way of searching for new hadrons (resonances). The study of analytical properties is performed using the language of Green functions and Feynman diagrams.

2.1 Complex variables and analytical functions

In this and subsequent Sections we remind ourselves of the properties of certain analytic functions which are necessary for further discussion.

Let us introduce the complex variable $z = x + iy$; this variables can be also written as $z = \rho\exp(i\varphi)$, see Fig. 2.1a. On this complex plane, analytic functions are defined.

The function $f(z) = 1/(z - a)$ provides a simple example. This function is analytic everywhere except of the point $z = a$. In this point a function has a pole singularity, see Fig. 2.1b.

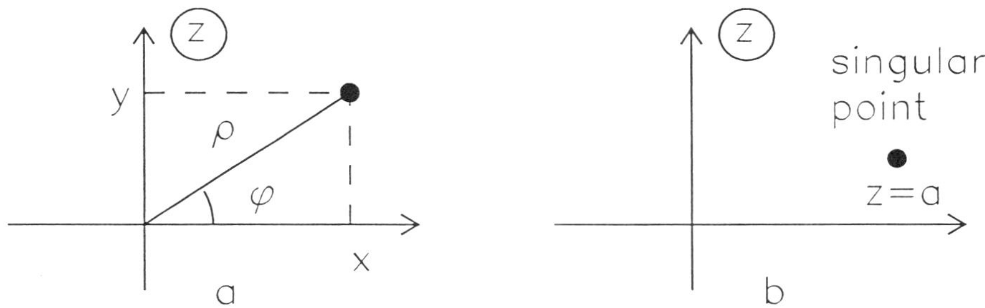

Fig. 2.1. a) Complex z-plane: the pion on plane corresponds to $z = x + iy$ or $z = \rho\exp(i\varphi) = \rho\cos\varphi + i\rho\sin\varphi$; b) The singular point $z = a$ of the analytical function $f(z) = 1/(z - a)$.

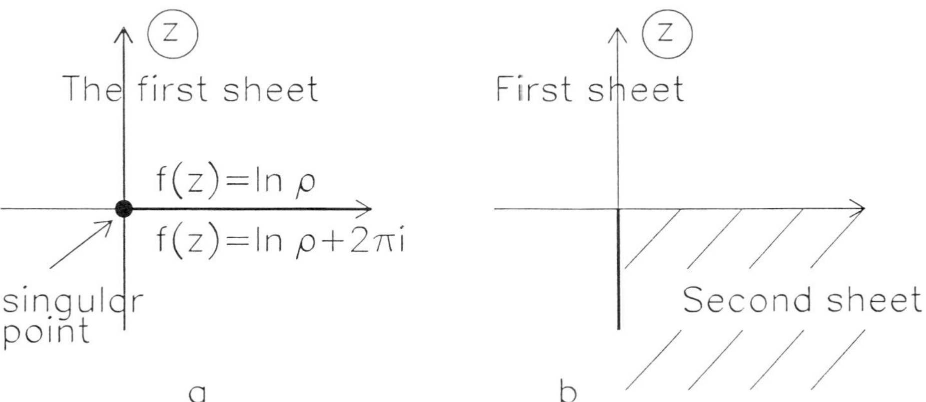

Fig. 2.2. Complex z-plane for $f(z) = \ln z$: a) The first sheet; b) The cut is located along the negative part of the imaginary axis so a part of the second sheet is seen.

Another example is given by the logarithmic function $f(z) = \ln z = \ln(\rho e^{i\varphi}) = \ln\rho + i\varphi$. This function is determined on an infinite number of sheets of the complex plane z. Let the first sheet be defined by $0 \le \varphi < 2\pi$; it is shown in Fig. 2.2a: there the cut related to a singular point $z = 0$ is directed along the real (positive) axis. On the upper edge of the cut $f(z) = \ln\rho$, while on the lower edge $f(z) = \ln\rho + 2\pi i$. The choice of the cut is arbitrary. Fig. 2.2b shows another cut position: in this figure one we can partly see the second sheet of the complex plane, with φ within the limits $-\pi/2 \le \varphi < 0$. The function $f(z) = \ln z$ has an infinite number of sheets $\ln z = \ln\rho + i\varphi + 2\pi i n$, where $0 \le \varphi < 2\pi$ and $n = 0, \pm 1, \pm 2, \ldots$. The first sheet is determined by $n = 0$.

The function $f(z) = \sqrt{z}$ has a square root singularity. This function is determined on two sheets $f(z) = \sqrt{z} = \sqrt{\rho \exp(i\varphi)} = \sqrt{\rho}\exp(i\varphi/2)$: the function $\sqrt{z}$ has different values at $-2\pi < \varphi \le 2\pi$. Let us determine the first sheet by $0 \le \varphi < 2\pi$ (see Fig. 2.3a):

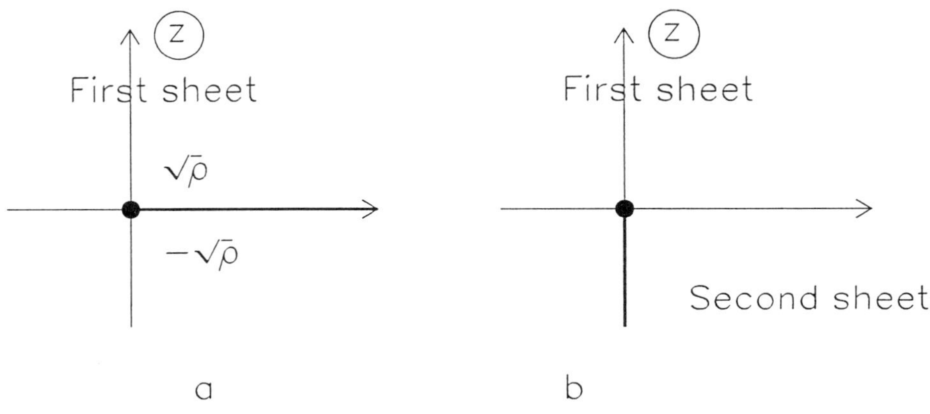

Fig. 2.3. Complex z-plane for $f(z) = \sqrt{z}$.

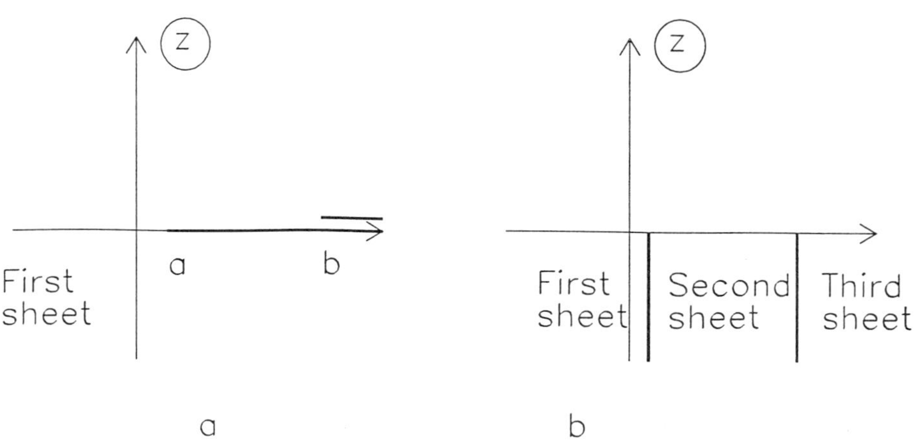

Fig. 2.4. Complex z-plane for $f(z) = \sqrt{z-a}/[1 + \sqrt{z-b}]$.

the cut is directed along the real positive axis. Fig. 2.3b shows a shifted cut.

A more complicated example is given by the function

$$f(z)=\frac{\sqrt{z-a}}{1+\sqrt{z-b}}$$

with two singular points, at $z=a$ and $z=b$. The singularities and the first sheet are shown in Fig. 2.4a, while Fig. 2.4b shows partly open second and third sheets.

2.2 Cauchy's theorem

Cauchy's theorem states that

$$\int_C \frac{dz'}{2\pi i}\frac{f(z')}{z'-z}=f(z), \tag{2.1}$$

if the contour of integration is of a circle-like type (see Fig. 2.5a) and the function $f(z)$ is analytic inside the contour. The proof of the Cauchy theorem is very simple: one takes as contour C a very small circle

$$z'-z=\epsilon e^{i\varphi}, \tag{2.2}$$

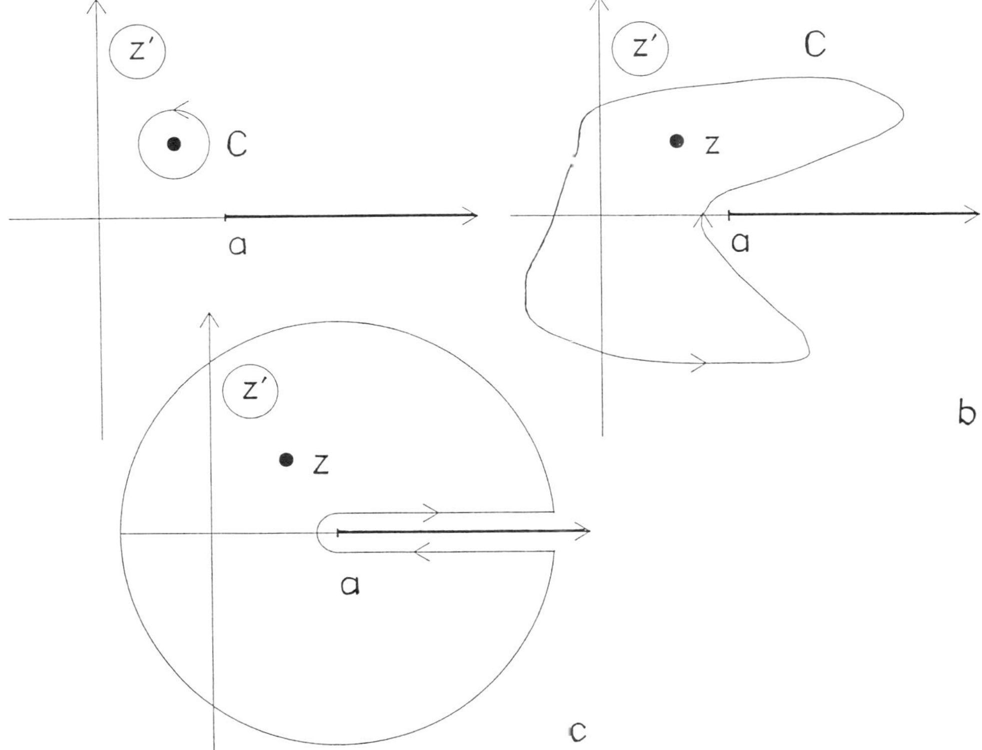

Fig. 2.5. Contour integration C in Eq. (2.1): its possible different positions.

with $\epsilon \to 0$. Then

$$\int\limits_C \frac{dz'}{2\pi i}\frac{f(z')}{z'-z} = \int\limits_C \frac{\epsilon d(e^{i\varphi})}{2\pi i}\frac{f(z+\epsilon e^{i\varphi})}{\epsilon e^{i\varphi}} = \int\limits_0^{2\pi} \frac{d\varphi}{2\pi} f(z) = f(z). \tag{2.3}$$

One may move the contour C so long as it does not cross singular points of $f(z)$, or the cuts, as is shown in Fig. 2.5b,c.

Fig. 2.5c shows the position of the contour C which is typical for the dispersion relation representation of amplitudes. The integral C can be represented as three integrals:

1. a big circle with radius $R \to \infty$,
2. the integral along the upper edge of the cut, from $z' = a$ till $z' = R$,
3. the integral along the lower edge of the cut.

Correspondingly,

$$f(z) = \int\limits_{C(R)} \frac{dz'}{2\pi i}\frac{f(z')}{z'-z} + \int_a^\infty \frac{dz'}{2\pi i}\frac{f(z'+i0)}{z'-z} + \int_\infty^a \frac{dz'}{2\pi i}\frac{f(z'-i0)}{z'-z}. \tag{2.4}$$

If $f(z') \to 0$ at $|z'| \to \infty$, then

$$f(z) = \int\limits_a^\infty \frac{dz'}{2\pi i}\frac{f(z'+i0)-f(z'-i0)}{z'-z}. \tag{2.5}$$

Let us denote

$$\frac{1}{2i}[f(z'+i0)-f(z'-i0)] \equiv disc\, f(z'). \tag{2.6}$$

Then Eq. (2.5) reads

$$f(z) = \int_a^\infty \frac{dz'}{\pi}\frac{disc\, f(z')}{z'-z}. \tag{2.7}$$

In the dispersion relation representation $disc\, f(z')$ is found from the unitarity condition.

2.3 Propagator function in quantum mechanics: the coordinate representation

For performing an analysis of the scattering amplitude, it is convenient to introduce the propagator function, or Green function. The propagator function determines the time evolution of the wave function:

$$\Psi(\vec r, t) = \int d^3r' K(\vec r, t; \vec r', t_0)\Psi(\vec r', t_0). \tag{2.8}$$

Here $\bar\Psi(\vec r', t_0)$ is the wave function determined at time t_0; $K(\vec r, t; \vec r', t_0)$ with $t \geq t_0$ is the propagator function. The propagator function should satisfy the boundary condition:

$$K(\vec r, t; \vec r', t_0)|_{t=t_0} = \delta(\vec r - \vec r'). \tag{2.9}$$

The propagator function allows us to find the wave function at any time t if the initial wave function at time t_0 is known, $(t_0 < t)$. It means that the function K determines the scattering amplitude $f(\theta)$.

The function K can be constructed if the full set of wave functions Ψ_n, which satisfy the Schrödinger equation (1.6), is known. Then

$$K(\vec{r},t;\vec{r}',t_0) = \sum_n \Psi_n(\vec{r},t)\Psi_n^*(\vec{r}',t_0), \tag{2.10}$$

where

$$\Psi_n(\vec{r},t) = \psi_n(r)e^{-iE_nt}, \tag{2.11}$$

and summation is performed over all eigenstates. The boundary condition (2.9) is equivalent to the completness condition of the set of wave functions being used:

$$\sum_n \psi_n(\vec{r})\psi_n^*(\vec{r}') = \delta(\vec{r}-\vec{r}'). \tag{2.12}$$

In the scattering process we deal with a continuous spectrum of states; therefore the summation over n should be replaced by an integration over states of the continuous spectrum. The interval d^3k contains $d^3k/(2\pi)^3$ quantum states, so we should replace in Eq.(2.12):

$$\sum_n \longrightarrow \int \frac{d^3k}{(2\pi)^3}. \tag{2.13}$$

Let us consider in detail the propagation function of a free particle described by the plane wave:

$$\Psi_k(\vec{r},t) = e^{i\vec{k}\vec{r}-i(k^2/2m)t}. \tag{2.14}$$

Then

$$K_0(\vec{r},t;0,0) = \int \frac{d^3k}{(2\pi)^3}\exp\left[i\vec{k}\vec{r} - i\frac{k^2}{2m}t\right] = \left(\frac{2m}{i\pi t}\right)^{3/2} e^{i(r^2m/2t)}. \tag{2.15}$$

It is taken into account here that the free particle propagation function $K_0(\vec{r},t;\vec{r}',t_0)$ depends on $\vec{r}-\vec{r}'$ and $t-t_0$ only, so we can put $\vec{r}'=t_0=0$.

The propagation function K describes the time evolution of a quantum state at $t > t_0$; it is convenient to use a propagator which is equal to zero at $t < t_0$. It is the Green function:

$$G(\vec{r},t;\vec{r}',t') = \theta(t-t')K(\vec{r},t;\vec{r}',t'). \tag{2.16}$$

Here $\theta(t)$ is the step function: $\theta(t) = 1$ at $t \geq 0$ and $\theta(t) = 0$ at $t < 0$. Because the K-function satisfies the equation

$$\left(i\frac{\partial}{\partial t} - \widehat{H}(r)\right)K(\vec{r},t;\vec{r}',t') = 0$$

$$\widehat{H}(r) = -\frac{\Delta}{2m} + V(r), \tag{2.17}$$

the Green function obeys the following equation

$$\left(i\frac{\partial}{\partial t} - \widehat{H}(r)\right)G(\vec{r},t;\vec{r}',t') = i\delta(t-t')\delta(\vec{r}-\vec{r}'). \tag{2.18}$$

Here the boundary condition (2.9) is used:

$$K(\vec{r},t;\vec{r}',t')\frac{\partial}{\partial t}\theta(t-t') = K(\vec{r},t;\vec{r}',t')\delta(t-t') = \delta(\vec{r}-\vec{r}')\delta(t-t'). \tag{2.19}$$

The Green function of a free particle is determined analogously by the function K_0:

$$G_0(\vec{r},t) = K_0(\vec{r},t;0,0)\theta(t). \tag{2.20}$$

Then, Eq.(2.15) gives

$$G_0(\vec{r},t) = \int \frac{d^3k}{(2\pi)^3}\theta(t)\exp\left[i\vec{k}\vec{r} - i\frac{k^2}{2m}t\right]. \tag{2.21}$$

This expression can be rewritten as an integral over the four-vector $(E,\vec{k})$:

$$G_0(\vec{r},t) = \int \frac{d^3k}{(2\pi)^3}\int\limits_{-\infty}^{+\infty}\frac{dE}{2\pi i}\cdot\frac{1}{-E+(k^2/2m)-i\epsilon}e^{\vec{k}\vec{r}-iEt}. \tag{2.22}$$

Here ϵ is infinitely small and positive: $\epsilon > 0$, $\epsilon \to 0$. For $t > 0$ the contour of integration over E is enclosed by the large circle in the lower half plane (see Fig. 2.6a): the factor $\exp[-iEt]$ guarantees an infinitesimally small contribution to the integral from this circle. The integral is equal to the residue at the pole $E = k^2/2m$, therefore we can replace in the case $t > 0$:

$$\left[-E + \frac{k^2}{2m} - i\epsilon\right]^{-1} \to 2\pi i\delta\left(E - \frac{k^2}{2m}\right). \tag{2.23}$$

For $t < 0$, the factor $\exp[-iEt]$ is infinitesimally small on the large circle in the upper half plane (see Fig. 2.6b): inside this enclosed contour there is no singularity so the integral (2.22) at $t < 0$ is equal to zero. We see that Eq.(2.21) exactly reproduces the definition of $G_0(\vec{r},t)$ given by Eq.(2.21). It should be pointed out that the factor $[-E+(k^2/2m)-i\epsilon]^{-1}$ in Eq. (2.22) is an operator $[-i\frac{\partial}{\partial t}+\hat{H}_0(r)]^{-1}$ in the momentum representation. It is important that the shift of the pole in the complex plane E which is determined by the value of $-i\epsilon$; just this shift suggests the evolution of the quark system in the positive-time direction.

The Green function $G(\vec{r},t;\vec{r}',t')$ satisfies the following integral equation:

$$G(\vec{r},t;\vec{r}',t') = G_0(\vec{r}-\vec{r}',t-t') \quad + \tag{2.24}$$

$$+\int d^3r''dt''G_0(\vec{r}-\vec{r}'',t-t'')(-i)V(r'')G(\vec{r}'',t'';\vec{r}',t').$$

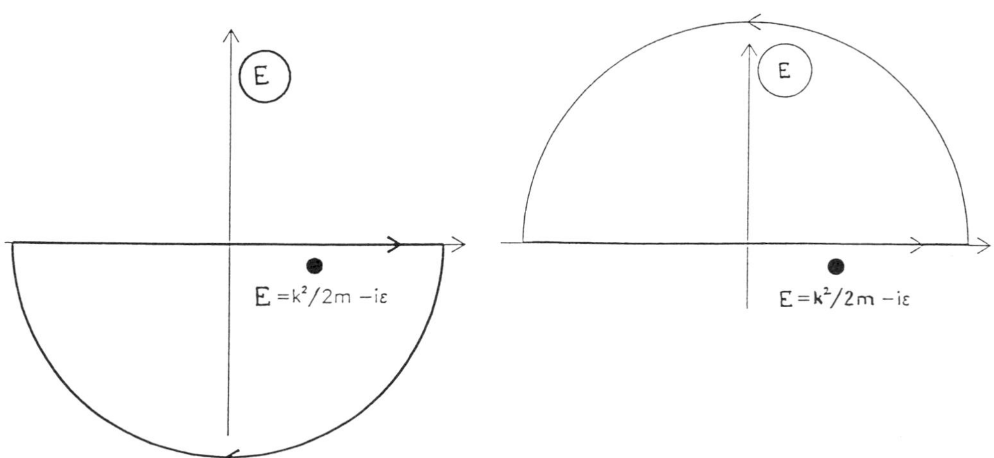

Fig. 2.6. The contours of integration over E in Eq. (2.22) for $t > 0$ and $t < 0$.

To justify Eq.(2.24) let us operate on Eq.(2.24) by $\{i(\partial/\partial t) - \hat{H}_0(r)\}$ where $\hat{H}_0 = -\Delta^2/2m$. We have as result

$$\left(i\frac{\partial}{\partial t} - \widehat{H}_0\right) G(\vec{r}, t; \vec{r}', t') = i\delta(\vec{r} - \vec{r}')\delta(t - t') \tag{2.25}$$

$$+ \int d^3r'' dt'' \delta(\vec{r} - \vec{r}'')\delta(t - t'')V(r'')G(\vec{r}'', t''; \vec{r}', t').$$

After performing the integration over $d^3r''dt''$ we have Eq. (2.18).

The Equation (2.24) can be written in a graphical form, shown in Fig. 2.7: thin lines correspond to free Green functions, G_0, while thick ones correspond to full Green functions, G.

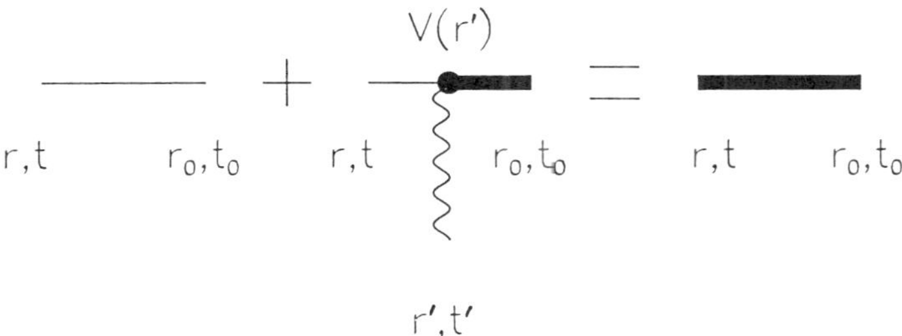

Fig. 2.7. Graphical form of Eq. (2.24) for the Green function.

The iteration of Eq. (2.24) demonstrates that the full Green function is an infinite set of diagrams of the type shown in Fig. 2.8. These diagrams describe the scattering of the effective particle on the field $V(r)$.

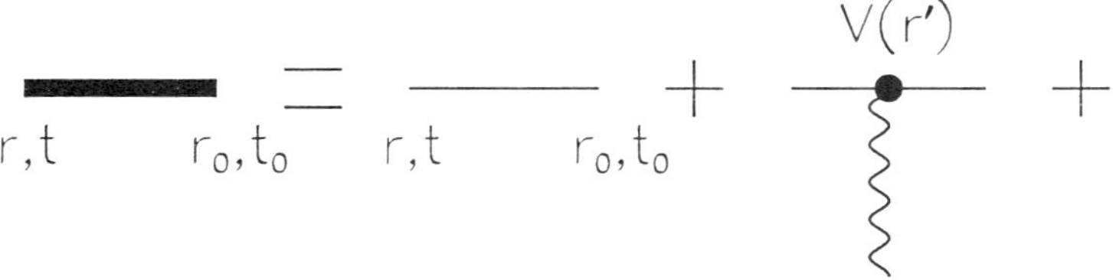

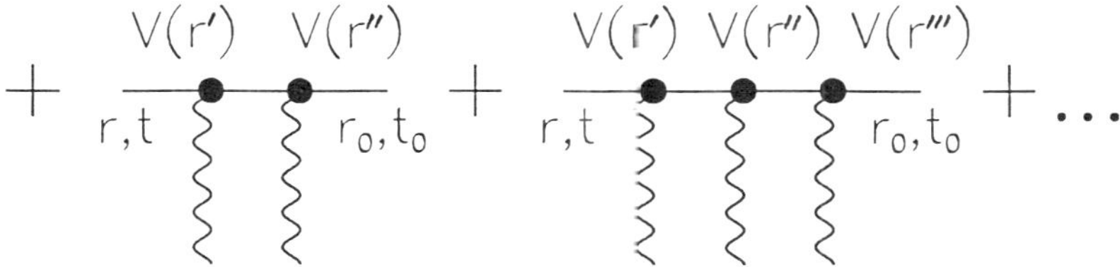

Fig. 2.8. Full Green function presented as an infinite set of scattering diagrams.

2.4 Propagator function in quantum mechanics: the momentum representation

Let us consider the Green functions in the momentum representation. The free Green function is determined as

$$G_0(k) = i\int d^3r dt\, G_0(\vec{r},t)e^{-i\vec{k}\vec{r}+iEt} = \frac{1}{-E+(\vec{k}^2/2m)-i\epsilon}. \tag{2.26}$$

The full Green function depends on two four-momenta:

$$G(k,p) = i\int d^3r dt \int d^3r' dt'\, G(\vec{r},t;\vec{r}',t')\exp[-i\vec{k}\vec{r}+iEt]\exp[i\vec{p}\vec{r}'-iE_pt']. \tag{2.27}$$

The Eq.(2.24) for Green function is rewritten in the momentum representation as

$$G(k,p) = (2\pi)^4\delta^{(4)}(k-p)G_0(k) - G_0(k)\int\frac{d^4k'}{(2\pi)^4}V(k-k')G(k',p). \tag{2.28}$$

Here the potential V in the momentum representation is defined as

$$V(q) = \int d^3r dt\, V(\vec{r},t)e^{-i\vec{q}\vec{r}+iq_0t}. \tag{2.29}$$

If $V(r)$ does not depend on t, the case which is discussed in lecture 1, then

$$V(q) = 2\pi\delta(q_0)\int d^3r\, V(\vec{r})e^{-i\vec{q}\vec{r}} = 2\pi\delta(q_0)\, V(\vec{q}). \tag{2.30}$$

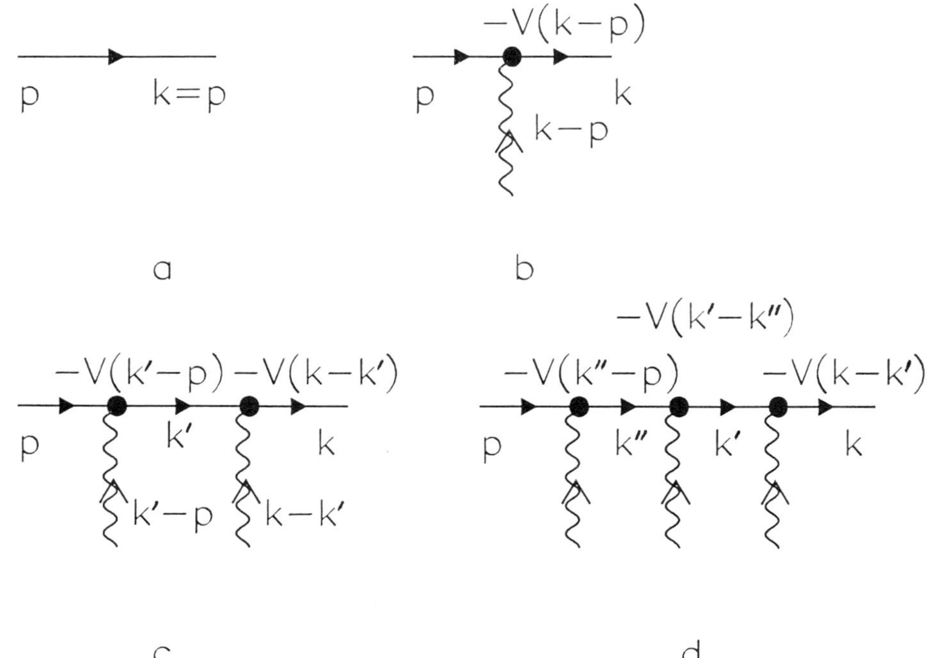

Fig. 2.9. Scattering diagrams for the full Green function in the momentum representation.

The iteration of Eq.(2.28) leads to the following representation of $G(k,p)$ in a series over V:

$$\begin{aligned} G(k,p) &= (2\pi)^4\delta^{(4)}(k-p)G_0(k) - G_0(k)V(k-p)G_0(p) \qquad (2.31)\\ &+ G_0(k)\int\frac{d^4k'}{(2\pi)^4}V(k-k')G_0(k')V(k'-p)G_0(p)\\ -G_0(k)\int\frac{d^4k'}{(2\pi)^4}&\frac{d^4k''}{(2\pi)^4}V(k-k')G_0(k')V(k'-k'')G_0(k'')V(k''-p)G_0(p)+\dots \end{aligned}$$

The expansion (2.31) corresponds to the set of diagrams shown in Fig. 2.9. These are Feynman diagrams for a scattering of the non-relativistic particle in field V.

The scattering amplitude $f(\theta)$, introduced by Eq.(1.20), is determined by the Green function by the relation

$$G(k,p) = (2\pi)^4\delta^{(4)}(k-p)G_0(k) + G_0(k)\frac{(2\pi)^3}{m}\delta(E-E_p)f(k,p)G_0(p). \qquad (2.32)$$

Here we redenote $f(\theta)$ as $f(k,p)$, namely $f(\theta)\equiv f(k,p)$.

2.5 Equation for the scattering amplitude $f(k,p)$

One can write an equation directly for the amplitude $f(k,p)$ keeping in mind that we consider here the case of the time-independent interaction. The equation for $f(k,p)$ may be easily derived substituting (2.32) into (2.28) (taking into account Eq. (2.30) as well). Then

$$f(k,p) = -\frac{m}{2\pi}V(\vec{k}-\vec{p}) - \int\frac{d^3k'}{(2\pi)^3}V(\vec{k}-\vec{k}')G_0(E,\vec{k}')f(k',p). \qquad (2.33)$$

Here $E'=E=E_p$ and the propagator of the free particle is rewritten in the form which underlines energy conservation in the intermediate states:

$$G_0(k')|_{E'=E} \equiv G_0(E,\vec{k}') = \frac{1}{-E+(k'^2/2m)-i\epsilon}. \qquad (2.34)$$

The amplitude $f(k,p)$ may be presented in a series over V:

$$f(k,p) = -\frac{m}{2\pi}V(\vec{k}-\vec{p}) + \frac{m}{2\pi}\int\frac{d^3k'}{(2\pi)^3}V(\vec{k}-\vec{k}')G_0(E,\vec{k}')V(\vec{k}'-\vec{p})$$

$$-\left(\frac{m}{2\pi}\right)^2\int\frac{d^3k'}{(2\pi)^3}\frac{d^3k''}{(2\pi)^3}V(\vec{k}-\vec{k}')G_0(E,\vec{k}')V(\vec{k}'-\vec{k}'')G_0(E,\vec{k}'')V(\vec{k}''-\vec{p})+\dots \qquad (2.35)$$

If the propagator is small, one may restrict oneself to a few terms on the r.h.s. of Eq. (2.35). If only the first term is taken into account, then one obtains

$$f(k,p) \simeq -\frac{m}{2\pi}V(\vec{k}-\vec{p}) \qquad (2.36)$$

This is the Born approximation for the scattering amplitude.

2.6 Propagators in the description of the two-particle scattering amplitude

Up to now our guideline was as follows: we considered the Schrödinger equation for two interacting particles, then we reduced it, in the c.m.s., to the equation for one particle scattered on the external field V. Just for this case we determined the scattering amplitude and the propagator of the non-relativistic particle. However, the technique which uses propagators allows us to calculate directly the scattering amplitude, without reducing the Schrödinger equation beforehand to the one-particle case. We can start with Eq. (2.35) and transform it into a form which manifests a propagation of two particles.

Let us consider the scattering of particles 1 and 2, which in the initial state have the four-momenta

$$k_1 = (\frac{k^2}{2m_1}, \vec{k}) \equiv (E_1, \vec{k}), \quad k_2 = (\frac{k^2}{2m_2}, -\vec{k}) \equiv (E_2, -\vec{k}). \tag{2.37}$$

The center-of-mass system is used here, as has been done in the previous consideration. The four-momenta of the final state are

$$p_1 = (\frac{k^2}{2m_1}, \vec{p}), \quad p_2 = (\frac{k^2}{2m_2}, -\vec{p}). \tag{2.38}$$

Energy conservation is taken into account here, for the potential is time-independent (see Eq. (2.30)).The total energy of particles 1 and 2 is equal to

$$E = \frac{k^2}{2m_1} + \frac{k^2}{2m_2} = \frac{k^2}{2m}, \tag{2.39}$$

where m is the reduced mass.

The equation (2.33) can be re-written using an explicit form for G_0:

$$f(k,p) = -\frac{m}{2\pi} V(\vec{k} - \vec{p}) - \int \frac{d^3k'}{(2\pi)^3} \frac{V(\vec{k} - \vec{k}') f(k', p)}{-E + k'^2/2m - i\epsilon}. \tag{2.40}$$

The propagator $[-E + k'^2/2m - i\epsilon]^{-1}$ stands for the free movement of the two-particle system; it may be presented as a product of free propagators of the particles 1 and 2:

$$\frac{1}{-E + k'^2/2m - i\epsilon} = \int_{-\infty}^{\infty} \frac{dE_1'/2\pi i}{(-E_1' + k'^2/2m_1 - i\epsilon)[-(E - E_1') + k'^2/2m_2 - i\epsilon]}. \tag{2.41}$$

The integration on the r.h.s. is performed according to the Cauchy theorem: the integration contour may be closed in the lower half-plane E_1' as was shown in Fig. 2.6*a*. Then $(-E_1' + k'^2/2m_1 - i\epsilon)^{-1} \to 2\pi i \delta(-E_1' + k'^2/2m_1)$.

Let us write Eq. (2.40), according to Eq. (2.41), as follows:

$$f(k,p) = -\frac{m}{2\pi} V(\vec{k} - \vec{p}) \tag{2.42}$$

$$- \int \frac{d^3k'}{(2\pi)^3} \frac{dE_1'}{2\pi i} V(\vec{k} - \vec{k}') \cdot \frac{1}{(-E_1' + k'^2/2m_1 - i\epsilon)(-E_2' + k'^2/2m_2 - i\epsilon)} \cdot f(k', p),$$

where $E_2' = E - E_1'$. The product of the two propagators on the r.h.s. of Eq. (2.42) clearly manifests the propagation of two particles in the intermediate state.

Eq. (2.42) is written in the c.m.s. of scattered particles 1 and 2, but it is easy to present it in an arbitrary system: the frame-independent consideration of the two-particle interaction amplitude is given in the next lecture, where the relativistic generalization of Eq. (2.42), the Bethe-Salpeter equation, is discussed.

2.7 Relativistic propagator for a free particle

The relativistic propagator of a free particle may be introduced analogously to the non-relativistic case. Namely, the wave functions of a non-relativistic particle are eigenstates of the Schrödinger operator $[\frac{i\partial}{\partial t} - \hat{H}_0]$, therefore the Green function is defined by the same operator:

$$[i\frac{\partial}{\partial t} - \frac{1}{2m}(\frac{\partial}{\partial \vec{r}})^2]G_0(\vec{r}, t) = i\delta(\vec{r})\delta(t). \tag{2.43}$$

The wave function of a free scalar particle obeys the Klein-Gordon equation

$$[(\frac{\partial}{\partial t})^2 - (\frac{\partial}{\partial \vec{r}})^2 + m^2]\varphi(x) = 0. \tag{2.44}$$

In the same way, the relativistic Green function is defined by the Klein-Gordon operator:

$$[(\frac{\partial}{\partial t})^2 - (\frac{\partial}{\partial \vec{r}})^2 + m^2]D(x) = i\delta^{(4)}(x). \tag{2.45}$$

So the propagator of a free relativistic particle in the momentum representation is equal to

$$D(k) = \frac{1}{m^2 - k^2 - i\epsilon}, \tag{2.46}$$

where k is the four-momentum of a particle with $k = (k_0, \vec{k})$ and $k^2 = k_0^2 - \vec{k}^2$. In the non-relativistic approximation $D(k)$ turns into the quantum mechanical propagator discussed in the previous sections. Let us introduce $E = k_0 - m$ and consider the case $E \ll m$. Then

$$\frac{1}{m^2 - k^2 - i\epsilon} = \frac{1}{m^2 - (m + E)^2 + \vec{k}^2 - i\epsilon} \tag{2.47}$$
$$\simeq \frac{1}{-2mE + \vec{k}^2 - i\epsilon} = \frac{1}{2m}\frac{1}{-E + (\vec{k}^2/2m) - i\epsilon}.$$

The r.h.s. of Eq.(2.47), up to factor $(2m)^{-1}$, coincides with Eq.(2.26) for a non-relativistic propagator in quantum mechanics. The relativistic Feynman propagator of Eq. (2.46) describes the propagation of a particle and its antiparticle:

$$\frac{1}{m^2 - k^2 - i\epsilon} = \frac{1}{2\sqrt{m^2 + \vec{k}^2}}\left[\frac{1}{-k_0 + \sqrt{m^2 + \vec{k}^2} - i\epsilon} + \frac{1}{k_0 + \sqrt{m^2 + \vec{k}^2} - i\epsilon}\right]. \tag{2.48}$$

The first term in square brackets corresponds to propagation of the relativistic particle with energy $\sqrt{m^2 + \vec{k}^2}$, while the second one describes propagation of the particle from the Dirac sea with negative energy $-\sqrt{m^2 + \vec{k}^2}$.

2.8 Mandelstam plane

Feynman diagrams provide information about analytic properties of amplitudes. The analytic properties of the scattering amplitude $1 - 2 \to 1' + 2'$ (see Fig. 2.10a) can be considered conveniently with use of the Mandelstam plane. Let us put for simplicity the masses of scattered particles in the process of Fig. 2.10a to be equal:

$$p_1^2 = p_2^2 = p_1'^2 = p_2'^2 = m^2. \tag{2.49}$$

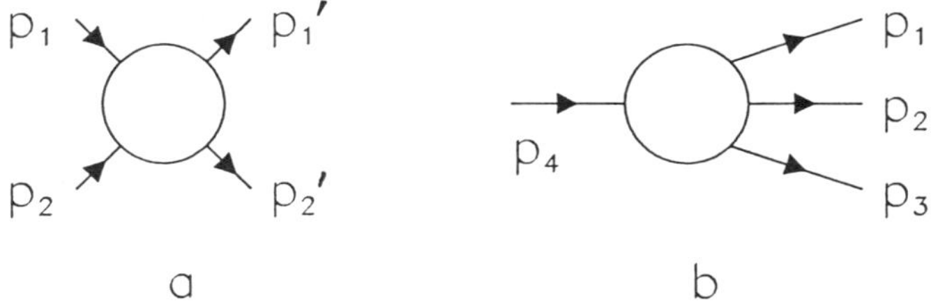

Fig. 2.10. Four-point amplitudes: a) scattering process $1 + 2 \to 1' + 2'$; b) decay $4 \to 1 + 2 + 3$.

The scattering amplitude of spinless particles depends on two independent variables. But for description of the scattering amplitude on the Mandelstam plane there are three variables:

$$\begin{aligned} s &= (p_1 + p_2)^2 = (p_1' + p_2')^2 \\ t &= (p_1 - p_1')^2 = (p_2 - p_2')^2 \\ u &= (p_1 - p_2')^2 = (p_2 - p_1')^2. \end{aligned} \tag{2.50}$$

These variables obey constraint

$$s + t + u = 4m^2. \tag{2.51}$$

The Mandelstam plane of the variables s, t and u is shown in Fig. 2.11. The physical regions are hatched. The physical region of the s-channel corresponds to the case shown in Fig. 2.10a: the incoming particles are 1 and 2, while particles $1'$ and $2'$ are outgoing;

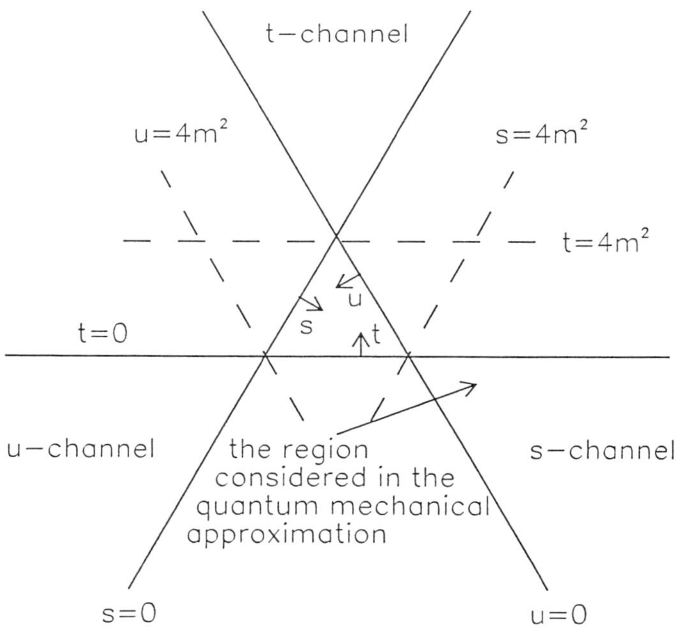

Fig. 2.11. The Mandelstam plane.

s is energy squared, t and u are momentum transfers squared. The physical region of the t-channel corresponds to the case when particles 1 and $1'$ collide, while the u-channel describes the collision of particles 1 and $2'$.

The Feynman diagram technique is a good guide for finding analytic properties of scattering amplitudes. Below typical singularities are considered as an example.

A) One-particle exchange diagrams are shown in Fig. 2.12a–c: they provide pole singularities of the scattering amplitude, which can be written as

$$\frac{g^2}{\mu^2 - t}, \qquad \frac{g^2}{\mu^2 - s}, \qquad \frac{g^2}{\mu^2 - u}, \tag{2.52}$$

where μ is the mass of a particle in the intermediate state, g is its coupling constant with external particles.

B) The two-particle exchange diagram is shown in Fig. 2.13a. It has square-root singularities in the s-channel, with a cut shown in Fig. 2.13b, and in the t-channel (see Fig. 2.13c). The cutting relates to the replacement of the Feynman propagator in the following way:

$$(k^2 - m^2)^{-1} \to \delta(k^2 - m^2), \tag{2.53}$$

thus giving the imaginary part of the diagram Fig 2.13a in the corresponding channel.

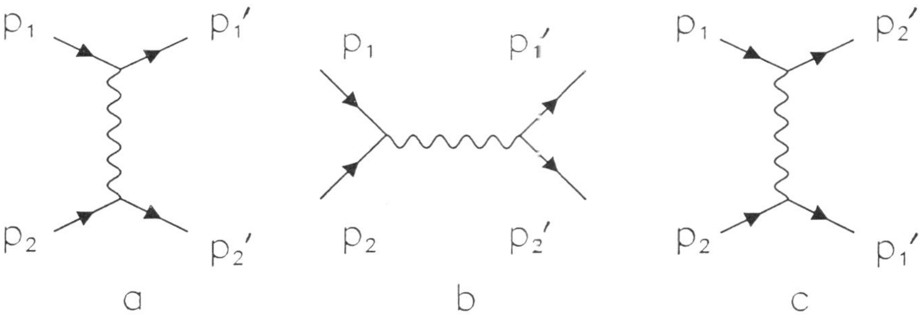

Fig. 2.12. One-particle exchange diagrams, which has pole singularities in: a) t-channel, b) s-channel, and c) u-channel.

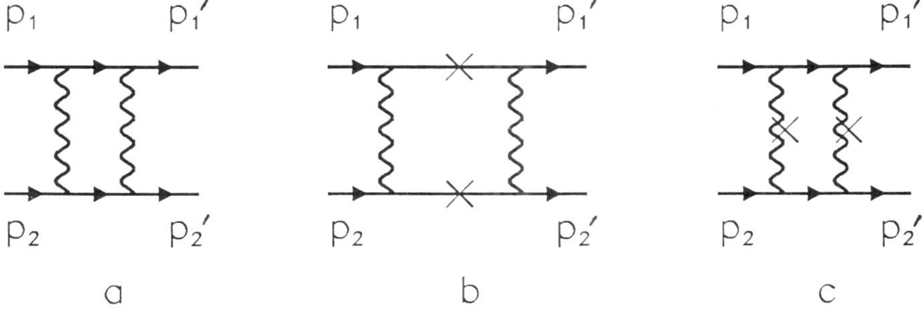

Fig. 2.13. Diagram with two-particle singularities in s- and t-channels.

The s-channel two-particle singularity is located at $s = (m_1+m_2)^2$; the type of singularity is as follows

$$\sqrt{s - (m_1 + m_2)^2}. \tag{2.54}$$

The t-channel singularity is located at $t = 4\mu^2$; it is of the type

$$\sqrt{t - 4\mu^2}. \tag{2.55}$$

C) Three-particle singularity in the s-channel relates to the diagram of Fig. 2.14a. The singularity is located at (see Fig. 2.14b):

$$s = (m_1 + m_2 + \mu)^2. \tag{2.56}$$

The type of singularity is as follows

$$\left[s - (m_1 + m_2 + \mu)^2\right]^2 \ln\left[s - (m_1 + m_2 + \mu)^2\right]. \tag{2.57}$$

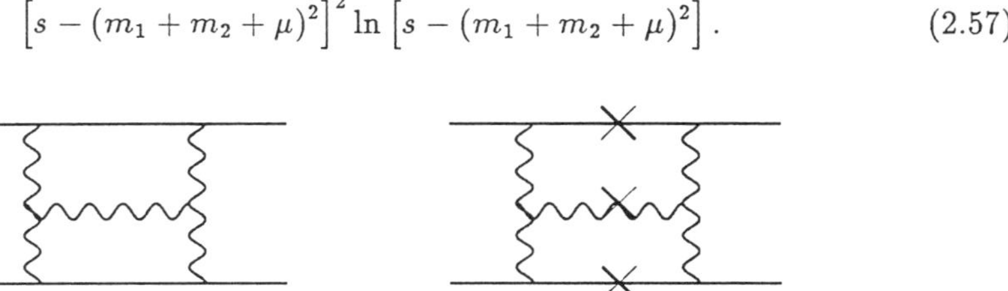

Fig. 2.14. Diagram with three-particle intermediate state.

2.9 Dalitz-plot

The four-point amplitude has an additional physical region in the case when particle masses (m_1, m_2, m_3) and m_4 are different and one of the masses is larger than the sum

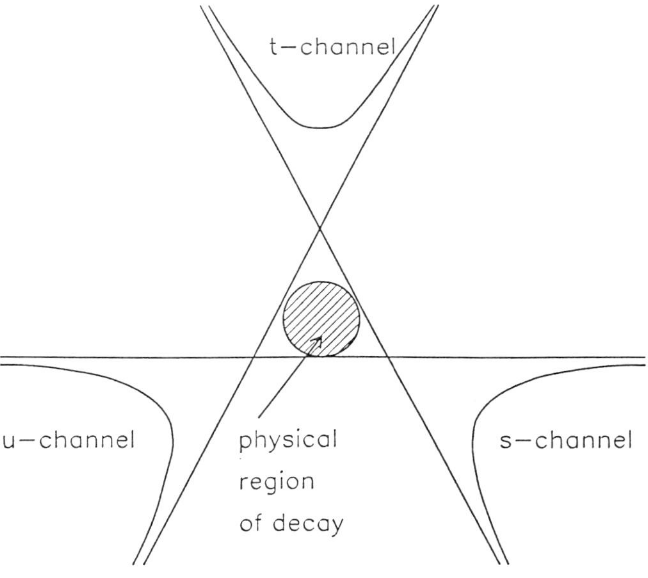

Fig. 2.15. The Mandelstam plane and the physical region for the decay $4 \to 1 + 2 + 3$.

of all others:

$$m_4 > m_1 + m_2 + m_3. \tag{2.58}$$

It leads to a possibility of the decay process (Fig. 2.10b)

$$4 \to 1 + 2 + 3. \tag{2.59}$$

The Mandelstam plane is shown for this case in Fig. 2.15. The physical region of the decay process is located in the center of the plane. This region of the decay process is shown separately in Fig. 2.16: it is called the Dalitz-plot.

The energies squared of the outgoing particles, $s_{ij} = (p_i + p_j)^2$, obey the constraint:

$$s_{12} + s_{13} + s_{23} - (m_1^2 + m_2^2 + m_3^2) = m_4^2. \tag{2.60}$$

The threshold singularities at

$$s_{ij} = (m_i + m_j)^2 \tag{2.61}$$

are touching the physical region.

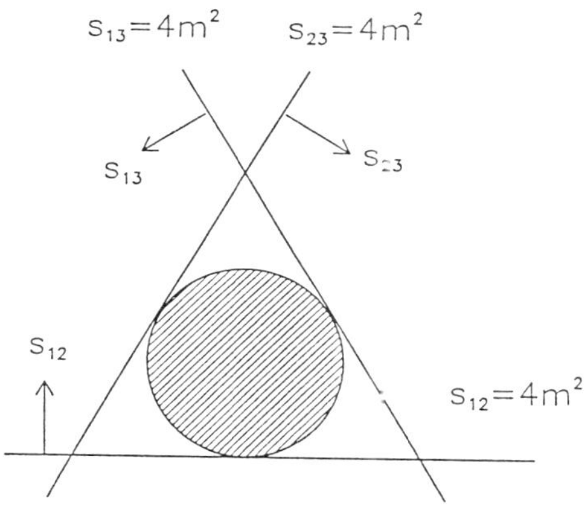

Fig. 2.16. Dalitz-plot of decay $4 \to 1 + 2 + 3$.

3 Dispersion relation N/D-method and Bethe-Salpeter equation

In this lecture the basic features of the dispersion integration method are considered for the scattering amplitude $1 + 2 \to 1' + 2'$. It is shown how the dispersion technique is connected with other methods, namely, the Feynman diagram and light cone variable approaches. Common and different features of various techniques are demonstrated by means of the Bethe-Salpeter equation.

3.1 N/D-method for one-channel scattering amplitude of spinless particles

Consider the analytic properties of scattering amplitudes for two spinless particles (with mass m) which interact through the exchange of another spinless particle (with mass μ).

This amplitude, $A(s,t)$, has s- and t-channel singularities. In the t-plane there are singularities at $t = \mu^2, 4\mu^2, 9\mu^2$ etc., which correspond to one- or many-particle exchanges. In the s-plane the amplitude has a singularity at $s = 4m^2$ (elastic rescattering) and singularities at $s = (2m + n\mu)^2$ where $n = 1, 2, \ldots$, corresponding to the production of n particles with mass μ. In a bound state with mass M there exists a pole singularity at $s = M^2$. If the mass of this bound state $M > 2m$, this is a resonance and the corresponding pole is located on the second sheet of the complex s-plane.

In the N/D-method we deal with partial wave amplitudes. Partial amplitudes in the s-channel depend on s only. They have all the s-channel righthand singularities of $A(s,t)$ at $s = M^2$, $s = 4m^2$, $s = (2m + \mu)^2, \ldots$ of Fig. 3.1.

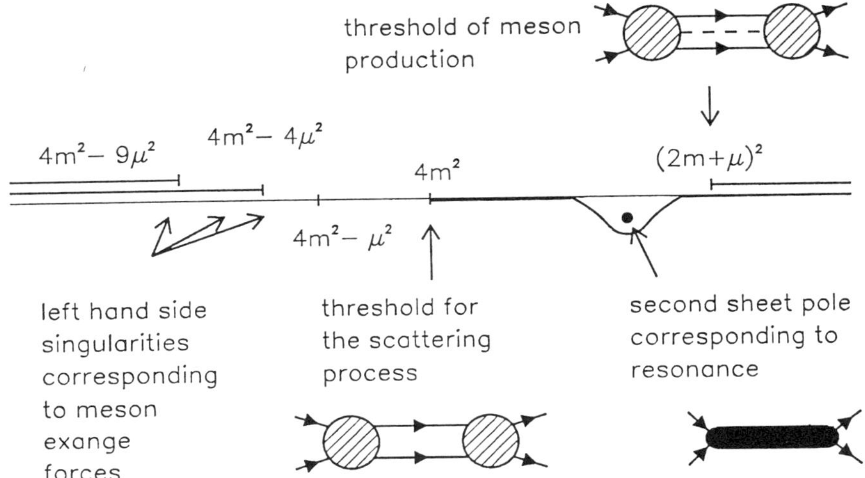

Fig. 3.1. Singularities of partial-wave amplitudes in the s-plane.

Lefthand side singularities of the partial amplitudes are connected with the t-channel exchanges contributing to $A(s,t)$. The S-wave partial amplitude is equal to

$$A(s) = \int_{-1}^{1} \frac{dz}{2} A(s, t(z)), \tag{3.1}$$

where $t(z) = -2(s/4 - m^2)(1 - z)$, and $z = \cos\theta$. L.h.s singularities correspond to

$$t(z = -1) = (n\mu)^2 \tag{3.2}$$

and are located at $s = 4m^2 - \mu^2, s = 4m^2 - 4\mu^2$, and so on.

The dispersion relation N/D-method provides the possibility to construct a relativistic two-particle partial amplitude in the region of low and intermediate energies.

Let us restrict ourselves to the consideration of the region in the vicinity of $s = 4m^2$. The unitarity condition for the partial wave amplitude (we consider the S-wave amplitude as an example) reads:

$$\text{Im } A(s) = \rho(s) \mid A(s) \mid^2 . \tag{3.3}$$

Here $\rho(s)$ is the two-particle phase space integrated at fixed s:

$$\begin{aligned} \rho(s) &= \frac{1}{2}\int d\phi_2(P; k_1, k_2) = \frac{1}{16\pi}\sqrt{\frac{s - 4m^2}{s}}, \\ d\phi_2(P; k_1, k_2) &= (2\pi)^4\delta^4(P - k_1 - k_2)\frac{d^3k_1}{(2\pi)^3 2k_{10}}\frac{d^3k_2}{(2\pi)^3 2k_{20}}. \end{aligned} \tag{3.4}$$

Here P is total momentum, $P^2 = s$; k_1 and k_2 are momenta of particles in the intermediate state. In the N/D-method the amplitude $A(s)$ is represented as

$$A(s) = \frac{N(s)}{D(s)}. \tag{3.5}$$

Here $N(s)$ has l.h.s singularities only, whereas $D(s)$ has only r.h.s ones. So the N-function is real in the physical region $s > 4m^2$. The unitarity condition can be rewritten as

$$\text{Im } D(s) = -\rho(s)N(s). \tag{3.6}$$

The solution of this equation is

$$D(s) = 1 - \int_{4m^2}^{\infty} \frac{d\tilde{s})}{\pi} \frac{\rho(\tilde{s})N(\tilde{s})}{\tilde{s} - s} \equiv 1 - B(s). \tag{3.7}$$

In Eq.(3.7) we neglect so called CDD-poles and normalize $N(s)$ by the condition $D(s) \to 1$ as $s \to \infty$.

Let us introduce the vertex function

$$G(s) = \sqrt{N(s)}. \tag{3.8}$$

We assume here that $N(s)$ is positive (the cases with negative $N(s)$ or if $N(s)$ changes sign need a special and more cumbersome treatment). Then the partial wave amplitude $A(s)$ can be expanded in a series

$$A(s) = G(s)[1 + B(s) + B^2(s) + B^3(s) + \cdots]G(s). \tag{3.9}$$

Its graphic interpretation is

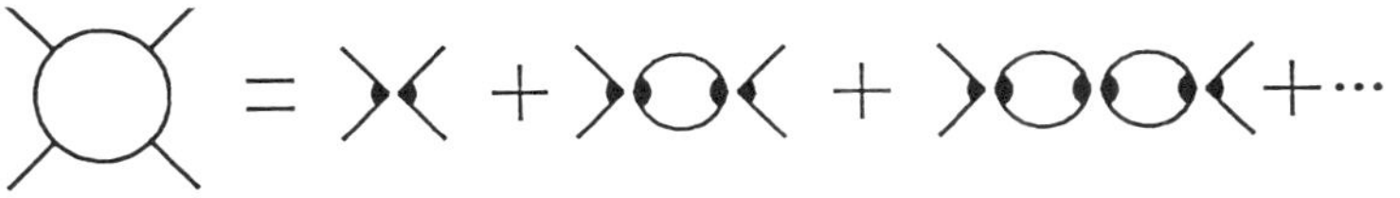

(3.10)

$B(s)$ is a loop-diagram

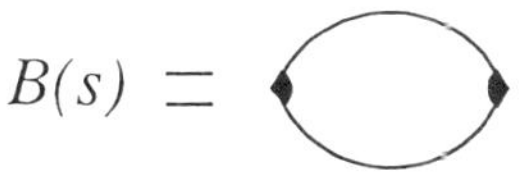

$B(s) =$ (3.11)

so the terms in Eq.(3.9) are amplitudes with different numbers of rescatterings.

3.2 N/D-method and K-matrix

As was shown above, in the N/D method, the amplitude A is written as

$$\begin{aligned} A(s) &= N(s) \Big/ \left[1 - \int_{4m^2}^{\infty} \frac{ds'}{\pi} \frac{N(s')\rho(s')}{s' - s}\right] \\ &= N(s) \Big/ \left[1 - P\int_{4m^2}^{\infty} \frac{ds'}{\pi} \frac{N(s')\rho(s')}{s' - s} - iN(s)\rho(s)\right]. \end{aligned} \tag{3.12}$$

The principal-value integral is real and does not contain the threshold singularity, so we have for the K-matrix representation

$$T(s) = \rho(s)A(s) = \frac{K(s)}{1 - iK(s)} \tag{3.13}$$

with

$$K(s) = \rho(s)N(s) \bigg/ \left[1 - P\int_{4m^2}^{\infty} \frac{ds' N(s')\rho(s')}{s' - s}\right]. \tag{3.14}$$

It is the K-matrix representation of the scattering amplitude for the one-channel case (see lecture 1, Eq.(1.45)).

An important point is that in the case, we have considered, the principal-value integral does not contain a threshold singular term: this is a property of the two-particle threshold singularity. A singular term related to the two-particle threshold exists in the semi-residue term only. We meet the opposite situation in the case of three particles in the intermediate state. This means that the K-matrix representation is not helpful for three-particle states.

3.3 Dispersion relation and Feynman diagrams: Light-cone variables

Loop diagram $B(s)$ plays the main role for the whole dispersion amplitude and we compare in detail the dispersion and Feynman expressions for $B(s)$.

The Feynman expression for $B_F(s)$, with a special choice of separable interaction $G(4K^2 + 4m^2)$, may be proved to be equal to the dispersion integral representation where the four-vector K is defined as

$$2K = k_1 - k_2 - \frac{k_1^2 - k_2^2}{P^2}P; \qquad K^2 = \frac{1}{P^2}(Pk)^2 - k^2; \qquad k = k_1. \tag{3.15}$$

In this subsection I use often the total momentum vector $P = k_1 + k_2$, so it is convenient for me to write here P^2 but not s.

The Feynman expression for the loop diagram reads

$$B_F(P^2) = \frac{1}{(2\pi)^4 i}\int \frac{d^4k\, G^2(4(Pk)^2/P^2 - 4k^2 + 4m^2)}{(m^2 - k^2 - i0)(m^2 - (P-k)^2 - i0)}. \tag{3.16}$$

Since a composite system is conveniently handled in the light-cone coordinates, they will be used hereafter:

$$k_- = \frac{1}{\sqrt{2}}(k_0 - k_z); \quad k_+ = \frac{1}{\sqrt{2}}(k_0 + k_z); \quad k^2 = 2k_+k_- - m_T^2; \quad m_T^2 = m^2 + k_T^2. \tag{3.17}$$

We choose the reference frame in which $P_T = 0$. Then

$$Pk = P_+k_- + P_-k_+ \tag{3.18}$$

and Eq. (3.16) takes the form for $G = 1$:

$$\begin{aligned} B_F(P^2) &= \frac{1}{(2\pi)^4 i} \times \\ &\times \int \frac{dk_+dk_-d^2k_T}{(2k_+k_- - m_T^2 + i0)(P^2 - 2(P_+k_- + P_-k_+) + 2k_+k_- - m_T^2 + i0)}. \end{aligned} \tag{3.19}$$

It should be mentioned that if $G \equiv 1$, one could have performed the integration over k_- right now, closing the integration contour around the pole

$$k_- = \frac{m_T^2 - i0}{2k_+} \tag{3.20}$$

and obtained the standard dispersion representation for a Feynman loop graph ($x = k_+/P_+$):

$$\frac{1}{(2\pi)^4 i}\int_0^1 \frac{dx d^2k_T(-2\pi i)}{2x(P^2 - m_T^2/x - P^2x + i0)} = \int \frac{ds}{\pi(s - P^2 - i0)}\int \frac{dx dk_T^2}{x(1-x)}$$
$$\times \quad \frac{\delta(s - m_T^2/[(x(1-x)])}{16\pi} = \int_{4m^2}^{\infty} \frac{ds \cdot \rho(s)}{\pi(s - P^2 - i0)}. \tag{3.21}$$

The variable x changes from 0 to 1, because for $x < 0$ and $x > 1$ both poles in k_- are located on the same side of the integration contour and the integral equals zero. The dispersion integral (3.21) is divergent at $s \to \infty$ because $G = 1$, and it is just G which provides the convergence of B_F in Eq.(3.16). The convergence of the integral (3.21) can be restored by a subtraction (or cutting) procedure.

For $G \neq 1$ some additional steps are required to obtain the dispersion representation; we introduce new variables ξ_+ and ξ_-

$$\begin{aligned} P_+k_- + P_-k_+ &= \sqrt{P^2}\xi_+ \\ P_+k_-P_-k_+ &= \sqrt{P^2}\xi_-. \end{aligned} \tag{3.22}$$

In terms of these variables, Eq.(3.16) takes the form

$$\begin{aligned} B_F(P^2) &= \frac{1}{(2\pi)^4 i} \times \\ &\times \int \frac{G^2(4(\xi_-^2 + m_T^2))d\xi_+ d\xi_- d^2k_T}{(\xi_+^2 - -\xi_-^2 - m_T^2 + i0)(P^2 - 2\sqrt{P^2}\xi_+ + \xi_+^2 - \xi_-^2 - m_T^2 + i0)} \\ &= \int_0^\infty 2d\xi_-\pi dk_T^2 G^2(4(\xi_-^2 + m_T^2)) \\ &\times \int_{-\infty}^{\infty} \frac{d\xi_+}{(\xi_+^2 - (\xi_-^2 + m_T^2) + i0)[(\xi_+ - \sqrt{P^2})^2 - (\xi_-^2 + m_T^2) + i0)]}. \end{aligned} \tag{3.23}$$

The integration over ξ_+ is performed, closing the integration contour in the upper half--plane, and two poles $\xi_+ = -\sqrt{\xi_-^2 + m_T^2} + i0$ and $\xi_+ = \sqrt{P^2} - \sqrt{\xi_-^2 + m_T^2} + i0$ contribute. The result of integration over ξ_+ is

$$\frac{2\pi i}{\sqrt{\xi_-^2 - m_T^2}(4(\xi_-^2 + m_T^2) - P^2)}. \tag{3.24}$$

The introduction of a new variable $s = 4(\xi_-^2 + m_T^2)$ yields

$$B_F(P^2) = \int_{4m^2}^{\infty} \frac{ds\, G^2(s)}{\pi(s - P^2)}\frac{1}{16\pi}\sqrt{1 - \frac{4m^2}{s}}, \tag{3.25}$$

that is just the dispersion representation (3.7).

3.4 Bethe-Salpeter equations in the momentum representation

Here I discuss the Bethe-Salpeter (BS) equation which is widely used for scattering processes and bound systems and compare it with a treatment of the same amplitudes based on a dispersion relation. BS-equation is a straightforward generalization of the non-relativistic Eq. (2.42) for the scattering amplitude.

The non-homogeneous BS equation in the momentum representation reads

$$A(p_1', p_2'; p_1, p_2) = V(p_1', p_2'; p_1, p_2) + \int \frac{d^4k_1\, d^4k_2}{i(2\pi)^4} A(p_1', p_2'; k_1 k_2) \times \frac{\delta^4(k_1 + k_2 - P)}{(m^2 - k_1^2 - i0)(m^2 - k_2^2 - i0)} V(k_1, k_2; p_1, p_2) \qquad (3.26)$$

or in graphical form:

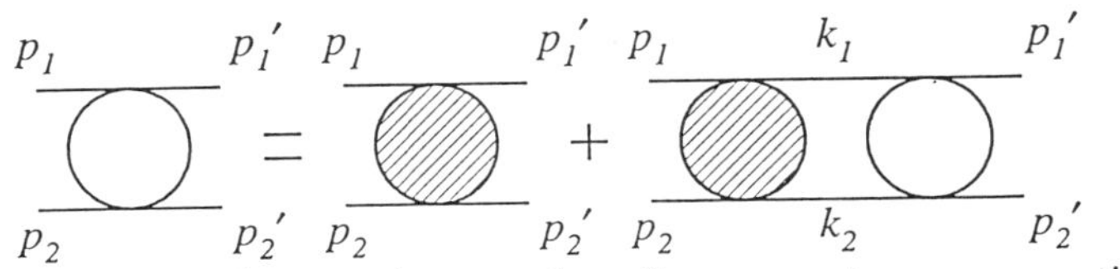

(3.27)

Here the momenta of the constituents obey the momentum conservation law

$$p_1 + p_2 = p_1' + p_2' = P, \qquad (3.28)$$

and $V(p_1, p_2; k_1, k_2)$ is a two-constituent irreducible kernel:

$$V(p_1, p_2; k_1, k_2) = \qquad (3.29)$$

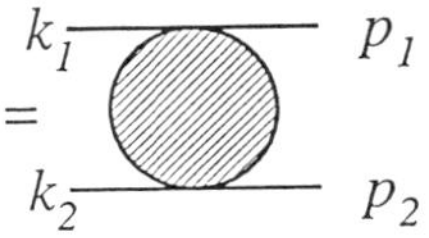

For example, it can be a kernel induced by a meson-exchange interaction

$$\frac{g^2}{\mu^2 - (k_1 - p_1)^2} \qquad (3.30)$$

In the general case $V(p_1, p_2; k_1, k_2)$ is an infinite sum of two-particle irreducible graphs

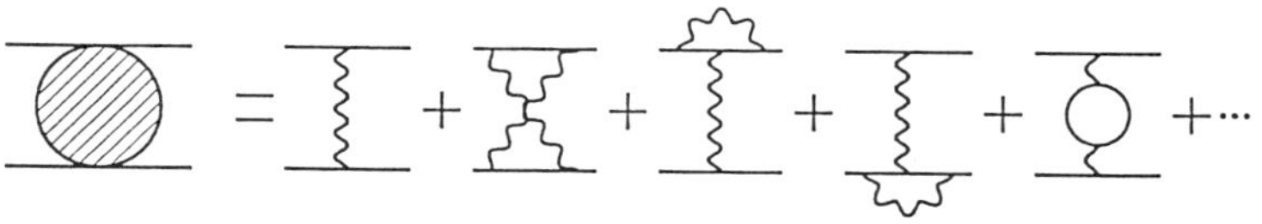

(3.31)

We would like to stress that the amplitude A, determined by the BS equation is the off-mass-shell amplitude. Even if we set $p_1^2 = p_1'^2 = p_2^2 = p_2'^2 = m^2$ in the l.h.s., the Eq.(3.26), the r.h.s. contains the amplitude $A(k_1', k_2; p_1', p_2')$ for $k_1^2 \neq m^2$, $k_2^2 \neq m^2$.

Restricting ourselves to one-meson exchange in the irreducible kernel V, by iterating Eq. (3.26) we come to infinite series of ladder diagrams

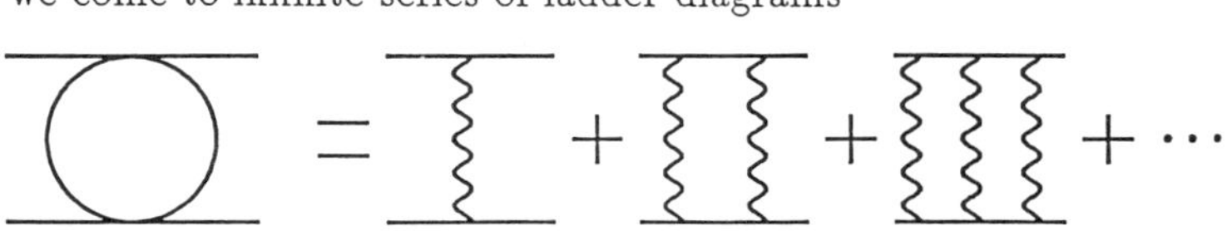

(3.32)

Let us investigate the intermediate states in these ladder diagrams. First of all, these diagrams have two-particle intermediate states which can appear as real states at c.m. energies $\sqrt{s} > 2m$. This corresponds to cutting the ladder diagrams across constituent lines:

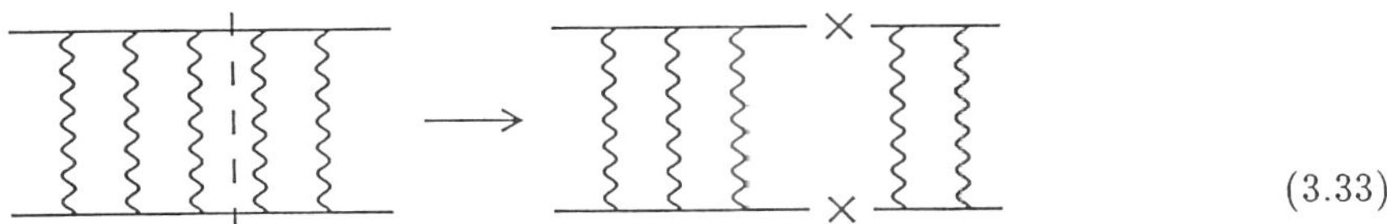

(3.33)

Such a two-particle state manifests itself as a singularity of the scattering amplitude at $s = 4m^2$. However the amplitude A considered as a function of s has not only this singularity but also an infinite set of singularities which correspond to the ladder diagram cuts across meson lines of the type:

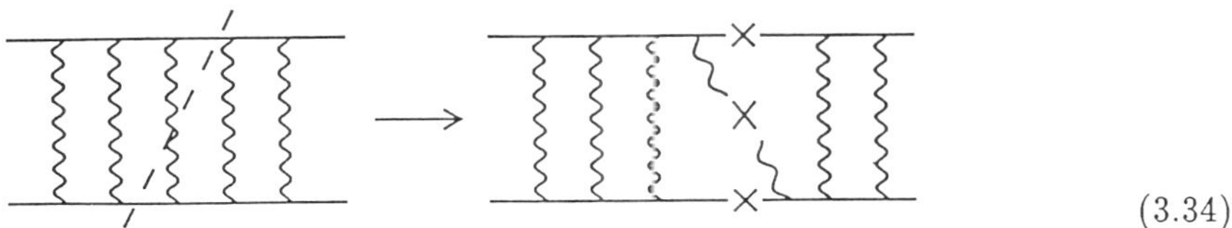

(3.34)

The diagrams which appear after this cut procedure are meson production diagrams, e.g., one-meson production diagrams:

$+$ $+$ (3.35)

Hence, the amplitude $A(p_1', p_2'; p_1, p_2)$ has the following cut singularities in the complex s-plane:

$$s = 4m^2 \tag{3.36}$$

which is related to the rescattering process. Other singularities are related to the meson production processes with cuts starting at

$$s = (2m + n\mu)^2; \qquad n = 1, 2, 3, \ldots \tag{3.37}$$

The four-point amplitude depends on six variables

$$\begin{aligned} & p_1^2,\ p_2^2,\ p_1'^2,\ p_2'^2 \\ s &= (p_1 + p_2)^2 = (p_1' + p_2')^2 \\ t &= (p_1 - p_1')^2 = (p_2 - p_2')^2. \end{aligned} \tag{3.38}$$

The seventh variable, $u = (p_1 - p_2')^2 = (p_1' - p_2)^2$, is not independent because of the relation

$$s + t + u = p_1^2 + p_2^2 + p_1'^2 + p_2'^2. \tag{3.39}$$

It is possible to decrease the number of variables in Eq.(3.26) if we consider an amplitude with definite angular momentum. For this purpose we consider Eq.(3.26) in the c.m.s. of particles 1 and 2 and expand the amplitude $A(p_1', p_2'; p_1, p_2)$ as well as the interaction term over the angular momentum states

$$\begin{aligned} \langle L'M'|A(p_1', p_2'; p_1, p_2)|\,LM\rangle &= A_L(s; p_1^2, p_2^2, p_1'^2, p_2'^2)\delta_{LL'}\delta_{MM'} \\ \langle L'M'|V(p_1', p_2'; p_1, p_2)|\,LM\rangle &= V_L(s; p_1^2, p_2^2, p_1'^2, p_2'^2)\delta_{LL'}\delta_{MM'}. \end{aligned} \tag{3.40}$$

For spinless particles the states $|LM>$ are spherical harmonics $Y_{LM}(\theta, \varphi)$. We thus get for the amplitude A_L the equation

$$A_L(s; p_1'^2, p_2'^2, p_1^2, p_2^2) = V_L(s; p_1'^2, p_2'^2, p_1^2, p_2^2)$$

$$+ \int \frac{d^4k}{i(2\pi)^4} V_L(s; p_1'^2, p_2'^2, k_1^2, k_2^2) \tag{3.41}$$
$$\times \frac{|Y_{LM}(\vec{k}/|\vec{k}|)|^2}{(m^2 - k_1^2 - i0)(m^2 - k_2^2 - i0)} A_L(s, k_1^2, k_2^2, p_1^2, p_2^2).$$

Here $k = k_1$, $k_2 = P - k$ and $P = p_1 + p_2 = (\sqrt{s}, 0, 0, 0)$.

If a bound state of the constituents exist, the scattering partial amplitude has a pole at $s = (p_1 + p_2)^2 = M^2$, where M is the mass of the bound state. This pole appears both in the on- and off-shell scattering amplitudes. This means that the infinite sum of diagrams of Fig. 3.2a type may be rewritten as a pole term of Fig. 3.2b plus some regular terms at $s = M^2$.

The left and right blocks in Fig. 3.2b, $\chi(p_1, p_2; P)$ and $\chi(p_1', p_2'; P)$, satisfy the homogeneous BS equation

$$\chi(p_1, p_2; P) = \int \frac{d^4k_1\, d^4k_2}{i(2\pi)^4} V(p_1, p_2; k_1, k_2)$$
$$\times \frac{\delta^4(k_1 + k_2 - P)}{(m^2 - k_1^2 - i0)(m^2 - k_2^2 - i0)} \chi(k_1, k_2; P), \tag{3.42}$$

whose graphical form is

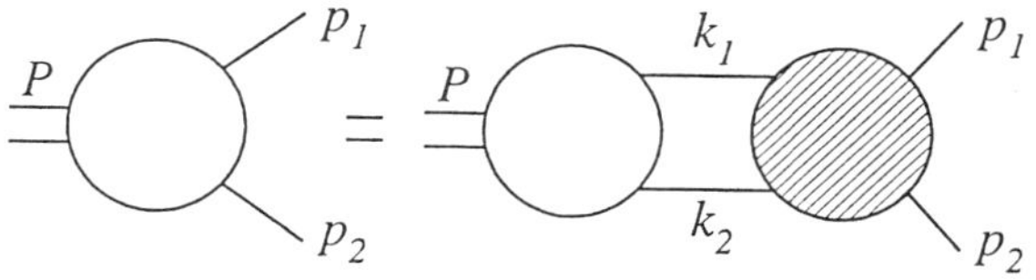

(3.43)

n iterations give

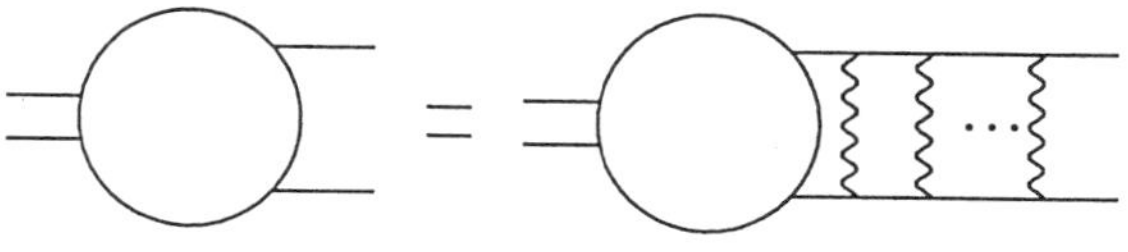

(3.44)

The same cutting procedure of the interaction block in the r.-h.-side of Eq.(3.44) shows us that the amplitude $\chi(p_1, p_2; P)$ contains all the singularities of the amplitude A given by Eqs. (3.33), (3.34).

The three-point amplitude $\chi(p_1, p_2; P)$ depends on three variables

$$P^2 \ (or \ \ s), \ \ p_1^2, \, p_2^2, \tag{3.45}$$

and again, as in the case of the scattering amplitude A, the BS equation contains the off-mass-shell amplitude $\chi(k_1, k_2; P)$; χ is a solution of the homogeneous equation, hence a normalization condition should be added independently.

The connection between χ and A in the pole has the sense of such a normalization condition. It is convenient to normalize χ as follows. For $P^2 \to M^2$

$$A(p_1, p_2; p_1', p_2') = \frac{\chi(p_1, p_2; P)\chi(P; p_1', p_2')}{P^2 - M^2} + Regular\ terms. \tag{3.46}$$

In the formulation of scattering theory, we start from a set of asymptotic states, containing constituent particles (with mass m) and mesons (with mass μ) only. We do not include in such a formulation of scattering theory composite particles as asymptotic

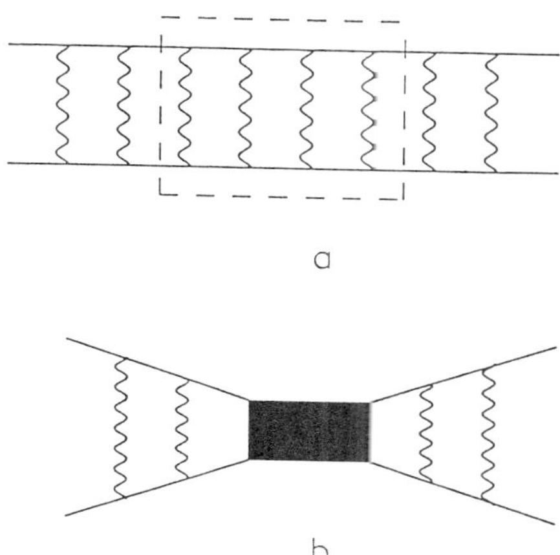

Fig. 3.2. a) Ladder diagram of the on-mass-shell scattering amplitude and the inner block which is the subject of consideration in eq.(3.27); b) Pole diagram which corresponds to composite particle and vertices of the transition "composite particle → constituents".

states; we simply cannot know beforehand whether such bound states exist or not. But if we consider processes of production of new constituent-anticonstituent outgoing pairs which can form bound states, the latter should be included into the set of asymptotic states.

3.5 Bethe-Salpeter equation with separable kernel

It was demonstrated that the Feynman diagram calculus of scattering amplitudes with separable interactions gives the same result as N/D dispertion relation method when the vertices in the c.m. system depend only on the space components of momenta. Here the BS equation with separable kernel is expessed in the form having the same solution of the scattering amplitude as is given by the N/D-method.

Let us write the scattering amplitude A, defined as an infinite sum of dispersion relation loop diagrams:

$$A(s) = s \rangle\langle s + s \rangle\bigcirc\langle s + s \rangle\bigcirc\bigcirc\langle s + \cdots \qquad (3.47)$$

The energy-off-shell amplitude emerges when the cut procedure of the series (3.47) is performed:

$$s \rangle\bigcirc_{\tilde{s}\,|\,\tilde{s}}\langle s + s \rangle\bigcirc_{\tilde{s}\,|\,\tilde{s}}\bigcirc_{\tilde{s}'}\langle s + \cdots \qquad (3.48)$$

This amplitude is also represented as an infinite sum of loop diagrams, but initial and final values $\tilde{s}$ and s are different

$$A(\tilde{s}, s) = \tilde{s} \rangle\langle s + \tilde{s} \rangle\bigcirc_{\tilde{s}}\langle s + \cdots \qquad (3.49)$$

It is the energy-off-shell amplitude which has to be considered in the general case. This amplitude satisfies the equation

$$A(\tilde{s},s) \;=\; G(\tilde{s})G(s)+G(\tilde{s})\int\limits_{4m^2}^{\infty}\frac{d\tilde{s}'}{\pi}\frac{G(\tilde{s}')\rho(\tilde{s}')A(\tilde{s}',s)}{\tilde{s}'-s}. \qquad (3.50)$$

Let us stress that in the dispersion approach we deal with amplitudes on the mass-shell, i.e. amplitudes for real constituents, whereas in the BS equation (3.26) we had off-mass-shell amplitudes. The apperance of the energy-off-shell amplitude in the dispersion method, Eq.(3.50), is the price we have to pay for keeping all the constituents on the mass-shell.

The diagrams representing (3.50) yield

$$A(\tilde{s},s) \;=\; G(\tilde{s}')\frac{G(s)}{1-B(s)}. \qquad (3.51)$$

In physical processes $\tilde{s}=s$ so the physical partial wave amplitude $A(s)$ is just $A(s)=A(s,s)$.

Let us consider the partial amplitude near the pole corresponding to the bound state. The pole condition is

$$1 \;=\; B(M^2) \;=\int_{4m^2}^{\infty}\frac{d\tilde{s}}{\pi}\frac{\rho(\tilde{s})G^2(\tilde{s})}{\tilde{s}-M^2}. \qquad (3.52)$$

In the vicinity of this pole

$$A(s)=G(s)\frac{1}{1-B(s)}G(s)\simeq\frac{G(M^2)}{\sqrt{B'(M^2)}}\cdot\frac{1}{M^2-s}\cdot\frac{G(M^2)}{\sqrt{B'(M^2)}}+\ldots \qquad (3.53)$$

Here we take into account that $1-B(\tilde{s})\simeq 1-B(M^2)-B'(M^2)(s-M^2)$. The homogeneous equation for the bound state vertex amplitude $G_v(s,M^2)$ reads

$$G_v(s,M^2) \;=\; G(s)\int\limits_{4m^2}^{\infty}\frac{d\tilde{s}}{\pi}G(\tilde{s})\frac{\rho(\tilde{s})}{\tilde{s}-M^2}G_v(\tilde{s},M^2). \qquad (3.54)$$

Here $G_v(s,M^2)$ is the analog of $\chi(p_1,p_2;P)$.

The only s-dependence on the right-hand-side of Eq.(3.54) is the factor $G(s)$, so

$$G_v(s,M^2) \;=\; G(s)N^{-1/2}(M^2). \qquad (3.55)$$

The normalization condition for $G_v(s,M^2)$ is the connection between $G_v(s,M^2)$ and $A(s,\tilde{s})$ in the vicinity of the pole. If we require

$$A(s,s) \;=\; \frac{G_v(M^2,M^2)G_v(M^2,M^2)}{M^2-s}+\textit{regular terms} \qquad (3.56)$$

then $N(M^2)=B'(M^2)$ and

$$G_v(s,M^2) \;=\; \frac{G(s)}{\sqrt{B'(M^2)}}. \qquad (3.57)$$

The vertex function $G_v(s,M^2)$ enters all processes containing the bound state interaction. For example, this vertex determines the form factor of a bound state. In more detail the problem of finding the form factor of the bound system is considered in lecture 5, where the deuteron is discussed.

4 Multichannel K-matrix and low-energy $\pi\pi$-scattering

Now we consider as an example a multichannel amplitude which has been investigated in detail for the realistic case of $\pi\pi$-scattering in the state $I, J^{PC} = 0, 0^{++}$. In the region near 1 GeV there is strong transition $\pi\pi \to K\bar{K}$, so the $\pi\pi$-amplitude presents us with a typical example of a two-channel amplitude.

4.1 K-matrix amplitude for the two-channel case

Let us present the amplitude in the form

$$\hat{A} = \frac{\hat{K}}{1 - i\hat{\rho}\hat{K}}, \tag{4.1}$$

where $\hat{A}, \hat{K}$ and $\hat{\rho}$ are 2×2 matrices:

$$\hat{A} = \begin{vmatrix} A_{11} & A_{12} \\ A_{21} & A_{22} \end{vmatrix}, \hat{K} = \begin{vmatrix} K_{11} & K_{12} \\ K_{21} & K_{22} \end{vmatrix}, \hat{\rho} = \begin{vmatrix} \rho_1 & 0 \\ 0 & \rho_2 \end{vmatrix}. \tag{4.2}$$

Here A_{ij} is the amplitude of the transition $i \to j$. For example, for the case $1 = \pi\pi$ and $2 = K\bar{K}$ the amplitude A_{11} is the scattering amplitude $\pi\pi \to \pi\pi$, and A_{12} is the transition amplitude $\pi\pi \to K\bar{K}$, and so on. The matrix $\hat{\rho}$ is constructed to describe the phase space factors of the states 1 and 2.

In strong interactions $A_{12} = A_{21}$ and $K_{21} = K_{12}$. Matrix elements K_{ij} are real and do not contain threshold singularities. But they may have pole singularities.

4.2 Example: $\pi\pi$-scattering amplitude $I, J^{PC} = 0, 0^{++}$ in the mass region near 1000 MeV

The S-wave $\pi\pi \to \pi\pi$ amplitude with $I = 0$ is described in the energy region $\sqrt{s} \leq$ 1100 MeV by the two-channel K-matrix as follows:

$$A_{\pi\pi}(s,t) = \frac{K_{\pi\pi} + i\rho_K[K_{\pi K}K_{K\pi} - K_{\pi\pi}K_{KK}]}{1 - i\rho_\pi K_{\pi\pi} - i\rho_K K_{KK} + \rho_\pi\rho_K[K_{\pi K}K_{K\pi} - K_{\pi\pi}K_{KK}]}. \tag{4.3}$$

Here the following notation is used for $\pi\pi$ and $K\bar{K}$ states:

$$|\pi\pi > \to \pi, \qquad |K\bar{K} > \to K, \tag{4.4}$$

so $K_{\pi\pi}$ refers to the scattering $\pi\pi \to \pi\pi$ while $K_{\pi K}$ is the K-matrix element for the transition $\pi\pi \to K\bar{K}$; ρ_π and ρ_K are phase space factors for $\pi\pi$ and $K\bar{K}$ states which are chosen in the forms:

$$\rho_\pi = \sqrt{1 - \frac{4m_\pi^2}{s}}, \qquad \rho_K = \sqrt{1 - \frac{4m_K^2}{s}}. \tag{4.5}$$

Also m_π and m_K are pion and kaon masses.

A two-pole K-matrix parametrization, with parameters given below, describes the experimental data:

$$
\begin{aligned}
K_{\pi\pi} &= \left[\frac{g_\pi^2}{M_1^2 - s} + \frac{G_\pi^2}{M_2^2 - s} + f_{\pi\pi}\right]\left(1 - \frac{m_\pi^2}{2s}\right) \\
K_{\pi\pi} &= \left[\frac{g_\pi g_K}{M_1^2 - s} + \frac{G_\pi G_K}{M_2^2 - s} + f_{\pi K}\right]\left(1 - \frac{m_\pi^2}{2s}\right) \\
K_{KK} &= \left[\frac{g_K^2}{M_1^2 - s} + \frac{G_K^2}{M_2^2 - s} + f_{KK}\right]\left(1 - \frac{m_\pi^2}{2s}\right).
\end{aligned}
\tag{4.6}
$$

Factors $(1 - m_\pi^2/2s)$ provide the Adler zero in the $\pi\pi$- scattering amplitude.

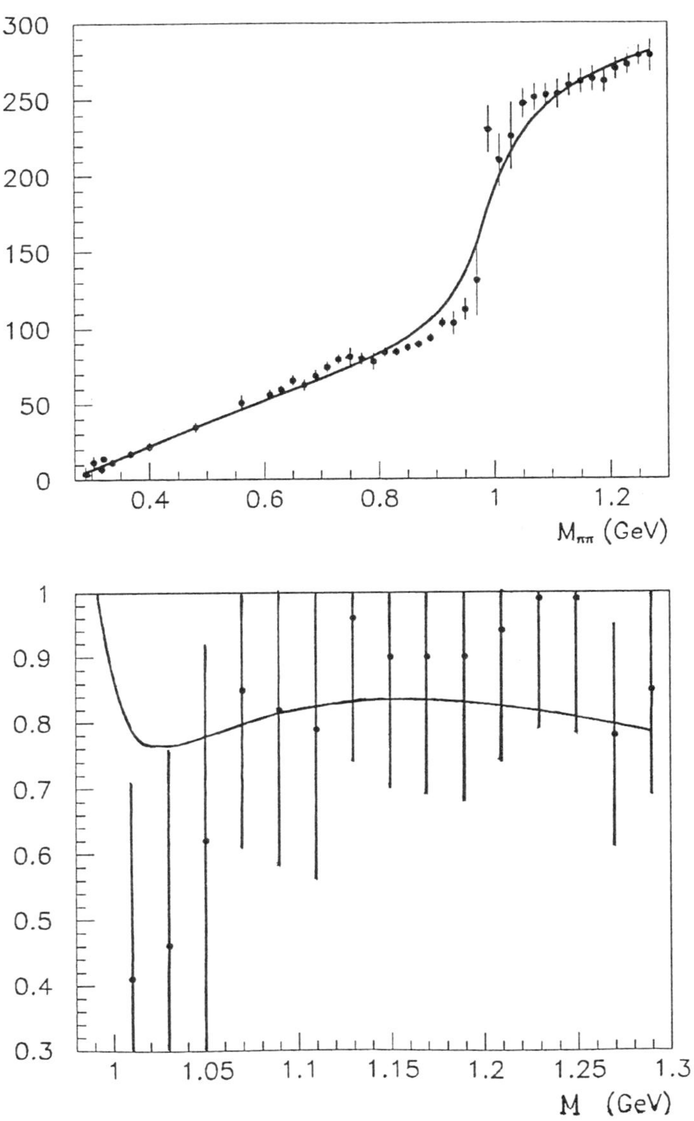

Fig. 4.1. a) Phase shift δ_S^0 of the $\pi\pi \to \pi\pi$ scattering amplitude. Here and further the curves show the fit results; b) Inelasticity η_S^0 of the $\pi\pi \to \pi\pi$ scattering amplitude.

The parameters of the K-matrix, which describe well the S-wave $\pi\pi$-amplitude, are as follows:

$$\begin{array}{ll} M_1 = 775 \text{ MeV} , & M_2 = 1170 \text{ MeV} \\ g_\pi = 860 \text{ MeV} , & G_\pi = 915 \text{ MeV} , \\ g_K = 445 \text{ MeV} , & G_K = -340 \text{ MeV} , \\ f_{\pi K} = 0.72 , & F_{\pi\pi} = f_{KK} = 0. \end{array} \tag{4.7}$$

The description of the experimental data using these parameters is demonstrated in Fig.4.1, where the $\pi\pi$ S-wave phase shift and the inelasticity parameter are shown.

The above expressions allow us to see the advantages and weak points of the K-matrix approach. The advantage is the simplicity in handling, which allows to explicity write down the formulae with the correct account of the r.h.s. singularities. The K-matrix may contain the poles, hence being able to determine the masses of the original bound states. However the K-matrix is unable to reproduce the analytic properties of the amplitude at $s \leq 0$, for it does not contain lefthand side singularities. It is clearly seen from from the above expression: the $\pi\pi$ amplitude defined in Eq. (4.3) contains on the first (physical) sheet the uncorrect singularity at $s = 0$. This means that the amplitude constructed in such a way is not valid at $s \leq 4m_\pi^2$.

4.3 K-matrix description of $\pi\pi$-production

Production of the $\pi\pi$-system is accompanied by the final state interaction of these pions. The corresponding final-state interaction factor is described by the K-matrix as well. We consider here, as examples, the reactions $\pi N \to \pi\pi N$ and $J/\Psi \to \pi\pi\phi$. At small momentum transfer to the nucleon, the reactions $\pi N \to \pi\pi N$ and $\pi N \to K\bar{K}N$ are described by the diagrams of Fig.4.2.

The amplitude of the diagram of Fig.4.2a (the S-wave $\pi\pi$ production) is written as:

$$A(\pi N \to (\pi\pi)_S N) = g(\bar{\Psi}\gamma_5\Psi)\frac{F(t)}{m_\pi^2 - t}A_{\pi\pi}(s,t). \tag{4.8}$$

Here Ψ is the nucleon wave function, $F(t)$ is the nucleon form factor for the pion-nucleon vertex and $A_{\pi\pi}(s,t)$ is the $\pi\pi$ production block.

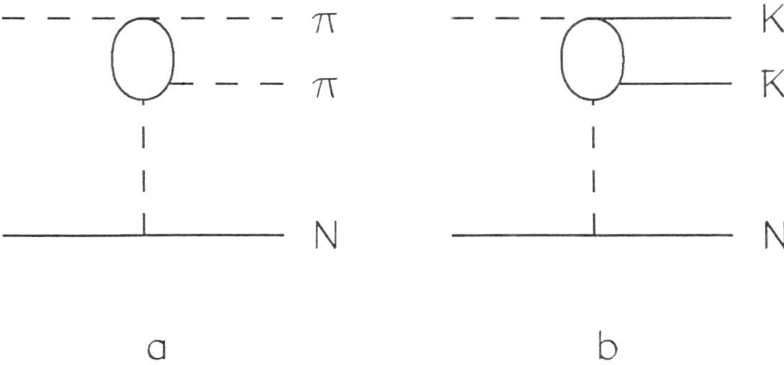

Fig. 4.2. Pole diagrams describing the $\pi\pi$ and $K\bar{K}$ production (the $K\bar{K}$ production block is included into the $a_{\pi\pi}$-amplitude (1) as an intermediate process).

Note that in the t-region, distant from $|t| \approx m_\pi^2$, the reggeization of the pion propagator may become necessary. The corresponding generalized formula is attained by the following replacement made in Eq.(4.8):

$$\frac{(\bar{\Psi}\gamma_5\Psi)}{m_\pi^2 - t} \to (\phi^+ \vec{\sigma}\vec{q}_\perp \phi) e^{-i(\pi/2)\alpha_\pi(t)} \frac{s_{\pi N}^{\alpha_\pi(t)}}{\sin(\pi\alpha_\pi(t)/2)}. \tag{4.9}$$

Here ϕ is the nucleon two-component spinor, $s_{\pi N}$ is the pion-nucleon energy squared, $\alpha_\pi(t)$ is the pion trajectory, and $\vec{q}_\perp$ is the component of the momentum transferred to the nucleon perpendicular to the pion beam direction ($q_\perp^2 \approx |t|$). The reggeized pion propagator takes into account the t-channel exchanges of the mesons belonging to the pion trajectory; the pion contribution is enlarged at small $|t|$ because $\alpha_\pi(m_\pi^2) = 0$. But it is not important for further anlysis which formula, (4.8) or (4.9), is being used. Main point is that in both cases the amplitude $A_{\pi\pi}$ can be factorized.

The S-wave $\pi\pi \to \pi\pi$ amplitude for the virtual pion is written in the following form:

$$A_{\pi\pi}(s,t) = \frac{K_{\pi\pi}(t) + i\rho_K[K_{\pi K}(t)K_{K\pi} - K_{\pi\pi}(t)K_{KK}]}{1 - i\rho_\pi K_{\pi\pi} - i\rho_K K_{KK} + \rho_\pi\rho_K[K_{\pi K}K_{K\pi} - K_{\pi\pi}K_{KK}]}. \tag{4.10}$$

K-matrix blocks, $K_{\pi\pi}(t)$ and $K_{K\pi}(t)$ are determined now by new coupling constants: $g_\pi(t)$, $G_\pi(t)$, $f_{\pi\pi}(t)$ and $f_{\pi K}(t)$. Namely,

$$\begin{aligned} K_{\pi\pi}(t) &= \left[\frac{g_\pi(t)g_\pi}{M_1^2 - s} + \frac{G_\pi(t)G_\pi}{M_2^2 - s} + f_{\pi\pi}(t)\right]\left(1 - \frac{m_\pi^2}{2s}\right) \\ K_{\pi K}(t) &= \left[\frac{g_\pi(t)g_K}{M_1^2 - s} + \frac{G_\pi(t)G_K}{M_2^2 - s} + f_{\pi K}(t)\right]\left(1 - \frac{m_\pi^2}{2s}\right). \end{aligned} \tag{4.11}$$

Using this technique the description of the two pion production in $\pi^- p$ collisions has been performed in ref. [12].

The analogous formula describes the final state interaction of pions in the reaction $J/\Psi \to \pi\pi\phi$. In this reaction the low energy S-wave is expressed in the form

$$\begin{aligned} A_{\pi\pi}(s) &= \frac{K_{\pi\pi}^* + i\rho_K[K_{\pi K}^* K_{K\pi} - K_{\pi\pi}^* K_{KK}]}{1 - i\rho_\pi K_{\pi\pi} - i\rho_K K_{KK} + \rho_\pi\rho_K[K_{\pi K}K_{K\pi} - K_{\pi\pi}K_{KK}]}, \\ K_{\pi\pi}^* &= \left[\frac{\Lambda_1 g_\pi}{M_1^2 - s} + \frac{\Lambda_2 G_\pi}{M_2^2 - s} + \Lambda_3\right]\left(1 - \frac{m_\pi^2}{2s}\right), \\ K_{\pi K}^* &= \left[\frac{\Lambda_1 g_K}{M_1^2 - s} + \frac{\Lambda_2 G_K}{M_2^2 - s} + \Lambda_4\right]\left(1 - \frac{m_\pi^2}{2s}\right), \end{aligned} \tag{4.12}$$

where Λ_i are constant complex parameters which describe the production amplitude and the three-meson final state interaction.

4.4 Analysis of the $(0, 0^{++})$ $\pi\pi$-amplitude pole structure in the 1000 MeV region

The $\pi\pi$-amplitude (4.3) can be easily investigated, in spite of its rather complicated form, by excluding in turn one or other interacting channel. This can be done by varying the G_a, g_a and f parameters.

If G_π, g_π, G_K, g_K and f tend to zero, the K-matrix poles become the poles of the scattering amplitude. Thus M_1 and M_2 would be the masses of the resonances, if their decay widths through the $\pi\pi$ and $K\bar{K}$ channels were small. The transitions

$$(\text{resonance 1}) \to \begin{pmatrix} \pi\pi \\ K\bar{K} \end{pmatrix} \to (\text{resonance 2}) \qquad (4.13)$$

lead to strong mixing of the original *states 1* and *2*. As a result, the mixed states acquire masses which are rather different from M_1 and M_2.

Let us first consider the original *states 1* and *2*. Their masses are found near 800 MeV and 1200 MeV. The coupling constants with $\pi\pi$ and $K\bar{K}$ channels are large. Such coupling constants would lead to the following widths:

$$\begin{aligned} \Gamma_1 &= \frac{g_\pi^2 \rho_\pi(M_1^2)}{M_1} \approx 900\ MeV\,, \\ \Gamma_2 &= \frac{G_\pi^2 \rho_\pi(M_2^2)}{M_2} + \frac{G_K^2 \rho_K(M_2^2)}{M_2} \approx 760\ MeV\,, \end{aligned} \qquad (4.14)$$

which are quite large. But when these two *states* mix only one keeps a large width, while the second one becomes narrow, $f_0(980)$. Let us consider the dynamics of the poles of the $\pi\pi$ scattering amplitude. To illustrate this, a simple method is used: we change the denominator of Eq.(4.3): $f \to xf$, $g_a \to xg_a$, $G_a \to xG_a$, and vary x from 0 (no mixing) up to 1 (the real case, corresponding to the description of the experimental data).

The pole movements with x are shown in Fig.4.3. When $x \to 0$, the transitions of the *particles 1* and *2* into $\pi\pi$ and $K\bar{K}$ *states* tend to zero, and the amplitude poles are located near the real axis, in the vicinity of $\sqrt{s} = M_{\pi\pi} = M_1 = 773$ MeV and $M_{\pi\pi} = M_2 = 1163$ MeV. With increasing x, the poles are shifted into the complex s-plane, and the real parts of the masses of both states drift to the $K\bar{K}$ threshold.

Let us discuss in more detail the pole positions for an intermediate x-value (e.g., $x = 0.4$). In this case we have the standard situation with two Breit-Wigner poles. The sheet-II pole, related to the first resonance, is located at $M_1^{II}(x = 0.4) \approx (880 - i140)$MeV, not far from the physical region (Fig.4.3a). The $K\bar{K}$ threshold singularity doubles the resonance poles: the second pole, which is also related to the first resonance, is located on sheet-III (Fig.4.3b), at $M_1^{III}(x = 0.4) \approx (865 - i185)$ MeV. The pole related to the second resonance is located on sheet-III, at $M_2^{III}(x = 0.4) \approx (1105 - i150)$ MeV (not far from the physical region, see Fig.4.3b), while its double, the sheet-II pole, is located at $M_2^{II}(x = 0.4) \approx (1125 - i145)$ MeV (Fig.4.3a).

In the final picture, when $x = 1$, both poles of the first resonance are located near the physical region and near the $K\bar{K}$ threshold:

$$\begin{aligned} M_1^{II} &= (987 - i40)\ \text{MeV}\,, \\ M_1^{III} &= (967 - i69)\ \text{MeV}\,. \end{aligned} \qquad (4.15)$$

The poles of the second resonance appear to be rather far from the physical region. The nearest one is located at

$$M_2^{III} = (1165 - i460)\ \text{MeV}\,. \qquad (4.16)$$

It is very important to find out what is the reason for the M_1 poles to appear in the region close to the physical axis. There are two possible explanations:

1) the existence of a strong $K\bar{K}$-component in the intermediate state, and

2) the mixing of the *states 1* and *2*.

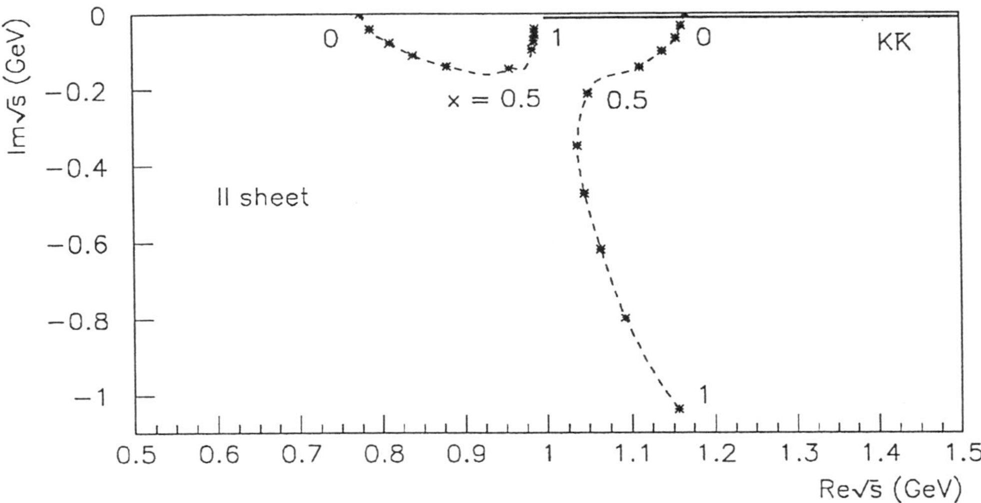

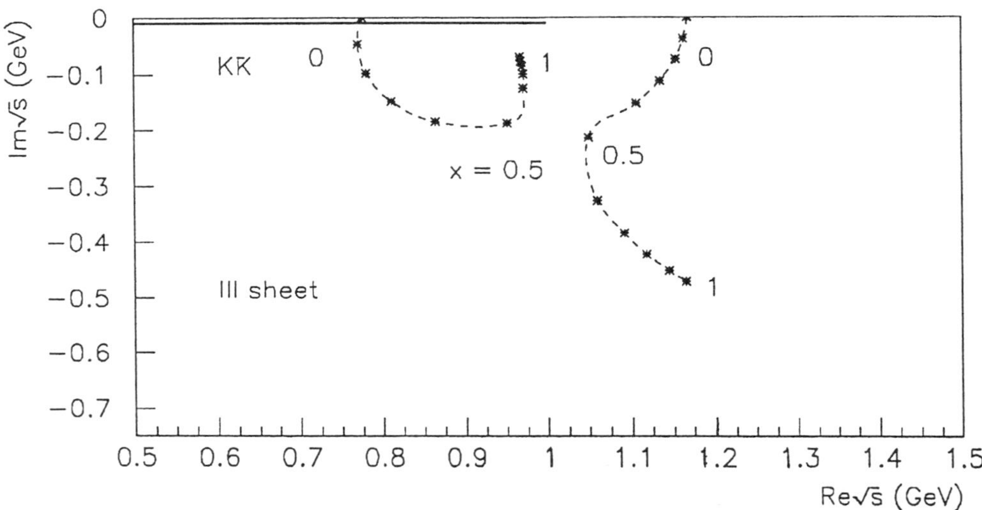

Fig. 4.3. "Dynamics" of the $\pi\pi$-amplitude poles with an increase of the g_π, g_K and G_π, G_K coupling constants. The position 0 corresponds to the K-matrix poles, and the position 1 corresponds to the physical values of coupling constants defined by (4.15).

To estimate the role of the $K\bar{K}$ component, let us exclude it by putting $g_K = G_K = f = 0$. In this case the $\pi\pi$-amplitude has poles at

$$M_1^{II}(g_K = G_K = f = 0) = (979 - i53)\ MeV\,,$$

and (4.17)

$$M_2^{II}(g_K = G_K = f = 0) = (1163 - i630)\ MeV\,.$$

Thus the pole M_1^{II} is shifted from (4.15) only slightly. Actually we expected this: both *state 1* and *state 2* are strongly related to the $\pi\pi$ channel and much more weakly to the $K\bar{K}$ channel:

$$\frac{g_K^2}{g_\pi^2} \approx \frac{G_K^2}{G_\pi^2} \approx \frac{1}{6}\,. \tag{4.18}$$

Using the same method, one can easily obtain that the mixing of *states 1* and *2* leads to a small $f_0(980)$ width. Let us switch off the contribution of the *resonance 2* by putting $G_\pi = G_K = 0$. Then we have for M_1^{II}:

$$M_1^{II}(G_\pi = G_K = 0) = (740 - i120)\ MeV\,. \tag{4.19}$$

This pole has a normal hadron width Γ = 240 MeV. It is located rather far from the $K\bar{K}$ threshold, although the $K\bar{K}$ singularity is present in the $\pi\pi$-amplitude due to the non-zero g_K and f coupling constants. Therefore our "experiments" with excluding the $K\bar{K}$ channel and the *state 2* show that the origin of the narrow resonance, $f_0(980)$, is linked to the *state 1* $\rightarrow \pi\pi \rightarrow$ *state 2* transition.

The two poles (4.15) are interpreted as the $f_0(980)$ resonance. This resonance is related to the $\pi\pi$ channel. Actually it is defined by this channel: the pole position changes only slightly after excluding the $K\bar{K}$ channel from the amplitude of Eq.(4.3). This fact, makes unconvincing the interpretation of this scalar meson as a $K\bar{K}$ molecule. The $f_0(980)$ resonance is a mixture of two original *states, 1* and *2*, which are strongly coupled to the $\pi\pi$ channel. It is very probable that one of these states is a P-wave $q\bar{q}$-meson while the other state is a long-awaited σ-meson which plays an important role in hadron and nuclear physics.

5 Fermion-fermion scattering

So far our discussion was concerned with spinless particles. In this lecture the dispersion technique is applied to fermion-fermion scattering. This case provides a good opportunity to demonstrate the method for the consideration of particles with spin. Construction of fermion-fermion states is done by the operator expansion, which is very helpful in connection with dispersion relation or light-cone variable techniques.

5.1 Four-component spinors, fermions and antifermions

Free particles with spin 1/2 are described by four-component spinors:

$$\Psi(p) = \begin{pmatrix} \Psi_1 \\ \Psi_2 \\ \Psi_3 \\ \Psi_4 \end{pmatrix}, \quad \text{or} \quad \Psi(p) = \begin{pmatrix} \varphi \\ \chi \end{pmatrix}, \tag{5.1}$$

where φ and χ are two-component spinors. Spinor field $\Psi(p)$ obeys the Dirac equation:

$$(\hat{p} - m)\Psi(p) = 0, \tag{5.2}$$

where $\hat{p} = p_\mu \gamma_\mu$. Matrices γ_μ ($\mu = 0, 1, 2, 3$) may be defined as

$$\gamma_0 = \begin{pmatrix} \mathbf{1} & 0 \\ 0 & -\mathbf{1} \end{pmatrix}, \qquad \vec{\gamma} = \begin{pmatrix} 0 & \vec{\sigma} \\ -\vec{\sigma} & 0 \end{pmatrix}, \tag{5.3}$$

where $\mathbf{1}$ is a unit 2×2 matrix, and $\vec{\sigma}$ are standard Pauli matrices

$$\sigma_1 = \begin{pmatrix} 0 & 1 \\ 1 & 0 \end{pmatrix}, \qquad \sigma_2 = \begin{pmatrix} 0 & -i \\ i & 0 \end{pmatrix}, \qquad \sigma_3 = \begin{pmatrix} 1 & 0 \\ 0 & -1 \end{pmatrix}. \tag{5.4}$$

The γ_5 matrix is defined as

$$\gamma_5 = i\gamma_1\gamma_2\gamma_3\gamma_0 = -\begin{pmatrix} 0 & 1 \\ 1 & 0 \end{pmatrix}; \tag{5.5}$$

γ-matrices satisfy the following commutation relations

$$\begin{aligned} \gamma_\mu\gamma_5 + \gamma_5\gamma_\mu &= 0 \\ \gamma_\mu\gamma_\nu + \gamma_\nu\gamma_\mu &= 2\delta_{\mu\nu} I. \end{aligned} \tag{5.6}$$

Here I is a unit 4×4 matrix, and $\delta_{\mu\nu}$ is the metric matrix

$$\delta_{\mu\nu} = \text{ diag } (1, -1, -1, -1). \tag{5.7}$$

The Dirac equation written for the two-component spinors φ and χ is

$$(p_0\gamma_0 - \vec{p}\vec{\gamma} - mI)\begin{pmatrix} \varphi \\ \chi \end{pmatrix} = p_0 \begin{pmatrix} \varphi \\ -\chi \end{pmatrix} - \vec{p}\begin{pmatrix} \vec{\sigma}\,\chi \\ -\vec{\sigma}\,\varphi \end{pmatrix} - m\begin{pmatrix} \varphi \\ \chi \end{pmatrix} = 0. \tag{5.8}$$

So, one has two equations for φ and χ:

$$\begin{aligned} p_0\varphi - \vec{p}\vec{\sigma}\chi - m\varphi &= 0, \\ -p_0\chi + \vec{p}\vec{\sigma}\varphi - m\chi &= 0. \end{aligned} \tag{5.9}$$

They can be rewritten in two equivalent forms:

$$\chi = \frac{\vec{p}\vec{\sigma}}{p_0 + m}\varphi,$$

$$(p_0^2 - \vec{p}^2 - m^2)\varphi = 0, \tag{5.10}$$

or

$$\varphi = \frac{\vec{p}\vec{\sigma}}{p_0 - m}\chi,$$

$$(p_0^2 - \vec{p}^2 - m^2)\chi = 0. \tag{5.11}$$

Let us define the conjugated spinor $\bar{\Psi}$ as

$$\bar{\Psi}(p) = \Psi^+(p)\gamma_0 = (\Psi_1^*, \Psi_2^* \; -\Psi_3^*, -\Psi_4^*). \tag{5.12}$$

The symbol $(+)$ denotes Hermitian conjugation $A^+ = (A_{ab})^+ = A_{ba}^*$, where $(*)$ means complex conjugation.

The spinor $\bar{\Psi}$ obeys the equation:

$$\bar{\Psi}(p)(\hat{p} - m) = 0. \tag{5.13}$$

To prove Eq.(5.13), one should perform Hermitian conjugation of Eq.(5.2), $\Psi^+(\hat{p}^+ - m) = 0$, and operate with γ_0 from the left-hand side: $\Psi^+(\hat{p}^+ - m)\gamma_0 = 0$, or $\Psi^+\gamma_0(\hat{p} - m) = 0$. We used here $\gamma_0^+ = \gamma_0$ and $\vec{\gamma}^+ = -\vec{\gamma}$.

a) Charge conjugation C. The Dirac equation has solutions with positive and negative energies which correspond to fermions and antifermions. Charge conjugation transforms the fermion solution into the antifermion one and vice versa. Let us demonstrate this starting from the Dirac equation for the field $\bar{\Psi}(p)$, Eq.(5.13). Let us transpose this equation,

$$(\hat{p}^T - m)\bar{\Psi}^T(p) = 0, \tag{5.14}$$

(here $A^T = (A_{ab})^T = A_{ba}$) and replace $p \to -p$:

$$(-\hat{p}^T - m)\bar{\Psi}^T(-p) = 0. \tag{5.15}$$

Let us introduce the matrix C which satisfies the relations:

$$C^{-1}C = 1, \qquad C^{-1}\gamma_\mu C = -\gamma_\mu^T. \tag{5.16}$$

It is

$$C = \gamma_2\gamma_0 = -\begin{pmatrix} 0 & \sigma_2 \\ \sigma_2 & 0 \end{pmatrix}, \tag{5.17}$$

and $C = C^{-1} = C^+ = -C^T$. Then one can rewrite Eq.(5.15)

$$C(-\hat{p}^T - m)C^{-1}C\bar{\Psi}^T(-p) = (\hat{p} - m)C\bar{\Psi}^T(-p) = 0. \tag{5.18}$$

The field $\Psi_C(p)$,

$$\Psi_C(p) = C\bar{\Psi}^T(-p), \tag{5.19}$$

satisfies the Dirac equation, and the solutions with positive energy stand for antifermions.

Using Eq.(5.10), one can represent $\Psi(p)$ as

$$\Psi(p) = \begin{pmatrix} \varphi \\ \chi \end{pmatrix} = \begin{pmatrix} \varphi \\ \frac{\vec{p}\vec{\sigma}}{p_0+m}\varphi \end{pmatrix}. \tag{5.20}$$

In the non-relativistic limit the two-component spinor φ is responsible for the description of the fermion:

$$\begin{pmatrix} \varphi \\ \frac{\vec{p}\vec{\sigma}}{p_0+m}\varphi \end{pmatrix} \longrightarrow \begin{pmatrix} \varphi \\ 0 \end{pmatrix} . \tag{5.21}$$

Non-relativistic antifermions are described by the spinor χ. Indeed,

$$\Psi_c(p) = -\begin{pmatrix} 0 & \sigma_2 \\ \sigma_2 & 0 \end{pmatrix} \begin{pmatrix} \frac{\vec{\sigma}^T\vec{p}}{p_0+m}\chi^* \\ -\chi^* \end{pmatrix} = \begin{pmatrix} \sigma_2\chi^* \\ \frac{\vec{\sigma}\vec{p}}{p_0+m}\sigma_2\chi^* \end{pmatrix} . \tag{5.22}$$

Introducing $\varphi_c = \sigma_2\chi^*$ we have a representation of $\Psi_c(p)$ which is analogous to Eq.(5.20)

$$\Psi_c(p) = \begin{pmatrix} \varphi_c \\ \frac{\vec{p}\vec{\sigma}}{p_0+m}\varphi_c \end{pmatrix} , \tag{5.23}$$

but with the replacement fermion $\rightarrow$ antifermion.

b) Space inversion P. Space inversion means a change of the sign of space coordinates, keeping the time unchanged:

$$\vec{x} \rightarrow -\vec{x}, \qquad t \rightarrow t. \tag{5.24}$$

Under the P-inversion:

$$\vec{p} \rightarrow -\vec{p}, \qquad p_0 \rightarrow p_0. \tag{5.25}$$

Angular momentum $\vec{M} = [\vec{r}, \vec{p}]$ does not change its sign, neither does the spin of a particle

$$\vec{\sigma} \rightarrow \vec{\sigma}. \tag{5.26}$$

Now let us operate on the spinor $\Psi(p)$ with P:

$$P\Psi(p) = P\begin{pmatrix} \varphi \\ \chi \end{pmatrix} = P\begin{pmatrix} \varphi \\ \frac{\vec{\sigma}\vec{p}}{p_0+m}\varphi \end{pmatrix} = \begin{pmatrix} P\varphi \\ -\frac{\vec{\sigma}\vec{p}}{p_0+m}P\varphi \end{pmatrix} = \begin{pmatrix} \varphi' \\ -\frac{\vec{\sigma}\vec{p}}{p_0+m}\varphi' \end{pmatrix} = \begin{pmatrix} \varphi' \\ -\chi' \end{pmatrix} . \tag{5.27}$$

One can see that spinors φ and χ behave in a different way under space inversion. This means that fermion and antifermion have opposite parities. If a fermion has positive parity, then

$$P = \gamma_0. \tag{5.28}$$

Different parities of fermion and antifermion are reflected in anticommutation of the operators P and C:

$$PC + CP = 0. \tag{5.29}$$

c) Bilinear forms. Treating particles with spin, the dispersion relation technique uses bilinear combinations $\bar{\Psi}(-p_2)\Psi(p_1)$, $\bar{\Psi}(-p_2)\gamma_5\Psi(p_1)$ and $\bar{\Psi}(-p_2)\gamma_\mu\Psi(p_1)$. Such bilinear forms describe the fermion state with momentum p_1 and that of an antifermion with momentum p_2. The space parities and charge parities of these bilinear forms are determined by the replacements

$$\Psi \rightarrow P\Psi \qquad \text{and} \qquad \Psi \rightarrow C\Psi, \tag{5.30}$$

where P and C are given by Eqs.(5.28) and (5.17). It is seen that with P-inversion

$$P: \begin{array}{lcl} (\bar{\Psi}\Psi) & \rightarrow & (\bar{\Psi}\Psi) \\ (\bar{\Psi}\gamma_5\Psi) & \rightarrow & -(\bar{\Psi}\gamma_5\Psi) \\ (\bar{\Psi}\gamma_\mu\Psi) & \rightarrow & -(\bar{\Psi}\gamma_\mu\Psi). \end{array} \tag{5.31}$$

Likewise C-inversion gives

$$C: \begin{array}{rcl} (\bar\Psi\Psi) & \to & (\bar\Psi\Psi) \\ (\bar\Psi\gamma_5\Psi) & \to & (\bar\Psi\gamma_5\Psi) \\ (\bar\Psi\gamma_\mu\Psi) & \to & -(\bar\Psi\gamma_\mu\Psi). \end{array} \tag{5.32}$$

It means that these bilinear combinations have the following quantum numbers J^{PC}:

$$\begin{array}{rcl} (\bar\Psi\Psi) & \to & 0^{++} \\ (\bar\Psi\gamma_5\Psi) & \to & 0^{-+} \\ (\bar\Psi\gamma_\mu\Psi) & \to & 1^{--}. \end{array} \tag{5.33}$$

In hadron physics, G-parity is also used. It is defined by a product of the two operations: $G = C \times P$. The states which are scalar in the isotopic space, ($I = 0$), do not change sign under isotopic inversion, hence for these states G-parity = C-parity. The states which are vectors in the isotopic state, ($I = 1$), change sign under isotopic inversion, so G-parity = $-C$-parity.

d) **Normalization.** The spinors used below have the following normalization

$$(\bar\Psi^j(p)\gamma_0\Psi_k(p)) = 2p_0\delta^j_k. \tag{5.34}$$

Here δ^j_k =diag$(1,1,1,1)$.
The spinors obey the completeness condition

$$\begin{array}{rcl} \sum\limits_{j=1,2} \Psi_{j\beta}(p)\bar\Psi^j_\alpha(p) & = & (\hat p + m)_{\beta\alpha} \\ \sum\limits_{j=3,4} \Psi_{j\beta}(p)\bar\Psi^j_\alpha(p) & = & (\hat p + m)_{\beta\alpha}. \end{array} \tag{5.35}$$

The completeness condition determines the propagator of the particle, see Eq.(2.10).

e) **Propagator.** The propagator of a free particle which obeys the Dirac equation (5.2), $G(p)$, is constructed by the same principle as is done for the spinless particle: it satisfies the equation $(\hat p - m)G(p) = -I$. Hence

$$G(p) = \frac{I}{-\hat p + m} = \frac{I}{-\hat p + m} \cdot \frac{\hat p + m}{\hat p + m} = \frac{\hat p + m}{-p^2 + m^2 - i\epsilon}. \tag{5.36}$$

In the above equation, the shift of the pole $m^2 \to m^2 - i\epsilon$ is done similarly to that in Eq. (2.46). The numerator $(\hat p + m)$ on the r.h.s. of Eq. (5.36) is due to the completeness condition for spinors, Eq. (5.35).

5.2 Spin structure of the fermion-fermion scattering diagrams

Now we determine the spin structure for the transition $NN \to NN$, where N is a fermion carrying spin and isospin. There exist two ways to write down the fermion--fermion scattering amplitude. The standard form of matrix elements for this transition is

$$\left(\bar\Psi(p_1')\tilde O^A\Psi(p_1)\right)\left(\bar\Psi(p_2')\tilde Q_A\Psi(p_2)\right) A(s,t). \tag{5.37}$$

Here p_1 and p_2 (p_1' and p_2') are the momenta of incoming (outgoing) fermions (see Fig. 5.1). The operators $\tilde O^A$ and $\tilde Q_A$ correspond to the state being considered. $A(s,t)$ is the part of the amplitude which depends only on the invariant energy and the invariant momentum-transfer.

However, for our purpose a more suitable choice is the spin-dependent factor in another form:

$$\left(\bar{\Psi}(p_1')O^A\Psi_c(-p_2')\right)\left(\bar{\Psi}_c(-p_2)Q_A\Psi(p_1)\right). \tag{5.38}$$

Here

$$\begin{aligned}\bar{\Psi}_c(-p_2) &= C\Psi(p_2)\\ \Psi_c(-p_2') &= -\bar{\Psi}(p_2')C.\end{aligned} \tag{5.39}$$

There is the following reason to use the spin-dependent part of the amplitude in the form of Eq. (5.38): we work with a series of loop-diagrams in the s-channel which is similar to a series (3.9) (or (3.10)). Calculations of the fermion-antifermion loop are rather easy to perform; just these fermion-antifermion s-channel loops appear when the charge-conjugated fields (5.39) have been introduced. This point may be clarified using proton-nucleon scattering as an example. Let us call a neutron an antifermion, and an antineutron a fermion. Then we may treat the proton-neutron loop diagram, in the spirit of Eq. (5.38), as a fermion-antifermion loop.

Following this prescription, first, we must construct the operators which correspond to definite NN states.

To this end, consider the bilinear form

$$\left(\bar{\Psi}_c(-p_2)Q_A\Psi(p_1)\right) \tag{5.40}$$

and introduce the momenta

$$P = p_1 + p_2, \qquad k = \frac{1}{2}(p_1 - p_2). \tag{5.41}$$

Let us concentrate first on the case with isospin $I = 1$. We choose Q_A in the form

$$\begin{array}{ll}\gamma_5 & for\ \ {}^1S_0\\ \gamma_5 K_{\mu\nu} & for\ \ {}^1D_2,\end{array} \tag{5.42}$$

where $K_{\mu\nu} = k_\mu k_\nu - k^2\delta^\perp_{\mu\nu}/3$ and $\delta^\perp_{\mu\nu} = \delta_{\mu\nu} - (P_\mu P_\nu)/P^2$. Remember that Ψ_c and Ψ have opposite parity, hence $(\bar{\Psi}_c\Psi)$ is pseudoscalar and $(\bar{\Psi}_c\gamma_5\Psi)$ is scalar.

For P-wave states, Q_A takes the form

$$\begin{array}{ll}1 & for\ \ {}^3P_0\\ \epsilon_{\mu abc}\gamma^a P^b k^c & for\ \ {}^3P_1\\ k_\mu\Gamma_\nu + k_\nu\Gamma_\mu - \frac{2}{3}(k\Gamma)\delta^\perp_{\mu\nu} & for\ \ {}^3P_2,\end{array} \tag{5.43}$$

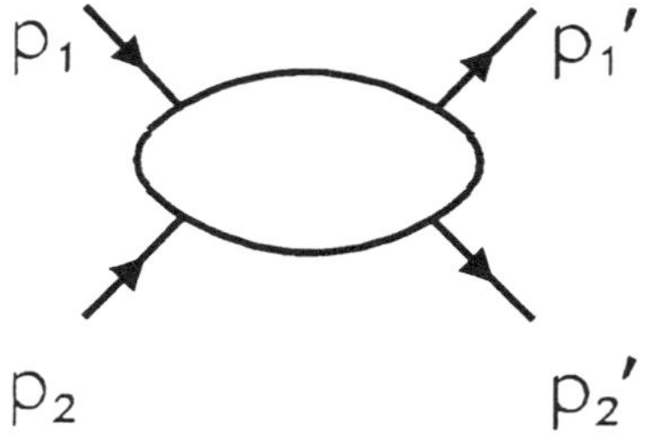

Fig. 5.1. Fermion-fermion scattering amplitude.

$\epsilon_{\mu abc}$ is the totally antisymmetric tensor, and the indices a, b, c run over 0,1,2,3. The operator

$$\Gamma_\mu = \gamma_\mu^\perp - \frac{k_\mu}{m + \sqrt{P^2}/2}$$

produces the bilinear form $(\bar{\Psi}\Gamma_\mu\Psi)$ for a pure S-wave state with spin 1. So the idea of the construction of the operators Q_A is rather simple: states with spins 0 or 1 are constructed using γ_5 or Γ_μ, while angular momentum states are generated by the tensors which are constructed from vector k : k_μ, $k_\mu k_\nu$, and so on. As an example, for F-states Q_A takes the form

$$\begin{array}{ll} {}^3F_2 & K_{\mu\nu a}\,\Gamma^a \\ {}^3F_3 & K^a_{\mu\nu}\gamma^b P^c \epsilon_{abc\lambda} + K^a_{\nu\lambda}\gamma^b P^c \epsilon_{abc\mu} + K^a_{\lambda\mu}\gamma^b P^c \epsilon_{abc\nu} \end{array} \tag{5.44}$$

Here

$$K_{\mu\nu\lambda} = k_\mu k_\nu k_\lambda - \frac{1}{5}k^2\left(k_\mu \delta^\perp_{\nu\lambda} + k_\nu \delta^\perp_{\lambda\mu} + k_\lambda \delta^\perp_{\mu\nu}\right). \tag{5.45}$$

The only state under discussion which can mix is ${}^3P_2$ (with ${}^3F_2$); for other states $Q^A = O^A$.

The amplitudes with $I = 0$ are constructed in the same way: Q_A takes the forms

$$\begin{array}{ll} {}^3S_1 & \Gamma_\mu \\ {}^3D_1 & K_{\mu a}\Gamma^a. \end{array} \tag{5.46}$$

These states can mix. Q_A for P-wave state is

$$ {}^1P_1 \qquad \gamma_5\, k_\mu. \tag{5.47}$$

The other states are constructed in an analogous way.

5.3 Fermion-fermion loop diagram calculations

Here we calculate loop diagrams with two fermions in the intermediate state. In lecture 3 it was shown how to calculate the loop diagram using different techniques, namely Feynman diagrams, light-cone variables or using the dispersion relation technique. Here, using the spin structures of the four-point amplitudes in the form of Eq. (5.38), we calculate the fermion loop diagrams in the framework of the dispersion relation technique.

Let us start from the loop diagram Fig.5.2. The phase space of fermions in the intermediate state is equal to

$$d\Phi(NN) = \frac{1}{2}(2\pi)^4\delta^4(P - k_1 - k_2)\frac{d^3k_1}{(2\pi)^4 2k_{10}}\frac{d^3k_2}{(2\pi)^3 2k_{20}}. \tag{5.48}$$

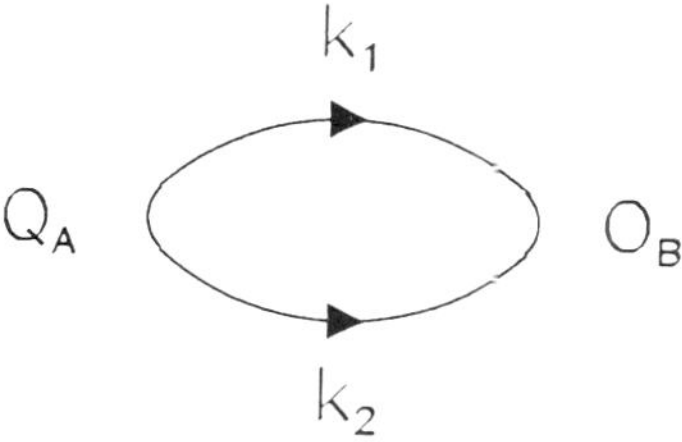

Fig. 5.2. Fermion-fermion loop diagram.

The matrix element for a real fermion transition reads

$$\sum_{spin\ states} \int d\Phi(NN) \left(\bar{\Psi}(p_1')O^B\Psi_c(-p_2')\right) \left(\bar{\Psi}_c(-k_2)O_B\Psi(k_1)\right)$$

$$\times \quad \left(\bar{\Psi}(k_1)Q^A\Psi_c(-k_2)\right) \left(\bar{\Psi}_c(-p_2)Q_A\Psi(p_1)\right). \tag{5.49}$$

For the sake of simplicity, the vertices of the loop diagram, G, are chosen to be equal to 1.

Summing over fermion spins in the intermediate state gives

$$\sum_a \Psi_a(k_1)\bar{\Psi}^a(k_1) = \hat{k}_1 + m, \quad \sum_a \Psi_{ca}(-k_2)\bar{\Psi}_c^a(-k_2) = -\hat{k}_2 + m, \tag{5.50}$$

so we have for the matrix element of Eq. (5.49):

$$\left(\bar{\Psi}(p_1')O^B\Psi_c(-p_2')\right) \rho_B^A \left(\bar{\Psi}_c(-p_2)Q_A\Psi(p_1)\right)$$

$$\hat{\rho}_B^A = \int d\Phi(NN)Tr\left[O_B(\hat{k}_1 + m)Q^A(-\hat{k}_2 + m)\right]. \tag{5.51}$$

The phase space matrix element ρ_B^A is responsible for the spin structure of the loop diagram.

The examples for ρ_B^A given below for the cases $I = 1$ and $J^P = 0^+, 1^-, 2^+$ (we also use a notion $\rho(J^P)$) read as follows

$$\begin{aligned} \hat{\rho}(^1S_0) &= \int d\Phi(NN)Tr\left[\gamma_5\left(\frac{1}{2}\hat{P} + \hat{k} + m\right)\right. \\ &\times \left.\gamma_5\left(-\frac{1}{2}\hat{P} + \hat{k} + m\right)\right] = 2s\frac{1}{16\,\pi}\sqrt{\frac{s - 4m^2}{s}}. \end{aligned} \tag{5.52}$$

Here $P^2 = s$ and $k^2 = (k_1 - k_2)^2/4 = m^2 - s/4$. The factor $2s$ appears because of the trace calculation, and $1/16\,\pi \cdot \sqrt{(s - 4m^2)/s}$ is the usual phase space value. Then

$$\begin{aligned} \hat{\rho}_\mu^{\mu'}(^3P_1) &= \int d\Phi(NN)\epsilon_{\mu abc}P^b k^c \epsilon^{\mu' a' b' c'} P_{b'} k_{c'} \\ &\times \quad \text{Tr}\left[\gamma^a\left(\frac{1}{2}\hat{P} + \hat{k} + m\right)\gamma_{a'}\left(-\frac{1}{2}\hat{P} + \hat{k} + m\right)\right] \\ &= \frac{4}{9}k^2(s + 2m^2)\epsilon_{\mu abc}\epsilon^{\mu a b' c}P^b P_{b'}\int d\Phi(NN) \\ &= -\frac{8}{9}s(s + 2m^2)\left(\frac{s}{4} - m^2\right)\delta_\mu^{\mu'\perp}\int d\Phi(NN). \end{aligned} \tag{5.53}$$

Here we used the result that $k_c k^{c'} \to k^2\delta_c^{c'\perp}/3$ because of the phase space integration. (Recall $\delta_{\mu\mu'} = diag(1, -1, -1, -1)$ and $\delta_\mu^{\mu'}$ =diag(1,1,1,1)). Furthermore

$$\hat{\rho}_{\mu\nu}^{\mu'\nu'}(^1D_2) = \int d\Phi(NN)K_{\mu\nu}\ \text{Tr}\left[(\hat{k}_1 + m)\gamma_5(-\hat{k}_2 + m)\gamma_5\right]K^{\mu'\nu'} \tag{5.54}$$

$$= \frac{2}{15}s\left(\frac{s}{4} - m^2\right)\left(\delta_\mu^{\perp\mu'}\delta_\nu^{\perp\nu'} + \delta_\mu^{\perp\nu'}\delta_\nu^{\perp\mu'} - \frac{2}{3}\delta_{\mu\nu}^{\perp}\delta^{\perp\mu'\nu'}\right)\int d\Phi(NN).$$

Here the substitution $k_\mu k_\nu k^{\mu'}k^{\nu'} \to k^4(\delta_{\mu\nu}^{\perp}\delta^{\perp\mu'\nu'} + \delta_\mu^{\perp\mu'}\delta_\nu^{\perp\nu'} + \delta_\mu^{\perp\nu'}\delta_\nu^{\perp\mu'})/15$ is used. Let us define

$$\hat{\rho}_B^A(J^P) = \rho_{BA}(s)\,I_B^A,$$

where operators I_B^A for the states considered above are equal to

$$
\begin{array}{ll}
{}^1S_0 & I = 1 \\
{}^3P_1 & I_\mu^{\mu'} = \delta_\mu^{\perp\mu'} \\
{}^1D_2 & I_{\mu\nu}^{\mu'\nu'} = \frac{1}{2}\left(\delta_\mu^{\perp\mu'}\delta_\nu^{\perp\nu'} + \delta_\mu^{\perp\nu'}\delta_\nu^{\perp\mu'} - \frac{2}{3}\delta_{\mu\nu}^{\perp}\delta^{\perp\mu'\nu'}\right).
\end{array}
$$

The phase space factors $\rho_{BA}(s)$ determine the dispersion relation expression for the loop diagram:

$$
B_{AB}(s) = \int_{4m^2}^{\infty} \frac{ds'}{\pi} \frac{G^B(s')\rho_{BA}(s')G_A(s')}{s'-s}. \tag{5.55}
$$

These loop diagrams are used for the construction of amplitudes in the same way as for the spinless case.

5.4 Example: Description of the deuteron in the dispersion relation technique

Now we turn to the consideration of the two-nucleon composite particle, the deuteron. The deuteron is a mixture of two neutron-proton states, S- and D-wave ones, which correspond to two bilinear fermion forms

$$
\begin{array}{ll}
\left(\bar\Psi_{\bar n}(-k_2)\, S_\mu \Psi_p(k_1)\right) & S - wave \\
\left(\bar\Psi_{\bar n}(-k_2)\, D_\mu \Psi_p(k_1)\right) & D - wave.
\end{array} \tag{5.56}
$$

Here Ψ_p and $\Psi_{\bar n}$ are fields of the proton and antineutron and operators S_μ and D_μ are equal, up to normalization constants, to the operators Γ_μ and $K_{\mu\nu}\Gamma_\mu$ which were introduced in the previous Section:

$$
S_\mu = n_1(s)\Gamma_\mu = -\frac{1}{\sqrt{2s}}\left[\frac{2k_\mu}{2m+\sqrt{s}} - \gamma_\mu^\perp\right] \tag{5.57}
$$

$$
D_\mu = n_2(s)K_{\mu\nu}\Gamma_\nu = -\frac{2}{s^{3/2}}\left[(m+\sqrt{s})k_\mu + \left(\frac{s}{4}-m^2\right)\gamma_\mu^\perp\right]. \tag{5.58}
$$

Here the normalization constants, n_1 and n_2 are defined by the condition (below $L_\mu = S_\mu$ or D_μ):

$$
\int \frac{1}{2} d\Phi(P;k_1,k_2) Tr\left[L_\mu(\hat k_1 + m)L^{\mu'}(-\hat k_2 + m)\right] = \rho_L(s)\delta_\mu^{\perp\mu'}; \tag{5.59}
$$

S-wave and D-wave phase space factors are

$$
\begin{aligned}
\rho_S(s) &= \frac{1}{16\,\pi}\sqrt{\frac{s-4m^2}{s}}, \\
\rho_D(s) &= \frac{1}{16\,\pi}\left(\frac{s-4m^2}{s}\right)^{5/2}.
\end{aligned} \tag{5.60}
$$

The amplitude $pn \to pn$ is a 2×2 matrix with elements

$$
\begin{aligned}
&\left(\bar\Psi_{\bar n}(-p_2)L_\mu\Psi_p(p_1)\right) A_{LL'}^{(s)} \left(\bar\Psi_p(p_1')L'^\mu\Psi_{\bar n}(-p_2')\right) \\
&\quad \equiv \langle L_\mu\rangle A_{LL'}^{(s)} \langle L'^\mu\rangle^+.
\end{aligned} \tag{5.61}
$$

The interaction block $\hat N$ has the same structure

$$\hat{N} = \begin{vmatrix} \langle S_\mu \rangle N_{SS}(s) \langle S^\mu \rangle^+ & \langle S_\mu \rangle N_{SD}(s) \langle D^\mu \rangle^+ \\ \langle D_\mu \rangle N_{DS}(s) \langle S^\mu \rangle^+ & \langle D_\mu \rangle N_{DD}(s) \langle D^\mu \rangle^+ \end{vmatrix} . \tag{5.62}$$

According to general prescriptions for the discussed technique, see as lecture 3, we should replace the interaction blocks by the vertices:

$$\begin{aligned} N_{SS} &\to G_{SS}(s)G^{SS}(s'), \qquad N_{SD} \to G_{SD}(s)G^{SD}(s') \\ N_{DS} &\to G_{DS}(s)G^{DS}(s'), \qquad N_{DD} \to G_{DD}(s)G^{DD}(s'). \end{aligned} \tag{5.63}$$

Remember that for the energy-off-shell transitions $s \neq s'$. The requirement $A_{DS} = A_{SD}$ leads to

$$G_{SD}(s) = G^{DS}(s). \tag{5.64}$$

In the general case, the interaction blocks are represented as a sum of vertices

$$N_{LL'} \to \sum_n G^n_{LL'}(s)G^{LL'}_n(s). \tag{5.65}$$

Here we discuss in detail the simplest case with $n = 1$ which allows to understand the principal point of the consideration. But we conclude a paragraph sketching the considerations for the general case.

Now we write down the equation for the amplitudes $A_{LL'}$. Let us represent $A_{LL'}$ as a sum of two amplitudes with different states before the last interaction (see Fig.5.3)

$$A_{LL'} = \sum_{L''=S,D} a_{LL''L'}G^{L''L'}. \tag{5.66}$$

$$A_{SS} = a_{SSS}\,G^{SS} + a_{SDS}G^{DS}. \tag{5.67}$$

For amplitudes $a_{LL'L}$ with $L = S$ we have the following system of equations

$$\begin{aligned} a_{SSS} &= G_{SS} + a_{SSS}b_{SSS} + a_{SDS}b_{DSS} \\ a_{SSD} &= G_{SD} + a_{SSS}b_{SSD} + a_{SDS}b_{DSD} \\ a_{SDS} &= a_{SDD}b_{DDS} + a_{SSD}b_{SDD} \\ a_{SDD} &= a_{SDD}b_{DDD} + a_{SSD}b_{SDD}. \end{aligned} \tag{5.68}$$

It is convenient now to characterize the loop diagram by three indices, i, L, j, i and j related to the vertex functions and L to the intermediate state:

$$b_{iLj} = \int_{4m^2}^{\infty} \frac{ds'}{\pi} \frac{G^{iL}(s')\rho_L(s')G_{Lj}(s')}{s' - s}. \tag{5.69}$$

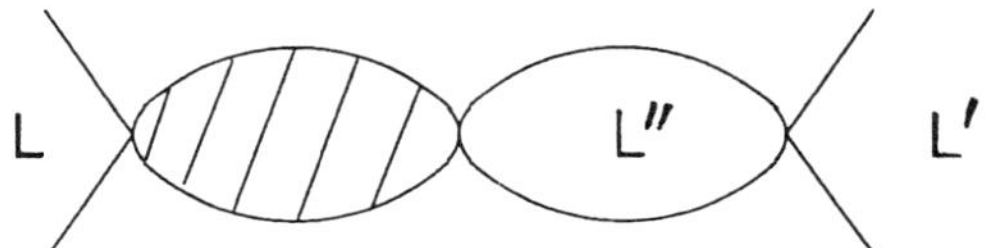

Fig. 5.3. Amplitude $A_{LL'}$ of Eq.(5.66).

Eq. (5.68) can be rewritten in a vector form

$$\begin{aligned}\vec{a}_s &= \vec{g}_s + \vec{a}_s\hat{B}\\ \vec{g}_s &= (G_{SS},\, G_{SD},\, 0,\, 0)\\ \vec{a}_s &= (a_{SSS}, a_{SSD}, a_{SDS}, a_{SDD}).\end{aligned} \tag{5.70}$$

Here $\hat{B}$ is the matrix

$$\hat{B} = \begin{vmatrix} b_{SSS} & b_{SSD} & 0 & 0\\ 0 & 0 & b_{SDS} & b_{SDD}\\ b_{DSS} & b_{DSD} & 0 & 0\\ 0 & 0 & b_{DDS} & b_{DDD}\end{vmatrix}. \tag{5.71}$$

The analogous equation can be obtained for $a_{LL'L}$ with $L = D$:

$$\begin{aligned}\vec{a}_d &= \vec{g}_d + \vec{a}_d\hat{B}\\ \vec{a}_d &= (a_{DSS},\, a_{DSD},\, a_{DDS},\, a_{DDD})\\ \vec{g}_d &= (0,\, 0\; G_{DS},\, G_{DD})\end{aligned} \tag{5.72}$$

Equations (5.70) and (5.72) can be written in a matrix form

$$\hat{a} = \hat{g} + \hat{a}\hat{B}. \tag{5.73}$$

Here

$$\begin{aligned}\hat{g} &= \begin{vmatrix}\vec{g}_s\\ \vec{g}_d\end{vmatrix} = \begin{vmatrix} G_{SS}, & G_{SD}, & 0, & 0\\ 0, & 0, & G_{DS}, & G_{DD}\end{vmatrix}\\ \hat{a} &= \begin{vmatrix}\vec{a}_s\\ \vec{a}_d\end{vmatrix} = \begin{vmatrix} a_{SSS}, & a_{SSD}, & a_{SDS}, & a_{SDD}\\ a_{DSS}, & a_{DSD}, & a_{DDS}, & a_{DDD}\end{vmatrix}.\end{aligned} \tag{5.74}$$

The solution of Eq. (5.73) is

$$\hat{a} = \hat{g}(1 - \hat{B})^{-1} \tag{5.75}$$

So, for the scattering amplitude we get the expression

$$A = \begin{vmatrix} A_{SS}(s) & A_{SD}(s)\\ A_{DS}(s) & A_{DD}(s)\end{vmatrix} = \hat{g}(1-\hat{B})^{-1}\hat{g}'. \tag{5.76}$$

Here

$$\hat{g}' = \begin{vmatrix} G^{SS} & 0\\ 0 & G^{SD}\\ G^{DS} & 0\\ 0 & G^{DD}\end{vmatrix}. \tag{5.77}$$

Of course, eq. (5.76) is a direct consequence of the unitarity condition if we suppose that the vertices $\hat{g}$ and $\hat{g}'$ have the form given by eqs. (5.74) and (5.77). Indeed, let us write the scattering amplitude in the form (5.76) and let the matrices $\hat{g}$, $\hat{g}'$ and $\hat{B}$ have no singularities both in the right and left-hand parts of the s-plane. The S-matrix is related to the scattering amplitude by

$$\hat{S} = I + 2i\sqrt{\hat{\rho}}\hat{A}\sqrt{\hat{\rho}}, \tag{5.78}$$

where the matrix $\sqrt{\hat{\rho}}$ refers to phase space volumes of S and D states, ρ_S and ρ_D:

$$\sqrt{\hat{\rho}} = \begin{bmatrix}\sqrt{\rho_S} & 0\\ 0 & \sqrt{\rho_D}\end{bmatrix}. \tag{5.79}$$

Then, from the unitarity condition for the $\hat{S}$ matrix ($\hat{S}^+\hat{S} = I$) one has for the amplitude:

$$(\hat{A} \; - \; \hat{A}^+) \; = \; 2i\hat{A}^+\hat{\rho}\hat{A}. \tag{5.80}$$

So, since the amplitude $\hat{A}$ is symmetric we can replace the operation of Hermitean conjugation by the complex one and rewrite Eq. (5.80) as:

$$\hat{g}\left((I-\hat{B})^{-1} - (I-\hat{B})^{*-1}\right)\hat{g}' = 2i\hat{g}(I-\hat{B}^*)^{-1}\hat{g}'\hat{\rho}\hat{g}(I-\hat{B})^{-1}\hat{g}'. \tag{5.81}$$

Evidently one can omit the matrices $\hat{g}'$ and $\hat{g}$ on the righthand and lefthand sides of those equation. Then, multiplying the expression by the matrix $(I - \hat{B})^*$ from the lefthand side and by the matrix $(I - \hat{B})$ from the righthand side one gets

$$\mathrm{Im}\ \hat{B} \; = \; \hat{g}'\hat{\rho}\hat{g}. \tag{5.82}$$

This relation gives us $b_{iLj}(s)$ in the form of Eq. (5.69), so we obtain the same result as in the diagram method.

Therefore, the diagram method is a modification of the conventional N/D method and can be obtained by means of the unitarity condition only. Moreover, this method can be easily represented by constituent rescattering diagrams which are very convenient for the study of the external fields interacting with the composite system.

The matrix elements of $\hat{A}$ contain the common factor $1/det|1-\hat{B}|$. It is convenient to separate it. For this end, let us introduce the matrix $\hat{\alpha}$:

$$\hat{\alpha} \; = \; \hat{A}\, det\,|1-\hat{B}|. \tag{5.83}$$

After having diagonalized the amplitude $\hat{A}$ (or $\hat{\alpha}$) one gets two eigenstates:

$$\begin{aligned} |d_\mu > \; &= \; \frac{\cos\Theta}{\sqrt{\rho_S}}\,|S_\mu > + \frac{\sin\Theta}{\sqrt{\rho_D}}\,|D_\mu > \\ |e_\mu > \; &= \; -\frac{\sin\Theta}{\sqrt{\rho_S}}\,|S_\mu > + \frac{\cos\Theta}{\sqrt{\rho_D}}\,|D_\mu > . \end{aligned} \tag{5.84}$$

The mixing angle Θ is determined by the condition $\langle d_\mu|A|e_\mu\rangle = 0$. It gives

$$(\cos^2\Theta - \sin^2\Theta)\frac{A_{DS}}{\sqrt{\rho_S\rho_D}} + \sin\Theta\cos\Theta\left(\frac{A_{DD}}{\rho_D} - \frac{A_{SS}}{\rho_S}\right) = 0. \tag{5.85}$$

The diagonal matrix elements are equal to

$$\begin{aligned} \langle d_\mu|\hat{A}|d^{\mu'}\rangle \; &= \; \frac{1}{det|1-\hat{B}|} \times \\ &\times \left(\frac{\cos^2\Theta}{\rho_S}\alpha_{SS} + \frac{2\cos\Theta\sin\Theta}{\sqrt{\rho_S\rho_D}}\alpha_{DS} + \frac{\sin^2\Theta}{\rho_D}A_{DD}\right)\delta_\mu^{\mu'} \\ \langle e_\mu|\hat{A}|e^{\mu'}\rangle \; &= \; \frac{1}{det|1-\hat{B}|} \times \\ &\times \left(\frac{\sin^2\Theta}{\rho_S}\alpha_{SS} - \frac{2\cos\Theta\sin\Theta}{\sqrt{\rho_S\rho_D}}\alpha_{DS} + \frac{\cos^2\Theta}{\rho_D}\alpha_{DD}\right)\delta_\mu^{\mu'}. \end{aligned} \tag{5.86}$$

The amplitude $\langle d_\mu|\hat{A}|d^{\mu'}\rangle$ has a pole at $s = M^2$, which corresponds to the deuteron bound state. This pole is related to the zero value of $\det|1-\hat{B}|$:

$$det|1-\hat{B}|_{s=M^2} \; = \; 0. \tag{5.87}$$

This pole is absent in the amplitude $\langle e_\mu|\hat{A}|e^{\mu'}\rangle$, which means that the numerator of this amplitude should be equal to zero at the point $s = M^2$:

$$\langle e_\mu \mid \hat{\alpha} \mid e^{\mu'} \rangle_{s=M^2} = 0. \tag{5.88}$$

It is possible to construct the deuteron vertices either
(I) in a direct way using the graph technique, or
(II) resolving Eq. (5.75) for the amplitude $\hat{a}$ in the limit $s \to M^2$.

I. Direct construction of deuteron vertices

The triangle block in the diagram Fig.5.4 gives us the deuteron form factors. The triangle block contains both transitions with orbital angular momentum conservation $L = L'$ and transitions with $L \neq L'$. To extract the deuteron form factors from the diagram of Fig.4 one should

(i) sum up the amplitudes of Fig.5.4 over all right and left loop diagrams, and

(ii) let the nucleon energies in the initial and the final states tend to the deuteron mass.

As a result, we have the double pole diagram of Fig.5.5 with deuteron form factors as the residue.

The diagram of Fig.5.4 with $L = L'$ can be obtained from the diagram of Fig.5.3, if one replaces the loop b_{iLj} with the triangle block Γ_{iLj}. The removal of the loop b_{iLj} is equivalent to differentiation of the scattering amplitude Fig.5.3 with respect to b_{iLj}. So the amplitude of Fig.5.4-type diagram with $L = L'$ is equal to

$$\begin{aligned} \langle d_\mu|\hat{T}|d^{\mu'}\rangle_{L=L'} &= -\langle d_\mu|\hat{\alpha}|d^{\mu'}\rangle(det|1-\hat{B}|)^{-2} \\ &\times \sum_{iLj}\Gamma_{iLLj}\frac{\partial}{\partial b_{iLj}}(det|1-\hat{B}|) + \ldots \end{aligned} \tag{5.89}$$

Here we write down only the terms contributing to the double pole as $s \to M^2$.
The amplitude $\langle d_\mu|\hat{\alpha}|d^{\mu'}\rangle$ in Eq.(5.89) is taken at $s = M^2$.
The determinant $\det|1-\hat{B}|$ is equal to

$$\begin{aligned} \det|1-\hat{B}| &= (1-b_{SSS})(1-b_{DDD})(1-b_{DSD}b_{SDS}) \\ &- (1-b_{SSS})b_{SDD}b_{DDS}b_{DSD} - (1-b_{DDD})b_{DSS}b_{SSD}b_{SDS} \\ &- b_{SSD}b_{DSS}b_{SDD}b_{DDS} \end{aligned} \tag{5.90}$$

so we have 8 terms in the r.h.s. of Eq. (5.89).

The same procedure can be used for the calculation of the diagrams of Fig.5.4-type with $L \neq L'$. However, in this case we should use differentiation with respect to $(b_{iLL'}b_{LL'j})$, see Fig.5.3. Eight combinations are possible

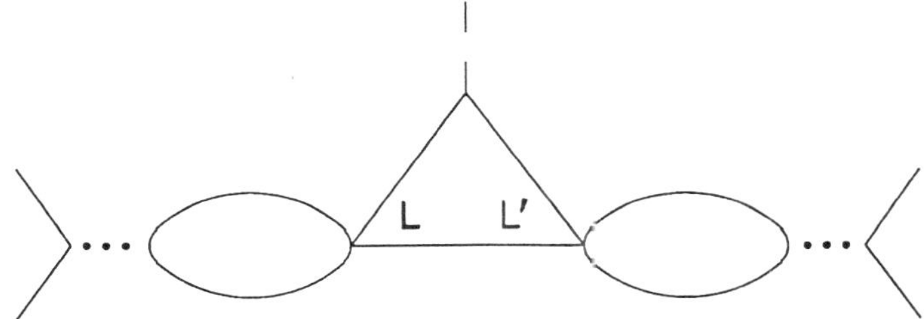

Fig. 5.4. Triangle graph for the amplitude of Eqs. (5.89) and (5.92).

$$(b_{DSD}b_{SDS}),\ (b_{SDS}b_{DSD}),\ (b_{DDS}b_{DSD}),\ (b_{DSD}b_{SDD})$$

$$(b_{SSD}b_{SDS}),\ (b_{SDS}b_{DSS}),\ (b_{SSD}b_{SDD}),\ (b_{DDS}b_{DSS}) \qquad (5.91)$$

So the corresponding amplitude is equal to

$$\begin{aligned} \langle d_\mu|\hat{T}|d^{\mu'}\rangle_{L\neq L'} &= -\langle d_\mu|\hat{\alpha}|d^{\mu'}\rangle(\ \det|1-\hat{B}|)^{-2} \\ &\times \sum_{iLL'j}\Gamma_{iLL'j}\frac{\partial(\ \det|1-\hat{B}|)}{\partial(b_{iLL'}b_{LL'j})} + \dots . \end{aligned} \qquad (5.92)$$

Eqs.(5.89) and (5.92) determine the weight of triangle diagrams $\Gamma_{iLL'j}$ in the form factor of the deuteron.

The double discontinuity of the triangle diagram $\Gamma_{iLL'j}$ is equal to

$$\text{disc}_{\tilde{s}}\ \text{disc}_{\tilde{s}'}\Gamma^{\mu}_{iLL'j}(\tilde{s},\tilde{s}') = \sum_{spin\ states}\int d\Phi(\tilde{P},\tilde{P}';k_1,k_1',k_2)$$

$$\times\ G^{iL}(\tilde{s})\langle L_\tau\rangle F^{\mu}\langle L'_\eta\rangle G_{L'j}(\tilde{s}')\xi^{\tau *}\xi^{\eta}. \qquad (5.93)$$

Here L_τ and L'_η are S- and D-wave operators, F^μ is the operator for the interaction of the electromagnetic field with the nucleon, and ξ^τ, ξ^η are deuteron polarizations. The phase space integration $d\Phi$ is the same as for the case of spinless particles with $\tilde{P}^2=\tilde{s}$, $\tilde{P}'^2=\tilde{s}'$ and $(\tilde{P}'-\tilde{P})^2=\tilde{q}^2=q^2$. The summation over spin states gives

$$disc_{\tilde{s}}disc_{\tilde{s}'}\Gamma^{\mu}_{iLL'j} = \int d\Phi(\tilde{P},\tilde{P}';k_1,k_1',k_2)G^{iL}(\tilde{s})\ \times$$

$$Tr\left[L_\tau(-\hat{k}_2+m)L'_\eta(\hat{k}'_1+m)F^\mu(\hat{k}_1+m)\right]G_{L'j}(\tilde{s}')\xi^{\tau *}\xi^{\eta}. \qquad (5.94)$$

Photon-nucleon interaction can be represented in the form:

$$F^\mu\ =\ \gamma^\mu f_1(q^2)\ +\ (k_1+k_1')^\mu f_2(q^2), \qquad (5.95)$$

where the functions f_1 and f_2 are expressed in terms of the conventional electric and magnetic nucleon form factors

$$\begin{aligned} f_1 &= f_M \\ f_2 &= 2m(f_E+f_M)/\,(4m^2-q^2). \end{aligned} \qquad (5.96)$$

The deuteron form factor discontinuity is a superposition of these triangle blocks $\Gamma_{iLL'j}$

$$\text{disc}_{\tilde{s}}\ \text{disc}_{\tilde{s}'}\ F^\mu(\tilde{s},\tilde{s}',q^2) = \sum_{iLL'j}\delta_{iLL'j}\ \text{disc}_{\tilde{s}}\ \text{disc}_{\tilde{s}'}\ \Gamma^{\mu}_{iLL'j}. \qquad (5.97)$$

The coefficients $\delta_{iLL'j}$ can be found from Eqs.(5.89) and (5.92). Direct calculation allows us to extract deuteron vertices, and Eq.(5.97) can be rewritten in the form

$$\text{disc}_{\tilde{s}}\ \text{disc}_{\tilde{s}'}\ F^\mu(\tilde{s},\tilde{s}',q^2) = \int d\Phi(\tilde{P},\tilde{P}';k_1,k_2,k_1')$$

$$\times\sum_{iL}\beta_{iL}G^{iL}(\tilde{s})\ \text{Tr}\left[L_\tau(-\hat{k}_2+m)L'_\eta(\hat{k}_1+m)F^\mu(\hat{k}_1+m)\right]$$

$$\times\qquad \sum_{L'j}G_{L'j}(\tilde{s'})\beta_{L'j}\xi^{\tau *}\xi^{\eta}. \qquad (5.98)$$

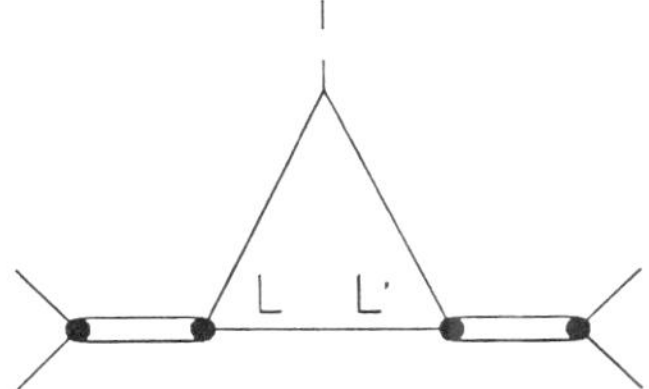

Fig. 5.5. The double pole deuteron triangle graph.

It should be mentioned that we have obtained vertex functions $G_{L'j}(\tilde{s}')\beta_{L'j}$ corresponding to the initial and final deuteron states; β_{Lj} are some new coefficients, which are determined below.

Now let us turn to the spin structure of the deuteron form factor

$$\begin{aligned} F_\mu(q^2) &= \Big[(P+P')_\mu \delta_{\tau\eta} F_1(q^2) + (q_\eta \delta_{\mu\tau} - q_\tau \delta_{\mu\eta}) F_2(q^2) \\ &- (P+P')_\mu \, q_\tau q_\eta \, \frac{F_3(q^2)}{2M^2}\Big] . \end{aligned} \tag{5.99}$$

Form factors F_1, F_2 and F_3 are related to the conventional electric, magnetic and quadrupole form factors by

$$\begin{aligned} F_1 &= F_E + \frac{q^2}{6M^2} F_Q \\ F_2 &= F_M \\ F_3\left(1-\frac{q^2}{4M^2}\right) &= F_M - F_E + \left(1-\frac{q^2}{6M^2}\right) F_Q . \end{aligned} \tag{5.100}$$

Let the r.h.s. of Eq. (5.98) be written in similar form:

$$\text{disc}_{\tilde{s}}\ \text{disc}_{\tilde{s}'}\ F^\mu(\tilde{s}, \tilde{s}', q^2) = \xi^{\tau *} \xi^\eta$$

$$\begin{aligned} \times & \Big[\left(\tilde{P}_\mu + \tilde{P}'_\mu - \frac{\tilde{s}'-\tilde{s}}{q^2}\tilde{q}_\mu\right)\delta_{\eta\tau} g_1(\tilde{s}, \tilde{s}', q^2) \\ + & (\tilde{q}_\eta \delta_{\mu\tau} - \tilde{q}_\tau \delta_{\mu\eta}) g_2(\tilde{s}, \tilde{s}', q^2) \\ - & \left(\tilde{P}_\mu + \tilde{P}'_\mu - \frac{\tilde{s}'-\tilde{s}}{q^2}\tilde{q}_\mu\right) \tilde{q}_\eta \tilde{q}_\tau \frac{g_3(\tilde{s}, \tilde{s}', q^2)}{2M^2}\Big] . \end{aligned} \tag{5.101}$$

The functions g_i are calculated using the explicit form of Eq. (5.98). These functions are the deuteron form factors F_i:

$$F_i(q^2) = \int_{4m^2}^{\infty} \frac{d\tilde{s} d\tilde{s}'}{\pi^2} \frac{g_i(\tilde{s}, \tilde{s}', q^2)}{(\tilde{s}-M^2)(\tilde{s}'-M^2)} . \tag{5.102}$$

Form factors F_i satisfy automatically the requirement $F_E(0) = 1$.

II. Construction the deuteron vertices resolving Eq.(5.75) in the limit $s \to M^2$

Now let us turn to another way of finding coefficients β_{iL} of Eq.(5.98) directly from Eq. (5.75). In the limit $s \to M^2$ we have $det|1-\hat{B}| \to 0$; this means that $\hat{a}$ is large and it is

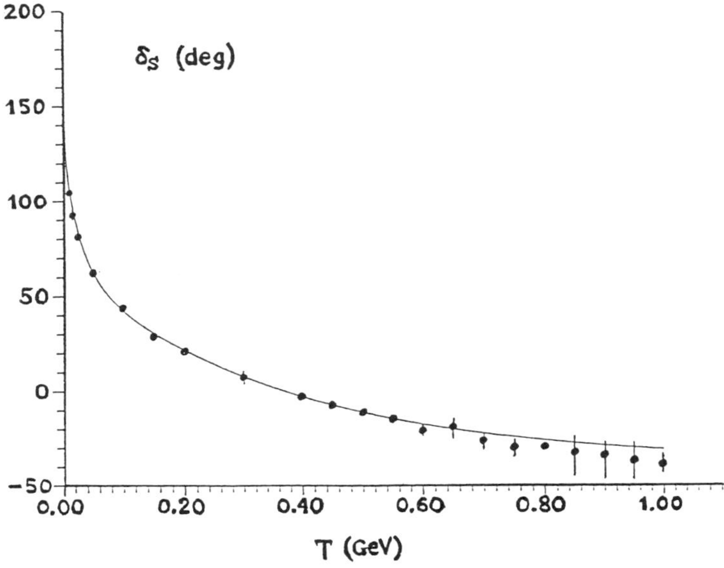

Fig. 5.6. Fit of the ${}^3S_1$ phase shift δ_S by the vertex functions.

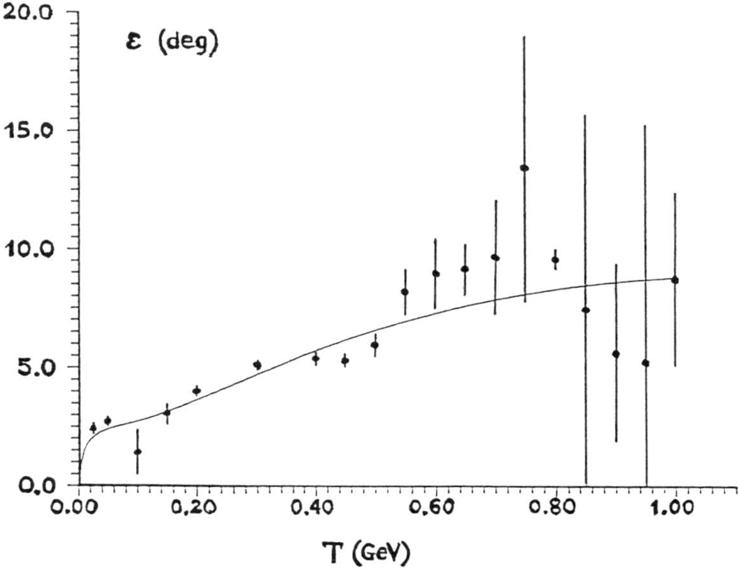

Fig. 5.7. Fit of the mixing parameter ϵ by the vertex functions.

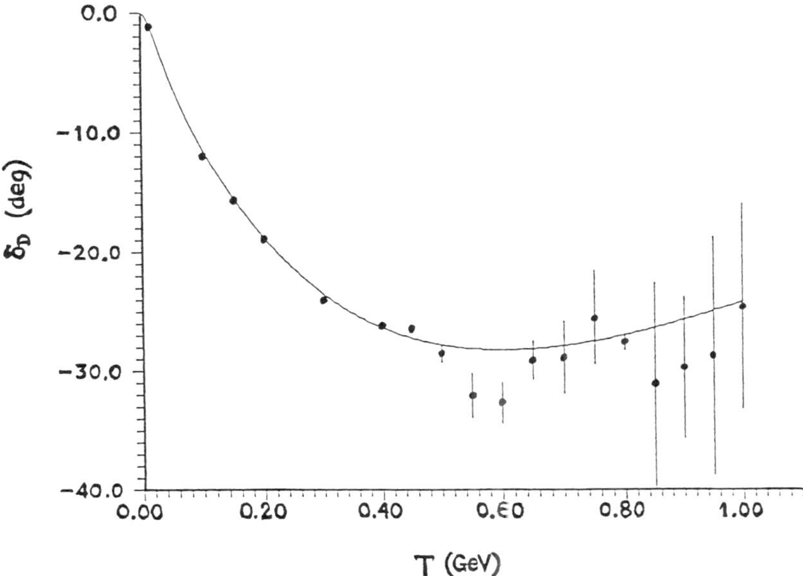

Fig. 5.8. Fit of the $^3D_1$ phase shift δ_D by the vertex functions.

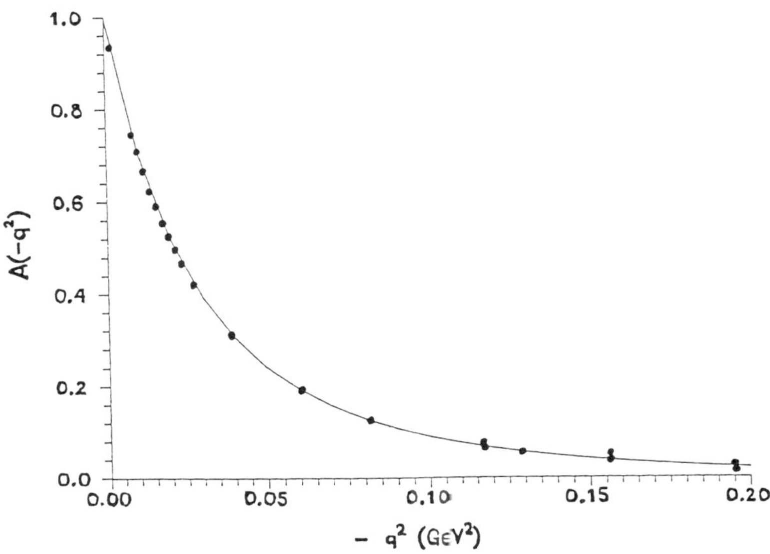

Fig. 5.9. Comparison of the form factor $A(-q^2)$ in the region $-q^2 = (0. - 0.2)$ GeV^2/c^2 with experimental data.

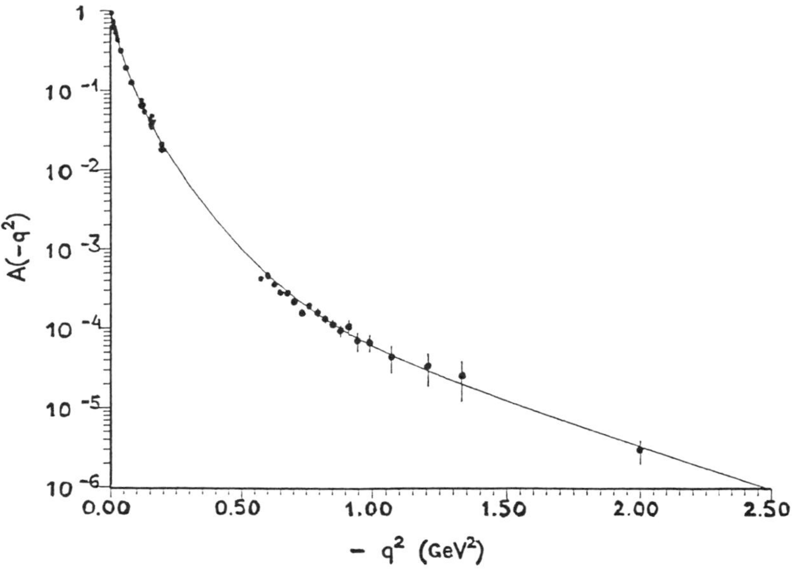

Fig. 5.10. Comparison of the form factor $A(-q^2)$ in the region $-q^2 = (0. - 2.0)$ $\mathrm{GeV}^2/\mathrm{c}^2$ with experimental data.

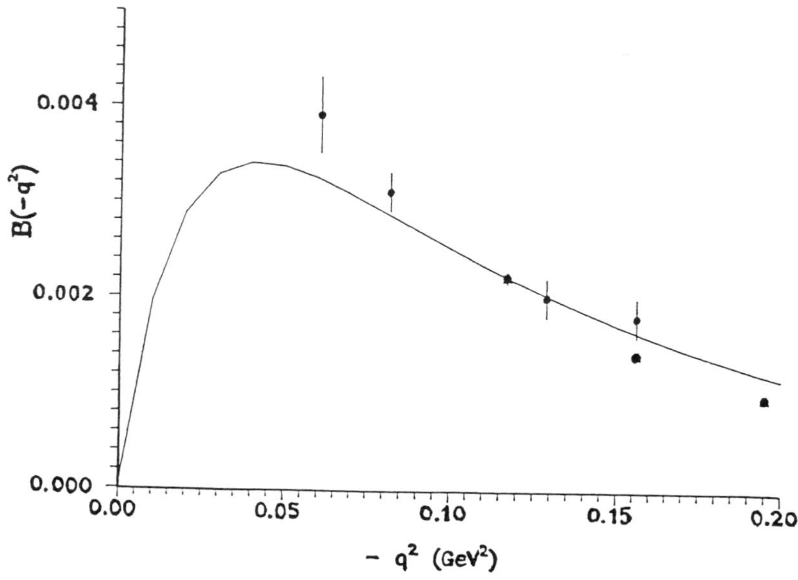

Fig. 5.11. Comparison of the form factor $B(-q^2)$ in the region $-q^2 = (0. - 0.2)$ $\mathrm{GeV}^2/\mathrm{c}^2$ with experimental data.

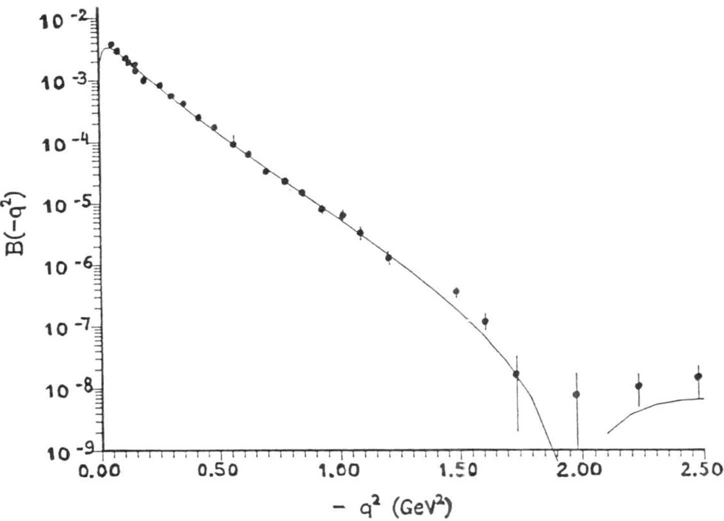

Fig. 5.12. Comparison of the form factor $B(-q^2)$ in the region $-q^2 = (0. \ - \ 2.0)$ GeV^2/c^2 with experimental data.

possible to neglect $\hat{g}$ in Eq. (5.75). At $s = M^2$ we have

$$\hat{a}(1 - \hat{B}) \; = \; 0 \tag{5.103}$$

This equation breaks into two identical ones for $\vec{a}_s$ and $\vec{a}_d$ (Eqs. (5.70),(5.72)); we denote the solutions of these equations as $\vec{a}_s^{(0)}$ and $\vec{a}_d^{(0)}$. As follows from Eq. (5.98), the coefficients β_{Lj}'s are determined by the projections of $\vec{a}_L$:

$$\begin{aligned} \vec{a}_s^{(0)} &= \text{Const } \Sigma_L \beta_{sL} \, \hat{L}_\mu \, \xi^\mu, \\ \vec{a}_d^{(0)} &= \text{Const } \Sigma_L \beta_{dL} \, \hat{L}_\mu \, \xi^\mu. \end{aligned}$$

Hence,

$$\begin{aligned} \beta_{SS} &= C \, a_{SSS}, \quad \beta_{SD} = C \, a_{SSD}, \\ \beta_{DS} &= C \, a_{SDS}, \quad \beta_{DD} = C \, a_{SDD}. \end{aligned} \tag{5.104}$$

The normalization constant C is determined from the requirement $F_E(0) = 1$.

This method was applied to the description of the electromagnetic properties of the deuteron. The S- and D-wave vertices were determined by the np scattering data in the energy region $T \leq 1.0$ GeV. (T is nucleon kinetic energy), or $\sqrt{s} \leq 2.35$ GeV. Description of the experimental data with reconstructed vertices are shown in Figs. 5.6, 5.7 and 5.8. In this energy region, inelastic processes are suppressed and because of this we do not include mesonic degrees of freedom explicity. The calculated values of the binding energy, the magnetic moment, and the quadrupole agree with data:

	Calculations	Experiment
μ_D	1.709 μ_B	1.715 μ_B
Q_D	25.0 e/fm^2	25.5 e/fm^2
ϵ	2.222 MeV	2.222 MeV
D-wave probability	5%	

The reconstruction of the deuteron vertices by fitting of the experimental data (Figs. 5.6–5.8) allows us to calculate the measured form factors $A(-q^2)$ and $B(-q^2)$:

$$\begin{aligned} A(-q^2) &= F_e^2(-q^2) + 2\left[\frac{q^2}{6M^2}\right]^2 F_Q^2(-q^2) - \frac{q^2}{6M^2} F_m^2(-q^2) \\ B(-q^2) &= -\frac{q^2}{6M^2}\left[1 - \frac{q^2}{4M^2}\right] F_m(-q^2). \end{aligned} \tag{5.105}$$

The description of the data is demonstrated by Figs. 5.9–5.12.

The dispersion relation technique takes into account relativistic effects which are important at large $-q^2$. It seems that the relativistic description of nucleons and the fulfilment of the gauge invariance requirement are the reasons for our successful treatment of the deuteron as a composite system of two nucleons only. Certainly, the two-nucleon description of the deuteron is invalid at small distances, or at large q^2. Indeed, we see that the calculated values of the form factors start to deviate from data in the region of $-q^2 \sim 2.0$ GeV^2/c^2. The region of application of perturbative QCD is restricted by values of the order of $-q^2 \geq 5$ GeV^2/c^2, so there is some room for non-nucleonic degrees of freedom such as $\Delta\Delta$-, $NN^*(1400)$-states or states containing mesons.

Acknowledgement. I thank D.V.Bugg, L.G.Dakhno and V.A.Nikonov for help, useful discussions and comments.

References

[1] R.P. Feynman, "Quantum Electrodynamics", W.A.Benjamin, New York, 1961.

[2] L.D. Landau and E.M. Lifshitz, "Quantum Mechanics", State Publishing House for Physics and Mathematics, Moscow, 1963.

[3] L.B. Okun, "Weak Interactions", State Publishing House for Physics and Mathematics, Moscow, 1963.

[4] G.F. Chew, "The Analytic S-Matrix", W.A. Benjamin, New York, 1966.

[5] V.N. Gribov, "Quantum Electrodynamics" (in russian), Leningrad Nuclear Physics Institute, Leningrad, 1974.

[6] V.V. Anisovich, M.N. Kobrinsky, J. Nyiri and Yu.M. Shabelski, "Quark Model and High Energy Collisions", World Scientific, Singapore, 1985.

[7] A.V. Anisovich and A.V. Sarantsev, Sov. J. Nucl. Phys. **55**, 1200, 1992.

[8] A.V. Anisovich and V.A. Sadovnikova, Sov. J. Nucl. Phys. **55**, 1483, 1992.

[9] V.V. Anisovich, D.V. Bugg and A.V. Sarantsev, Nucl. Phys. **A537**, 501, 1992.

[10] V.V. Anisovich, M.N. Kobrinsky, D.I. Melikhov and A.V. Sarantsev, Nucl. Phys. **A544**, 747, 1992.

[11] V.V. Anisovich, D.I. Melikhov, B.Ch. Metsch and H.R. Petry, Nucl. Phys. **A563**, 549, 1993.

[12] V.V.Anisovich, A.A.Kondashov, Yu.D.Prokoshkin, et al. Phys. Lett., in press.

GAMS EXPERIMENTS AND RESULTS

Madjid Boutemeur

Montréal University, Montréal,
H3C 3J7, Quebec, Canada

1. INTRODUCTION

GAMS, which is the acronym of an electromagnetic calorimeter, designates also the collaboration * which has in recent years focussed attention on the detection of neutral decays of mesons to understand more thoroughly the light meson spectrum and to detect possible non-$q\bar{q}$ (exotic) mesons.

The production of neutral mesons is studied both in π^- charge exchange reactions and in the central region of high energy hadron collisions. The study of reactions with neutral final states that decay eventually to photons has been of much interest with the emergence of QCD as a theory of strong interactions and has triggered the search for exotic states.

At the Institute for High Energy Physics (IHEP, Russia), systematic studies of detectors for high energy photons ($E_\gamma > 100$ MeV), begun in 1973, have resulted in the conception of the lead-loaded glass multi-cell calorimeters GAMS-4000 and GAMS-2000. GAMS-type calorimeters allow the simultaneous measurement of the impact coordinates and energy of a large number of photons and the reconstruction, with high accuracy, of the mass and momentum of decaying particles.

In this report, we will first give a general survey of the experimental setups and their underlying principles with a focus on photon detection. We will then describe some partial wave analysis methods as used by GAMS to analyse the data. Finally, some of the physics results will be discussed.

1.1: The GAMS physics program

The GAMS program is based on the following two exclusive reactions

* IHEP(Protvino, USSR), KEK(Tsukuba, Japan), LANL(Los Alamos, USA), LAPP (Annecy, France), INFN(Pisa,Italy) and IISN(Brussels, Belgium)

Hadron Spectroscopy and the Confinement Problem
Edited by D.V. Bugg, Plenum Press, New York, 1996

a) charge exchange reaction:

$$\pi^- p \longrightarrow M^0 n \ , \text{ with } M^0 \longrightarrow k\gamma \tag{1.1}$$

b) central production :

$$hp \longrightarrow M^0 \text{ h p} \ , \text{ with } M^0 \longrightarrow k\gamma, \tag{1.2}$$

where M^0 is a neutral state decaying to k photons. The first reaction is measured with 38 GeV/c pions at IHEP and 100 GeV/c (also 230 GeV/c) pions at CERN. The second reaction is measured at CERN with 300 GeV/c pions and 450 GeV/c protons incident on nucleons of an active scintillator target.

The M^0 states that decay into some combination of π^0, η, η', or ω mesons, like $2\pi^0$, $4\pi^0$, $\pi^0\eta$, $\pi^0\omega$, $\eta\eta$, $\eta\eta'$, $\eta\omega$, $\omega\omega$, have been studied. The fact that the ρ^0 meson has no open pure neutral decay channel turns out to be an advantage since it removes an overwhelming background coming from the vector sector.

In the quark-model framework, mesons are $q\bar{q}$ bound states with a total spin S equal to 0 or 1 and with a relative angular momentum L. The parity P and the charge conjugation C of the $q\bar{q}$ system obey the relations : $P = (-1)^{L+1}$ and $C = (-1)^{L+S}$. Thus, certain J^{PC} like $0^{--}, 1^{-+}, 2^{+-}, \ldots$ are not allowed by the quark-model and they are called exotic quantum numbers. Most of the channels studied by GAMS are potential sources of exotic states, since final states with possible exotic quantum numbers like $\pi^0\pi^0(J^{PC} = 0^{++},\ 2^{++})$ and $\pi^0\eta(J^{PC} = 1^{-+})$, are copiously produced. However, most of the possible non-$q\bar{q}$ states have quantum numbers the same as $q\bar{q}$ mesons, so they can mix with these. Glueballs are flavourless. They ought to have zero isospin and strangeness, but apart from this, they have no clear-cut signature. Therefore their indentification is difficult and requires a lot of systematic work. Isovector non $q\bar{q}$ mesons, namely hybrids ($q\bar{q}g$) and four-quark states ($q\bar{q}q\bar{q}$) can be identified in the exotic quantum number sector.

Exclusive measurements offer the advantage that kinematic constraints can be applied to each event. This greatly reduces the errors on measured quantities such as the M^0 invariant mass. Furthermore, the M^0 allowed angular momentum states are often restricted, in particular when the t-distribution of the produced M^0 is limited to small values (OPE), which improves one's ability to make spin assignments to the observed states.

The total cross section for π^- p (h p) interactions is nearly constant in the energy domain from about 20 GeV to 450 GeV (24mb (40mb)). The sum of the measured differential cross sections [ref. 1.1] for $\pi^- p \longrightarrow M^0 n$, with $M^0 = \pi^0$, η, ω , which are by far the most copiously produced neutral states, accounts for less than 10^{-4} of the total cross section. This implies that less than one neutral state is produced in 10^4 pion interactions. Moreover, we are interested in reactions with cross sections more than an order of magnitude smaller. With a total cross section of 24mb (40mb), only 6% (10%) of the incident pi^- (h) beam interacts in the 60 cm long liquid-hydrogen target of the CERN GAMS-4000 (NA12) experiment. To get reasonable rates of interesting events, one needs very high intensity hadron beams. In the CERN NA12 experiment, the π^- beam intensity was $2.5 \times 10^7 \pi^-/s$. One of the main problems then is to get a good rejection power against the majority of charged particles.

2. The Experimental setups

2.1 The GAMS electromagnetic calorimeters

Two large GAMS-type calorimeters have been built: GAMS-2000 and GAMS--4000. The lead-glass cells in both detectors have the same dimensions: $38 \times 38 \times 450$ mm^3. The former has 48 x 48 -1 = 2303 cells, one cell being left out in the centre to provide a hole for the passage of the unscattered incident particles. GAMS-4000 is a matrix of 64 x 64 - 4 =4092 cells, the hole consisting of 2x2 cells in this case. GAMS-2000 was installed on the 4B beam line at the Institute for High Energy Physics in Protvino where it has been used by the members of the GAMS collaboration in the framework of the 6th Joint CERN-IHEP Experiment. GAMS-4000, also referred to as the NA12 experiment, was located at CERN in the North area on the H8 beam line.

Table 2.1 Main parameters of GAMS-2000 and GAMS-4000

	GAMS-2C00	GAMS-4000
Radiation length X_0(cm)	2.6	2.9
Lead-glass cell dimensions (mm^3)	$38 \times 38 \times 450$	$38 \times 38 \times 450$
Total number of cells	$48 \times 32 - 1 = 1535$	$64 \times 64 - 4 = 4092$
Useful area (m^2)	2.2	5.0
- Coordinate accuracy (σ_x(mm))		
at 25 GeV/c	1.3	2.0
at 200 GeV/c	-	1.0
- σ_E/E for single γ(%)		
at 25 GeV/c	1.8	2.2
at 200 GeV/c	-	1.5
Maximum number of simultaneous γ	10	20

Since the two experimental setups are similar, we will be decribing only the GAMS-4000 (NA12) apparatus. A view of the GAMS-4000 electromagnetic calorimeter is shown in figure 1. It consists of a light-tight box containing the lead-glass matrix, supported in a mainframe by two screws that allow to move the whole box up and down. The mainframe itself can be moved in the transverse direction relative to the beam line. Both movements are used to bring each individual cell into the beam line for calibration purposes.

Detailed descriptions of the GAMS detectors, their photomultiplier gain tuning systems and data acquisition systems are given in [ref. 2.1], [ref. 2.2] and [ref. 2.3]. A summary of the main characteristics of GAMS-2000 and GAMS-4000 calorimeters are given in table 2.1.

2.2 The NA12 experimental layout

A scheme of the NA12 experimental layout used to study the charge exchange reaction is shown on figure 2. The 100 GeV π^- were produced by striking a beryllium target with 450 GeV protons of the CERN SPS machine. They were transported to the north area (NA) through the H8 beam line. The main contaminations to the π^-

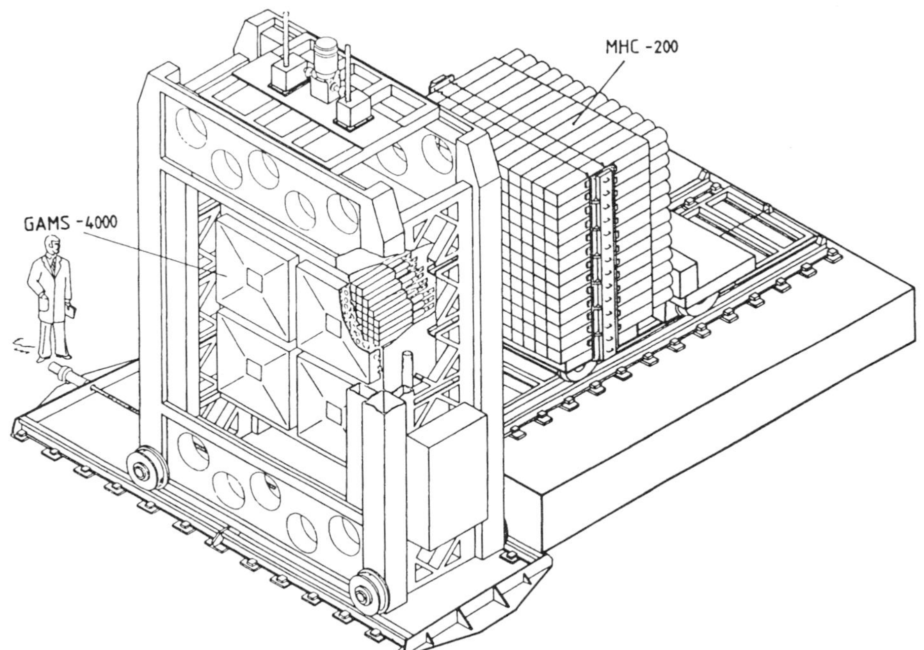

Figure 1.

beam were from K^- (6%), electrons (3%) and antiprotons (2%). In high transmission mode, $\Delta P/P = \pm 3\%$ in the H8 beam. The SPS machine cycle is 12 seconds with a 1.5 second flat top. The maximum acceptable beam intensity in the NA12 setup was 2.5×10^7 particles/second.

The triggerering logic is quite simple since it involves rejecting events whenever and wherever a charged particle is detected in the final state. Coincident pulses from the beam-telescope counters S_i made of thin scintillator and placed upstream of the target, signal the arrival time of a particle on the 60 cm long liquid hydrogen target. The target is surrounded with a guard system made of two parts : 1) thin scintillator counters (AC, AH_i, GS-scintillators) that veto events with charged particles emitted in the reaction and also the unscattered incident beam particles (94%); 2) a cylindrical lead-glass barrel placed right after the thin scintillators. It consists of eight 10 cm thick crowns of twelve cells with a truncated prism shape and by lead-scintillator sandwiches (SW_i, Albedo).

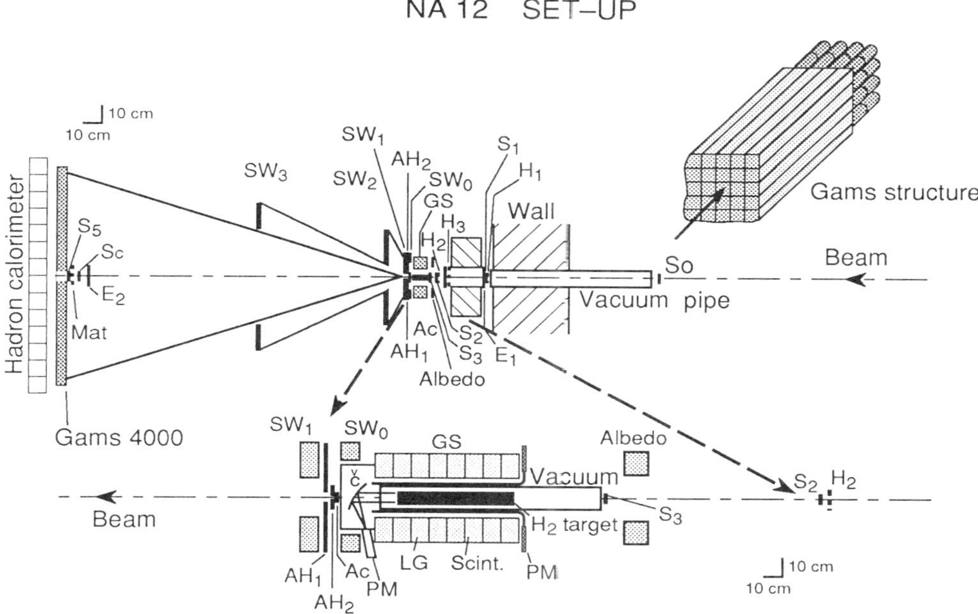

Figure 2.

The purpose of the lead-glass barrel is primarily to veto (offline) events with a recoil N^* or $\Delta \longrightarrow n\pi \longrightarrow 2\gamma n$ while the role of the sandwiches is to supress events with photons emitted outside the geometrical acceptance of GAMS-4000.

The most critical part of the trigger is the veto function in the very forward direction. Therefore, several counters share this function : 1) the AC counter, 6 cm in diameter, made of fast scintillator (NE103), heavily loaded by the unscattered incident particles and the scatterred ones up to an angle of 4.5 degrees; 2) the AH_1 counter, with a 19 cm external diameter and a 5 cm hole in the centre covering particles scattered up to 11 degrees; 3) the AH_2 counter, made of four scintillator sheets, covers scattering angles up to 33 degrees (larger angles are covered by the 12 GS-scintillator counters). All these counters are equipped with fast amplifiers (100 MHZ) working at a rather low High Voltage, this being made possible due to the use of low level (2 mV) discriminators [ref. 2.4].

2.2.1 The beam position hodoscopes

A system of three hodoscopes H_0, H_1, H_2, placed upstream of the liquid hydrogen target, allows the localisation and the measurement of the size and the angular divergence of the incident π^- beam. Each hodoscope is made of two detections planes X and Y. A plane consists of 15 (70×3 mm^3) NE110-type scintillators placed such that there is a 1 mm overlap between adjacent scintillators. The transverse coordinates (X,Y) of the beam particle are then obtained by playing with the coincidences of the photomultiplier signals coupled to the fired scintillators.

2.3 The liquid hydrogen target

The liquid hydrogen target is contained in a 60 cm long and 6 cm diameter cylinder maintained at a temperature of 20.6 K. The ends of the cylinder are closed with a thin 175μm layer of transparent mylar.

A 100 GeV π^- crossing the target parallel to the beam axis emits Cherenkov light at an angle of 26 degrees relative to the π^- direction. The light collection system is made of : 1) a thin 100 μm aluminised layer covering the inner surface of the cylinder; 2) a conical mirror placed upstream of the target directing the produced light to an inclined concave mirror (250 μm aluminium). The latter focusses the light towards a photomultiplier XP2262 located at its focal point.

The Cherenkov light yield depends on the distance covered in the target by the beam particle before the interaction. It is also affected by the light reflection and absorption in the target. The light intensity spectrum for a 100 GeV π^- traversing the entire target without interaction has a Gaussian shape with 30% resolution.

2.3.1 Longitudinal coordinate of the interaction point

The distribution of the Cherenkov light intensity N(I) produced by a π^- that has interacted in the target is parametrised as follows :

$$N(I) = N_0 e^{-0.0334\ I}. \tag{2.1}$$

The Z coordinate of the interaction point in the target can be related to the produced Cherenkov light intensity by requiring it to have a reconstructed uniform distribution. The obtained relation is as follows :

$$Z = L_t \frac{1 - e^{-0.0334\ I}}{1 - e^{-0.0334\ I_{max}}} \tag{2.2}$$

where L_t = 60 cm is the total target length and I_{max} is the average light intensity produced by the traversing pions. The rather poor resolution ±5 cm on Z is sufficient given that the electromagnetic GAMS calorimeter is placed at 15 m (or 20 m) away from the target.

2.4 The electromagnetic GAMS detector

In this section, after a brief reminder of the main processes involoved in the electromagnetic shower developement [ref. 2.5] [ref. 2.6] [ref. 2.7], we will discuss the choice of the GAMS cell dimensions, the cell response, the shower profile and the reconstruction of the photon impact points.

2.4.1 Electromagnetic showers

At high energies the combined effect of photon radiation and pair production gives rise to a spectacular effect : a cascade shower. An initial electron radiates photons, which convert to pairs, which radiate and produce other pairs in turn, the number of particles increasing exponentially with depth in the medium. A mathematical formulation of showers, even restricted to average behaviour [ref. 2.8], is difficult. Quantitatively accurate treatments of shower development are usually performed using Monte Carlo techniques. Computer code has been developed that takes into account the energy dependence of the cross section, the lateral spread of the shower due to multiple scattering, statistical fluctuations and other complications, see [ref. 2.9] and [2.10], for example.

2.4.2 : Radiation loss of electrons : 'Bremsstrahlung'

High energy electrons lose energy mainly by radiating photons while being decelerated in the Coulomb field of the atomic nuclei of the traversed medium. In a medium of thickness dx, the energy loss by radiation is given by [ref. 2.6] :

$$\frac{dE}{dx}\Big|_{rad} = -\frac{E}{X_0}, \tag{2.3}$$

where X_0 is the radiation length defined as :

$$\frac{1}{X_0} = 4\alpha r_e^2 N_0 \rho \frac{Z(Z+1)}{A} \ln(\frac{128}{Z^{\frac{1}{3}}}), \tag{2.4}$$

where $\alpha = 2\pi e^2/hc$ is the fine structure constant, $r_e = e^2/m_e c^2$ the electron classical radius, N_0 the Avogadro number and ρ, Z, A are the density, the atomic number and the atomic mass of the radiator material respectively.

For high energy electrons, the energy loss by ionisation is quite constant with energy, i.e. :

$$\frac{dE}{dx}\Big|_{ionization} = constant\ with\ E, \tag{2.5}$$

while the energy loss by radiation is proportional to energy :

$$\frac{dE}{dx}|_{rad} \propto E. \tag{2.6}$$

The critical energy E_c is defined when the energy loss by radiation equals the energy loss by ionisation,

$$\frac{dE}{dx}|_{ionization} = \frac{dE}{dx}|_{rad},$$

and it can be fairly approximated by :

$$E_c \simeq \frac{600}{Z} \text{ MeV}. \tag{2.7}$$

In the ultrarelativistic limit, where energy loss by ionisation can be neglected, the mean energy $< E >$ of an electron with initial energy E_0 after having traversed a thickness x is given by integrating relation (2.6) :

$$< E >= E_0 e^{-\frac{x}{X_0}}. \tag{2.8}$$

Therefore, the radiation length may be defined simply as the thickness of the medium which reduces the mean energy of a beam of high energy electrons by a factor e.

For $E_e >> m_e c^2$, the mean angle of the radiated photon is independent of the photon energy and it can be approximated by :

$$< \theta_\gamma > \simeq \frac{m_e c^2}{E_e}. \tag{2.9}$$

Thus, most of the radiation lies inside a narrow cone with its axis along the momentum vector of the electron. (example : for $E_e > 500$ MeV, $< \theta_\gamma >$ is less than 1 mrad).

2.4.3 Absorption of gamma rays in matter

Gamma rays interact with matter via three processes : photoelectric absorption, Compton scattering and pair production. For $E_\gamma > 100$ MeV, the pair production process is dominant with a cross section essentially independent of energy. A beam of high energy photons of intensity I_0 is attenuated by pair production in a thickness x of absorber according to :

$$I = I_0 e^{-\frac{7x}{9X_0}}. \tag{2.10}$$

So, in the thickness of $\frac{9}{7}X_0$, which is called the conversion length, the intensity is reduced by a factor e. Also, a radiation length X_0 corresponds to an absorber thickness where pair production happens with a probability $P = 1 - \exp(-7/9) = 54\%$. For $E_\gamma >> m_e c^2$, the electron and positron (the created pair) travel approximately in the same direction as the gamma ray. The average angle between the direction of motion of the created electron (positron) and the gamma ray is :

$$< \theta_e > \simeq \frac{m_e c^2}{E_\gamma}, \tag{2.11}$$

which is very similar to relation (2.9).

2.4.4 The choice of the cell dimensions

From relations (2.3) and (2.6), we can see that one needs high-Z materials with small X_0 to contain individual showers in a small volume. In lead-loaded glass ($\approx 50\%$PbO, $\approx 50\%SiO_2$) calorimeters, the shower energy is measured through the Cherenkov light emitted by the electrons (positrons).

The Cherenkov light intensity produced by a charged particle is :

$$I \propto L \sin^2 \theta_c, \tag{2.12}$$

where L is the total track length of the charged particle in the shower and θ_c is half the opening angle of the Cherenkov cone with its axis along the electron trajectory:

$$\cos \theta_c = \frac{1}{n\beta} \simeq n^{-1}, \tag{2.13}$$

n being the refractive index of the glass and $\beta = v/c \simeq 1$ for relativistic electrons. The number of photons N_{ph} emitted per centimeter of path length in the visible spectrum is approximately :

$$N_{ph} \approx 500 \sin^2 \theta_c/\text{cm}. \tag{2.14}$$

(Example : with n=1.62 for a typical lead glass, $\vartheta_c = 52°$ and $N_{ph} \approx 300/\text{cm}$.)

The Cherenkov light produced in each cell is measured with a photomultiplier fixed at one end of the cell. The cell dimensions have to be chosen in such a way as to optimise energy and coordinate reconstruction accuracy.

2.4.5 Lateral dimension of a cell

For a calorimeter made of parallelepipedic cells each having a lateral dimesion d, the localisation of the impact point of the photon on the detector is in principle better for smaller d (i.e high granularity). However, below a certain size the resolution on the position is affected by: 1) statistical fluctuation of the number of photons produced in the cell; 2) the losses by reflections on the sides of the cell. Additionally, one should have in mind that the detector cost scales as d^2.

Measurements made at IHEP with electrons on lead glass have shown that the resolution σ_x on the reconstructed coordinate for $d > 3$cm varies like :

$$\sigma(E_0, d) \approx \frac{1}{\sqrt{E_0}} e^{\frac{d}{d_0}}, \tag{2.15}$$

where E_0 is the photon energy and $d_0 = 26$mm for F8-00 lead glass. For $d < 3$cm, σ_x depends only weakly on d [ref. 2.11].

Considering various constraints, of which some are mentioned above, a lateral cell dimension d = 38 mm has been adopted for GAMS, the corresponding σ_x being 2 mm at 25 GeV and 1 mm at 100 GeV.

2.4.6 Optimal cell length

In principle, the longer the cell (many X_0), the better is the shower containement and also the energy resolution. However, since we are measuring the energy via the

Cherenkov light yield, we should worry about the light absorption in the cell. The light absorption probability increases with the cell length. Thus, a very long cell would deliver no light at the end.

Measurements, also made at IHEP, with 5 and 25 GeV electrons on F8-00 lead glass have shown [ref. 2.12] that the energy resolution first decreases fast by increasing the cell length up to about 35 cm and then slowly flattens out to reach a minimum at about 45 cm.

A length of 45 cm has been adopted for the GAMS cells. This corresponds to about 16 radiation lengths and 97% shower containement at 100 GeV for F8-00 lead glass. The relative energy resolution (F.W.H.M) $\frac{\Delta E}{E}$ is :

$$\frac{\Delta E}{E} = 0.027 + \frac{0.125}{\sqrt{E(GeV)}}, \tag{2.16}$$

where the contributions to the first term are mainly from the lack of full-containement of the shower and spatial non uniform attenuation of Cherenkov light inside the cell. The second term corresponds to fluctuations in photoelectron statistics.

2.5 GAMS calibration

The energy deposited in a cell k of GAMS is related to the response A_k of the cell by the calibration coefficient factor α_k :

$$E_k = \alpha_k A_k. \tag{2.17}$$

The electromagnetic shower developed by an electron being practically identical to the one developed by a photon with the same energy, the calibration coefficients are measured using a 100 GeV monoenergetic electron beam. The beam is defocussed in order to irradiate uniformly a large number of cells (3x4 cells) at the same time. The automatic vertical and horizontal movements of GAMS allow the alignement of each lead glass cell relative to the incident beam axis. The calibration coefficients are determined using an iterative procedure described below. The procedure can be adopted to calibrate some other multi-cell calorimeters. It does not necessitate the measurement of the impact point on the cell of beam particles used for the calibration.

2.5.1 The calibration method

The electromagnetic shower developed by an electron generally spreads over many cells. Therefore, it is natural that the calibration coefficient for a given cell depends on the calibration coefficients of the surounding cells. If E_0 is the total energy of the shower, then :

$$E_0 = \sum_k E_k \tag{2.18}$$

and the calibration coefficient for cell k can then be written as :

$$\alpha_k = \frac{E_0 - \sum_{i \neq k} \alpha_i A_i}{A_k}. \tag{2.19}$$

Let $\rho_k = \frac{E_k}{E_0}$ be the energy fraction of E_0 seen by the cell k; relation (2.17) becomes :

$$\frac{1}{\alpha_k} = \frac{A_k}{\rho_k E_0}. \qquad (2.20)$$

The fraction ρ_k depends on the impact coordinates of the incident electron and therefore varies from one event to another *.

An estimate of $\frac{1}{\alpha_k}$ can be made as an average for n events :

$$\frac{1}{{\alpha_k}^n} = < \frac{1}{\alpha_k} >^{(n)} = \frac{\sum_{i=1}^n w_k \frac{A_k^i}{\rho_k^i E_0}}{\sum_{i=1}^n w_k^i}. \qquad (2.21)$$

A weight w_k^i is introduced to take into account the amount of energy deposited in the cell k for the event i. The best choice for w_k^i is found to be :

$$w_k^i = \left(\frac{E_k^i}{E_{tot}^i}\right)^2 = (\rho_k^i)^2, \qquad (2.22)$$

where E_{tot}^i is the total energy deposited in the detector for the event i.

Combining relation (2.21) with relation (2.22), we obtain the following iterative algorithm :

$$\begin{aligned} \frac{1}{\alpha_k^{(n+1)}} &= < \frac{1}{\alpha_k} >^{(n+1)} \\ &= \frac{1}{\sum_{i=1}^{n+1} w_k^i}\left(\sum_{i=1}^{n} w_k^i \frac{A_k^i}{\rho_k^i E_0} + \frac{A_k^{(n+1)}}{\rho_k^{(n+1)} E_0}\right) \\ &= \frac{1}{\sum_{i=1}^{n+1} w_k^i}\left(\frac{1}{\alpha_k^{(n)}} \sum_{i=1}^{n} w_k^i + w_k^{(n+1)} \frac{A_k^{(n+1)}}{\rho_k^{(n+1)} E_0}\right) \end{aligned} \qquad (2.23)$$

Since the impact coordinates of the calibrating particle are unknown, $\rho_k^{(n+1)} = \frac{E_k^{(n+1)}}{E_0}$ is replaced with the estimator :

$$\hat{\rho}_k^{(n+1)} = \frac{\hat{E}_k^{(n+1)}}{\hat{E}_{tot}^{(n+1)}} = \frac{\alpha_k^{(n)} A_k^{(n+1)}}{\sum_i \alpha_i^{(n)} A_i^{(n+1)}}. \qquad (2.24)$$

This implies that the energy $E_k^{(n+1)}$ seen by the cell k and the total energy $E_{tot}^{(n+1)}$ deposited in the detector for the current iteration (n+1), are evaluated using the calibration coefficients found for the previous iteration (n).

Inserting (2.24) in (2.23) , we obtain the final iterative algorithm:

$$\frac{1}{\alpha_k^{(n+1)}} = \frac{1}{\alpha_k^{(n)}}\left(1 + \frac{w_k^{(n+1)}}{\sum_{i=1}^{n+1} w_k^i}\left[\frac{E_{tot}^{(n+1)}}{E_0} - 1\right]\right). \qquad (2.25)$$

---

* For GAMS, ρ_k varies from 23% to 80 % depending on whether the impact point is on the edge or at the centre of the fired cell.

The iteration is stopped when $E_{tot}^{(n+1)} = \sum_i \alpha_i^{(n)} A_i^{(n+1)} = E_0 = E_{beam}$, i.e. when $\alpha_k^{(n+1)} \simeq \alpha_k^{(n)}$.

In practice, 500 electrons per cell are necessary to calibrate GAMS. The whole calibration procedure lasts 10 hours and it is performed at the begining and the end of each run. The calibration coefficients for the cells surrounding a dead cell are automatically overestimated. This helps in compensating the energy loss in the dead cell but it does affect the resolution on the impact coordinate determination.

A flash lamp is used to monitor the calibration coefficients during the run and to correct them to account for a possible drift in the photomultiplier gains. Light pulses produced by a flash lamp are distributed through optical fibres over all the GAMS cells. The light output of the lamp is monitored by a comparison to the light yield of a temperature stabilised photodiode.

2.6 Photon impact coordinate reconstruction on GAMS

Here we assume that the cluster or the electromagnetic shower measured by the detector is generated by a single photon. We will see later how to assign a cluster to a single or to many photons depending on its profile. The barycentre (centre of gravity) of the shower is defined by :

$$x_0 = \frac{\sum_{i,j} i \times E_{i,j}}{\sum_{i,j} E_{i,j}}, \quad y_0 = \frac{\sum_{i,j} j \times E_{i,j}}{\sum_{i,j} E_{i,j}}, \tag{2.26}$$

where (i,j) are the coordinates in cell units (i.e. i is the line number(0,63) and j is the column number (0,63) for the considered cell). The barycentre coordinates (x_0, y_0) are biased estimates of the true impact point coordinates (x_{true}, y_{true}); they coincide only when the impact point is right at the centre of a cell.

To get a better estimate, we need to find the relationship between the barycentre and the true coordinates. Two methods were used to obtain this. The first method involves the use of a 1 mm pitch hodoscope in front of a cell, that allows to measure, with 1 mm accuracy, the effective true impact coordinates which can then be compared to the barycentre ones. The second method is based on the uniform irradiation of a cell, the reconstructed barycentre coordinates are then corrected such as to obtain a uniform distribution of the impact point on the cell. The two methods were found to be equivalent and the correction function $x_{true} = f(x_0)$ is found to be an odd polynomial of the ninth order.

2.6.1 Shower profile

The knowledge of the shower profile is crucial for photon reconstruction in multi-cellular electromagnetic calorimeters. It is defined as the mean response function $R(E, x, y)$ which depends on the the energy E and the impact coordinates (x,y) of the incident photon.

In the case of the GAMS calorimeter, it has been verified with 100 GeV and 200 GeV electrons with a normal incidence on the cells, that the shower profile scales linearly with energy, i.e.:

$$R(E, x, y) = E \times G(x, y), \tag{2.27}$$

where G(x,y) is the shower profile function which relates the average fraction of deposited energy to the impact coordinates (x,y) of the incident photon and has to be measured experimentally. For GAMS, a sample of 200 000 electron induced showers was used to map out G(x,y) in steps of $\Delta x \delta y = 2$mm. The average profile was then stored as a table. However, to allow fast online reconstruction of photons, G(x,y) has been parametrised as :

$$G(x,y) = F(x)F(y) + \epsilon(x)\epsilon(y),)$$

where the shapes of the function F and ϵ are shown on figure 3.

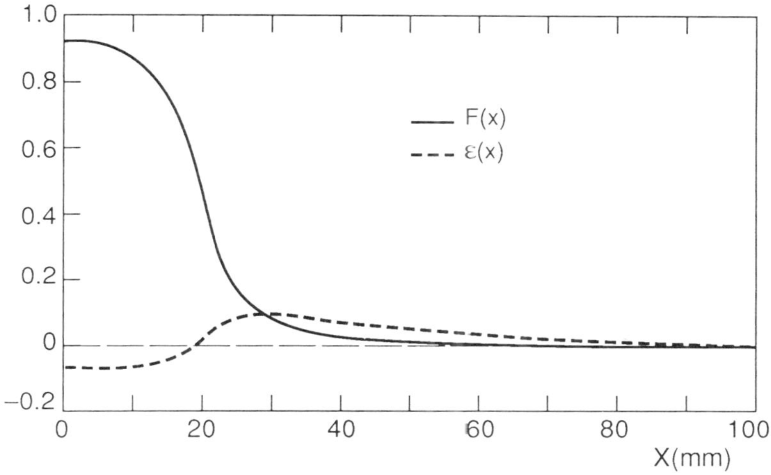

Figure 3.

2.6.2 Photon reconstruction

2.6.2.1 Case of a single photon

In the simple case where the measured electromagnetic shower is induced by a single photon, the energy E and the coordinates (x,y) of the photon can be obtained by comparing the measured shower to the shower profile function R(E,x,y). Assuming that the cell energy resolution is purely statistical $\sigma_E \propto \sqrt{E}$, (i.e. $\sigma_E = A\sqrt{E}$ where A is a constant), a χ^2 can be constructed as follows:

$$\chi^2 = \sum_i (\frac{E_i - E_i^R}{AE})^2, \tag{2.28}$$

where the summation runs over all the shower cells and E_i^R is the expected energy fraction for cell i as given by the shower profile function. To account for the small amplitudes with non Gaussian fluctuations delivered by the cells located on the edges of the shower (far from the impact point), a weight $w_i = 1 - \frac{E_i}{E_{shower}}$ is introduced into the χ^2 expression :

$$\chi^2 = \sum_i \left(\frac{E_i - E_i^R}{AE_i w_i + B} \right)^2 . \tag{2.29}$$

The constant B is introduced as an empirical fixup, chosen to obtain the usual distribution for the χ^2. The corrected barycentre coordinates and the total shower energy released are used as the starting point for the minimisation of χ^2. The best estimate of the photon (E,x,y) are then obtained as the final output of the procedure.

2.6.2.2 Multi-photon reconstruction

Often, a measured electromagnetic cluster is composed of many overlapping showers, and the separation between the contributing showers necessitates an accurate measurement of the shower profile in the detector. The following procedure was used to search for photons in GAMS :

1): Search for clusters; a cluster is defined as a group of adjacent cells surrounded with all zero amplitude cells, and two clusters are separated if they have no more that one corner of a cell in common.

2): Search and list all local maxima in the cluster, where a maximum is defined as a cell having an amplitude greater than its surrounding cell amplitudes.

3): If more than one maximum is found, define sub-clusters, each being the 3×3 cells around the maximum.

4): Perfom a fit of each sub-cluster to the shower profile as in the case of a single photon (see previous subsection). This procedure evaluates the contribution of the shared cells between sub-clusters to the energy.

5): Redo the fit of each individual sub-cluster to the shower profile using all the contributing cells. The final energy sharing is obtained by iterating the procedure, taking into account the reconstructed coordinates and the energy of the sub-cluster.

6): The fitted sub-cluster is considered as a photon if it has a chisquare value (relation 2.29) greater than 3, otherwise a fit of the sub-cluster considering more than one contributing photon is attempted.

Using the above procedure, the GAMS performance is such that two overlapping photon showers separated by 15 mm can be disentangled with 50% efficiency if $E_{\gamma_1} \approx E_{\gamma_2}$.

2.6.3 Fake photons

In order not to miss the reconstruction of any photon, the parameters of the shower reconstruction procedure are tuned such that even dubious photons are accepted. Conditions, like the separation between two photons correlated with their relative invariant mass, can later be applied to recombine two wrongly separated photons into the original single photon. Other fake photons can come from the accidental occurrence of a second interaction while the GAMS ADCs gate is open (60 ns). These can be found by looking for a π^0 peak in the 3 photon invariant mass spectrum. Figure 4, shows fully reconstructed 8 photons event from the $\eta\eta$ channel, where the first η decayed into 2 photons and the second η decayed into $3\pi^0$, each decaying into 2 photons.

2.7 Channel separation and mass-constrained fit

The channels are separated depending on the final photon multiplicity. A final state with 4 photons can originate from one of the following channels : $\pi^0\pi^0$, $\pi^0\eta$,

$\pi^0\eta'$, $\eta\eta'$, $\omega\gamma$, $\pi^0\gamma\gamma$, ..., for example. Invariant masses of all the pairs that can be formed with the recontructed photons are calculated. A pair is assigned to a candidate particle π^0, η, or η' if its invariant mass lies in [100, 170] MeV, [470,630] MeV or [780,1170] MeV, respectively, where the mass windows were defined as the regions in the photon pair invariant mass spectrun, where the signal to noise ratio seems reasonable. For a given photon multiplicity, there is usually more than one possible channel assignment or solution. The best possible solution is chosen according to the smallest χ^2 obtained from a mass constrained fit.

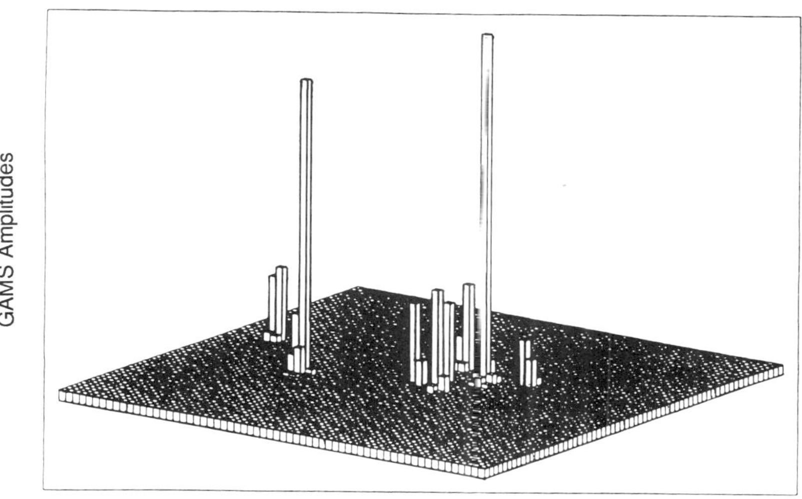

Figure 4.

2.7.1 Mass constrained fit

For the exclusive charge exchange reaction $\pi^- p \longrightarrow M^0 n$, $M^0 \longrightarrow k\gamma$, a constrained χ^2 based on each photon measured variables E,x,y and the correponding errors σ_E, σ_x, σ_y, can be formed as follows :

$$\chi^2 = \chi_0^2 - 2\lambda_n F_n(x_i, y_i, E_i) - \sum_{(ij)}^{\text{all pairs}} 2\lambda_{(ij)} F_{ij}(x_i, y_i, E_i), \tag{2.30}$$

where

$$\chi_0^2 = \sum_i \left[\left(\frac{x_i^F - x_i}{\sigma_{x_i}} \right)^2 + \left(\frac{y_i^F - y_i}{\sigma_{y_i}} \right)^2 + \left(\frac{E_i^F - E_i}{\sigma_{E_i}} \right)^2 \right],$$

is the non constrained χ^2, with x_i^F, y_i^F and E_i^F being the final variables that minimise the χ^2, and

$$F_n(x_i, y_i, E) = M_n^2 - M_{n_{PDG}}^2 \quad \text{and}$$

$$F_{ij}(x_i, y_i, E_i) = M_{ij}^2 - M_{ij_{PDG}}^2,$$

are the mass contraints expressed in terms of the photon measured variables, where M_n and M_{ij} are the calculated invariant masses of the recoiling neutron and the photon pair, respectively. The subscript PDG indicates the Particle Data group mass value of the considered particle and λ_n and λ_{ij} are Lagrange multipliers. The total number of constraints is equal to the total number of included pairs in the fit plus one for the recoiling neutron.

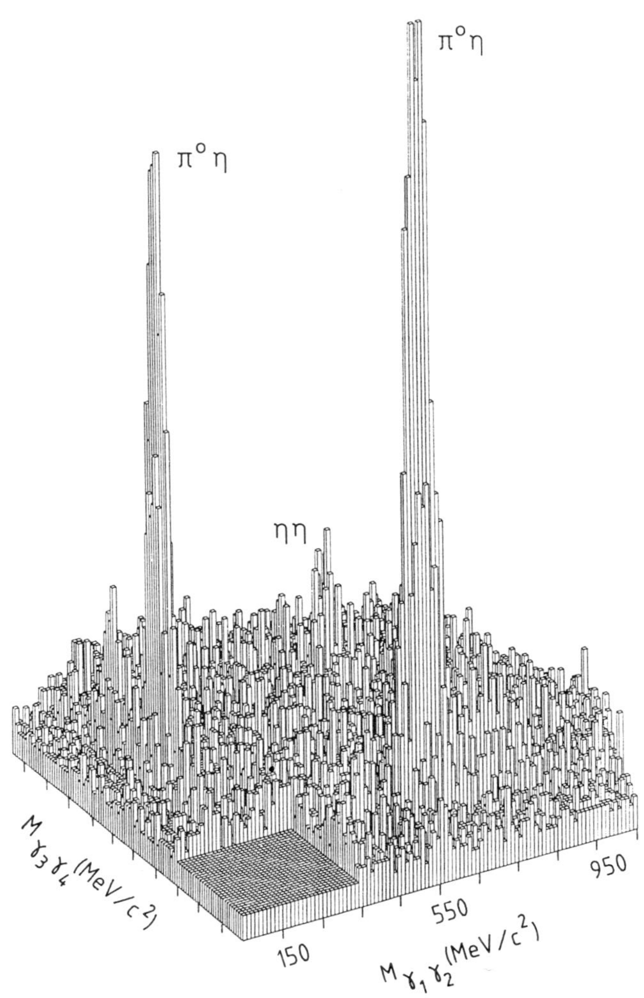

Figure 5.

The χ^2 is minimised iteratively. The final corrected variables (E,x,y) for each photon are found to enhance the invariant mass resolution for the M^0 state. It is worth mentioning here, that the iterative procedure used to minimise the χ^2, can be used to study the PULL variables [ref. 2.13] to check and possibly correct the errors σ_E, σ_x and σ_y associated with each measured photon variable (E,x,y).

A lego plot of all recombinations of the $4-\gamma$ data into pairs is shown in figure 5;

$\eta\pi^0$ and $\eta\eta$ peaks (channels) are seen over a small background. The $\pi^0\pi^0$ events which contribute to a peak about ten times higher than the $\pi^0\eta$ one, have been left out. Using the mass constrained fit and the associated χ^2 value for each fitted solution, the $\pi^0\pi^0$, $\pi^0\eta$, and $\eta\eta$ channels can be separated among the 4-gamma event samples with more than 90% purities.

3: Partial wave analysis

Three partial wave analysis methods for 2-body decays of the neutral state M^0, produced in charge exchange reactions, are described. The first two methods consist of energy independent and energy dependent fits based on spherical harmonic moments. The moments are directly related to the amplitudes and the relative phases of the helicity partial waves. The third method consists of a direct fit of the total production amplitude expression to the observed angular distributions.

3-1: Generalities

For an unpolarised target and without measuring the recoiling neutron polarisation, the differential intensity of events produced in the reaction

$$\pi^- + p \longrightarrow M^0 + n; \quad M^0 \longrightarrow a + b$$

can be written [ref. 3.1] as :

$$I(t, M, \theta, \Phi) = \frac{\partial^4 \sigma}{\partial t \partial M \partial \cos\theta \partial \Phi} = \frac{1}{2} \sum_{\lambda_p, \lambda_n} |H_{\lambda_p,\lambda_n}(t, M, \theta, \Phi)|^2, \tag{3.1}$$

where t is the momentum transfer between the incident π^- and the M^0 system, M is the M^0 invariant mass, λ_p, λ_n are the proton and neutron helicities and θ, Φ are the polar and azimuthal angles of either particle a or particle b in the Gottfried-Jackson (GJ) frame *.

The full helicity amplitude $H_{\lambda_p,\lambda_n}(t, M, \theta, \Phi)$ which is the sum of the terms corresponding to all possible intermediate spin j and helicity m states of M^0 is given by :

$$\sqrt{4\pi} H_{\lambda_p,\lambda_n}(t, M, \theta, \Phi) = \sum_{j=0}^{\infty} \sum_{m=-j}^{j} \sqrt{2j+1} H^j_{\lambda_p,\lambda_n,m} d^j_{m,0}(\theta) e^{im\Phi} \tag{3.2}$$

and where $d^j_{m,0}(\theta) = < j, m | e^{-i\theta J_Y} | j, 0 >$ is the matrix element of the rotation operator, for an angle θ around the Y axis, between the helicity states m=1 and m=0. J is the total angular momentum, and the factor $\sqrt{2j+1}$ is introduced to get equal coefficients for the contributions $|H^j_m|^2$ in the integrated intensity.

Inserting the relation (3.2) in the relation (3.1) we get :

$$\begin{aligned} 4\pi I(\theta, \Phi) = \frac{1}{2} \sum_{\lambda_p,\lambda_n} \sum_{j_2,m_2} \sum_{j_1,m_1} \sqrt{2j_1+1}\sqrt{2j_2+1} e^{i(m1-m2)\Phi} \\ \times H^{*j_1}_{\lambda_p,\lambda_n,m1} H^{j_2}_{\lambda_p,\lambda_n,m2} d^{j_1}_{m1,0}(\theta) d^{j_2}_{m2,0}(\theta). \end{aligned} \tag{3.3}$$

* The Gottfried-Jackson frame is the M^0 rest frame rotated such that its Z axis is parallel to incident π^- direction.

The helicity amplitude products of relation (3.3) can be expressed in terms of the density matrix elements :

$$\rho^{j_1 j_2}_{m_1 m_2} = \frac{1}{2N} \sum_{\lambda_p,\lambda_n} H^{*j_1}_{\lambda_p,\lambda_n,m1} H^{j_2}_{\lambda_p,\lambda_n,m2}, \tag{3.4}$$

where

$$N = \frac{\partial^2 \sigma}{\partial t \partial M}.$$

Therefore relation (3.3) can then be rewritten as :

$$\begin{aligned} I(\theta,\Phi) = \frac{N}{4\pi} & \sum_{j_1 m_1 j_2 m_2} \sqrt{2j_1+1}\sqrt{2j_2+1}\rho^{j_1 j_2}_{m_1 m_2} \\ & \times d^{j_1}_{m1,0}(\theta) d^{j_2}_{m2,0}(\theta) e^{i(m1-m2)\Phi} \end{aligned} \tag{3.5}$$

and its integral gives the trace equality :

$$\sum_{jm} \rho^{jj}_{mm} = 1. \tag{3.6}$$

As the strong interaction conserves parity, for the process $1 + 2 \longrightarrow 3 + 4$, the helicity amplitudes satisfy the relation:

$$H_{\lambda_3,\lambda_4,\lambda_1,\lambda_2} = \eta(-1)^k H_{-\lambda_3,-\lambda_4,-\lambda_1,-\lambda_2}, \tag{3.7}$$

where η is the intrinsic parity product of particles 1, 2, 3 and 4 and $k = \sum_i (s_i + \lambda_i)$ is the sum of their spins and helicities. Combining (3.7) and (3.4) we get the relation

$$\rho^{j_1 j_2}_{m_1 m_2} = (-1)^{(m_1+m_2)} \rho^{j_1 j_2}_{-m_1,-m_2}, \tag{3.8}$$

which allows a reduction in the total number of density matrix elements.

3-2: Spherical harmonic moments and helicity amplitudes

In any reference frame where the Z axis is in the production plane, the most general angular distribution of the M^0 system can be written, according to parity conservation, as a sum of the real parts $Re(Y^m_l)$ of the spherical harmonic moments (we will write Y^m_l instead of $Re(Y^m_l)$ for simplicity):

$$I(t,M,\theta,\Phi) = N \sum_{l=0}^{\infty} \sum_{m=-1}^{1} < Y^m_l > Y^m_l(\theta,\Phi), \tag{3.9}$$

where $< Y^0_0 >= \frac{1}{4\pi}$ and the coefficients $< Y^m_l >$ are the normalised moments.

Using standard relations amongst rotation matrix products, the $< Y^m_l >$ coefficients can be expressed as a sum of the real parts of the elements $\rho^{j_1 j_2}_{m_1 m_2}$:

$$< Y^m_l >= \frac{1}{4\pi} \sum_{j_1,m_1,j_2,m_2} C(l,j_1,j_2,m,m_1,m_2) Re\rho^{j_1 j_2}_{m_1 m_2}, \tag{3.10}$$

where

$$C(l,j_1,j_2,m,m_1,m_2) = -1^m\sqrt{(2l+1)(2j_1+1)(2j_2+1)}\times \begin{pmatrix} l & j_1 & j_2 \\ m & -m_1 & m_2 \end{pmatrix}\begin{pmatrix} l & j_1 & j_2 \\ 0 & 0 & 0 \end{pmatrix}.$$

This expression illustrates the fact that in the abscence of polarisation measurement, only combinations of the density matrix elements are measureable.

In the t channel, the most interesting combinations of the density matrix elements are those relative to a well defined exchanged parity. If M^0 has natural parity (i.e. $P = -1^J$ where J is its total angular momentum), then the corresponding helicity amplitudes can be expressed as [ref. 3.2] :

$$H^{j\pm}_{\lambda_p,\lambda_n,m} = \frac{1}{\sqrt{2}}[H^j_{\lambda_p,\lambda_n,m} \mp -(-)^m H^j_{\lambda_p,\lambda_n,-m}], \tag{3.11}$$

with

$$H^{j-}_{\lambda_p,\lambda_n,0} = H^j_{\lambda_p,\lambda_n,0} \quad \text{and} \quad H^{j+}_{\lambda_p,\lambda_n,0} = 0.$$

The corresponding density matrix element are :

$$\rho^{j_1j_2\pm}_{m_1m_2} = \frac{1}{2}(\rho^{j_1j_2}_{m_1m_2} \mp (-1)^{m_2}\rho^{j_1j_2}_{m_1,-m_2}), \tag{3.12}$$

where

$$\rho^{j_1j_2\pm}_{m_1m_2} = \frac{1}{8N}\sum_{\lambda_p\lambda_n} H^{*j_1\pm}_{\lambda_p,\lambda_n,m_1} H^{j_2\pm}_{\lambda_p,\lambda_n,m_2}.$$

Table 3.1

Expression of moments $< Y^m_l >$ in terms of helicity amplitudes for $L_{MAX} = 2, M_{MAX} = 1$

(Sums over the helicity indices of the proton and the neutron are implicit)

$$\sqrt{4\pi}N < Y^0_0 > = S_0^2 + P_0^2 + P_-^2 + P_+^2 + D_0^2 + D_-^2 + D_+^2$$
$$\sqrt{4\pi}N < Y^0_1 > = 2\ S_0P_0 + 1.789\ P_0D_0 + 1.549(P_-D_- + P_+D_+)$$
$$\sqrt{4\pi}N < Y^1_1 > = 1.414\ S_0P_- + 1.095\ P_0D_- - 0.632\ P_-D_0$$
$$\sqrt{4\pi}N < Y^0_2 > = 2\ S_0D_0 + 0.894\ P_0^2 - 0.447\ (P_-^2 + P_+^2) + 0.639\ D_0^2 + 0.319(D_-^2 + D_+^2)$$
$$\sqrt{4\pi}N < Y^1_2 > = 1.414\ S_0D_- + 1.095\ P_0P_- + 0.452\ D_0D_-$$
$$\sqrt{4\pi}N < Y^2_2 > = 0.548(P_-^2 - P_+^2) + 0.391(D_-^2 - D_+^2)$$
$$\sqrt{4\pi}N < Y^0_3 > = 1.757\ P_0D_0 - 1.014(P_-D_- + P_+D_+)$$
$$\sqrt{4\pi}N < Y^1_3 > = 1.171\ P_0D_- + 1.014\ P_-D_0$$
$$\sqrt{4\pi}N < Y^2_3 > = 0.926(P_-D_- - P_+D_-)$$
$$\sqrt{4\pi}N < Y^0_4 > = 0.857\ D_0^2 - 0.571(D_-^2 + D_+^2)$$
$$\sqrt{4\pi}N < Y^1_4 > = 1.107\ D_0D_-$$
$$\sqrt{4\pi}N < Y^2_4 > = 0.452(D_-^2 - D_+^2)$$
$$N = \frac{\partial^2\sigma}{\partial t\ \partial M} \quad and \quad LL' = Re(LL'^*).$$

We can see from relation (3.11) that the amplitudes H^+ and H^-, corresponding to natural parity exchange and unnatural parity exchange respectively, do not interfere in the expression of the moments $< Y_l^m >$.

After summing over the proton and neutron helicities and using the spectroscopic notation, relation (3.11) can be rewritten for a natural parity M^0 system as :

$$L_{\lambda\pm} = \frac{1}{\sqrt{2}}[H_\lambda^L \mp (-1)^\lambda H_{-\lambda}^L] \quad \lambda \neq 0, \tag{3.13}$$

with $L_{0-} = H_0^L$ and $L_{0+} = 0$.

The moments $< Y_l^m >$ can be written then as bilinear products of the amplitudes $L_{\lambda\chi}$ of the system $M^0 \longrightarrow a + b$ with spin L and helicity λ, produced by natural parity exchange (NPE)$\chi = 1$ or by unnatural parity exchange (UPE)$\chi = -1$. Table (3.1) gives the relations between the moments $< Y_l^m >$ and the helicity amplitudes considering only the waves S, P and D (i.e. $L_{MAX} = 2$ ($l_{MAX=4}$)).

4: The moments $< Y_l^m >$

Including the factor N in the coefficients $< Y_l^m >$ of relation (3.9), we obtain a new set of real coefficients t_l^m which depend on the momentum transfer t and the mass M of the M^0 system. The angular distribution of the produced events can then be written as :

$$I_{prod}(\theta, \Phi, t, M) = \sum_{l=0}^{\infty} \sum_{m=0}^{l} t_l^m Y_l^m(\theta, \Phi), \tag{4.1}$$

where the summation over m is done from m=0 to m = l instead of -l to l ($Y_l^m = Y_l^{-m}$).

The produced angular distribution in a small enough interval of t and M, where we assume that the coefficients t_l^m are constant, is normalised to the total number of events produced in that interval :

$$\int_\Omega I_{prod}(\Omega) d\Omega = N_{prod} = \sqrt{4\pi} t_0^0, \tag{4.2}$$

and the normalised spherical harmonics are given by :

$$< Y_l^m > = \frac{1}{N_{prod}} \int_\Omega I_{prod}(\Omega) Y_l^m(\Omega) d\Omega. \tag{4.3}$$

Inserting the relation (4.1) in relation (4.3) and using the usual normalisation formulae for spherical harmonics, we obtain a simple relation between the real coefficients t_l^m and the moments $< Y_l^m >$:

$$< Y_l^m > = \frac{1}{N_{prod}} \epsilon_l^m t_l^m, \tag{4.4}$$

with

$$\epsilon_l^m = 1 \quad \text{if } m = 0 \quad \text{and } \frac{1}{2} \quad \text{if } m \neq 0.$$

Now, the problem is firstly to determine the coefficients t_l^m which give the unnormalised moments $\sqrt{4\pi}N_{prod} < Y_l^m >$, and secondly to solve the equations of table 3-1 to get the amplitudes and the relative phases of helicity partial waves.

The angular distribution of the produced events is related to the angular distribution of the observed events through the efficiency and acceptance function $A(\Omega)$:

$$I_{obs}(\Omega) = A(\Omega)I_{prod}(\Omega) = A(\Omega)\sum_{\mu=0} t_\mu Y_\mu(\Omega). \tag{4.5}$$

The index μ replaces indices (l,m) where $\mu = 0$ for (l=0,m=0) and $\sum_\mu \equiv \sum_l \sum_{m\geq 0}$. The efficiency and acceptance function should be known for all $M, t, \cos\theta$, and Φ bins. It is determined using a full Monte Carlo simulation of the studied process including all the detector effects and geometry. In the following subsection, we will describe some methods of extracting the coefficients t_l^m from the measured data.

4-1: Fit by χ^2 minimisation

This method consists [ref. 4.1] of extracting the coefficients t_l^m by a fit of the distribution $I_{obs}(\Omega)$ (relation 4.5) to the measured angular distribution. For a given mass M and in a particular momentum transfer bin *, the measured events are grouped into $\delta\Omega = (\Delta\cos\theta, \Delta\Phi)$ bins where the size of a bin is chosen depending on the resolution in $\cos\theta$ and Φ and on available statistics.

Let w_k be the weight of the total number of measured events in bin k and w_k^{exp} the weight of the expected number of events in the same bin. The χ^2 to be minimised is then given by :

$$\chi^2 = \sum_k (\frac{(w_k - w_k^{exp})^2}{(\sigma_k)^2}, \tag{4.6}$$

where

$$w_k^{exp} = \frac{\Delta\Omega}{\sqrt{4\pi}} A(\Omega_k) \sum_\mu Y_\mu(\Omega_k). \tag{4.7}$$

Here σ_k is the error on w_k and $A(\Omega_k)$ is the average value of $A(\Omega)$ in the bin k.

One can prove [ref. 4.1] that the variance of the total expected weight is equal to :

$$\sigma_k^2 = \frac{< w_k^2 >}{< w_k >} w_k^{exp}. \tag{4.8}$$

The χ^2 minimization can be done by solving the system of equations :

$$\frac{\partial\chi^2}{\partial t_\mu} = 0. \tag{4.9}$$

The error matrix $E^{-1}(t_\alpha)$ for the fitted coefficients t_μ is then given by the relation :

$$[E^{-1}(t_\alpha)]_{\mu\mu'} = -\frac{1}{2}\frac{\partial^2\chi^2}{\partial t_\mu \partial t_{\mu'}}. \tag{4.10}$$

* Depending on the available statistics, and under some assumptions(!), the measured angular distributions are sometimes integrated over all the accepted t range.

4-2: Fit using a maximum likelihood

In order to apply the maximum likelihood [ref. 4.2] method to the observed angular distribution, the corrections for acceptance and efficiency need a particular parametrisation. The observed angular distribution defined in relation (4.5) is normalised to the total number of observed events W for a given mass and t bin, $\Delta M \times \Delta t$ as follows :

$$\int_{\Omega} I_{obs}(\Omega) d\Omega = W \ , \tag{4.11}$$

and the acceptance and efficiency function $A(\Omega)$ can also be expanded as a series of spherical harmonics:

$$A(\Omega) = \sum_{\mu} a_{\mu} Y_{\mu}(\Omega). \tag{4.12}$$

Combining (4.5), (4.11) and (4.12) we obtain a linear relation between the coefficient t_{μ} :

$$t_0 = \frac{1}{a_0}[W - \sum_{\mu>0} \epsilon_{\mu} a_{\mu} t_{\mu}], \tag{4.13}$$

with

$$\epsilon_{\mu} \equiv \epsilon_l^m = 1 \text{ for } m = 0 \text{ and} \frac{1}{2} \text{ for } m \neq 0.$$

The normalised angular distribution of the produced events is simply the probability of producing an event in the bin $\Delta\Omega$ in the considered $\Delta M \times \Delta t$ interval, and it is obtained by combining the relations (4.12), (4.13) and (4.1):

$$I_{prod}(\Omega) = \frac{Y_0}{a_0} W + \sum_{(\mu>0)} \left[Y_{\mu}(\Omega) - \frac{Y_0}{a_0} a_{\mu} \epsilon_{\mu} \right] t_{\mu}. \tag{4.14}$$

If we take out all the constant factors then the ith event, with a weight p_i, will contribute to the likelihood function F with a factor P_i such that :

$$P_i = [I_{prod}]^{p_i}. \tag{4.15}$$

The best set of coefficients t_{μ} is then obtained by minimising the quantity :

$$-\log(F) = \sum_{i}^{N} \log(P_i). \tag{4.16}$$

The minimisation can be done using the MINUIT package which also provides routines for the estimation of the associated errors. Special care should be taken when estimating the errors mainly because we are assigning individual weights to each event.

This method, besides the fact that it is slow, has some major advantages over the the χ^2 fit method. One of the advantages is the number of free parameters in the fit. Since the parameter t_0 is expressed as a linear combination of the other free parameters (relation 4.13), the maximum likelihood method has one free parameter less than the χ^2 fit method for any given set of contributing helicity partial waves. Another advantage is the sensitivity to regions in $\cos\theta$ and Φ where the acceptance is nil. The expected weight w_k^{exp} in relation (4.7) is equal to zero for $A(\Omega) = 0$, and that might affect the determination of the coefficient t_μ especially when the region where the acceptance is zero is wide. However, it is necessary to study in detail the effects of zero acceptance on the results and the associated errors by applying both methods on a Monte Carlo sample of events.

4.3 The helicity partial waves

4.3.1 Energy independent fit of the moment

The helicity partial waves : S_0, P_0, P_-, P_+, D_0, D_-,..., each being a complex number, are calulated in each $\Delta M \times \Delta t$ interval by solving the equation system given in table 3.1.

Since the equations do not contain products of the type L_-L_+, and L_0L_+, we can fix a reference phase for each of the natural parity and unnatural parity helicity partial waves. This decreases the number of unknowns by 2. The remaining free parameters can then be obtained by minimising the quantity :

$$\chi^2 = \sum_{l=0}^{l_{max}} \sum_{m=0}^{l} \left[\frac{(Y_l^m)_{meas} - (Y_l^m)_{theo}}{\sigma_{Y_l^m}} \right]^2_{\Delta M, \Delta t}, \tag{4.17}$$

where $(Y_l^m)_{meas}$ is the measured normalised moment value, $\sigma_{Y_l^m}$ its associated measurement error and $(Y_l^m)_{theo}$ is the normalised moment expression as given in table 3.1. The equations of table 3.1 might have more than one possible solution in each $\Delta M \times \Delta t$ bin. These solutions can be found either by numerical techniques, which involve forcing the minimisation to look for other minima, or by analytical techniques like the method of Barrelet [ref. 4.3] [ref. 4.4]. Once all the solutions are found for each $\Delta M \times \Delta t$ bin, they have to be linked from one bin to the next one in order to get the mass spectra of the helicity partial wave amplitudes. This last step is not trivial and necessitates some assumptions mainly on how a given partial wave amplitude would vary from one bin to the next.

4.3.2 Energy dependent fit of the moment

This method of extracting the helicity partial waves from the moments can be very efficient if the number of the contributing waves is small. It consists of assuming an analytical expression for each partial wave amplitude L_α. The expression,in general, is a sum of of one or more Breit-Wigner shapes and a polynomial background :

$$L_\alpha = \sum_j BW^j_{L_\alpha} e^{i\phi L_\alpha} + P^n(M), \tag{4.18}$$

with

$$BW^j_{L_\alpha} = \frac{c_j^{L_\alpha}}{(M^j_L - M - i\frac{\Gamma^j_L}{2}} \tag{4.19}$$

and $L_\alpha = [L_0,\ L_+,\ L_-]$, M^j_L, Γ^j_L are the mass and the width of the jth Breit-Wigner shape of the partial wave L_α, ϕ_{L_α} is the relative phase, and $P^n(M)$ is a nth order polynomial of M.

For a given t interval, the partial wave helicity amplitude expressions have to be introduced into the moment equations of table 3.1 and the free parameters, which consist of the coefficients associated to the Breit-Wigners and the backgound, can then be extracted for the entire studied mass range by minimising the χ^2:

$$\chi^2 = \sum_{\Delta M^0} \sum_{l=0}^{l_{max}} \sum_{m=0}^{l} \left[\frac{(Y_l^m)_{meas} - (Y_l^m)_{theo}}{\sigma_{Y_l^m}} \right]^2_{\Delta M, \Delta t} \tag{4.20}$$

which looks like the χ^2 given in relation (4.17) but summed over all the mass bins. The number of free parameters can be reduced by fixing some of the parameters for the known resonances. This method, even though tedious, has the merit of being a good tool for evaluating the weight of a given resonance in describing the measured normalised moment spectra.

4.3.3 Direct fit of the measured angular distributions

In the case where the available statistics is not suffucient to perform a good moment analysis, the measured angular distributions are fitted directly to the total production amplitude expressed in terms of the contributing helicity partial waves.

In each $\Delta M \times \Delta t$ interval, the number of expected events in the solid angle bin $\Delta\Omega = \Delta \cos\theta_{GJ} \times \Delta\Phi_{GJ}$ is given by :

$$N_{exp}(\Delta\Omega) = \frac{\Delta\Omega}{4\pi} \epsilon(\Delta\Omega) \frac{\partial\sigma}{\partial\Omega}, \tag{4.21}$$

where $\epsilon(\Delta\Omega)$ is the everage of the acceptance and efficiency function in $\Delta\Omega$ and $\frac{\partial\sigma}{\partial\Omega}$ is the differential cross section written in terms of the helicity waves S_0, P_0, $P_\pm$,... Assuming the Poisson Law for populating the bin $\Delta\Omega$, the probablity distribution can be written as :

$$P(\Delta\Omega) = \frac{e^{N_{exp}(\Delta\Omega)}[N_{exp}(\Delta\Omega)]^{n(\Delta\Omega)}}{[n(\Delta\Omega)]!}, \tag{4.22}$$

with $n(\Delta\Omega)$ being the observed number of events in $\Delta\Omega$. The most probable values of the helicity partial wave amplitudes and phases for the considered mass bin are obtained by minimising the function :

$$F = -\log(\prod_i P(\Delta\Omega_i)) = -\sum_i \log P(\Delta\Omega_i), \tag{4.23}$$

which corresponds to the maximum of the likelihood function. Here again we have to search for all the minima of the function F in order to get all the possible solutions. For smaller statistics, one can still integrate the angular distributions over the Φ_{GJ} angle. However, it will not be anymore possible to disentengle the natural L_+ and the unnatural L_- partial waves.

5 Physics results

Most results come from studies of exclusive production of neutral final states in the charge exchange reaction (1.1). The studied processes include two-body cross sections of known mesons, rare decay modes of established mesons, reconfirmation and measurement of branching ratios of less well established mesons, and the search for new states. Table 5.1 gives a summary of the states discovered recently with GAMS, and the references for already published results. Most of the newly discovered states have been seen in $\eta\eta$ and $\eta\eta'$ channels and fall into already filled SU(3)-nonets (mainly 0^{++} and 2^{++}). On the other hand, it is believed [ref 5.11][ref. 5.12] that the η and the η' couple strongly to gluon pairs. This is probably (!) supported by the large branching ratios for radiative J/Ψ decays into the η and particularly into the η'. So, most of the newly observed states are potential glueball candidates.

Table 5.1. New states observed with GAMS

Name	Decay mode(s)	J^{PC}	Energies GeV/c	References
r(2520)	$\in \pi^0 \to 4\gamma$	6^+	38	[5.1]
G(1590)	$\eta\eta \to 4\gamma,\ 8\gamma$	0^{++}	38, 100, 230	[5.2], [5.3]
	$\eta\eta' \to 4\gamma$		38	[5.4]
	$4\pi^0 \to 8\gamma$		38, 100, 230	[5.5]
X(1850)	$\eta\eta \to 4\gamma,\ 8\gamma$	2^{++}	100	[5.2]
	$4\pi^0 \to 8\gamma$		100	[5.5]
X(1750)	$\eta\eta \to 4\gamma,\ 8\gamma$	0^{++} or 2^{++}	38	[5.6]
X(1910)	$\eta\eta' \to 4\gamma$	?	38	
X(2100)	$\eta\eta \to 4\gamma$	?	100	
X(2220)	$\eta\eta' \to 4\gamma$	2^{++} or 4^{++}	38, 100	[5.7]
X(1640)	$\omega\omega \to 6\gamma$	2^{++}	38	
X(1960)	$\omega\omega \to 6\gamma$	2^{++}	38	
X(2140)	$\omega\pi^0 \to 5\gamma$	1^{--}	38 100	
$\rho_5(2330)$	$\omega\pi^0 \to 5\gamma$	1^{--}	38 100	
M(1400)	$\eta\pi^0 \to 4\gamma$	1^{-+}	100	[5.8]
S(1300)	$\eta\pi^0 \to 4\gamma$	0^{++}	100	[5.9]

(The above table does not include the exclusive measurement $\pi^- p \longrightarrow M^0 \pi^- p$ at 300 GeV [ref. 5.10].)

5.1 The G(1590) 0^{++}

It was first observed in the partial wave analysis of the $\eta\eta$ events produced in reaction (1.1) with 38 GeV/c π^-. The G(1590) was seen as a clear peak in the S-wave with a measured mass of (1592 ± 25) MeV/c^2 and a width of (210 ± 40) MeV/c^2. A similar peak has been observed in the $\eta\eta'$ mass spectrum. This peak is quite near the threshold, but it cannot be accounted for by phase space or threshold effects. The partial wave analysis of the $\eta\eta$ data at 100 GeV/c has also been performed. Two peaks have been seen in the S-wave (figure 6), they have been identified as the $f_0(1300)$ (ex $\epsilon(1300)$) and the G(1590) mesons. Similar results have been obtained with 230 GeV/c pions.

At 38, 100 and 230 GeV/c π^- beam momenta, the most striking $\eta\eta$ resonance produced in reaction (1.1) is the G(1590). At these energies its decay into $2\pi^0$ has not been observed. It has also not been observed by an other experiment [ref. 5.13] looking for the production of $K^0\bar{K}^0$. The measured branching ratios for the G(1590) into $\eta\eta$ and $\eta\eta'$ are: $BR(G \rightarrow \eta\eta) = 0.12 \pm 0.03$ [ref. 5.4], and $BR(G \rightarrow \eta\eta') = 0.35 \pm 0.1$ [ref. 5.1]. It has been found that $BR(G \rightarrow \pi^0\pi^0) < 5\%$. The suppression of the $\pi^0\pi^0$ decay mode makes it difficult to identify the G(1590) as a normal $q\bar{q}$ state. The G(1590) has also been observed to decay into $4\pi^0$[ref. 5.5]. This decay mode is expected to be rare for $q\bar{q}$ mesons in comparison with other 4π decay modes (ρ -dominance). A mechanism of gluon decoloration has been proposed [ref. 5.14] which would favor the decay of glueballs preferably into η and η' mesons rather than π^0 or K^0 mesons. This mechanism gives branching ratios for the G(1590), assumed to be a glueball, which are in agreement with measured values. It has been shown [ref. 5.14][ref. 4.15] that the ratio $BR(G \rightarrow 4\pi^0)/BR(G \rightarrow \text{all } 4\pi) \simeq 1/5$ if the gluon decoloration mechanism is the main mechanism at work in this decay.

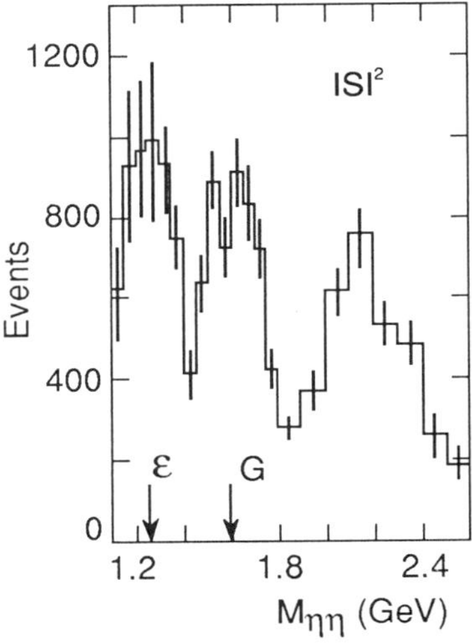

Figure 6.

Another strong argument to give a glueball assignment to the G(1590) comes from its observation in the exclusive central production reaction : $\pi^- p \longrightarrow M^0\pi^- p$ with 300 GeV/c pions [ref. 5.10]. The G has been observed as a peak directly in the $\eta\eta$ mass spectrum. Compared to the well known $f_2(1270)$ meson, it has been found in the charge exchange reaction: $\sigma(\pi^- p \rightarrow Gn)/\sigma(\pi^- p \rightarrow f_2 n) \simeq 1/25$ with 40 and 100 GeV/c pions, and in central collisions that : $\sigma(\pi^- p \rightarrow G\pi^- p)/\sigma(\pi^- p \rightarrow f_2\pi^- p) \simeq 10$ with 300 GeV/c pions. Neither G nor f_2 are detected in central collisions with 40 GeV/c π^- at the sensitivity of the experiment. These observations, made in the frame of supposedly preferred gluonic processes, support the glueball hypothesis.

At this school, we have heard about other interpretations of the G(1590):

1) The G(1590) (cf. Prof. E. Palano) is only observed by GAMS experiment(s) and only in the partial wave analysis. It was never obrserved as a convincing clear peak in the $\eta\eta$ mass spectrun. Therefore, one should not take so seriously the G(1590) as an established resonance unless another experiment confirms it.
2) The $0^{++}(1500)$ state observed by the Crystal Barrel collaboration (cf. Prof. M. Faessler) is the same as the G(1590) and it is a good glueball candidate. In this case one has to explain the mass difference between the two observations. J.M. Frere has proposed an explanation based on phase space consideratons.
3) There might be some evidence (cf. Prof. M. Faessler) for a 0^{++} resonance at 1590 MeV in the $K\bar{K}$ channel of the Crystal Barrel data. This makes room for the G(1590) simply being an $s\bar{s}$ state.

At this stage, one cannot have a definitive conclusion; further data and analyses are needed to settle the interpretation of G(1590).

5.2 The M(1400)1^{-+}

The $\pi^0\eta$ data produced in reaction (1.1) with 100 GeV/c π^- beam has been analysed. They are shown on figure 7. A strong forward-backward asymmetry has been observed at $M_{\pi^0\eta} \simeq 1400$ MeV in the Gottfried-Jackson frame of the $\pi^0\eta$ system. This asymmetry, which has been noticed previously [ref. 5.16], may only arise through the presence of an odd wave interfering with the dominant $a_2(1320)$ peak [ref. 5.17]. Three methods (cf. section 3 and 4) have been used for the partial wave analysis. The first two methods consist respectively of an energy independent and energy dependent fit of the helicity partial wave amplitudes to the spherical harmonic moments. The moments have been evaluated from the measured angular distributions in each 35 MeV mass bin using a maximum likelihood method . The third method consists of a direct fit of the differential cross section, expressed in terms of the partial wave amplitudes to the $\eta\pi^0$ experimental angular distributions. Results of the three methods has been found to converge to a unique interpretation of the observed $\eta\pi^0$ angular distributions; the production of a non-$q\bar{q}$, $J^{PC} = 1^{-+}$, $I^G = 1^-$ meson has been observed in the P_0^2 amplitude at $M_{\eta\pi^0} = 1406 \pm 20$ MeV with a width of 180 ± 20 MeV. Being an isovector, this exotic state cannot be a glueball. Two interpretations have been proposed: that it is a hybrid state containing a valence gluon, $q\bar{q}g$ [ref. 5.18], or that it is a four-quark state [5.19].

Data on the $\eta\pi^-$ system obtained in the reaction $\pi^- p \longrightarrow \eta\pi^-$ at 6.3 GeV beam momentum at the KEK 12 GeV Proton Synchrotron also exhibits [ref. 5.20] a strong forward-backward asymmetry. The partial wave analysis showed a clear enhancement of the P_+ wave which can be described with a Breit-Wigner shape with a of mass 1323 MeV and a width of of 143 MeV. This may correspond to the M(1400), though the mass value is lower and the width is narrower than that reported by GAMS.

There are several arguments against the GAMS and KEK P-wave shapes. The major one comes from the acceptance problem; both detectors have almost zero acceptance for $|\cos\theta_{GJ}| > 0.8$ making the partial wave analysis very hard. Another argument comes from the dominance of the $a_2(1320)$ peak which sits right at the same place as the P-wave and suggests that the shape of the P-wave is somehow due to a leakage from the D-wave. Furthermore, the backgound around the $a_2(1320)$ D-wave peak is about the same size as the peak-amplitude of the P-wave.

New results with higher statistics of the $\eta\pi$ system are expected to come out soon from the BNL experiment; this will hopefully bring more consclusive results.

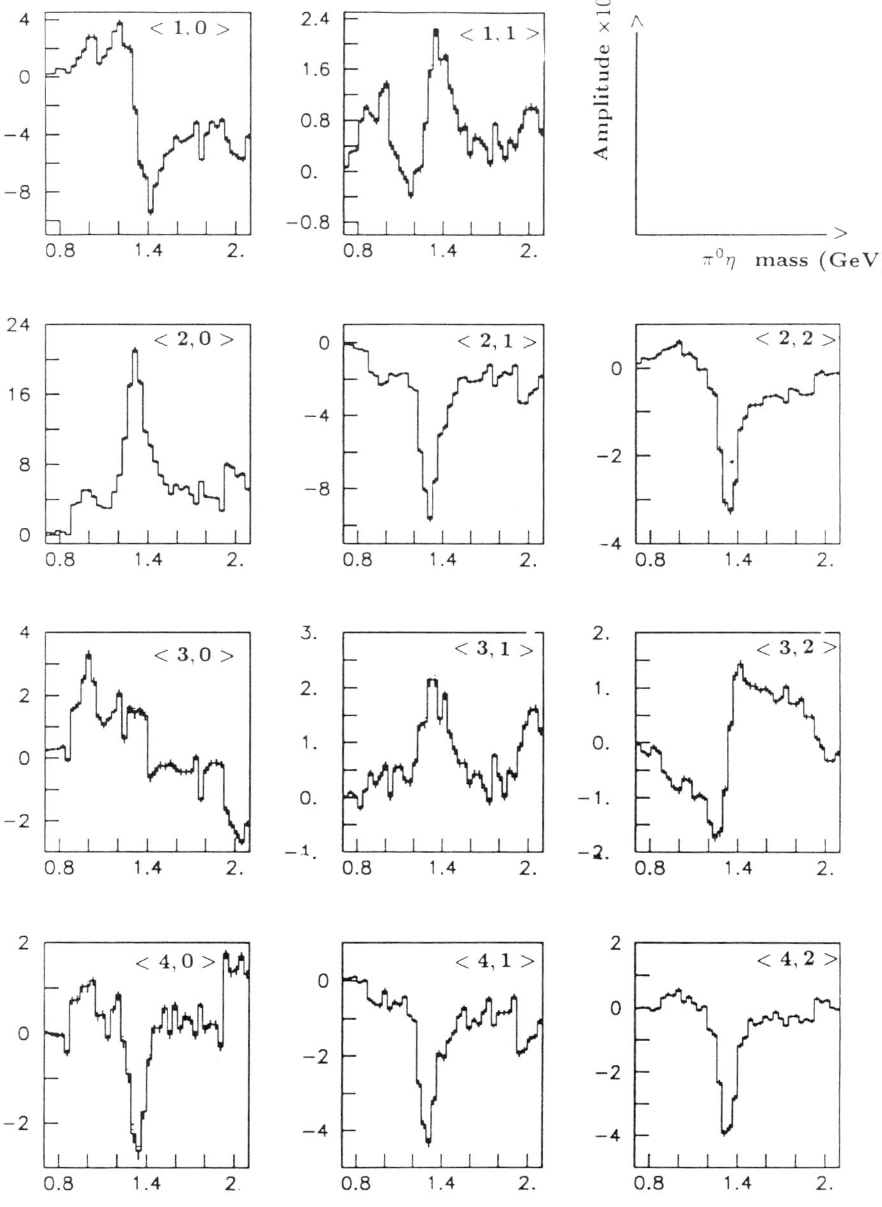

Figure 7.

5.3 Some other results

A new structure, X(1300), has also been observed in the S-wave ($J^{PC} = 0^{++}$) partial wave analysis of $\eta\pi^0$ (100GeV)). It could be the right a_0 member instead of the $a_0(980)$ (ex $\delta(980)$) in the $^3P_0$ nonet. The $a_0(980)$ has been proposed as a $K\bar{K}$ molecule [5.20].

In the $\eta\eta'$(38GeV), a peak at $M_{\eta\eta'} = 1910$MeV with a width of $\Gamma = 90 \pm 35$ MeV was observed [ref.5.22]. The low statistics does not allow the spin assignement. However, the absence of the symmetric decay modes $\pi^0\pi^0$, $\eta\eta$ and $K^0_sK^0_s$ as well as the t-dependance (indicating a non-OPE production mechanism), could be explained if the observed state had exotic quantum numbers $J^{PC} = 1^{-+}$.

In the charge exchange reaction (1.1), where the recoiling baryon is an N^* and not a neutron, a narrow peak was observed in the $\eta\eta$(38GeV) mass spectrum [5.23] with a mass of 1740 MeV and a width of $\Gamma < 80$ MeV. This particle shows exotic charcteristics that differ from those of ordinary mesons : its production is accompanied by isobars and not neutrons, suppression of the OPE mechanism, very small production cross section and narrow width.

6 CONCLUSION

Hadron spectroscopy is still far from being completely understood. New steps have been made through the studies of the neutral decays of light mesons produced in hadron interactions.

The emergence of numerous states that do not find their place either in SU(3) or in simple gluonic schemes, shows the need for a systematic exploration for a thorough knowledge of the light meson spectrum. One may hope that out of the present complexity of the meson spectrum, theory will be able to refine its methods and bring QCD on the same footing as QED.

The contents of this report are far from being exhaustive; either on the GAMS detection systems nor on GAMS physics results. A lot of GAMS data is still not analysed yet !.

ACKNOWLEDGEMENTS

I would like to thank Prof. D. Bugg for inviting me to such a pleasant and interesting school, the GAMS collaboration for the contents of this report and my officemate P. Vikas for helping me writing down the report.

References

[1.1] W.A. Apel et. al., Nucl. Phys. B151 (1979), B152(1979) 1; Phys. Lett. B83 (1979) 181.

[2.1] F. Binon et. al., Nucl. Inst. Meth. A248(1986)86.

[2.2] D. Alde et. al., Nucl. Inst. Meth. A240(1985)343.

[2.3] D. Alde et. al., Int. Rep. CERN-EP/87-88, february 1987.

[2.4] D. Brahy and E. Rossa, Nucl. Inst. and Meth. 192 (1982)359.

[2.5] D. H. Perkins, Introduction to High Energy Physics, Addison-Wesley Pub. Co. Inc.

[2.6] E. Segré, Nuclei and Particles, 2nd ed., W. A. Bejamin, Inc. (1977).

[2.7] K. Kleinknecht, Detectors for particle Radiation, Cambridge Univ. Press (1986).
[2.8] B. Rossi, hight Energy Particles, Prentice Hall pub. co. (1952).
[2.9] W. R. Nelson et al., The EGS code system, SLAC Report 265, (1985).
[2.10] R. Brun, GEANT3, CERN repport DD/EE/84-1 (1987).
[2.11] F. Binon et al., Nucl. Inst. and methods, A248 (1986) and references therein.
[2.12] F. Binon et al., Nucl. Inst. and methods. 188 (1981)507.
[2.13] R. K. Bock et al.,editors, Formulae and methodes in experimental data evaluation, volume 3, published by the EPS, Computational Physical Group.
[3.1] B.R. Martin, D. Morgan, G. Shan, Pion Pion Interactions in Partcile Physics, academic Press, 1976.
[4.1] G. Costa et. al., Nucl. Phys. B175 (1980) 402.
[4.2] G. Grayer et al. Nucl. Phys. B75 (1974) 189.
[4.3] A. Gersten, Nucl. Phys. B12 91969) 537.
[4.4] E. Barrelet, Nuovo Cim. 8A(1972) 331.
[5.1] F. Binon et. al., Lett. Nuovo Cim. 39(1984)41.
[5.2] F. Binon et. al., Nuovo Cim. 78A(1983)313.
[5.3] F. Binon et. al., Nuovo cim. 80A(1984)363.
[5.4] D. Alde et. al., Nucl. Phys. B269(1986)485.
[5.5] D. Alde et. al., CERN-EP/87-161 sept. 1987.
[5.6] D. Alde et. al., Phys. Lett. 182B(1986)105.
[5.7] D. Alde et. al., Phys. Lett. 177B(1986)120.
[5.8] D. Alde et. al., Phys. lett. 205B(1988).
[5.9] M. Boutemeur, AIP Conf. Proc. 185, SUH-URH Chung Ed. p. 389 Brookhaven (1988).
[5.10] D. Alde at al., Phys. Lett. 201B(1988)160.
[5.11] J. F. Donoghue, AIP conf. Proc . 81, C. A. Heusch and W. T. Kirk Ed p. 97, am. Inst. Phys. New-York (1982).
[5.12] V. Novikov at. al., Nucl Phys. B191(1981)301.
[5.13] B. Cohen et. al., Phys. Rev. D25(1982)1786.
[5.14] S. S. Gershtein et. al., Z. Phys. C24(1984)305.
[5.15] S. S. Gershtein, Prep. IHEP 87-42, Serpukhov 1987.
[5.16] W. D. Apel et. al., Nucl. Phys. B193(1981)269.
[5.17] T. Barnes, UTPT (Toronto) 85-21.
[5.18] M. S. Chanovitz, Phys. Lett. B187(1987)409.
[5.19] F. Close and H. Lipkin, Phys. Lett. B169(1987)245.
[5.20] H. Aoyagi et al., Phys. Lett. B (1993)246-254
[5.21] J. Weinstein, AIP Conf. Proc. 185, SUH-URH Chung Ed. p. 400 Brookhaven(1988).
[5.22] D. Alde et al., Phys. Lett. B276 (1992) 375-378
[5.23] D. Alde et al., Phys. Lett. B284 (1992) 457-460

PHYSICS OF 0^{++} AND 2^{++} MESONS

T. H. Burnett

Physics Department
University of Washington
Seattle WA 98195-1560

INTRODUCTION

These lectures present an introduction to the study of scalar and tensor mesons, with an emphasis on an introduction to the phenomenology of spectroscopy. Specifically I examine why one uses Breit-Wigner resonance functions to characterize the parameters of the particles, and how the simple function must be modified, sometimes dramatically, in order to understand and interpret the experimental results. I apply this to the lowest mass tensors, where the situation is well understood, then to the scalars, showing that the quark model can account for the observed scattering in both cases. I then study some recent results leading to the conclusion that non quark model states clearly exist.

Since the major goal of study of the mesons is to look for more states than can be accounted for by the quark model, I remind you that scalar ($J^{PC}=0^{++}$) and tensor ($J^{PC}=2^{++}$) mesons result from quark-antiquark bound states with L=1, and the quark spins parallel or anti-parallel. A classic example, about which there is no controversy, is afforded by the charmonium system shown in Figure 1.

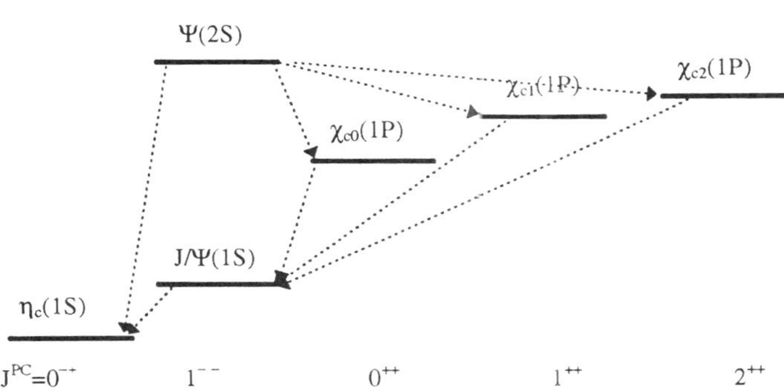

Figure 1. The familiar charmonium level scheme. Dotted lines are photon transitions.

Hadron Spectroscopy and the Confinement Problem
Edited by D.V. Bugg, Plenum Press, New York, 1996

In the quark model, scalar and tensor states are naturally associated, with masses differing only by the spin-orbit coupling, 140 MeV in this case. They also couple to the same final states, especially to two pseudoscalars (e.g., $\pi\pi$, $\eta\eta$, KK, etc.)—an important experimental problem for untangling contributions.

Table 1. Masses and widths of charmonium scalar and tensor states

State	Mass (MeV)	Width (MeV)
$\chi_{c0}(1P)$	3415.1±1.0	13.5±3.3±4.2
$\chi_{c2}(1P)$	3556.17±0.13	2.00±0.18

For example, both χ_0 and χ_2 decay to $2(\pi\pi)$ at about the same branching ratio, 3%, but the widths are quite different; the ratio of the widths is in fact well understood if the spin assignments and the charmonium model are accepted. A theme of these lectures is the interpretation of observed widths and masses: how they are determined from data, and how then one relates the measurements to dynamical quantities.

The quark model works quite well for charmonium (see lectures by T. Barnes[1]). A prerequisite for looking for mesons with gluonic degrees of freedom is to understand fully the predictions of the quark model. This is also a theme of these lectures.

Finally, let me state without apology that this is not a complete review. I will concentrate on some interesting aspects for the beginning student, and refer to the some excellent recent reviews[2].

An Initial Look At The Spectrum

I start with tables of accepted, likely, and some not-so-likely states. The particle names are in boldface if the particle is in the PDG 1994[3] summary (PDG94), and so is considered to be well established, and italic if it does not appear at all, that is, represents a new claim. The column headed "QM" is a sometimes tentative quark model assignment for the observed resonance. I'll return to consider evidence in selected channels later. When there are multiple entries for masses and widths, it is to emphasize the existence of different experiments perhaps supporting the result, or to allow the reader to judge whether the two claims are consistent.

Table 2. I=0 scalar candidates. References can be found in PDG94 or in the text. Names in boldface appear in the PDG summary, quantities in italics have appeared since the PDG94.

Name	mass (MeV)	width (MeV)	QM	Comments
$\mathbf{f_0(980)}$	980±10	40 - 400	1P	Controversial!
$\mathbf{f_0(1300)}$	1000-1500	150 - 400	1P′	
$f_0(1370)$				"represents broad pp elastic resonances in this region"
	1340±40	*255±50*		Crystal Barrel: $\pi^0\pi^0$, ηη
	1374±38	*375±61*		Crystal Barrel: $\rho^+\rho^-$
$f_0(1450)$	*1445±5*	*65±10*		WA91: 4π
$f_0(1525)$	*1505±20*	*150±20*		Crystal Barrel: $\pi^0\pi^0$, ηη
	1488±10	*148±17*		E760: ηη (J^{PC} not determined)
	1505	*148*		J/ψ →γ4π reanalysis
$f_0(1590)$	1581±10	180±17		GAMS: ηη′, ηη, $4\pi^0$
$f_0(1750)$	*1748±10*	*264±25*		E760: ηη (J^{PC} not determined)
	1750	*160*		J/ψ →γ4π reanalysis
X(1950)	1952±14	238±35		LASS: K*K*
	1964±35	225±50		E760: $\pi^0\pi^0$
$f_0(2104)$	2086±30	210±63	2P′	MkIII: J/ψ →γ2π;
	2104±35	*203±10*		E760: ηη (J^{PC} not determined)
	2104	*203*		J/ψ →γ4π reanalysis

Table 3. I=0 tensor candidates.

Name	mass (MeV)	width (MeV)	QM	Comments
$\mathbf{f_2(1270)}$	1275±5	185±20	1P	
$f_2(1430)$	1421	30		various
$f_2(1520)$	1510±8	120±5		Crystal Barrel. This was the A_X, now the $f_0(1525)$. ππ, ηη
$\mathbf{f_2'(1525)}$	1525±5	76±10	1P′	
$f_2(1640)$	1647	58		GAMS ωω
	1640	*140*		J/ψ →γ4π reanalysis
$\mathbf{f_J(1710)}$	1709±5	140±12		The original θ discovered in J/ψ radiative decays. Central production favors spin 2
$f_2(1900)$	1918±12	390±60		WA91: 4π
X(1910)	1810-1920	90-250		GAMS: $4\pi^0$, ωω, ηη'
$f_2(2010)$	2011±70	202±60	2P′	BNL φφ
$f_2(2150)$	2130±35	130±30		GAMS ηη
$f_2(2300)$	2297±28	149±40	1F′	BNL φφ
$f_2(2340)$	2339±60	319±80	3P′	BNL φφ

TENSORS: THE $F_2(1270)$ AND $F_2'(1515)$

We now look at some of the cleanest evidence for the existence, parameters, and spin assignments of the $f_2(1270)$ and $f_2'(1515)$. It is of course well established that these are the I=0 components of the tensor 1P quark model nonet, along with the $a_2(1320)$ and K2*(1430). Figure 2 shows the $\pi\pi$ mass distribution for the decays $J/\psi \rightarrow \omega\pi\pi \rightarrow 5\pi$ as observed by DM2[4]. The curve is two simple Breit-Wigner functions on top of a simple polynomial background. For the $f_2(1270)$, I've annotated the plot to show the assumed

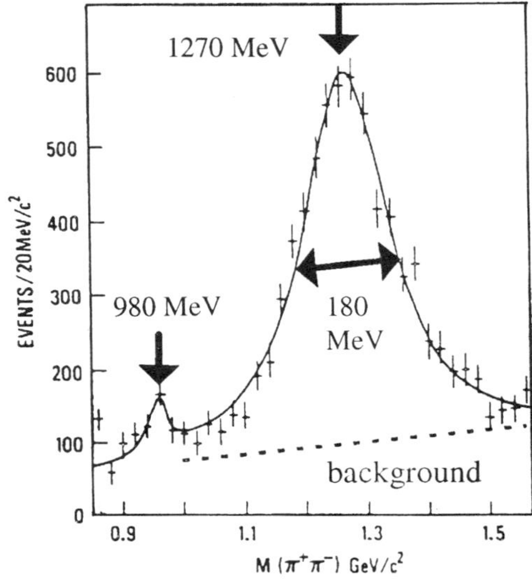

Figure 2. Observation of the $f_2(1270)$ and $f_0(980)$ in $J/\psi \rightarrow \omega\pi\pi$. Data from DM2[4].

Table 4. $B(J/\psi \rightarrow VT)$, in units of 10^{-4}.

V\T	$f_2(1270)$	$f_2'(1515)$
ω	43±6	<2.2
ϕ	<4	8±4

background (mostly $J/\psi \rightarrow b_1\pi$) and how the parameters correspond to the peak and full-width at half maximum (FWHM). This is one example of a group of related decays, the general $J/\psi \rightarrow VT$. Since the vector particle V, ϕ or ω, is well known to be composed of strange and non-strange quarks, and since the decays obey the OZI rule, a system recoiling against an (ω,ϕ) is expected to be dominantly composed of (non-strange, strange) quarks. This is confirmed by the branching fractions in Table 4.

How do we know that the f_2 has spin 2?

I now address the obvious difference between scalar and tensor mesons: the spin. What does it mean to have spin? Consider the amplitude for $f_2 \rightarrow \pi\pi$ in the rest frame of the f_2. Since the pions have no spin, the f_2 having spin J=2 means that this amplitude must have an index that takes 2J+1=5 values, and has specific transformation properties under rotation. In this case, the amplitude must take the form[5]

$$A_\lambda(\theta,\phi) = C \cdot d^2_{\lambda 0}(\theta) e^{i\lambda\phi}$$

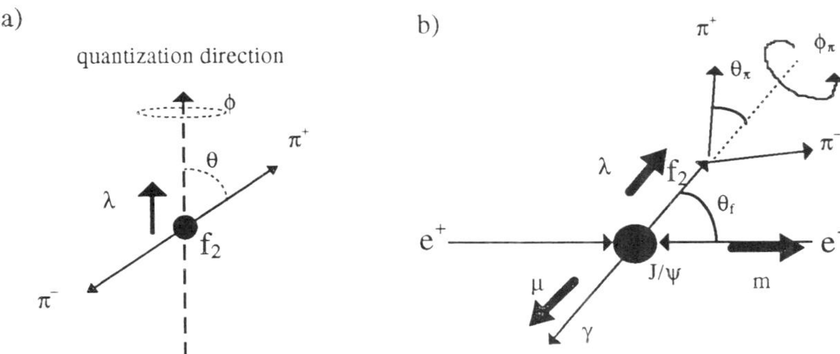

Figure 3. Diagrams showing (a) the decay $f_2 \to \pi\pi$ at rest, and (b) the cascade $e^+e^- \to J/\Psi \to \gamma f_2$, $f_2 \to \pi\pi$. Particles, angles, and helicity indices are labeled. In (b), note that θ_π refers to the helicity angle.

where λ is the polarization of the f_2, and θ and ϕ are the polar angles of the π^+ with respect to the polarization direction, see Figure 3a.
But we measure angles, not indices. Consider $J/\psi \to \gamma f_2$, Figure 3b.The amplitude must have the form

$$T_{m\mu} = \Sigma_\lambda\ A_\lambda d^1_{m\lambda}(\theta_f)\ d^2_{10}(\theta_\pi)\ \exp(i\lambda\phi_\pi)$$

where $m=\pm1$ and $\mu=\pm1$ are the J/ψ and γ polarizations, indices of the amplitude because they represent external particles. Note that θ_π and ϕ_π are defined in the helicity frame, the rest frame of the f_2 reached by a boost along the f_2 direction in the J/ψ frame. Since the f_2 is internal, the total amplitude is a coherent superposition of the possible values, which are $\lambda=m$, $m\pm1$. Finally, parity requires $A_\lambda = A_{-\lambda}$, so there are three possible amplitudes corresponding to f_2 helicity 0, 1, or 2. One often sees the definitions $x=A_1/A_0$ and $y = A_2/A_0$. The final rate is an incoherent sum over m and λ.

So, we know that the f_2 has spin 2 partly since the distributions of $\cos\theta_\pi$ and ϕ_π are not consistent with being flat, as would be expected for spin 0. In practice, the joint distribution is fit to the parameters x and y. For example, Figure 4 shows the distribution of $\cos\theta_\pi$ for the events in the $f_2(1270)$ region. The line is a fit roughly corresponding to $x=1$, $y=0$, and it is clear that the distribution is not consistent with the uniform distribution expected for spin 0.

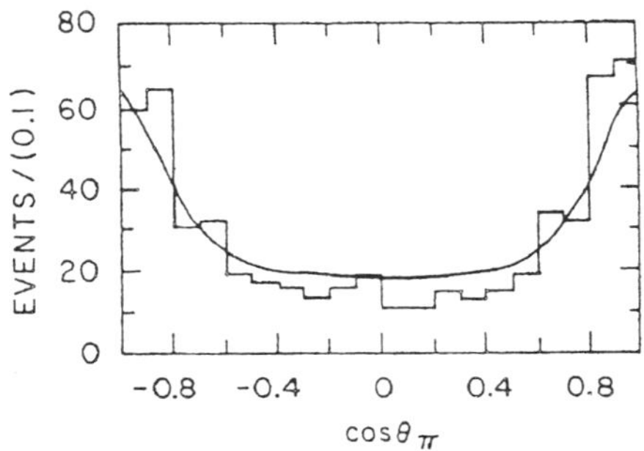

Figure 4. Distribution of $\cos\theta_\pi$ for events in the f_2 region, Mark III[6] data.

In Figure 5 we see dramatic evidence from Mark III data on (a) $J/\psi \to \gamma\pi\pi$ and (b) $J/\psi \to \gamma KK$ of $f_2(1270) \to \pi\pi$ and $f_2'(1515) \to K^+ K^-$, which are the dominant decays of each. Two other objects of interest are visible, $f_J(1710)$ seen in both, and $f_0(2100) \to \pi\pi$.

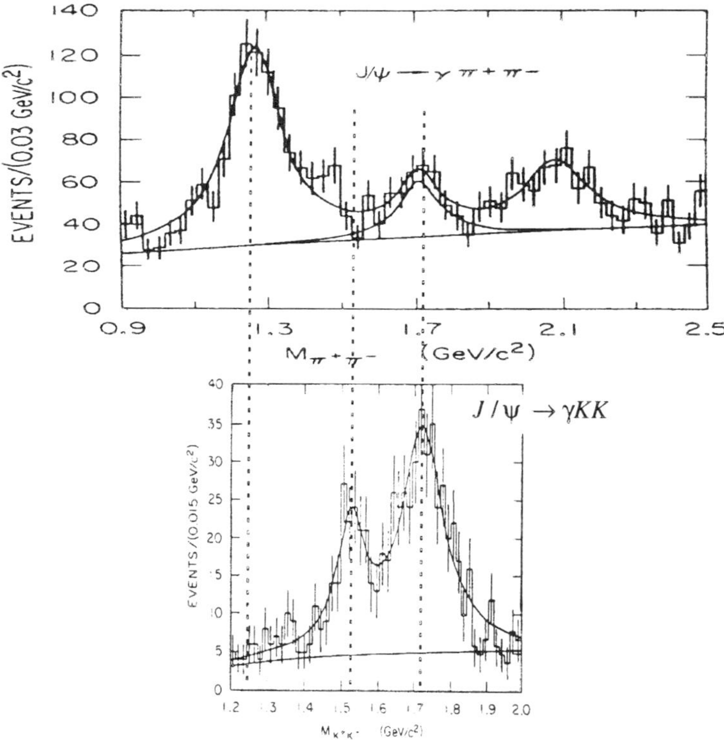

Figure 5. Plots of $J/\psi \to \gamma\pi\pi$ and $J/\psi \to \gamma KK$, from Mark III, Ref. 6.

What is a Breit-Wigner?

Now we examine the function used to find the mass and width of resonances like the $f_2(1270)$, the famous Breit-Wigner (henceforth BW). According to the PDG[7]:

$$a_\lambda(s) = \frac{m\Gamma_{el}}{m^2 - s - im\Gamma_{tot}}$$

is the elastic partial wave scattering amplitude as a function of mass squared s for spin λ, corresponding to a resonance with mass m, width Γ_{tot} and elasticity Γ_{el}/Γ_{tot}. This is represented pictorially, for a resonance that couples to $\pi\pi$, in Figure 6(a): The internal line implies a pole of the scattering amplitude, at the complex value $s=m^2$ -i $m\Gamma$. A resonance corresponds to a pole in the amplitude.

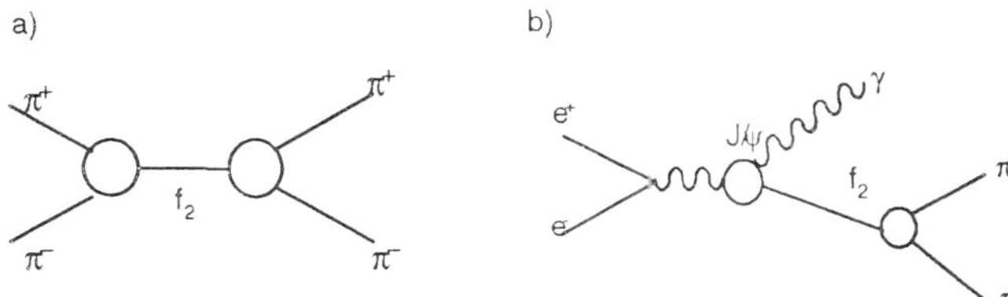

Figure 6. Diagrams showing the f_2 as an internal line in (a) elastic $\pi\pi$ scattering, and (b) J/ψ radiative decay

But this is elastic $\pi\pi$ scattering, not the production and decay to $\pi\pi$ that we are studying. Consider instead J/ψ production and decay, Figure 6(b). This amplitude, as a function of the $\pi\pi$ invariant mass, must have the same pole, even though the production mechanism is completely different. Thus, as a function of the $\pi\pi$ mass, we expect to see the same peak.

In Figure 7, I show figures drawn by the program MathCad[8], a very easy (and recommended) way to study properties of functions and to do analysis. The equations use a ":=" symbol to represent definition. All of the computations in these lectures were performed by MathCad, and many of the equations were imported from the worksheet that computed the plots shown.

You see the definition of the BW function using the same notation as above, and various plots using the $f_2(1270)$ parameters. The phase shift is defined as shown: see the lectures by Anisovich[9] for a derivation of this quantity using non-relativistic quantum mechanics. Notice that the maximum phase variation and peak in the magnitude occur at the same place.

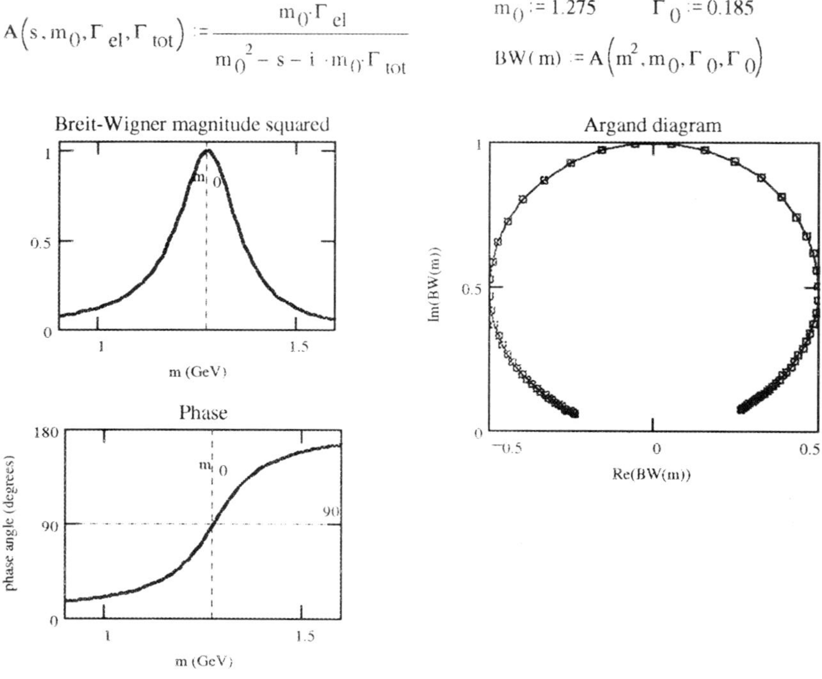

Figure 7. MathCad formulas and plots showing properties of the simple Breit-Wigner function

Interpretation of the widths: why the difference?

The width, or energy uncertainty is simply related to the decay rate by the uncertainty principle. The quark model interpretation of the f_2 and f_2', that they are the hidden non-strange and strange members of a nonet, was clearly supported by J/ψ decays. Yet these very similar objects have quite different total widths. I show a simple calculation of $\Gamma(f_2' \to KK)/\Gamma(f_2 \to \pi\pi)$: the phase space for a spin 2 object with mass M to decay to two particles of mass *m* is $\Phi(M, m) := \frac{k(M^2, m, m)^5}{M}$, where k is the CM momentum. Then we expect a ratio $\frac{2 \cdot \Phi(1.515, m_K)}{3 \cdot \Phi(1.270, m_\pi)} = 0.383$, where 2/3 is the ratio of the SU(3) Clebsch-Gordan coefficients. This is to be compared with the measured $\frac{76}{185} = 0.411$ Figure 8 shows the effect of the phase space factor. This is not bad, indicating the importance of the phase space and angular momentum factors, but there are some missing details:

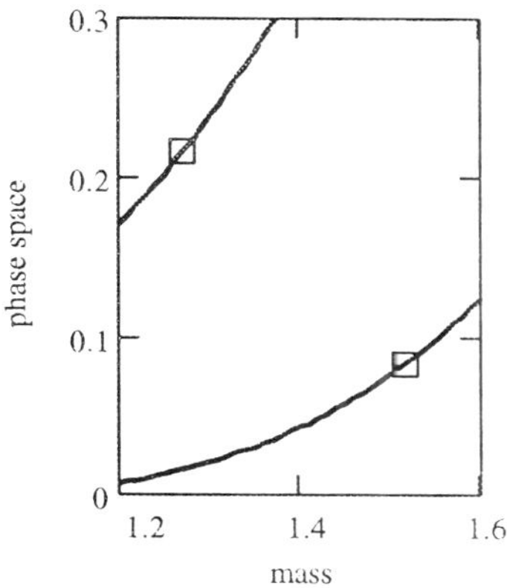

Figure 8. Phase space *vs.* mass for the f_2 and f_2', in a simple model. The two lines are for $\pi\pi$ and KK, with the boxes marking the positions of the f_2 and f_2'.

- other channels, like $\eta\eta$, 4π, etc.
- energy dependence, including "barrier factor". In PDG94 we find the comment[7]:

> "A better form incorporates the known kinematic dependencies, replacing $m\Gamma_{\text{tot}}$ by $\sqrt{s}\Gamma_{\text{tot}}(s)$ where $\Gamma_{\text{tot}}(s)$ is the width the resonance particle would have if its mass were $\sqrt{s}$ ..."

The Mass-Dependent Width BW

So let's modify our simple BW according to this prescription. PDG94 doesn't mention another necessary ingredient, a form factor that corresponds to the wave function for two pions to form an f_2 for example. A model for this is non-relativistic scattering by a spherical square well, the source of the Blatt-Weiskoff barrier penetration factor. Another possible parametrization is an exponential, which I'll in fact use later. Using the f_2 parameters $\Gamma_f := 0.185$, $m_f := 1.270$ and using the spin 2 Blatt-Weiskoff barrier penetration factor function $v_2(x) := \dfrac{x^2}{x^2 + 3\cdot x + 9}$, the modified width is

$$G_f(s) := \sqrt{s}\cdot\Gamma_f\cdot\frac{k_{\pi\pi}(s)\cdot v_2\left[\left(k_{\pi\pi}(s)\cdot R\right)^2\right]}{k_{\pi\pi}\left(m_f^2\right)\cdot v_2\left[\left(k_{\pi\pi}\left(m_f^2\right)\cdot R\right)^2\right]} \tag{1}$$

and the resulting modified BW function

$$B_2(s) := \frac{G_f(s)}{m_f^2 - s - i\cdot G_f(s)} \tag{2}$$

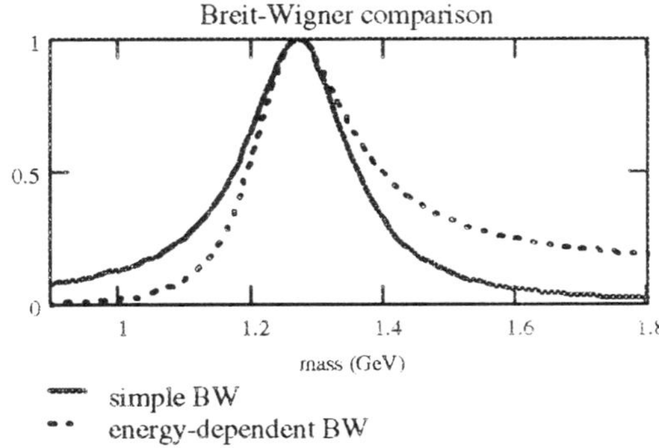

Figure 9. Comparison of a simple BW with the energy-dependent version for L=2.

The magnitude squared for the simple and energy-dependent BW functions are shown in Figure 9, using a barrier radius R=3 GeV^{-1}, about 0.6 fm. Note how prominent the high-mass tail is. For fits like the DM2 fit to $J/\psi\to\omega\pi\pi$ shown above, use of the energy-dependent form would clearly have required a different background, and may have resulted in different fit parameters.

How is this different? Either can be expressed as $A(s)=1/[f(s)-i]$ with f a real function. Any function in this form automatically satisfies the so-called elastic unitarity relation $\text{Im}[A(s)]=|A(s)|^2$, which is the Argand normalization. But this function is related to the actual scattering amplitude $T(s)$ by division by a phase space factor: $A(s) = \rho(s)\ T(s)$. The two-body phase space factor, including the angular momentum barrier factor, is

$$\rho(s) = \frac{k(s)^{2\lambda+1}}{\sqrt{s}} \tag{3}$$

where $k(s)$ is the cm momentum for the two-body decay. Then

$$\text{Im}\ T(s) = \rho(s)\ |T(s)|^2.$$

We see that the function $f(s)$ above is not arbitrary, and must in fact have a denominator factor of $\rho(s)$. The BW form with a mass dependent width satisfies this, while the simple form does not.

Another way of looking at the above is that unitarity requires that the function $f(s)$ have a branch point corresponding to the square root singularity implied by the factor $k(s)$.

Analyticity: The Dispersive part

Having shown that the simple BW is in fact not consistent with unitarity, I now discuss another requirement; there is a relation between the absorptive (imaginary) and dispersive (real) parts of an amplitude: we expect the denominator function to be a real analytic function with right-hand branch cuts corresponding to the thresholds.
This function is known as the vacuum polarization, $\Pi(s)$. For a BW it has only an imaginary part:

$$\Pi(s) = -i\sqrt{s}\Gamma(s).$$

But the real part cannot be constant: analyticity demands that it satisfy a *dispersion relation:*

$$\mathrm{Re}\,\Pi(s) = \frac{1}{\pi}\int_{thresh}^{\infty} \frac{\mathrm{Im}\,\Pi(s')}{s'-s} ds'$$

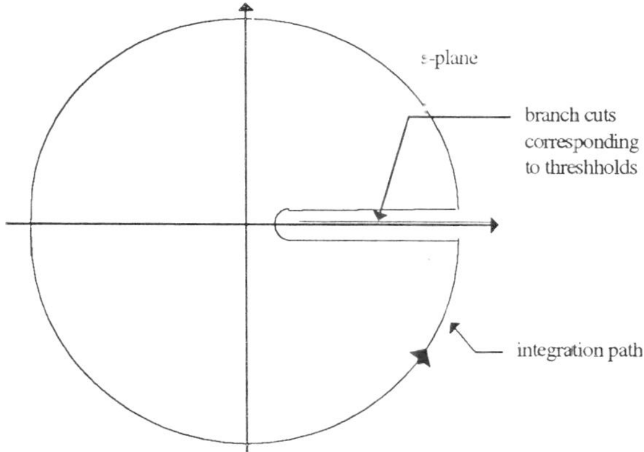

Figure 10. Integration path for the dispersion integral in the complex s-plane. The radius of the circle is taken to infinity, where (for properly-behaved functions the contribution vanishes), leaving the contribution from the branch cut: since Π is a real analytic function, the discontinuity is the same as the imaginary part.

Thus the integral is along the real axis, starting with the first threshold, with the principal value taken. (See the contribution of Anisovich[9] to these lectures.) So we must replace $-i\sqrt{s}\Gamma(s)$ by the full Π(s) in the BW denominator, which becomes

$$m_0^{\;2} + \mathrm{Re}\,\Pi(s) - s + i\,\mathrm{Im}\,\Pi(s);$$

m_0 is interpreted as a "bare" mass, the mass in the absence of final-state corrections. In practice it cannot be separated from constant contributions to Re Π, for example a subtraction.

Apply to f_2 and f_2'

Now lets use both effects, proper unitarity and analyticity, to study the effect on the $f_2(1270)$ line shape. First, I adopt the simpler exponential form for the barrier, defining a vertex, or phase space function

$$f(s,k) := \frac{k^5}{\sqrt{s}}\cdot\exp\left(-\frac{k^2}{k_0^{\;2}}\right), \qquad k_0 := 0.7 \tag{4}$$

The function for Im Π(s) must allow for all accessible decay channels, the full width. I include the channels ππ, KK, and ηη (but note that the f_2 has 10% BF to 4π).

$$\mathrm{Im}\Pi(s) := -\gamma^2\cdot\left(3\cdot f\left(s,k_{\pi\pi}(s)\right) + f\left(s,k_{KK}(s)\right) + s_P^{\;2}\cdot f\left(s,k_{\eta\eta}(s)\right)\right)$$

where γ is a coupling constant, and the coefficients are SU(3) factors. The ππ scattering amplitude is

$$A(s) := \frac{\gamma^2 \cdot 3 \cdot f(s, k_{\pi\pi}(s))}{\Pi(s) - s}$$

where ReΠ is determined by a numerical integration. Figure 11 shows the real and imaginary parts of $\Pi(s)$, and the amplitude squared. The two constants γ and m_0 were adjusted to 1.46 and 1.47, to give the observed mass and width.

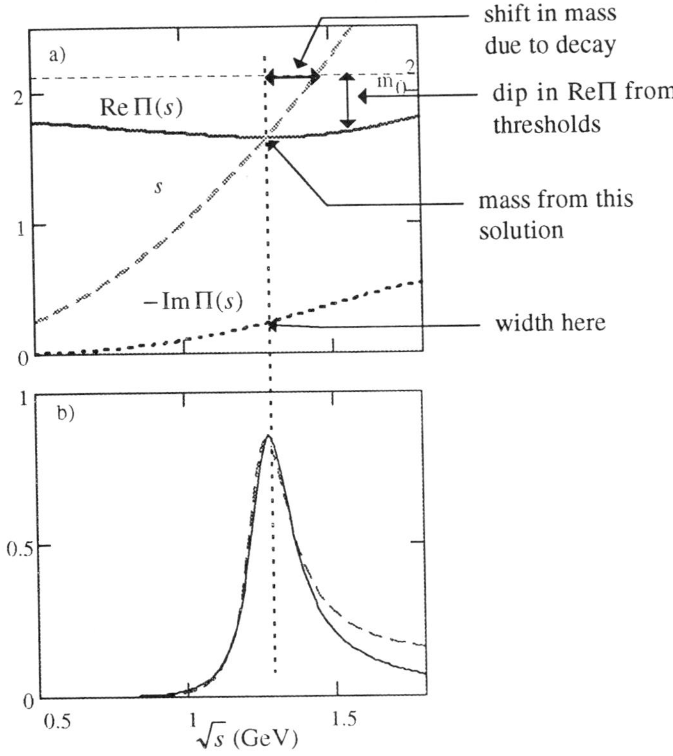

Figure 11. The vacuum polarization and line width for the simple model of f_2 discussed here. (a) has the vacuum polarization, with annotations about the features, and (b) the resulting magnitude squared in the solid, compared with the simple modified BW function, dashed.

Fitting two values with two free parameters may not seem very profound, but when I apply the same procedure to the f_2' but with a bare mass of m_0+2m_s where m_s accounts for the additional mass due to each strange quark, I get the observed mass and width with m_s=76 MeV. See Törnqvist[10] for a more complete treatment of the tensor nonet.
However, comparing the two curves for the line shape, we see that all this effort has little effect on the line shape in Figure 11, perhaps justifying use of the simple BW. But wait until you see the scalars, coming up.

Multi-resonance Vacuum Polarization

Here I summarize the formalism for the modifications necessary to the simple BW when one has strong coupling to decay channels, and more than one resonance. We generalize the vacuum polarization to allow for more than one resonance with the same quantum numbers (*e.g.* f, f′) in which case we have to allow for transitions among the states: the following diagram

α ——○—— β
i

represents the transition $\alpha \to \beta$ with an intermediate two-body state *i*, ππ for example. (Following Törnqvist[11], I let Greek letters represent the single-particle states.) Such a transition amplitude would be large only if both states coupled strongly to the same decay channel *i* which is not the case for f_2 and f_2' as we have seen.

Now to generalize our previous discussion, I introduce the vertex, or threshold, function $G_{\alpha i}(s)$,

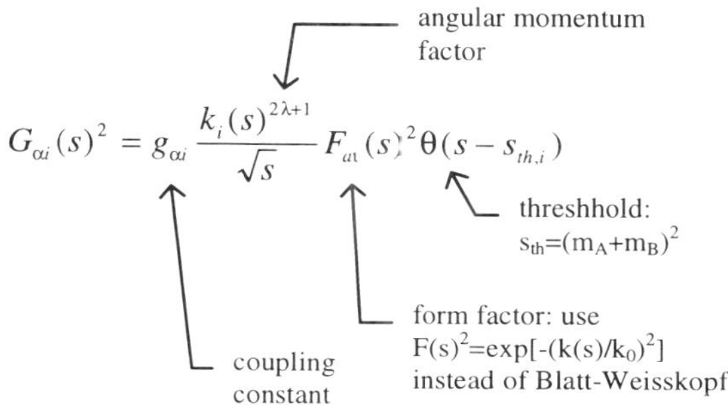

which represents the coupling and the square root of the phase space and threshold factors, as annotated on the equation.
Then unitarity requires that Π "know" about all two body (at least) channels:

$$\operatorname{Im}\Pi_{\alpha\beta}(s) = -\sum_i G_{\alpha i}(s)\, G_{\beta i}^{\,*}(s). \tag{5}$$

The sum must include *all* channels. We make the approximation that only low-lying two-body channels are important in affecting the shape. Analyticity is as above:

$$\operatorname{Re}\Pi_{\alpha\beta}(s) = \frac{1}{\pi}\int \frac{\operatorname{Im}\Pi_{\alpha\beta}(s')}{s'-s}\,ds' \tag{6}$$

where the principal value is taken.

Finally, the amplitude for j→i is

$$A_{ij}(s) = \sum_{\alpha\beta} G^{\dagger}(s)_{i\alpha}\, P(s)_{\alpha\beta}\, G_{\beta j}(s),$$

$$P(s)_{\alpha\beta} = \left[\left(m_0^{\,2} - s\right)\delta_{\alpha\beta} + \Pi_{\alpha\beta}(s)\right]^{-1}. \tag{7}$$

And pictorially

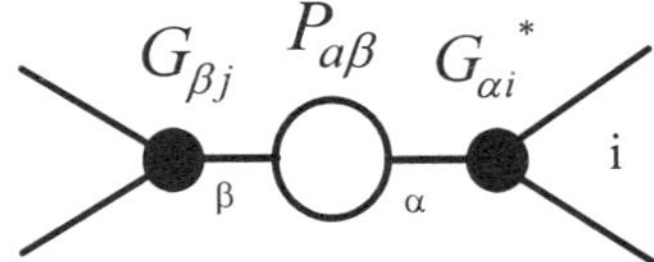

Note the matrix inversion. *P* is the propagator, corresponding to the BW denominator factor.
As a consequence of Eq. (5), the amplitude, satisfies the unitarity equation

$$\frac{A - A^{\dagger}}{2i} = AA^{\dagger}. \tag{8}$$

THE LOW-LYING SCALARS

Introduction

The CERN-Munich ππ phase shift data, inferred from the reaction $\pi^{-}p \rightarrow \pi^{-}\pi^{+}n$, show a dramatic variation that cannot be produced by a single resonance. The most rapid phase motion is near 180°, where the amplitude has a sharp dip, rather than a peak, at a mass just below the KK threshold. This is the $f_0(980)$. Such behavior must be the result of interference between two components. The rapidly varying phase can only be due to a pole close to the real axis. Since the other, background, component is slowly varying, it is not clearly due to a pole; but if so, the pole must be far off the real axis. The crucial question for the scalars is: what is the dynamical origin of these two components?

Some History

I give a brief list of several attempts to interpret the ππ scattering data:

- Au, Morgan and Pennington[12]: This pioneering effort fit the ππ scattering data, especially central production, to a parametrization. They then examined that function to determine the number and positions of poles. The result: 3 poles in the KK threshold region!
- Morgan and Pennington[13]: This subsequent analysis used Yost functions instead, and has a quite complete review. Figure 12, from that paper, shows the variety of values for the (sheet II) pole near the threshold.
- Zou and Bugg[14]: pole plus background modeled by t-channel ρ exchange.
- Anisovitch[9]: two-pole K-matrix approach, described at this school. The two input K-matrix poles result in two poles in the amplitude, one of course near the KK threshold.
- Törnqvist[11]: I will follow this approach to demonstrate how the modified BW can in fact account for the data.

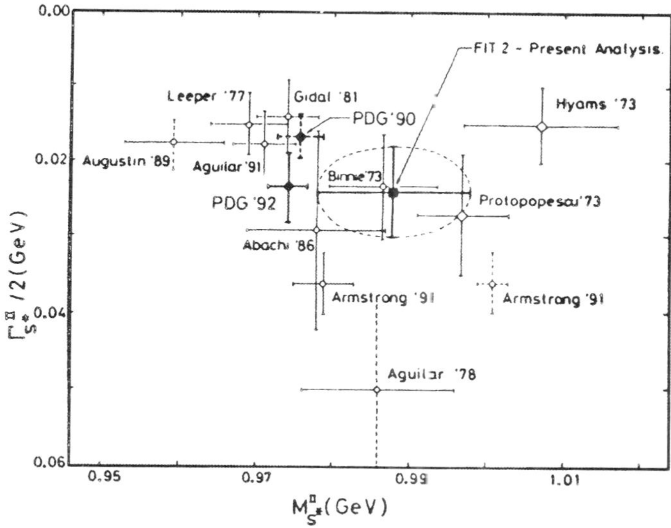

Figure 12 The variety of pole positions for the $f_0(980)$, from Morgan and Pennington ref. (13).

Big difference: L=0 for ππ rather than L=2

I note the difference that the factor of k^4 makes: Figure 13 compares the threshold functions for scalar and tensor decays to ππ. Both have the square root singularity, but the scalar function has a dramatic cusp that, as we'll see, can have a very noticeable effect on the real part of the amplitudes.

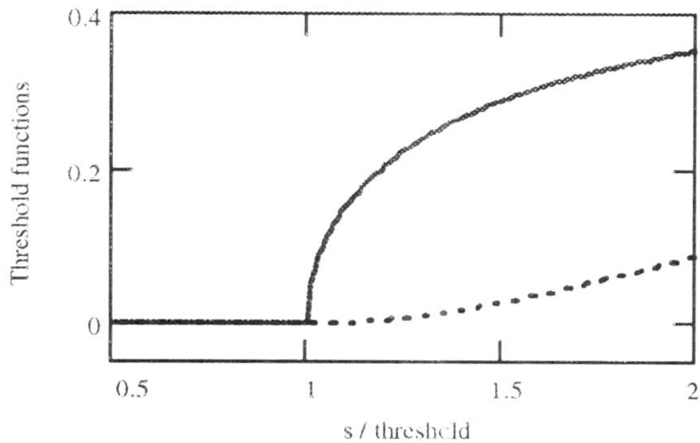

Figure 13. plot showing the difference between spin 0 and 2 threshold functions.

I=1/2: the K_0* (1430)

To understand the scalars, we start with a state that must have quark content, and is not (like the $a_0(980)$) mixed up with the KK threshold and possible molecular interpretations.

This state is clearly seen in the phase shift analysis of $K^-p \to K^-\pi^+n$, in which the K*(890) and $K_2(1430)$ are each seen as a clean BW, and account for all of the 1^{-+} and 2^{++}, respectively. So it is a challenge to understand the very distorted shape seen by LASS[15], shown in Figure 14

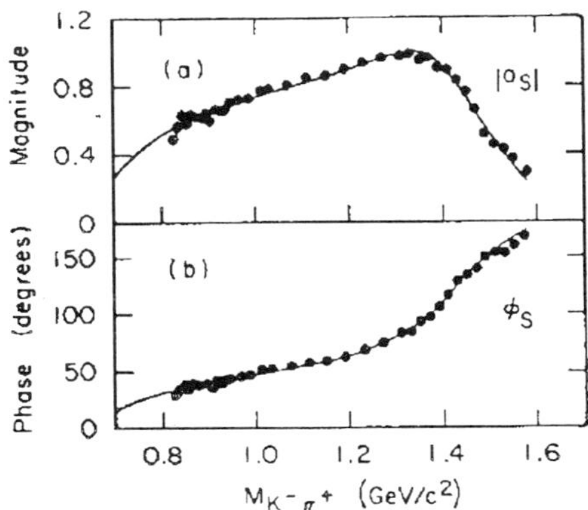

Figure 14. Kπ S-wave scattering data from LASS.

The PDG assigned the mass and width to be 1430 and 287 MeV from the fit shown, which is worth examining. A simple BW was added to a coherent background parametrized by an effective range formula. The relative phase was adjusted to maintain unitarity in the form $\mathrm{Im}\,A = |A|^2$. The parameters are those of the BW. This procedure, known as the Dalitz-Tuan[16] method, is of course quite *ad hoc,* a point made by in the LASS paper. It cannot be consistent with analyticity. An alternate, and simpler hypothesis, is that *all* of the data is due to the scalar quark model state, just as the other two states dominate their respective channels. This is exactly what Törnqvist tried.

A requirement that amplitudes for scalar decays to two pseudoscalars must obey is a zero near s=0, known as the Adler[17] zero. This is enforced by the following vertex function, corresponding to G^2 :

$$F(s, s_A, k) := (s - s_A)\cdot\frac{k}{\sqrt{s}}\cdot\exp\left(-\frac{k^2}{k_0^2}\right) \tag{9}$$

where s_A is the Adler zero, k is the CM momentum including the threshold theta function. Then our I=1/2 vacuum polarization function is

$$\mathrm{Im}\Pi(s) := -\gamma^2\cdot\left(1.5\cdot F\left(s, s_{A.K\pi}, k_{K\pi}(s)\right) + 1.5\cdot F\left(s, 0, k_{K\eta'}(s)\right)\right) \tag{10}$$

The numerical factors are SU(3) Clebsch-Gordan coefficients; the Kη coupling is zero due to the particular value for the ηη′ mixing. Using this function, Törnqvist fit the data above, finding values for the parameters as shown in Table 5.

Table 5. Parameters for Törnqvist fit to the Kπ scalar scattering data.

Parameter	description	fit value
k_0	form factor damping	0.56 GeV
γ	coupling constant	1.1395
m_0+m_s	bare mass	1.520 GeV
$s_{A.K\pi}$	Adler zero for Kπ	-0.46 GeV^2

The bare mass is an overall bare mass for the multiplet, m_0, plus m_s, the difference due to the heavier strange quark. This is strongly coupled with the form factor parameter k_0.

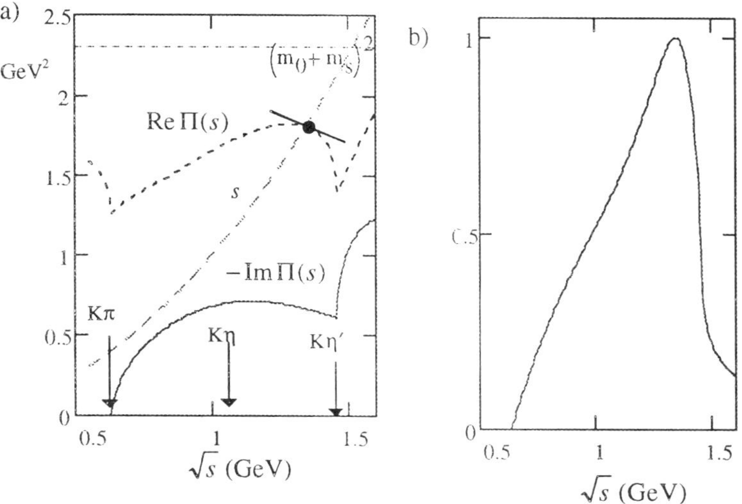

Figure 15. (a) Scalar I=1/2 vacuum polarization, with Tornqvist parameters; (b) corresponding Kπ intensity.

Figure 15 shows the resulting vacuum polarization and Kπ amplitude squared. Note that the mass (the intersection of Re Π and s, indicated by the dot in the figure) is only shifted by a few hundred MeV from m_0+m_s. Törnqvist calls this the BW mass, m_{BW}=1.346 GeV to distinguish it from the different position of the pole. Note also the effect of the Kη′ threshold on Re Π, resulting in a noticeable slope. This has the effect of changing the apparent width. The BW denominator factor can be expanded near the zero in the real part as

$$m_0^2 + \mathrm{Re}\,\Pi(s) - s + i\,\mathrm{Im}\,\Pi(s) \approx \left(m_{\rm BW}^2 - s\right)\left(1 - \left.\frac{d\,\mathrm{Re}\,\Pi(s)}{ds}\right|_{m_{\rm BW}^2}\right) + i\,\mathrm{Im}\,\Pi(m_{\rm BW}^2)$$

$$\propto \left(m_{\rm BW}^2 - s\right) - i m_{\rm BW}\Gamma_{\rm BW} \qquad (11)$$

where m_{BW} and Γ_{BW} are basically defined by the zero in the real part, or a phase shift of 90° and the rate of change of the phase there:

$$\mathrm{Re}\,\Pi(m_{\rm BW}^2) + m_0^2 = m_{\rm BW}^2,$$

$$\Gamma_{\rm BW} = \frac{-\mathrm{Im}\,\Pi(m_{\rm BW}^2)}{\left(1 - \frac{d\,\mathrm{Re}\,\Pi(s)}{ds}\right) m_{\rm BW}}. \qquad \mathbf{(12)}$$

Thus an approaching threshold, like Kη′ in this case, *decreases* the effective width, here to 406 MeV. It is not a large effect for the K_0*, but is for other members of the nonet. This can be interpreted as a mixing of the state with the virtual 2-particle state.

I=1: the $a_0(980)$

Now we examine the situation for the I=1 sector. The prominent, narrow $a_0(980)$ is quite dramatic in Figure 16, from Crystal Barrel[18] study of $\bar{p}p \to \pi^0\pi^0\eta$:

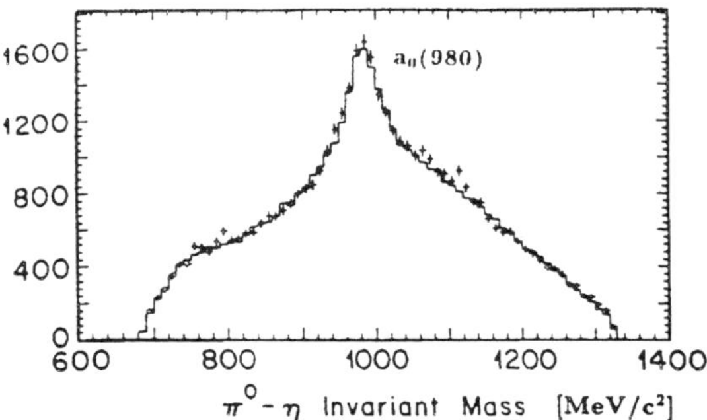

Figure 16. Distribution of $\pi^0\eta$ from the Crystal Barrel.

A BW fit to this figure results in M=982±2, Γ=54±10, the best measurements so far. A more recent approach[19] gives, M=1024+5-26. What do these parameters mean? Is a simple BW the correct way to parametrize this effect?

We instead apply the modified BW. There are three decay channels. $\pi\eta$, KK, and $\pi\eta'$. The imaginary part of Π is

$$\mathrm{Im}\Pi(s) := -\gamma^2 \cdot \left(2 \cdot s_P^{\,2} \cdot F\left(s, 0, k_{\pi\eta}(s)\right) + F\left(s, 0, k_{KK}(s)\right) + 2 \cdot c_P^{\,2} \cdot F\left(s, 0, k_{\pi\eta'}(s)\right)\right)$$

where $s_P = \sin\delta_p$, and $c_P = \cos\delta_p$ where $\delta_p = -54°$ is the pseudoscalar mixing angle.

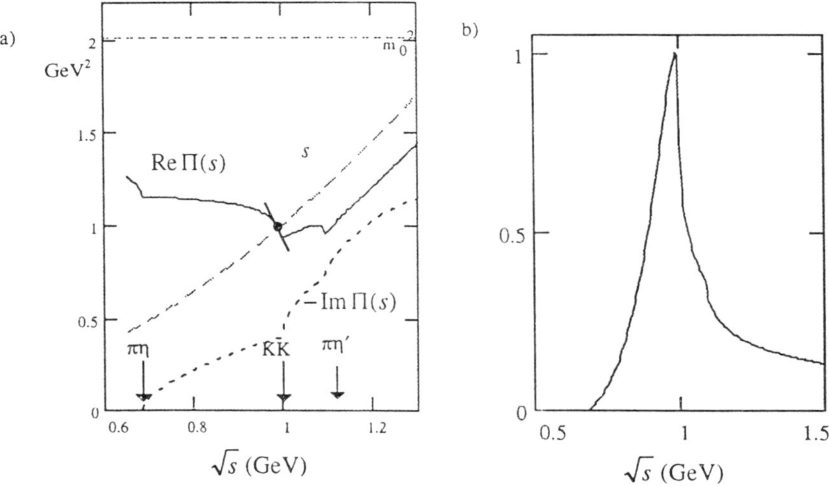

Figure 17: (a) I=1 scalar vacuum polarization function; (b) corresponding $\pi\eta$ intensity.

One parameter was adjusted for this case: m_s was determined to be 100 MeV, so the bare mass in this case is m_0=1420 MeV. Note, in, Figure 17a, how dramatically larger the shift in mass is, 400 MeV to just below the KK threshold.

Here the slope of Re Π is much more negative than the I=1/2 case, so that the apparent width is greatly reduced (see Figure 17a), where the FWHM is 140 MeV,. An interpretation is that the a_0 wave function has a large component of virtual KK. This is perhaps equivalent to the molecule of Isgur and Weinstein[20]. The BW mass and width for this channel are 991 and 109 MeV respectively.

I=0: $f_0(980)$ & $f_0(1300)$

Now to the bottom line: can this simple procedure account for the complicated ππ phase shifts? Figure 18 shows a representation of the ππ S-wave phase shift data used for Crystal Barrel[22] fits, a two-pole K-matrix. Note how the phase goes through 180^0 and keeps going, with extremely rapid behavior at the KK threshold.

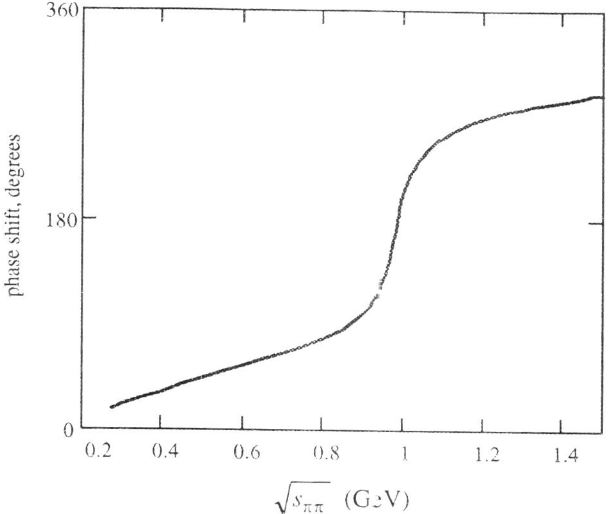

Figure 18: ππ S-wave phase shift parametrization, see text.

I=0 Vacuum Polarization

We've seen how important it is to include the effects of the dispersion relation in the real part of Π. Now we see the effect of another important consequence of the modified BW formalism. That is, the fact that the I=0 sector has two quark states that I'll label

$$n \equiv \left(u\bar{u} + d\bar{d}\right)/\sqrt{2}\,, \quad s \equiv s\bar{s}.$$

Recall that this was unimportant for the tensors since the coupling to common decay channels was small, with the decay channels in fact having little effect. The coupling constants are determined by SU(3) and can be represented as a matrix g with a column index taking five values for the decay channels ππ, KK, ηη, ηη'. and η'η'.

$$g := \gamma \cdot \begin{bmatrix} \sqrt{3} & 1 & s_P^{\,2} & -\sqrt{2}\cdot s_P \cdot c_P & c_P^{\,2} \\ 0 & \sqrt{2} & \sqrt{2}\cdot c_P^{\,2} & 2\cdot s_P \cdot c_P & \sqrt{2}\cdot s_P^{\,2} \end{bmatrix} \quad (13)$$

The overall coupling γ remains the same as for I=1/2 and I=1. Then we define a vector of the vertex functions for the decay channels:

$$F_{thresh}(s) := \begin{bmatrix} F\left(s, 0.5\cdot m_\pi^{\,2}, k_{\pi\pi}(s)\right) \\ F\left(s, 0, k_{KK}(s)\right) \\ F\left(s, 0, k_{\eta\eta}(s)\right) \\ \beta^2 \cdot F\left(s, 0, k_{\eta\eta'}(s)\right) \\ F\left(s, 0, k_{\eta'\eta'}(s)\right) \end{bmatrix} \quad (14)$$

were the $\pi\pi$ function has an Adler zero at position predicted by current algebra, and the $\eta\eta'$ channel has a special enhancement factor β. Finally, $\mathrm{Im}\Pi$ is a symmetric 2x2 matrix:

$$\mathrm{Im}\Pi(s) := -g \cdot \mathrm{diag}\left(F_{thresh}(s)\right) \cdot g^T \quad (15)$$

The one new parameter, β, is set by Törnqvist to 1.6 to fix higher mass behavior, but could be accounting for a missing SU(3) singlet channel, as discussed below.

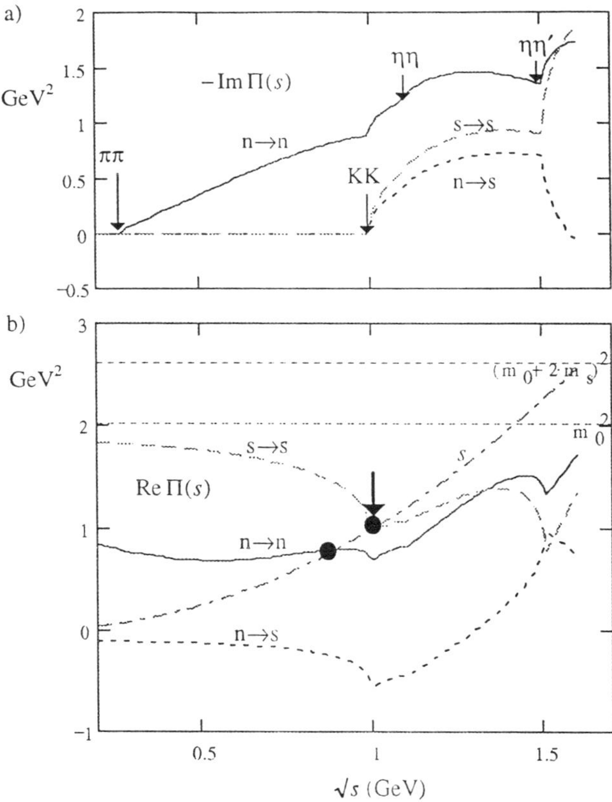

Figure 19. Imaginary (a) and real (b) parts of the scalar I=0 vacuum polarization function.

Figure 20a shows the resulting $\pi\pi$ phase shift, compared with the K-matrix approach mentioned earlier. The intensity, Figure 20b, shows the sharp dip and secondary peak at 1.2 GeV.

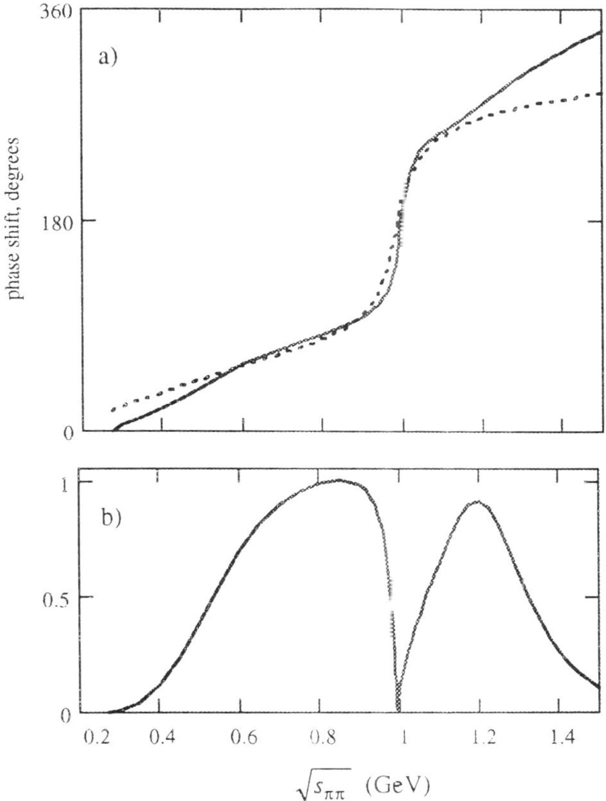

Figure 20 (a) Phase shift from quark model (solid) compared with K-matrix (dashed). (b) Magnitude squared of the $\pi\pi$ amplitude..

Diagonalize the Mass matrix:

To find the actual states, one must diagonalize the propagator P(s) to find the eigenstates. Figure 21 shows the real and imaginary parts of the mixing angle. The interpretation of the angle is that δ=0 corresponds to no mixing, clearly the case for low masses, while the line at −35.26 degrees marks ideal mixing, where the eigenstates correspond to SU(3) singlet and octet, the case above the KK threshold. See Törnqvist[11] for an extensive discussion.

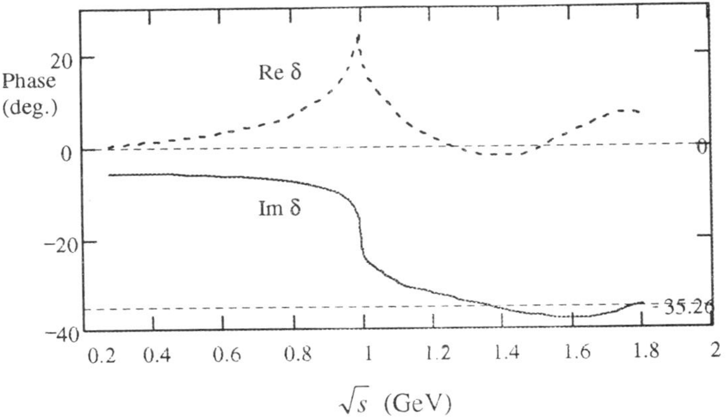

Figure 21. Results of diagonalization of the I=0 propagator matrix.

RECENT EXPERIMENTAL EVIDENCE

I have presented a summary of what I think is compelling evidence that the low mass scalar scattering data, to 1.2 GeV in I=0, 1.5 GeV or so for I=1/2, 1 can be accounted for by the quark model. (But note a different point of view from the conference organizer![21]). Then, given that the quark model prediction for the first radial excitations are typically above 2 GeV, we can examine the data with some confidence that states in this region must correspond to non quark-antiquark states. A review of Table 2 shows that there are in fact many candidates for glueballs or whatever. Due to lack of space (and time) I will concentrate on evidence for the $f_0(1525)$, which should perhaps be renamed $f_0(1505)$.

Crystal Barrel

The Crystal Barrel has made giant contributions to meson spectroscopy lately, studying the annihilation of stopping antiprotons in liquid hydrogen. Final states include $\pi^0\pi^0\pi^0$, $\pi^0\pi^0\eta$, and $\pi^0\eta\eta$. A simultaneous analysis[22] of the first two reactions yielded evidence for $f_0(1375)$ and $f_0(1525)$ in $\pi^0\pi^0$ and $\eta\eta$, the parameters summarized in Table 2. The Dalitz plot for $\pi^0\pi^0\pi^0$ is shown in Figure 22.. Since it has six-fold symmetry, each event is entered six times. Concentrate on the top corner of the triangle, where the arrow labeled "A" points at a horizontal band of events across the plot, which is the evidence for $f_0(1525)$. (The blob "B" is a complication that led to a tensor interpretation, the A_X, in previous analyses; it results from interference of the $f_0(980)$ in the crossed channels.)

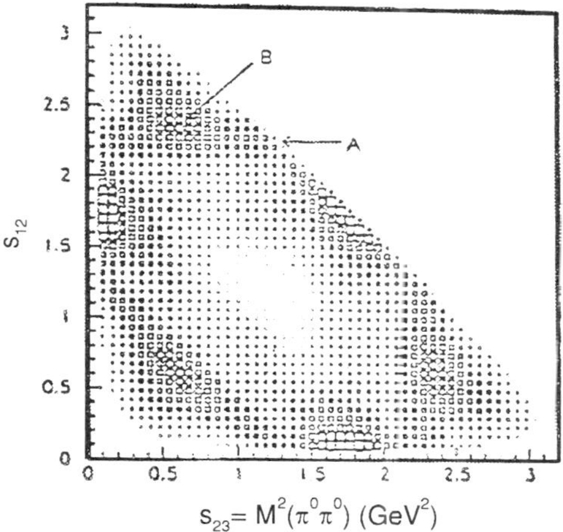

Figure 22. The Dalitz plot for $\bar{p}p \rightarrow \pi^0\pi^0\pi^0$

The evidence for quite broad $f_0(1375)$ in this analysis is not as visible. However a new analysis[23] of the $\pi^0\eta\eta$ final states shows very clearly the need for an enhancement in the 1300 MeV region.

E760 at FNAL: $\bar{p}p$ Annihilation in Flight

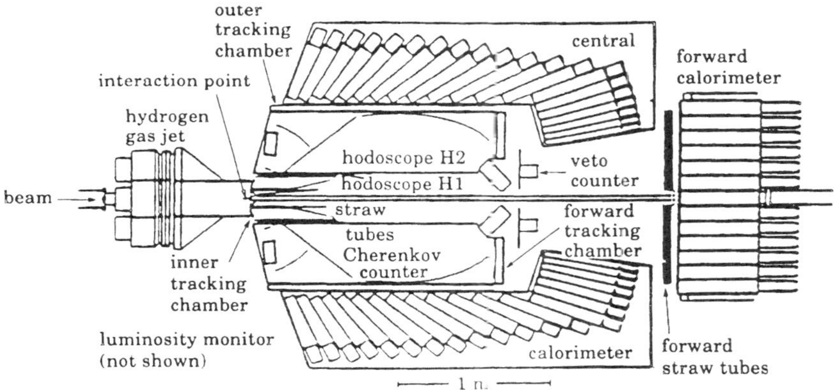

Figure 23 The E760 apparatus

Experiment E760 at the antiproton accumulator at Fermilab, shown in Figure 23, has studied the reaction $\bar{p}p \rightarrow \eta\eta\pi^0$ and observed three clear peaks at two beam energies, Figure 24. A spin-parity analysis has not yet been published, but the peaks can only correspond to even spin particles. The fit values are shown in the E760 entries in Table 2.

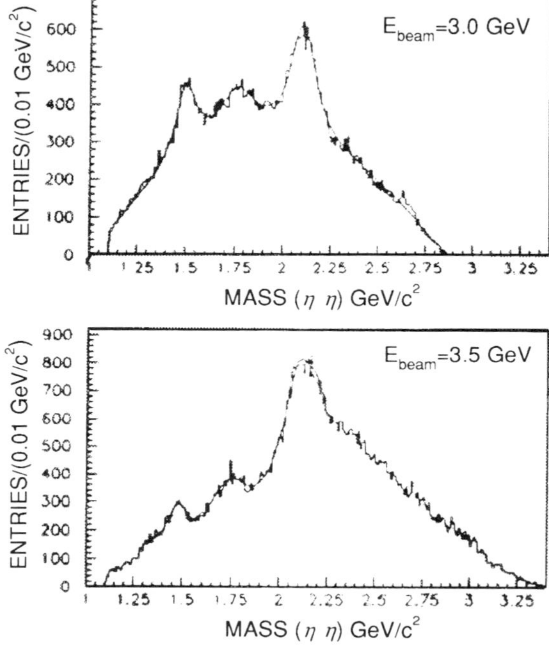

Figure 24. $\eta\eta$ mass spectrum from E760.

Reanalysis of Mark III J/$\psi \to \gamma$ 4π

The nature of J/ψ radiative decays is to enhance states with gluonic degrees of freedom. The largest single channel is to four pions, so these data received a lot of attention a decade ago when the data was collected by the Mark III and DM2 experiments. The Mark III data[24] shown in Figure 25 shows three peaks that seem to correspond to the E760 distributions. A spin-parity analysis identified a large component of $\rho\rho$ 0^{-+}, probably the tail of the η(1490), but more that half was not clearly identified, or especially, associated with the peaks. DM2[25] analyzed a larger similar data set, using the same technique and concluded that all three peaks pseudoscalar resonances. The PDG accepted η(1760) as a candidate.

Neither analysis allowed for S-wave interactions among the final state pions. A recent reanalysis by Bugg and several of the lecturers at this school, this author included[26] used a larger data set corresponding to 4.9×10^6 produced J/ψ taken by Mark III in 1983 and 1985. (The published[27] analysis used data corresponding to 2.7×10^6 produced J/ψ taken 1982 and 1983)

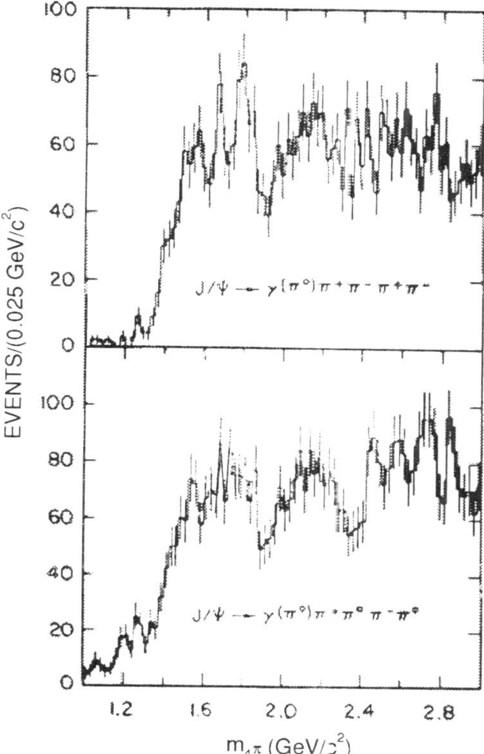

Figure 25. Mark III data

It was found that the data could be accounted for by four components: pseudoscalar consistent with the tail of the $\eta(1440)$; three scalar resonances coupling to $\sigma\sigma$; where "σ" corresponds to the $\pi\pi$ scattering amplitude; a tensor state at 1640; and incoherent phase space background. The primary result, the scalar, is shown in Figure 26: with the resonance parameters used shown in Table 2.

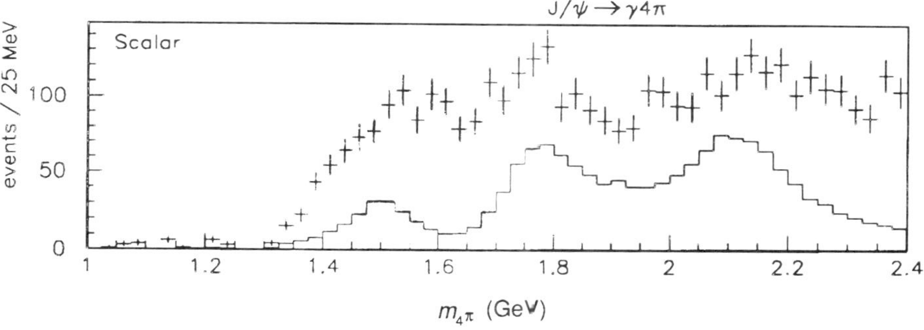

Figure 26. The scalar component (solid line) of the fit to Mark III data (crosses).

SUMMMARY, FURTHER QUESTIONS

I've presented a description of what I think is the most important evidence for the existence of a relatively narrow (Γ=150 MeV) scalar state at 1500 MeV. As we saw, the two I=0 quark-antiquark states have bare masses in this region as well, but coupling to final states produced the dramatic changes that seem to account for the rather complicated scattering data. A narrow width implies that such effects are rather smaller, but especially that the origin of the state is different. For a slightly different analysis, see Amsler and Close[28].

What then is the $f_0(1370)$? Remember that its parameters come from the assumption of a simple BW form to describe $\pi\pi$ scattering. As we have seen this is at best a convenient parametrization, unlikely to be related to an actual state. In fact, the $\pi\pi$ scattering function shown in Figure 20 has a broad peak in this region, it is possible that what is called the $f_0(1370)$ is simply a manifestation of this peak.

A related question is, why doesn't the $f_0(1520)$ itself show up in the $\pi\pi$ scattering data? There is a potential answer to this: Bugg[29] and collaborators have reanalyzed the peripheral production data on which the $\pi\pi$ phase shifts are based, and found an acceptable solution that includes a resonance at 1500, which shows up as a dip in the scattering amplitude. The same analysis does not require the $f_0(1370)$.

Questions

I finish with a list of questions and points that I did not cover:

- Can the simple model of Törnqvist be modified to account for the $f_0(1520)$, presumably with SU(3) singlet couplings to the same final states? Would it then obviate the need for the parameter β? Can it account for $J/\psi \to \phi\pi\pi$ and $J/\psi \to \phi KK$?
- A Crystal Barrel analysis[30] of $\bar{p}p \to \eta\pi^0\pi^0$ indicates the need for an I=1 scalar at 1450 MeV. This is clearly inconsistent with the quark-model explanation for the low-lying scalars presented here, and needs confirmation, or perhaps an alternate explanation.
- Are the states claimed by WA91[31] (1445 MeV) GAMS[32] (1590 MeV) consistent with $f_0(1520)$, in spite of the stated errors?
- Is the $J/\psi \to \gamma\pi\pi$ data consistent with $J/\psi \to \gamma 4\pi$ and expectations of the $2\pi/4\pi$ branching ratios? (This is roughly unity for the Bugg[29] analysis.)
- I did not comment on the tensors above the $f_2(1270)$ and $f_2'(1525)$. There are clearly non-quark model candidates here, with some central open questions being: Is the $f_J(1710)$ scalar or tensor? How reliable is the evidence for $f_2(1640)$ in the Bugg $J/\psi \to \gamma 4\pi$ reanalysis, and is it the same as the GAMS $\omega\omega$ resonance? It was not seen in $J/\psi \to \gamma\omega\omega$.

REFERENCES

[1] T. Barnes, lectures presented at this school.
[2] L. Montanet, *Nucl. Phys. B* (Proc. Suppl.) 39B,C (1995) 281; A. Palano, *ibid.*, 287.
[3] Review of Particle Properties; *Phys. Rev.* **D50** (1994)1173.
[4]Augustin et al, (DM2) *Nucl. Phys.* **B320** (1987) 369.
[5] Ref. 3, pg 1287.
[6] R. Baltrusaitis et al (Mark III) *Phys. Rev.* **D35** (1987) 2077.
[7] Ref. 3, page 1293:, eq. 23.57.
[8] *MathCad 6.0 User's Guide* MathSoft Inc.,101 Main Street, Cambridge MA 02142.
[9] V. I. Anisovitch, lectures presented at this school.
[10] N. A. Törnqvist, *Phys. Rev. D* **29,** 121 (1984).
[11] N. A. Törnqvist, "Understanding the Scalar Meson qq-bar Nonet", Helsinki preprint HU-SEFT R 1995-05.
[12] K. L. Au, D. Morgan, and M. R. Pennington, *Phys. Rev.* D **35** (1987).1633.
[13] D. Morgan, and M. R. Pennington, *Phys. Rev.* **D48** (1993) 1185.
[14] B. S. Zou and D. V. Bugg, *Phys. Rev.* **D50** (1994) 591.
[15] D. Aston et al (LASS collaboration), *Nucl. Phys.* **B296** (1988) 493.
[16] R. H. Dalitz and S. Tuan, *Ann. Phys.* **10** (1960). 307.
[17] S. L. Adler, *Phys. Rev.* **137B** (1965) 102; **139B** (1965) 1638.
[18] C. Amsler et al (Crystal Barrel collaboration) *Phys. Lett* **B291** (1992) 347.
[19] D.V. Bugg, V.V. Anisovich, A. Sarantsev, B.S. Zou *Phys. Rev.* **D50** (1994) 4412.
[20] J. Weinstein and N. Isgur, *Phys. Rev.* **D41** (1990).2236. See also Ref. 1.
[21] D. Bugg, summary talk presented at this school.
[22] Crystal Ball Collaboration. (V.V. Anisovich et al.).. *Phys. Lett.* **B323** (1994) 233-241.
[23] S. Spanier, presentation at Hadron 95, Manchester UK (1995).
[24] Baltrusaitis et al,(Mark III) *Phys. Rev.* **D33** (1986) 1222.
[25] Bisello et al (DM2), *Phys. Rev.* **D39** (1989) 701.
[26] D.V. Bugg, I. Scott, B.S. Zou, V.V. Anisovich, A.V. Sarantsev, T. Burnett, and S. Sutlief, "Further Amplitude Analysis of $J/\psi \rightarrow \gamma\ (\pi^+,\pi^-,\pi^+,\pi^-)$", accepted by *Phys. Lett.*
[27] Baltrusaitis et al,(Mark III) *Phys. Rev.* **D33** (1986) 1222.
[28] C. Amsler and F. E. Close, Rutherford preprint RAL-TR-95-003, "IS F0 (1500) A SCALAR GLUEBALL?" Submitted to *Phys. Rev.* D.
[29] D. V. Bugg, B. S. Zou, V.V. Anisovich, A.V. Sarantsev, presentation at Hadron 95, Manchester UK (1995).
[30] C. Amsler at al, (Crystal Barrel) *Phys. Lett.* **B333** (1994) 277
[31] WA91 Collaboration (F. Antinori et al) i. **B353** (1995) 589.
[32] GAMS Collaboration (Alde et al) *Phys. Lett.* **B201** (1988) 160.

PHYSICS OF 0^-, 1^+ AND 2^- MESONS

Antimo Palano

Dipartimento di Fisica dell'Universitá and Sezione INFN
Bari, Italy

1 INTRODUCTION

Mesons are quark-antiquark bound states. For a fermion-antifermion pair like $q\bar{q}$ having orbital angular momentum L and total spin S, the P-parity and the C-parity are related to L and S by:

$$P = (-1)^{L+1}, \quad C = (-1)^{L+S}.$$

If J is the total spin, J^{PC} multiplets can be identified, in terms of nuclear spectroscopy notation, as:

$$N^{(2S+1)}L_J$$

where N is the principal quantum number. States having N$\geq$2 are called radial excitations. Light mesons occur in multiplets of nine members having the same J^{PC} configuration: three have isospin one, four have isospin 1/2 and carry strangeness, two have isospin zero.

The present status of the $q\bar{q}$ multiplets is summarized in Table 1 [1]. We notice the presence of empty cases, but this is a problem which could be solved by making new experiments with high statistics and good data analysis. The real source of concern in the field of light meson spectroscopy is the fact that several mesons multiplets have not very well defined members. In addition, meson resonances exist which have no place in the above multiplets.

Large experimental and theoretical efforts have been made in the last few years in order to solve the spectrum of the light mesons. The main problem is the possibility of the existence, among the discovered meson resonances, of non $q\bar{q}$ mesons such as states made only by gluons (glueballs), by mixtures of quarks and gluons (hybrids) or multiquark states. The existence of these states is expected from QCD, the theory which describes the interactions of quarks and gluons.

The search for these new resonances is now 15 years old. It has been performed in a large variety of processes and it has been pursued insistently and systematically. The basic ideas on which this search has been carried on are the following:

– a) Compare meson spectroscopy from different production mechanisms;

Hadron Spectroscopy and the Confinement Problem
Edited by D.V. Bugg, Plenum Press, New York, 1996

Table 1: Status of the lower J^{PC} multiplets

$N^{(2S+1)}L_J$	J^{PC}	$u\bar{d}, u\bar{u}, d\bar{d}$ I=1	$u\bar{u}, d\bar{d}, s\bar{s}$ I=0	$s\bar{u}, \bar{s}d$ I=1/2	θ_{quad}
1^1S_0	0^{-+}	π	η, η'	K	-10^0
1^3S_1	1^{--}	ρ	ω, ϕ	$K^*(892)$	39^0
1^1P_1	1^{+-}	$b_1(1235)$	$h_1(1170), h_1(1380)$	K_{1B}	
1^3P_0	0^{++}	$a_0(980)$	$f_0(1300), f_0(980)$	$K_0^*(1430)$	
1^3P_1	1^{++}	$a_1(1260)$	$f_1(1285), f_1(1510)$	K_{1A}	
1^3P_2	2^{++}	$a_2(1320)$	$f_2(1270), f_2'(1525)$	$K_2^*(1430)$	28^0
1^1D_2	2^{-+}	$\pi_2(1670)$		$K_2(1770)$	
1^3D_1	1^{--}	$\rho(1700)$	$\omega(1600)$	$K^*(1680)$	
1^3D_2	2^{--}			$K_2(1820)$	
1^3D_3	3^{--}	$\rho_3(1690)$	$\omega_3(1670), \phi_3(1850)$	$K_3(1780)$	29^0
1^3F_4	4^{++}	$a_4(2040)$	$f_4(2050), f_4(2220)$	$K_4^*(2045)$	
2^1S_0	0^{-+}	$\pi(1300)$	$\eta(1295)$	$K(1460)$	
2^3S_1	1^{--}	$\rho(1450)$	$\omega(1420), \phi(1680)$	$K^*(1410)$	
2^3P_2	2^{++}		$f_2(1810), f_2(2010)$	$K_2^*(1980)$	
3^1S_0	0^{-+}	$\pi(1770)$	$\eta(1760)$	$K(1830)$	

- b) Use reactions which can tag the flavour content or the quantum numbers of the produced states;
- c) Compare with expectations from the quark model.

The relativistic Godfrey-Isgur (GI) model [3] is one often used in these comparisons since it offers predictions for meson spectroscopy from the pion to the upsilon.

Scalar and tensor resonances have been discussed by T. Burnett [2]; in these lectures we will discuss the present status of the $J^P = 0^-$, 1^+ and 2^- mesons.

2 PSEUDOSCALAR MESONS

The pseudoscalar nonet is obtained (in the notations of the previous section) with L=0 and S=0 so that J=0, P=–1 and C=+1. The established members of the nonet are summarized in table 1.

In the framework of SU(3), with three flavours u,d,s, there are nine $q\bar{q}$ combinations so that states having the same J^{PC} are grouped in an octet and a singlet. If we label X_8 the I=0 member of the octet and X_1 the singlet, their wavefunctions are given by:

$$X_8 = (u\bar{u} + d\bar{d} - s\bar{s})/\sqrt{6}$$

$$X_1 = (u\bar{u} + d\bar{d} + s\bar{s})/\sqrt{3}.$$

However, there is no reason to expect that the I=0 physical states correspond to X_1 and X_8. They can mix, and their flavor content is related to the OZI rule.

The OZI (Okubo, Zweig, Iizuka) rule states that "processes in which the initial quark pairs cannot appear as part of the final state particle, are suppressed". In QCD, OZI suppression is related to the suppression of the annihilation quark lines into hard gluons. The annihilation into n hard gluons therefore corresponds to a suppression of n powers of α_S.

The physical states X_0 and X_0' are expected to be mixed states of X_1 and X_8.

$$X_0 = X_8 cos\theta - X_1 sin\theta$$

$$X_0' = X_8 sin\theta + X_1 cos\theta,$$

where θ is the mixing angle. The mixing angles for the different nonets as computed by the quadratic Gell-Mann Okubo mass formula are summarized in Table 1. When the OZI rule works well, instead of X_8 and X_1 we expect ideal mixing. In this case, for $\theta = 35.3^0$:

$$X_0 = (u\bar{u} + d\bar{d})/\sqrt{2}$$

$$X_0' = s\bar{s}.$$

When there are large OZI violating forces then the physical states will be nonideal and will approximate X_1 and X_8. In general, the mixing will be intermediate between these extremes.

We observe a remarkable peculiarity of the pseudoscalar nonet: it has the largest deviation from ideal mixing. In the latter case one would expect:

$$m_{(I=1)} = m_{X^0}, \quad 2m_{(I=1/2)} = m_{X^0} + m_{X_0'}.$$

For the nearly ideally mixed nonet with $J^{PC} = 1^{--}$ we have:

$$m_{\rho(770)} \cong m_{\omega(772)},$$

$$2m_{K^*(892)}(= 1784) \cong m_{\omega(782)} + m_{\phi(1020)}(= 1802).$$

However, in the pseudoscalar nonet we have:

$$m_{\pi(135)} \neq m_{\eta(548)}$$

$$2m_{K(494)}(= 988) \neq m_{\eta(548)} + m_{\eta'(958)}(= 1506).$$

One question immediately arises: what is the origin of the large deviations from ideal mixing in the pseudoscalar nonet?

In the framework of the possible existence of gluonium states one can allow the possibility of mixing between glueballs, hybrids and $q\bar{q}$ states. Glueballs are I=0 singlets; therefore in the presence of a pseudoscalar glueball, one could expect them to mix with the I=0 members of the nonet. We could then expect to find a sizeable glue component in η and η' mesons.

3 MEASUREMENT OF THE NON-$q\bar{q}$ COMPONENT OF η AND η' MESONS

The measurement of the glue component in the I=0 members of the pseudoscalar nonet has been made in J/ψ decays. Inputs to this problem have also been obtained in the study of radiative meson decays. In order to measure the possible non-$q\bar{q}$ component of η and η', experiments DM2 (at Orsay) [4] and Mark III (at SLAC) [5] have performed a systematic study of hadronic J/ψ decays to:

$$J/\psi \rightarrow Vector + Pseudoscalar.$$

Following Rosner [6], the wave functions of the pseudoscalar mesons can be written:

$$|\eta> = X_\eta|u\bar{u} + d\bar{d}> + Y_\eta|s\bar{s}> + Z_\eta|other \quad components>$$

$$|\eta'> = X_{\eta'}|u\bar{u} + d\bar{d}> + Y_{\eta'}|s\bar{s}> + Z_{\eta'}|other \quad components>,$$

where Z measures the amount of the possible non-$q\bar{q}$ component and

$$X_\eta^2 + Y_\eta^2 + Z_\eta^2 = X_{\eta'}^2 + Y_{\eta'}^2 + Z_{\eta'}^2 = 1.$$

For the hypothesis Z=0, if η and η' were unmixed octet and singlets, we would expect:

$$X_{\eta'} = -Y_\eta = \sqrt{2/3}$$

$$X_\eta = Y_{\eta'} = \sqrt{1/3}.$$

3.1 A model for J/ψ decays into Meson Pairs

The diagrams contributing to J/ψ decay into meson pairs are shown in Fig. 1.

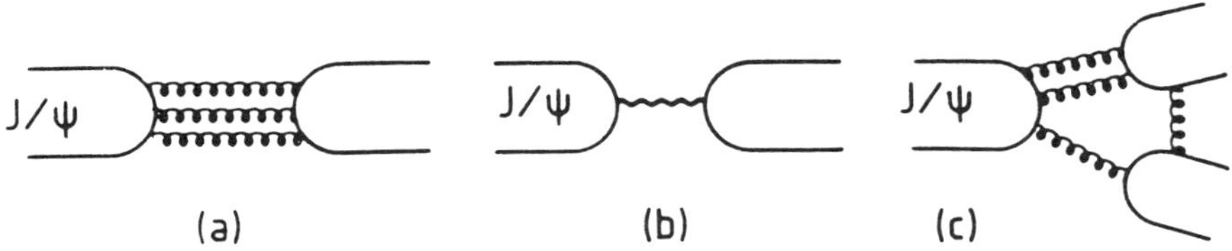

Figure 1: J/ψ decay diagrams: a) SOZI suppressed strong decays, b) electromagnetic decays and c) DOZI suppressed decays into mesons.

An important key in this problem is the presence of both single disconnected diagrams, labelled SOZI (fig. 1a) and double disconnected diagrams, labelled DOZI (fig. 1c). The amplitudes describing J/ψ decay to two mesons can be written as follows [7]:

- a) A singly OZI violating amplitude (SOZI) g;
- b) A doubly OZI violating amplitude (DOZI) r, expressed relative to g;
- c) An electromagnetic amplitude e;
- d) A factor (1 - s_g) for every strange quark contributing to g.

The list of the amplitudes describing the different J/ψ decays and the corresponding measured branching fractions are summarized in Table 2.

We notice that the reduced branching fractions $\tilde{B} = B/p^3$ have been used in order to take into account the P-wave matrix element for the decay of J/ψ into the two final state mesons. The results from the fits are shown in Table 3. The phase of e relative to g θ_e has also been included in the fit. We notice that $|s|$, $|e|$ and $|r|$ are rather small, being 16%, 11% and 10% of $|g|$ respectively.

In the case of the DM2 fit, the assumption that Z=0 has been used. Taking instead the Mark III results only,

$$X_\eta^2 + Y_\eta^2 = 1.00 \pm 0.16$$

$$X_{\eta'}^2 + Y_{\eta'}^2 = 1.44 \pm 0.25,$$

a result which shows that there is no need for additional non-$q\bar{q}$ componenents in η and η' mesons. These results are summarized in the Rosner plot shown in Fig. 2.

One essential ingredient of this study is the presence of the DOZI amplitudes. The absence of such contributions gave indications of positive evidence for the presence of

Table 2: Amplitudes and Branching ratios used in the analysis of $J/\psi \to PV$

Process $J/\psi \to PV$	SOZI Amplitude	DOZI Correction	$\tilde{B}$ (Mark III) (Units of 10^{-3})	$\tilde{B}$ (DM2) (Units of 10^{-3})
$\rho^+\pi^-, \rho^0\pi^0, \rho^-\pi^+$	$g+e$		1.556 ± 0.161	
$K^{*+}K^-, K^{*-}K^+$	$g(l-s_g)+e$		1.017 ± 0.061	
$K^{*0}\bar{K}^0, \bar{K}^{*0}K^0$	$g(l-s_g)-2e$		0.836 ± 0.055	
$\omega\eta$	$(g+e)X_\eta$	$\sqrt{2}rg(\sqrt{2}X_\eta + Y_\eta)$	0.632 ± 0.058	0.531 ± 0.052
$\omega\eta'$	$(g+e)X_{\eta'}$	$\sqrt{2}rg(\sqrt{2}X_{\eta'} + Y_{\eta'})$	0.079 ± 0.010	0.091 ± 0.024
$\phi\eta$	$(g(l-2s_g)-2e)Y_\eta$	$rg(\sqrt{2}X_\eta + Y_\eta)$	0.287 ± 0.031	0.278 ± 0.026
$\phi\eta'$	$(g(l-2s_g)-2e)Y_{\eta'}$	$rg(\sqrt{2}X_\eta' + Y_\eta')$	0.182 ± 0.025	0.239 ± 0.027
$\rho\eta$	$3eX_\eta$		0.071 ± 0.010	0.070 ± 0.018
$\rho\eta'$	$3eX_\eta'$		0.054 ± 0.009	0.039 ± 0.014
$\omega\pi^0$	$3e$		0.159 ± 0.017	0.169 ± 0.016
$\phi\pi^0$	0		≤ 0.0026	

Table 3: Results from the fits in the determination of the quark contents of η and η' mesons

Coupling	Mark III	DM2
$\lvert g\rvert$	1.21 ± 0.05	1.120 ± 0.041
$\lvert s_g\rvert$	0.21 ± 0.04	0.176 ± 0.030
$\lvert e\rvert$	0.13 ± 0.01	0.137 ± 0.006
θ_e	1.26 ± 0.11	1.170 ± 0.090
r	-0.14 ± 0.02	-0.088 ± 0.019
X_η	0.67 ± 0.05	0.637 ± 0.044
Y_η	-0.74 ± 0.10	-0.771 ± 0.037
$X_{\eta'}$	0.58 ± 0.06	0.436 ± 0.044
$Y_{\eta'}$	1.05 ± 0.12	0.900 ± 0.021
$\chi^2/d.o.f.$	0.02	0.047

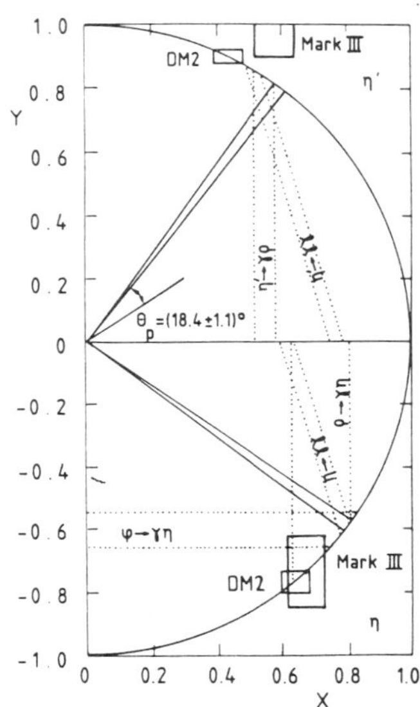

Figure 2: Rosner plot for η and η'. The results of the Mark III and DM2 experiments are compared with radiative widths of pseudoscalar mesons. X and Y represent the non-strange and strange fractions of the states. The circle correspond to $X^2 + Y^2$=1. In the DM2 fit this condition is imposed.

glue in the η' meson [8], but this was ruled out by a later analysis. Dropping the DOZI amplitudes, one obtains:

$$X_\eta^2 + Y_\eta^2 = 1.1 \pm 0.2$$

$$X_{\eta'}^2 + Y_{\eta'}^2 = 0.65 \pm 0.18,$$

which shows that the η' would not be saturated with $q\bar{q}$ only. However, this solution gives a poor description of the data: $\chi^2/NDF = 25/2$.

3.2 The pseudoscalar mixing angle

The $\eta - \eta'$ mixing angle θ_P is related to the measured fractions X and Y by:

$$X_\eta = Y_{\eta'} = \sqrt{1/3}cos\theta_P - \sqrt{2/3}sin\theta_P$$

$$X_{\eta'} = -Y_\eta = \sqrt{1/3}sin\theta_P + \sqrt{2/3}cos\theta_P,$$

from which one obtains:

$$tan\theta_P = -\frac{\sqrt{2}X_\eta + Y_\eta}{X_\eta - \sqrt{2}Y_\eta}.$$

The two experiments obtain respectively:

$$\theta_P = -19.0 \pm 1.0^0 \quad MarkIII$$

$$\theta_P = -20.0 \pm 1.6^0 \quad DM2.$$

Additional inputs to the measurement of the pseudoscalar mixing angle come from the measurement of the $\gamma\gamma$ widths of the pseudoscalar mesons. The invariant $\gamma\gamma$ mass spectrum from the process $\gamma\gamma \to \gamma\gamma$ from Crystal Ball (at DESY) is shown in Fig. 3 [9], where clear π^0, η and η' signals can be seen.

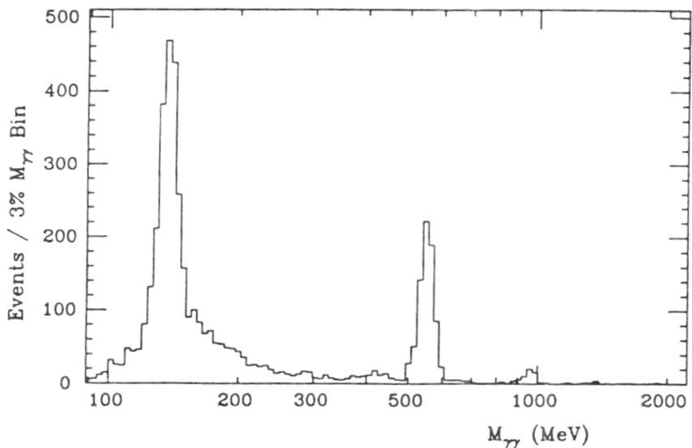

Figure 3: $\gamma\gamma$ effective mass in two photon interactions from Crystal Ball.

The $\gamma\gamma$ widths of these mesons have been computed to be:

$$\Gamma_{\gamma\gamma}(\pi^0) = C\frac{M^3_{\pi^0}}{f^2_\pi}$$

$$\Gamma_{\gamma\gamma}(\eta) = C\frac{M^3_\eta}{3f^2_\pi}\left[\frac{f_\pi}{f_8}cos\theta_P - \sqrt{8}\frac{f_\pi}{f_1}sin\theta_P\right]^2$$

$$\Gamma_{\gamma\gamma}(\eta') = C\frac{M^3_{\eta'}}{3f^2_\pi}\left[\frac{f_\pi}{f_8}sin\theta_P + \sqrt{8}\frac{f_\pi}{f_1}cos\theta_P\right]^2,$$

where $C = \alpha^2/(64\pi^3)$, f_1 and f_8 are the singlet and octet decay constants and from theory one fixes $f_\pi/f_8 = 0.8$. The resulting pseudoscalar mixing angle is:

$$\theta_P = -22.2 \pm 1.2^0,$$

in good agreement with the results from J/ψ decays.

3.3 Gluonium in the pseudoscalar nonet?

The previous section has shown that there is little evidence for a possible mixing of the pseudoscalar mesons with a hypothetical gluonium resonance. However this does not rule out the possible existence of glueballs or hybrids with η and η' quantum numbers. Bag models predict, in fact, a pseudoscalar glueball in the 1-2 GeV region [10], while more recent lattice calculations give a higher mass, around 2.2 GeV [11]. Isoscalar hybrids, on the other hand, are expected between 1.0 and 1.5 GeV.

The task of identifying gluonium is rather hard, since meson multiplets are not well defined, nor is it completely defined by the structure of the radial excitations. In addition, several meson resonances with different quantum numbers, but having the same decay modes, are close together in mass. It is then clear that no single experiment can discover non $q\bar{q}$ mesons, so that the only chance to find them is to compare meson spectroscopy from different reactions.

4 THE E/ι PUZZLE

One classical example of the difficulties encountered in identifying glueball states can be found in the so called "E/ι puzzle" which can be stated as follows. One, two, three or even more states have been observed decaying to $K\bar{K}\pi$, $\eta\pi\pi$ or $\rho^0(770)\gamma$ in the mass region around 1.45 GeV; some experiments find $J^{PC} = 0^{-+}$ while other experiments find $J^{PC} = 1^{++}$. The interest in this region is that these objects could be glueballs, hybrids, molecules or normal $q\bar{q}$ states. Large data sets have been collected in order to study this subject during the last 15 years. However, still now, it is not easy to fit all the results into a clear picture.

5 RESULTS FROM J/ψ RADIATIVE DECAYS.

The starting point in gluonium spectroscopy can be dated with the discovery, in radiative J/ψ decay (fig. 4), from experiments Mark II [12] and Crystal Ball [13] (at SPEAR), of a surprisingly large signal in the $K\bar{K}\pi$ effective mass from the decay:

$$J/\psi \to \gamma(K\bar{K}\pi).$$

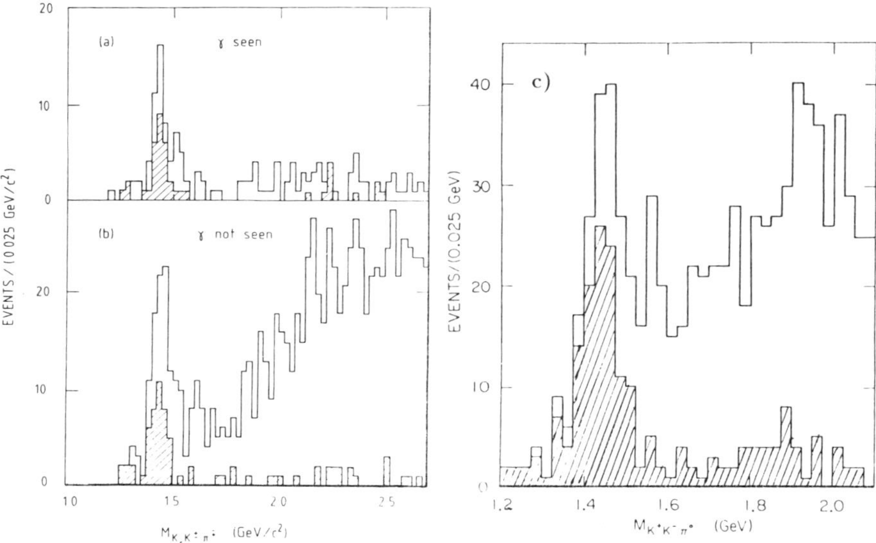

Figure 4: Observation of the $\iota(1450)$ by MarkII in $J/\psi \to \gamma K_S^0 K^{\pm}\pi^{\mp}$ with the radiative γ a) detected, b) undetected. c) Observation of ι by Crystal Ball in the $K^+K^-\pi^0$ final state.

The new resonance was labelled $\iota(1450)$ and had the following parameters:

$$m = 1454 \pm 3 \quad MeV$$

$$\Gamma = 98 \pm 7 \quad MeV$$

$$BR(J/\psi \to \gamma\iota) = 4.5 \pm 0.4 \times 10^{-3}.$$

The quantum numbers were determined by the Crystal Ball experiment to be $J^{PC} = 0^{-+}$, by performing a partial wave analysis of the $K^+K^-\pi^0$ system. Its decay mode was determined to be entirely $a_0(980)\pi$. The branching fraction for $J/\psi \to \gamma\iota(1450)$ was shown to be the largest of the J/ψ radiative decays, of the same size as the $J/\psi \to \gamma\eta'$ radiative decay (see Fig. 5).

The first measurement of the spin of the $\iota(1450)$ was made using an isobar model, i.e. assuming intermediate resonance production. In this specific case, $a_0(980)\pi$ and

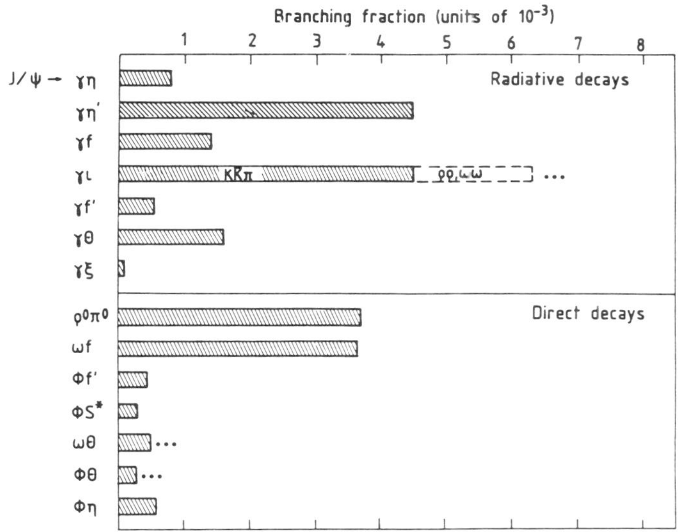

Figure 5: Branching fraction for several resonances observed in J/ψ decays.

$K^*\bar{K}$ intermediate states were considered. The analysis was repeated with increased statistics [14], but using the helicity formalism, i.e. without making use of the assumptions of the isobar model. This was in order to avoid complications involved in describing the $a_0(980)$ resonance, an object poorly understood. The analysis confirmed the results obtained by the Crystal Ball group, i.e. that in the hypothesis of one single resonance we are dealing with a pseudoscalar meson. The discriminating variable used in this analysis is θ_ι, the angle formed by the $\iota(1450)$ with respect to the e^+e^- beam-line. However this angle suffers strongly from losses due to geometric acceptance of the apparatus, expecially in the regions where different spin amplitudes can be well discriminated.

A further indication that something new is happening in the $K\bar{K}\pi$ mass spectrum produced in J/ψ radiative decay, is the absence (or rather weak) production of $f_1(1285)$, a $J^{PC} = 1^{++}$ state. This suggests that, in this particular process, spin are suppressed. This reminds us of Yang's [15] theorem, which forbids two massless gluons to produce spin one objects (see Fig. 6a).

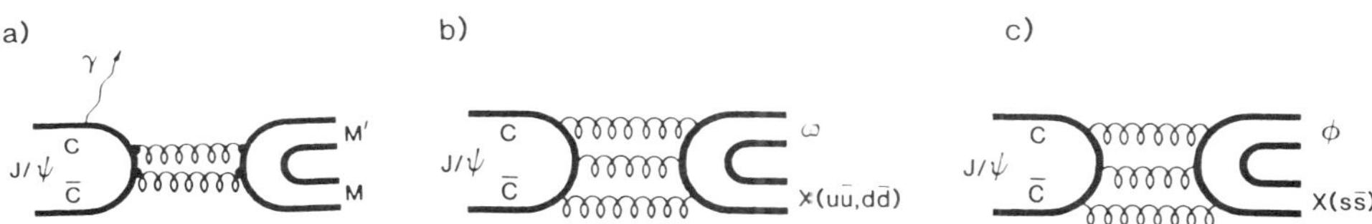

Figure 6: Diagrams of J/ψ decays: a) Radiative J/ψ decay; Hadronic decays to b) $J/\psi \to \omega X$ and c) $J/\psi \to \phi X$.

Since the pseudoscalar nonet is already full, and the $\iota(1450)$ is produced with very high rate in J/ψ radiative decay, mediated by gluons, then the $\iota(1450)$ is a candidate for being the first glueball [16].

5.1 Problems

However, some problem begin to arise. First, with increasing statistics, the $\iota(1450)$ lineshape appears not to be fitted with a single Breit-Wigner (fig. 7).

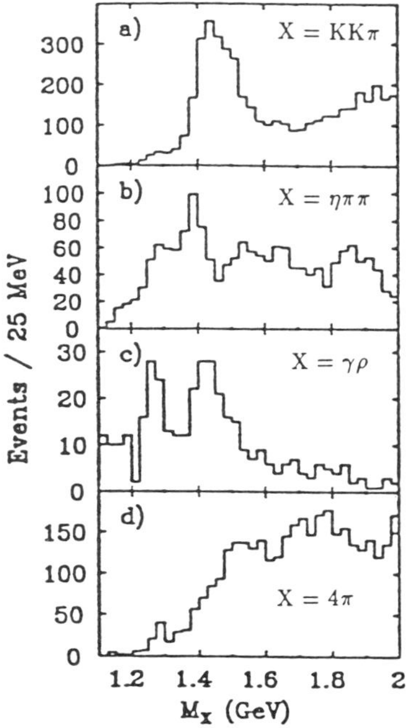

Figure 7: Summary of mass spectra from radiative J/ψ decays (Mark III).

Secondly, the $\iota(1450)$ decay mode appears to be inconsistent; it is entirely due to $a_0(980)\pi$ decay, since a clear $K^*(\rightarrow K\pi)$ signal appears in the high mass region of the $K\bar{K}\pi$ mass spectrum. In addition, if the decay of $\iota(1450)$ proceeds mostly through $a_0(980)\pi$ then one would expect a strong $\iota(1450)$ signal in the $\eta\pi\pi$ decay mode, since $a_0(980)$ goes mostly through $\eta\pi$. This is not the case.

5.2 $a_0(980)$

The $a_0(980)$ resonance, an I=1 state with $J^P = 0^+$, is observed as a peak in the $\eta\pi$ mass spectrum, Fig. 8a [17], and as a threshold enhancement in the $K\bar{K}$ mass spectrum, Fig. 8b [18].

The two effects are assumed to belong to the same state. One simple way to describe this two channel resonance is due to Flatté [19]; he uses the formalism of a coupled channel Breit-Wigner amplitude:

$$\frac{d\sigma}{dm} = C\left|\frac{m_R\sqrt{\Gamma_0\Gamma_i}}{m_R^2 - m^2 - im_R(\Gamma_{\pi\eta} + \Gamma_{K\bar{K}})}\right|^2$$

$$\Gamma_{\eta\pi} = g_\eta q_\eta$$

$$\Gamma_{K\bar{K}} = \begin{cases} g_K\sqrt{(1/4)m^2 - m_K^2} & \text{above threshold} \\ ig_K\sqrt{m_K^2 - (1/4)m^2} & \text{below threshold,} \end{cases}$$

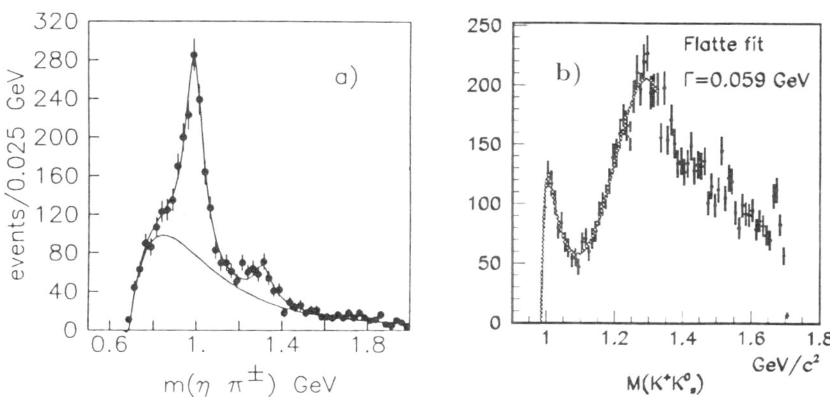

Figure 8: Observation of $a_0(980)$: a) $a_0(980) \rightarrow \eta\pi^{\mp}$ centrally produced in the reaction $pp \rightarrow p_f(\eta\pi^+\pi^-)p_s$ by Omega/WA76, b) $a_0(980) \rightarrow K^0_S K^{\pm}$ in $\bar{p}p \rightarrow K^0_S K^{\pm}\pi^{\mp}$ at rest from OBELIX.

where q is the decay c.m. momentum, and g_K and g_η are the couplings to the $K\bar{K}$ and $\eta\pi$ final states respectively. The $a_0(980)$ parameters are the following:

$$m_R = 983 \quad MeV$$

$$\Gamma_{\eta\pi} = 57 \quad MeV.$$

The ratio g_K/g_η is usually fixed to the SU(3) prediction: $g_K/g_\eta = 3/2$

A plot of the resulting lineshape obtained from the above parametrization is shown in Fig. 9a). Fig. 9b) shows a threshold enhancement in the $K\bar{K}$ mass spectrum from $\iota(1450)$ decay which suggests the presence of the $a_0(980)\pi$ decay mode.

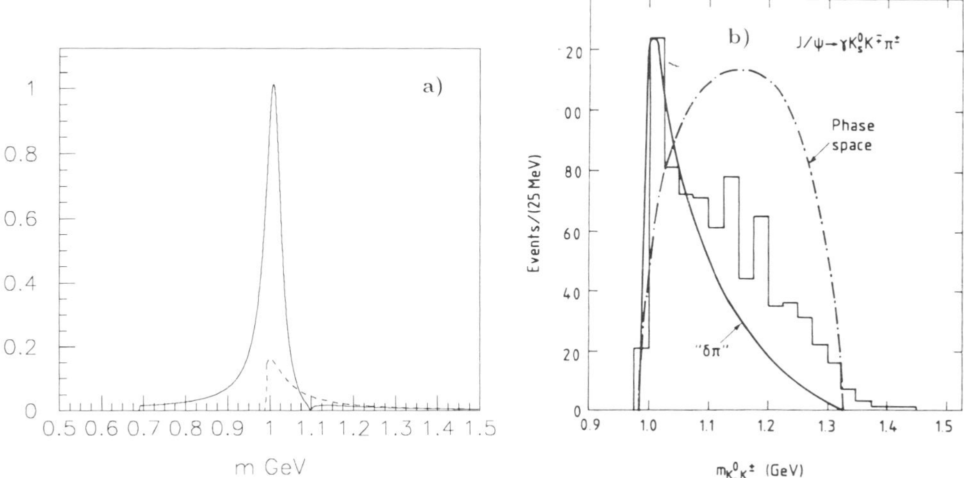

Figure 9: a) $a_0(980)$ lineshape in the $\eta\pi$ (solid) and $\bar{K}K$ (dashed) channels from a coupled channel Breit-Wigner, b) $K^0_S K^{\pm}$ effective mass in the $\iota(1450)$ decay from DM2.

5.3 The present status of $\iota(1450)$

More recent results from J/ψ radiative decays have been able to shed a little more light on the $\iota(1450)$ puzzle. The results from different spin-parity analyses of the $K\bar{K}\pi$ and $\eta\pi\pi$ systems from different experiments [20] indicate that the $\iota(1450)$ peak is in fact due to at least two resonances: a $J^{PC} = 0^{-+}$ state and a $J^{PC} = 1^{++}$ state at about the same mass, Fig. 10.

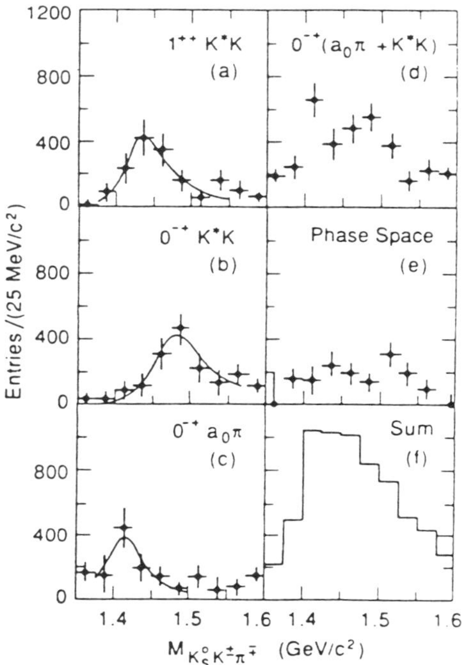

Figure 10: Results from the Partial Wave Analysis of the $K\bar{K}\pi$ system in the $\iota(1450)$ region by Mark III.

The pseudoscalar state could be due to a superposition of two different resonances, one decaying to $a_0(980)\pi$ and the other decaying to $K^*\bar{K}$. The first of these two possible pseudoscalar resonances matches with the corresponding one obtained from the $\eta\pi\pi$ study.

Further input to the $\iota(1450)$ problem is the reported evidence of a radiative $\iota(1450)$ decay to $\rho^0(770)\gamma$:

$$B(J/\psi \to \gamma\iota)B(\iota \to \rho^0\gamma) = 0.64 \pm 0.12 \pm 0.07 \times 10^{-4}.$$

The Mark III results are summarized in fig. 7. We observe $\iota(1450)$ signals in the $K\bar{K}\pi$, $\eta\pi\pi$ and $\rho^0\gamma$ mass spectra. We also notice the presence of a rather weak, but significant, signal of $f_1(1285)$ in the $K\bar{K}\pi$, $\eta\pi\pi$, $\rho^0\gamma$ and 4π mass spectra.

6 INFORMATION FROM J/ψ HADRONIC DECAYS

A powerful tool which could help in understanding the nature and the quark composition of a given resonance comes from the comparison of J/ψ radiative and hadronic decays:

$$J/\psi \to \omega(780)X^0$$

$$J/\psi \to \phi(1020)X^0,$$

where $X^0 = \eta\pi\pi, K\bar{K}\pi$. Under the hypothesis that diagrams shown in Fig. 6 dominate, we would expect to observe strangeonium resonances associated to ϕ production and $u\bar{u} + d\bar{d}$ states associated with an ω signal. In order to be a little more quantitative, we define [7]:

$$R_V = \frac{\Gamma(J/\psi \to \omega X)}{\Gamma(J/\psi \to \phi X)}$$

and

$$R_\gamma = \frac{\Gamma(J/\psi \to \gamma X)}{\Gamma(J/\psi \to (\omega,\phi)X)}.$$

We expect

$$R_V = 2 \text{ if X is an SU(3) singlet}$$

$$R_V = 0.5 \text{ if X is an SU(3) octet.}$$

For a glueball, we expect $R_V > 1$ and $R_\gamma >> 1$. For comparison, η and η' production gives:

$$\eta : R_V = 2.4 \pm 0.4, \quad R_\gamma = 0.6 \pm 0.1$$

$$\eta' : R_V = 0.5 \pm 0.1, \quad R_\gamma = 11.4 \pm 2.3.$$

The results from this study are summarized in Fig. 11.

Statistics here are rather limited, as is the quality of background subtraction. In addition, no detailed partial wave analyses have been performed on most of these spectra. We notice, however, the presence of the $J^{PC} = 1^{++}$ $f_1(1420)$ in the $K\bar{K}\pi$ mass spectrum associated to the ω, and of a weak $f_1(1285)$ associated to ϕ. The $\eta\pi\pi$ spectrum, on the other hand, shows indications for $f_1(1285)$ in all the mass spectra and possibly some $\iota(1450)$.

As one can see from the above results, J/ψ decays alone are not able to solve the gluonium puzzle in the pseudoscalar sector. We should then look to other sources of information such as $p\bar{p}$ annihilations, hadron induced reactions, central production or $\gamma\gamma$ collisions.

7 RESULTS FROM $\bar{p}p$ ANNIHILATIONS

Thirty years ago, $\bar{p}p$ annihilations have been extensively studied in bubble chambers; today there is a new revival with modern technologies. Large data sets are being collected and analyzed, expecially at CERN LEAR. We expect the annihilation region of quark and antiquarks in this process will be a source of $q\bar{q}$ states, but also of exotics such as glueballs, hybrids or multiquark resonances.

7.1 The "E" resonance

Historically, the pseudoscalar state labelled $\iota(1450)$ in the previous section was in fact discovered in 1964 in $\bar{p}p$ annihilations at rest in a bubble chamber in the reaction [21]:

$$\bar{p}p \to (K^0_S K^\pm \pi^\mp)\pi^+\pi^-. \qquad (1)$$

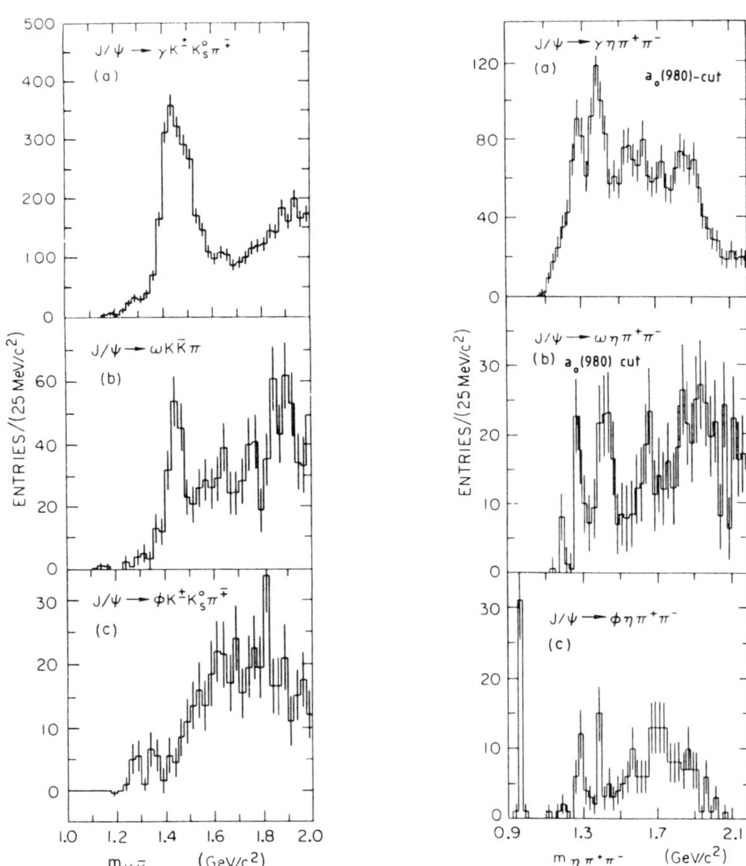

Figure 11: Results from J/ψ decays to $VK_S^0K^{\pm}\pi^{\mp}$ and $V\eta\pi\pi$ where $V = \gamma, \omega, \phi$. The data are from Mark III.

The resonance has a mass of $m = 1425 \pm 7$ MeV and a width $\Gamma = 80 \pm 10$ MeV. Its spin-parity was found to be $J^P = 0^-$ and its decay was found to proceed through 50 % $K^*\bar{K}$ and 50% $a_0(980)\pi$. This resonance was labelled "E".

Further confirmation of this effect has been reported 20 years later by ASTERIX at LEAR [22] in the study of $\bar{p}p$ annihilations at rest in gaseous hydrogen; further confirmation comes from the OBELIX collaboration at LEAR [23]. Fig. 12a) shows the evidence for this resonance in liquid hydrogen. Notice that the background on the left side of the peak is combinatorial, not physical. This experiment has high statistics but less complete angular coverage compared with the bubble chamber experiment. They also confirm the pseudoscalar nature of the resonance, but the data can only be fitted using a superposition of two states: a lower one decaying to $a_0(980)\pi$ and a higher one decaying to $K^*\bar{K}$, thus confirming the results from radiative J/ψ decays.

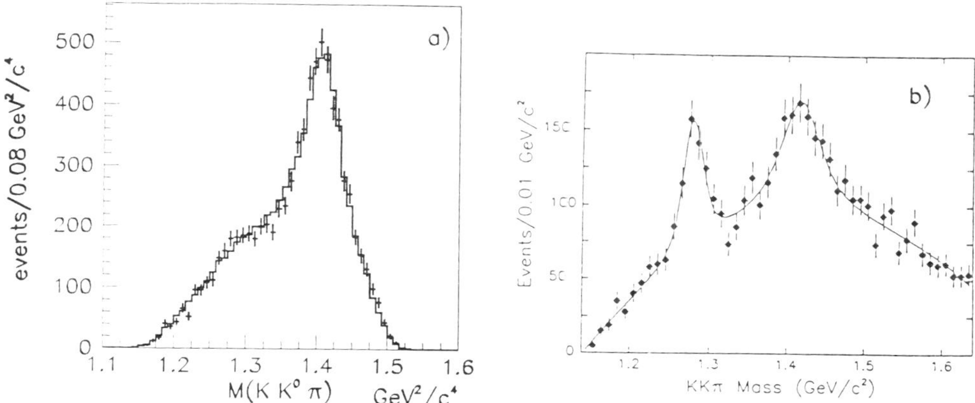

Figure 12: a) Observation of the "E" meson in $\bar{p}p \to K^0K^\pm\pi^\mp\pi^+\pi^-$ at rest by OBELIX, b) $K^0_SK^+\pi^-$ effective mass distribution from $\bar{p}p$ annihilations at 8 GeV/c.

One interesting observation which can be made is the absence, in the $K\bar{K}\pi$ mass spectrum, of the axial meson $f_1(1285)$; this resembles the suppression already described in radiative J/ψ decay of spin one objects. In the case of $\bar{p}p$ annihilations at rest, the explanation is relatively simple.

In reaction(1), the dipion system in the outgoing system is mostly in an S-wave, and its angular momentum with respect to the $K\bar{K}\pi$ system is zero, due to the small available phase space. Thus, for S and P initial states, the only possibilities are the following:

$$^1S_0 \quad \bar{p}p \to 0^{-+}(\pi\pi)$$

$$^3P_1 \quad \bar{p}p \to 1^{++}(\pi\pi).$$

In conclusion, annihilations in S-wave give $J^{PC} = 0^{-+}$ while annihilations in P-wave give $J^{PC} = 1^{++}$.

Concerning the contributions of S and P waves, they depend on the hydrogen density. All the models give a dominance of P-wave at low pressure, which decreases as the density increases until in liquid hydrogen the amount of P-wave is reduced to 10%.

We thus expect production of 0^{-+} resonances in $\bar{p}p$ annihilations at rest in liquid hydrogen. Evidence for this is the absence of the axial meson $f_1(1285)$. As the $\bar{p}p$ annihilation occurs in flight, this rule does not hold any more, so axial mesons should show up. This is in fact the case, as can be seen in Fig. 12b) [24] from the E771 experiment at BNL.

Recent data from $\bar{p}p$ annihilations at rest in the reaction:

$$\bar{p}p \rightarrow (\eta\pi\pi)\pi\pi \qquad (2)$$

confirm these results: a pseudoscalar resonance is observed in the 1400 MeV region with no evidence for axials. Reactions (1) and (2) at rest behave therefore as good filters to select pseudoscalar mesons.

8 RESULTS FROM PERIPHERAL HADRONIC INTERACTIONS

Hadron interactions from incident π and K (see fig. 13a)) have been extensively used for meson spectroscopy, giving a huge amount of information. These reactions

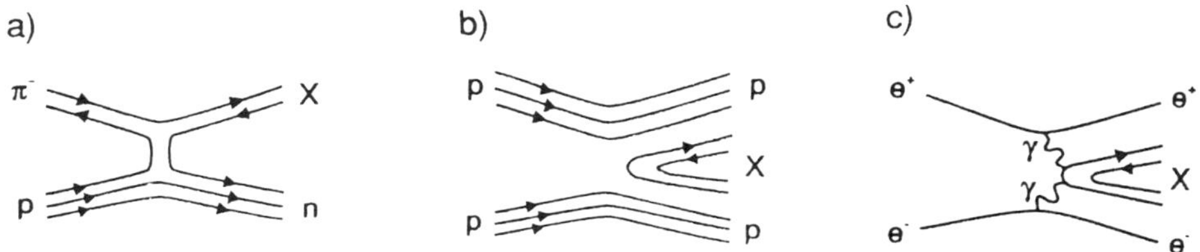

Figure 13: Diagrams of reactions used to study light meson spectroscopy: a) hadron induced "peripheral" reactions, b) central hadronic collisions, c) $\gamma\gamma$ collisions.

are dominated by the production of $q\bar{q}$ mesons, but glueballs and hybrids can also be produced. We label these reactions as "peripheral", to distinguish them from the "central" ones which will be discussed in sect. 9. When the energy of the incident hadron is low one cannot really draw real peripheral graphs, i.e. diagrams where the four momentum transfer from the recoiling target is small.

8.1 Results from π induced reactions

Relevant contributions to the E/ι puzzle have been obtained in the study of $K\bar{K}\pi$, $\eta\pi\pi$ and $\rho^0(770)\gamma$ final states from π incident reactions at different energies.

The reaction:

$$\pi p \rightarrow K\bar{K}\pi n \qquad (3)$$

has been studied in bubble chambers and with hadron spectrometers at the Brookhaven AGS and at KEK. Results from these experiments are not in complete agreement. The CERN bubble chamber experiment [25], studying reaction (3) at 3.95 GeV/c, reported evidence for two axial mesons (see fig. 14a), $f_1(1285)$ and $f_1(1420)$.

We notice here the presence of peaks at the positions of these two resonances with relative sizes quite different from what is observed in J/ψ radiative decay to $\bar{p}p$ annihilations at rest. This suggests that no strong selection rules are active in these reactions. A Dalitz plot analysis of the $K\bar{K}\pi$ spectrum gives $J^P = 1^+$ as quantum numbers of the resonance at 1420 MeV, with little pseudoscalar contribution. The E-771 experiment, at BNL, studied the same final state at 8 GeV/c, with much higher

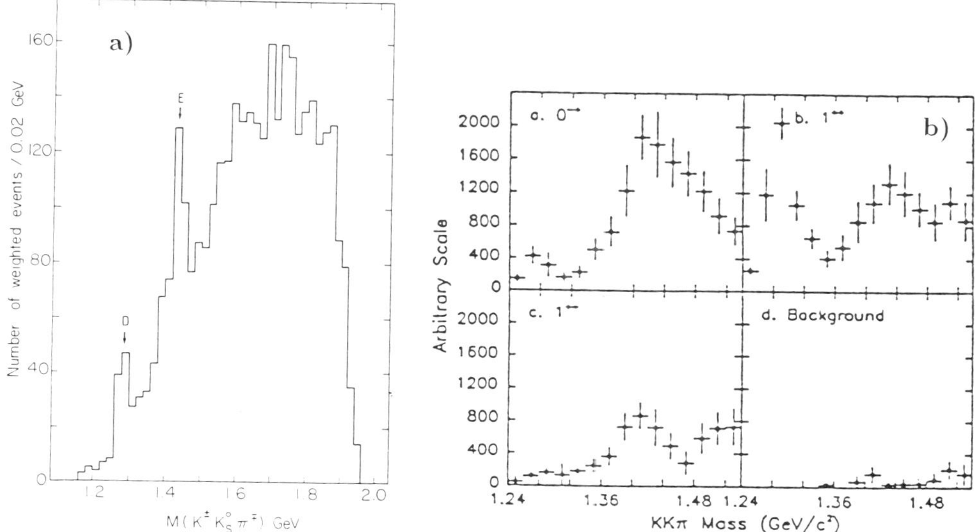

Figure 14: a) $K_S^0 K^\pm \pi^\mp$ effective mass from $\pi^- p$ interactions at 3.95 GeV/c; b) Partial Wave Analysis of the $K_S^0 K^+ \pi^-$ system in $\pi^- p$ interactions at 3 GeV/c.

statistics [26]. The results obtained are more complex, Fig. 14b). They find the $K\bar{K}\pi$ mass spectrum dominated by $J^{PC} = 0^{-+}$ with some contribution from 1^{++} and 1^{+-}. An inspection of the Dalitz plot in the 1420 MeV region, similar in both experiments, shows asymmetric K^* bands. While the CERN experiment interpreted this asymmetry as due to background from incoherent K^* production, the BNL experiment interpreted it as due to interference between different C-parity resonances.

Further information has been obtained in the study of the reaction:

$$\pi p \rightarrow \eta \pi \pi n \qquad (4)$$

at Argonne, KEK and CERN [27]. All these experiments find evidence for a pseudoscalar resonance in the 1420 MeV region, thus confirming the results from the study of reaction (3). They also find evidence for still another pseudoscalar state $\eta(1295)$, placed just below the axial meson $f_1(1285)$. The $\eta\pi\pi$ mass spectrum from GAMS [28] is shown in Fig. 15.

The VES experiment at Serpuhkov has also studied the $\eta\pi\pi$ system in the reaction:

$$\pi^- N \rightarrow \pi^-(\eta\pi^+\pi^-)N \quad at \quad 37 \quad GeV/c.$$

However, the partial wave analysis of this system does not show any evidence for $\eta(1295)$.

A further contribution to the E/ι puzzle comes from the study of the $\rho^0\gamma$ final state, since radiative decays of mesons give information on the quark content of mesons. The experiment Lepton-F (at Serpukohv) [29] finds evidence for a $\phi\gamma$ decay mode of $f_1(1285)$, but no evidence for a $\rho^0\gamma$ decay mode, with rates:

$$BR(f_1(1285) \rightarrow \phi\gamma) = 0.9 \pm 0.2 \times 10^{-3}$$

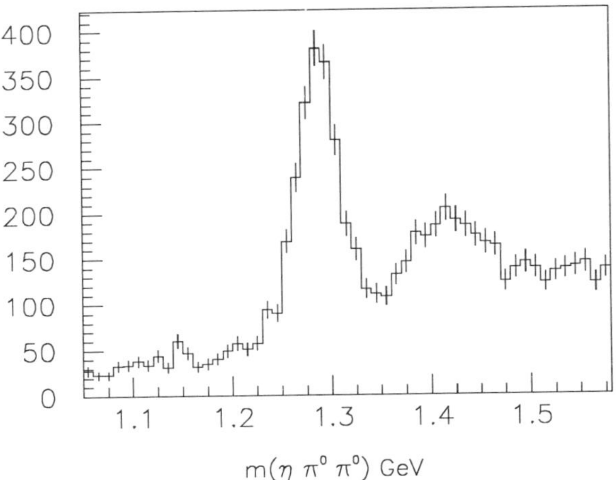

Figure 15: $\eta\pi^0\pi^0$ effective mass distribution from $\pi^- p$ interactions at 100 GeV/c. The data are from GAMS.

$$BR(f_1(1285) \rightarrow \rho^0\gamma) < 0.05.$$

No evidence is found for $f_1(1420) \rightarrow \phi\gamma$. These results are surprising and they conclude that $f_1(1285)$ and $f_1(1420)$ mesons do not belong to the same nonet. The VES experiment [30] has different results, but with higher statistics:

$$BR(f_1(1285) \rightarrow \phi\gamma) < 0.9 \times 10^{-3}$$

$$BR(f_1(1285) \rightarrow \rho^0\gamma) = 2.8 \pm 0.7 \times 10^{-2}.$$

8.2 Results from K$^-$p peripheral interactions

Kaon induced reactions test the strangeness content of the produced mesons, since we would expect strangeonium resonances to be strongly produced in these reactions. Two experiments have given contributions relevant to the question of the E/ι puzzle.

The ACNO collaboration [31], studying the reaction

$$K^- p \rightarrow K\bar{K}\pi\Lambda, \qquad (5)$$

found evidence for yet another axial meson, $f_1(1510)$ and little or no evidence for $f_1(1420)$. This result has been confirmed at SLAC by the LASS experiment [32] which, in a more complete PWA of reaction (5), found, in addition to $f_1(1510)$, evidence for the still missing member of the $J^{PC} = 1^{+-}$ multiplet: $h_1(1380)$, Fig. 16. The presence of the two mesons, lying in the same mass region, was demonstrated by the interference pattern observed between the two $J^P = 1^+$ mesons which produced an asymmetry in K^* production. We therefore have the problem of having 10 candidates for the axial meson multiplet when we expect only 9. In addition $f_1(1420)$ seems not to be strongly coupled to $s\bar{s}$.

9 RESULTS FROM CENTRAL PRODUCTION

In this class of reactions, sketched in Fig. 13b), the incident hadron interacts with the target proton at low 4-momentum transfer. The two hadrons leave the interaction

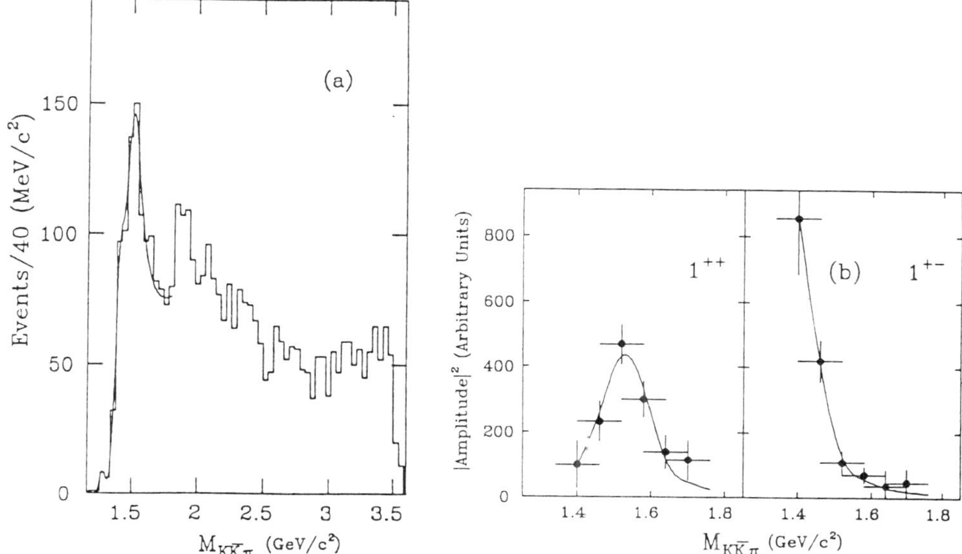

Figure 16: $K_S^0 K^{\pm}\pi^{\mp}$ from $K^- p$ interactions at 11 GeV/c from LASS.

region nearly undisturbed, producing a system which is almost at rest in the centre of mass. The resulting double exchange diagram could originate from different production mechanisms; but at very high energies, contributions from Double Pomeron Exchange (DPE) are expected. Since the Pomeron is understood in QCD in terms of a multigluon system, DPE reactions could enhance the production of glueball or hybrid states.

This class of reactions has been studied at the CERN ISR (SFM and AFS experiments) and Omega/WA76 [33]. Results related to the E/ι puzzle come from experiment WA76 only [34]. The results are summarized in fig. 17.

The WA76 experiment has studied the $K\bar{K}\pi$, $\eta\pi\pi$, 4π and $\rho^0\gamma$ systems with good statistics and negligible background. The data show evidence for $f_1(1285)$ in all these channels. A clear peak is also observed in the 1.42 GeV region of the $K\bar{K}\pi$ effective mass which, in the spin parity analysis, is associated to the $J^{PC} = 1^{++}$ $f_1(1420)$ meson with little pseudoscalar contribution ($\approx 10\%$). No evidence is found for other decay modes of this meson. The spin-parity analysis of the $\eta\pi\pi$ system also shows no evidence for pseudoscalars: no $\eta(1295)$, no $\iota(1450)$. This suggests that the dynamics related to this process are different from those observed either in J/ψ decays or in hadron induced reactions. The 4π mass spectrum shows evidence for structure in the 1450 MeV region which it is not associated to an axial meson.

On the other hand, the $\rho^0\gamma$ spectrum above the η' shows evidence for $f_1(1285)$ and no evidence for $f_1(1420)$. A comparison with the same spectrum observed in radiative J/ψ decays shows that indeed the structure observed there is related to $\iota(1450)$ and not to $f_1(1420)$.

In summary, the central production process definitely shows enhanced axial meson production and suppressed pseudoscalars. A strong $f_1(1420)$ signal is observed to decay only to $K^*\bar{K}$ with little production of pseudoscalar $\iota(1450)$.

10 RESULTS FROM $\gamma\gamma$ COLLISIONS

This class of reactions is sketched in Fig. 13c) and resembles the central production process, but the incident particles are here electrons and the exchanged particle virtual

γ's. According to the vector dominance model, the hadronic interactions of γ's proceed through the conversion of the γ into a virtual $q\bar{q}$ pair with $J^{PC} = 1^{--}$, a vector meson. We would then expect gluonium to be strongly suppressed in $\gamma\gamma$ collisions. Hybrids and 4-quarks states are allowed, on the other hand. These reactions can therefore be considered as "anti-glueball" filters.

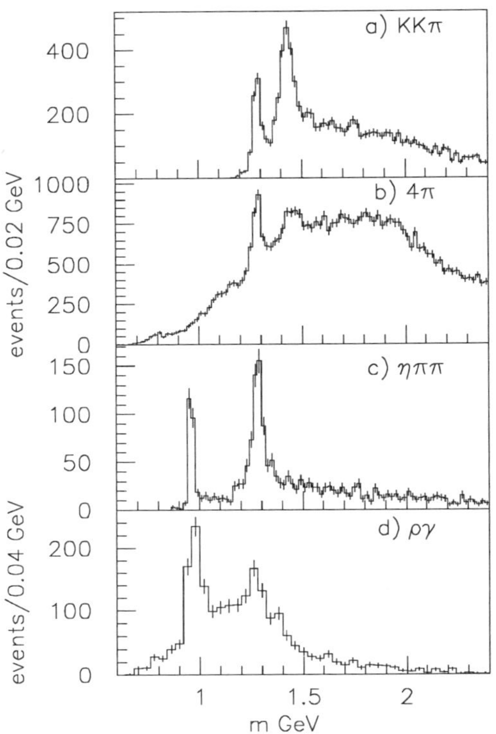

Figure 17: Summary of the results from Omega/WA76. Centrally produced $K\bar{K}\pi$, $\eta\pi^+\pi^-$, $2\pi^+2\pi^-$ and $\rho^0\gamma$ mass distributions in pp interactions at 300 GeV/c.

What also makes these reactions particularly interesting is the possibility of using the Yang theorem to discriminate between different spin configurations of the produced meson system. In particular, events are here classified as "no tag" and "tag". In the first case, the scattered electrons have a very small four momentum transfer, so that they are lost in the beam pipe and are undetected. Kinematically this means that the exchanged virtual γ has almost zero mass so that it is nearly real. Since the Yang theorem states that two massless photons do not couple to spin one, in this class of events spin zero particles are allowed but spin one are forbidden.

In the "tag" events, one of the two electrons has a relatively high four momentum transfer, so that it is able to escape from the beam pipe and is detected. In this case the γ becomes massive and highly virtual (γ^*) so that the Yang theorem does not hold any more and both 0 and 1 spins are allowed. Evidence for this is given in Fig. 18 [35], where the $\eta\pi\pi$ mass spectrum is shown in the two configurations. In the first case only the spin 0 η' is visible while in the second case also $f_1(1285)$ is produced. It is surprising that no $\eta(1295)$ is visible in the 'no tag' events. It is also interesting that no $\iota(1450)$ signal is produced in these reactions.

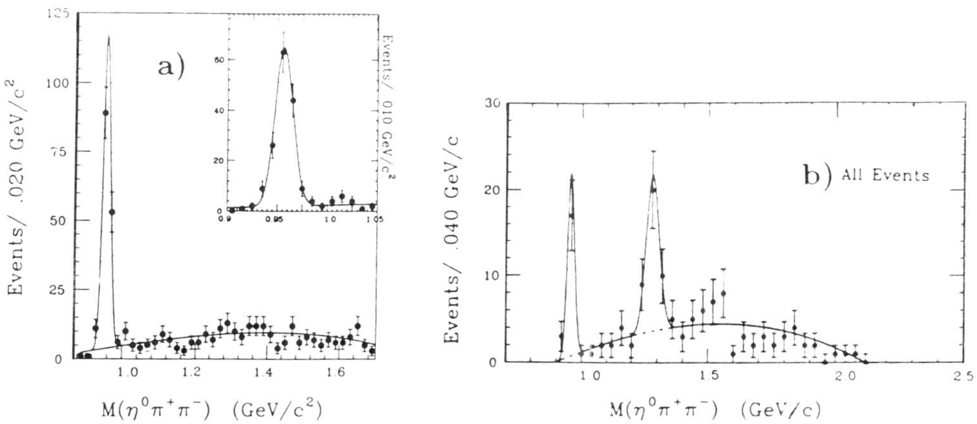

Figure 18: $\gamma\gamma \to \eta\pi\pi$ for a) "no tag" and b) "tag" events. Data from Mark II.

The $K\bar{K}\pi$ mass spectrum for the 'tag' events is shown in Fig. 19 [36] where we notice the production of $f_1(1285)$ and $f_1(1420)$. No signals are observed in the 'no-tag' events, confirming therefore the axial meson nature of these two states.

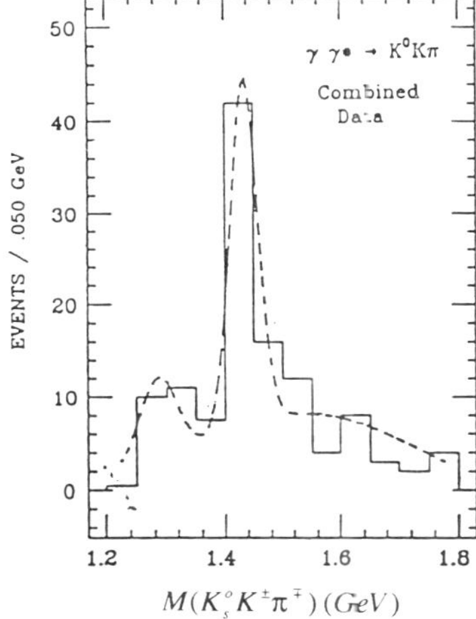

Figure 19: $\gamma\gamma \to K\bar{K}\pi$ for "tag" events. The spectrum is obtained adding the results of all experiments.

11 DISCUSSION OF THE PSEUDOSCALAR MESONS

In order to disentangle the pseudoscalar puzzle and have indications on possible non-$\bar{q}q$ candidates in this sector, we should compare the observed physical states with what is expected from the quark model. The comparison is made in Fig. 20a).

We notice that the GI model contains two predictions for the radial excitations which are labelled as P1 and P2. This is done in order to accomodate the experimental evidence for $\eta(1295)$, a candidate for a radial excitation of the η meson. The previous

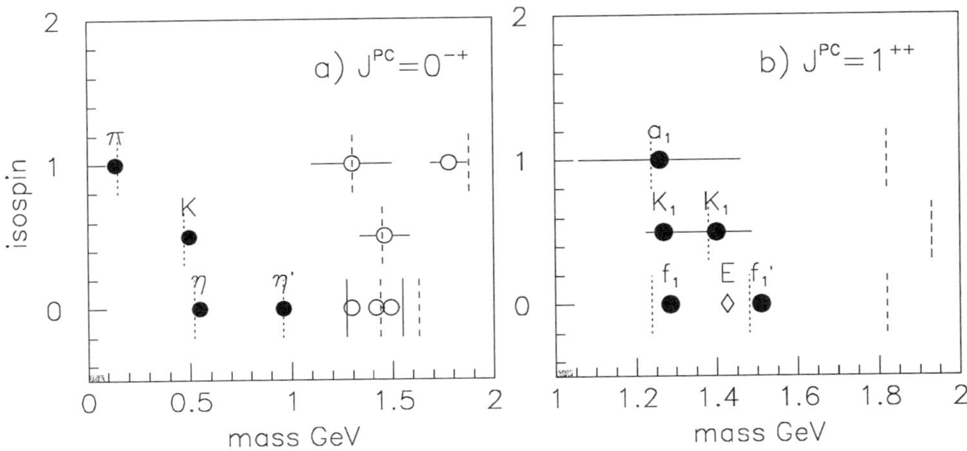

Figure 20: Comparison between measurements and expectations from the GI model for meson masses. Symbols represents masses, horizontal bars their full widths. Vertical lines show the expectations from the GI model. Dotted lines: ground states, dashed: radial excitations. a) $J^{PC} = 0^{-+}$, b) $J^{PC} = 1^{++}$. The solid lines in a) represent the expectations from the P1 model.

sections have shown that in addition to (π, K, η and η') mesons several sources report evidence for still other pseudoscalar resonances: $\eta(1295)$ and $\iota(1450)$.

The evidence for $\eta(1295)$ comes only from a PWA of the $\eta\pi\pi$ system in πp interactions. It is not directly visible in the mass spectrum since it is hidden over the overwhelming $f_1(1285)$ signal. It is not produced in $\gamma\gamma$ collisions, J/ψ decays, central production or $\bar{p}p$ annihilations. It is also not seen by VES in the high statistics study of the $\pi(\eta\pi\pi)$ system. In $\gamma\gamma$ collisions or $\bar{p}p$ annihilations at rest, the dynamics can be such that axial mesons are forbidden so that this meson could show up, but this is not the case. One could ask how solid is the evidence for this meson.

Turning to $\iota(1450)$, the experimental evidence for this state is becoming more and more real. The actual status can be summarized as follows:

- a) A resonance (two states?) is observed in the following decay channels:

$$\begin{aligned}\iota(1450) \; &\to K^*\bar{K} \\ &\to \rho^0\gamma \\ &\to \pi a_0(980) \;\to K\bar{K}\pi \\ &\qquad\qquad\quad\;\; \to \eta\pi\pi,\end{aligned}$$

- b) It is produced in J/ψ radiative (hadronic?) decays, $\bar{p}p$ annihilations, and πp interactions,
- c) It is not produced in $\gamma\gamma$ collisions or central production.

The $\iota(1450)$ is a candidate for being a glueball, a hybrid, or a radial excitation.

The strong radiative decay of $\iota \to \rho^0\gamma$ would seem against the glueball interpretation and in favour of the radial excitation. However, this seems not to be the case since a positive but low rate for the decay of a pseudoscalar glueball is expected by several models [7]. Radial excitations on the other hand are expected to have a rather small radiative decay width.

The information obtained from hadronic J/ψ decays is not clear and more data are needed. Assuming the ι to be a glueball, Seiden et al [37] have computed the following ratios:

$$B(J/\psi \to \phi\iota) = 0.83 \times 10^{-4}$$

$$B(J/\psi \to \omega\iota) = 2.4 \times 10^{-4}.$$

Experiments give:

$$B(J/\psi \to \phi\iota) < 2.1 \times 10^{-4}$$

$$B(J/\psi \to \omega"E") = 6.8 \pm 1.7 \times 10^{-4},$$

where "E" is the signal observed in the $K\bar{K}\pi$ mass spectrum, which however seems to be related to the axial meson $f_1(1420)$, rather than to the pseudoscalar $\iota(1450)$. One could nonetheless allow for some pseudoscalar contribution, so that the above branching ratios are not in disagreement. Hadronic J/ψ decays to $\eta\pi\pi$ do not give too much information, since no spin parity analysis of the possible structures observed in the 1.4 GeV region has been published. However, the structure observed in $J/\psi \to \omega\eta\pi\pi$ in the 1.4 GeV region would suggest the effective presence of some $\iota(1450)$ signal, since $f_1(1420)$ does not decay to $\eta\pi\pi$.

In conclusion, the number of resonances hiding below the $\iota(1450)$ signal is still not established. It is also not clear what is its (their) nature: radial excitation, glueball, or hybrid?

12 RADIAL EXCITATIONS

Radial excitations are usually difficult to detect. They have low cross sections and they are broad. Their rates can be estimated to be one order of magnitude lower than the ground states.

A recent high statistics experiment, VES at Serpukhov [38], has provided a good measurement of the I=1 pseudoscalar radial excitations, $\pi(1300)$ and $\pi(1800)$ shown in Fig. 21.

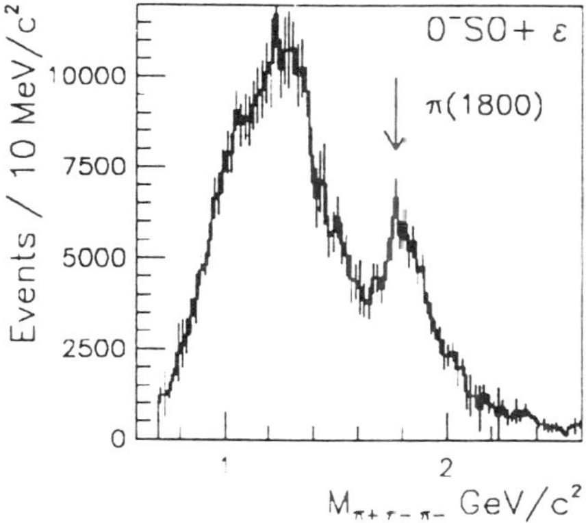

Figure 21: Results from PWA of the 3π system in the pseudoscalar projection in $\pi^- N$ interactions at 37 GeV/c from VES.

An inspection of Fig. 20a) shows that $\iota(1450)$ falls exactly where the GI model predicts a radial excitation. However, under the hypothesis that $\iota(1450)$ is an excited

η', it is not clear why it should be so strongly produced in radiative J/ψ decays, a process which is supposed to be mediated by gluons.

13 DISCUSSION OF $f_1(1420)$

The status of the axial mesons is shown in Fig. 20b). We observe a fair agreement with expectations from the GI model and also that two states, $f_1(1420)$ and $f_1(1510)$, are competing for being the $s\bar{s}$ member of this nonet.

The question is difficult to answer because there are too many uncertainties. The I=1/2 strange members are not C,G-eigenstates: $K_1(1280)$ ($J^{PC} = 1^{+-}$) and $K_1(1400), (J^{PC} = 1^{++})$ can mix to form C,G-eigenstates Q_+ and Q_-. The mixing angle is close to 45^0. The experimental determinations of the masses of these two states are:

$$m_{Q_+} = 1.31 \pm 0.015 \quad GeV$$

$$m_{Q_-} = 1.37 \pm 0.020 \quad GeV.$$

In addition, we have to deal with the poor determination of the mass of the a_1 meson which leads to the a_1 problem (see sect. 15).

Using the GMO quadratic mass formula:

$$tg^2\theta = \frac{4Q_+^2 - 3f_1'^2 - a_1^2}{3f_1^2 - 4Q_+^2 + a_1^2},$$

one obtains $\theta = 53.2^0$ for $f_1(1420)$ and $\theta = 62.0^0$ for $f_1(1510)$. We can therefore compute the strange contribution in these two mesons, under the hypothesis that they belong to the axial meson nonet.

$$X_f' = \sqrt{1/3}sin\theta + \sqrt{2/3}cos\theta.$$

We obtain 90% $s\bar{s}$ composition for $f_1(1420)$ and 80% for $f_1(1510)$.

We consider now the results obtained from $\gamma\gamma$ collisions which measured the $\gamma\gamma$ width of $f_1(1285)$ and $f_1(1420)$ to be respectively 2.2 ± 0.4 and 1.5 ± 0.3 [40]. The prediction for the ratio of production of the I=0 members of the axial mesons is:

$$\Gamma_{f_1}/\Gamma_{f_1'} = \frac{m_{f_1}}{m_{f_1'}} \left(\frac{sin\theta + \sqrt{8}cos\theta}{\sqrt{8}sin\theta - cos\theta} \right)^2.$$

We expect this ratio to be 2.0 for $f_1(1420)$, a value which is not inconsistent with a measured ratio of 1.5 ± 0.5. For $f_1(1510)$ the above expression gives 1.0 but the $\gamma\gamma$ width of this meson has never been measured.

From the above considerations, if $f_1(1420)$ belongs to the axial nonet, we expect it to have 90% $s\bar{s}$ composition. However this contradicts the results obtained from hadronic J/ψ decays, where this meson is observed opposite to the ω rather than the ϕ. In addition, a strangeonium state should be copiously produced in K induced reactions. However, in this case $f_1(1510)$ is observed with little evidence for $f_1(1420)$.

All these observations go in the direction that, most likely, $f_1(1420)$ does not belong to the axial nonet. It could therefore be an extra state and it is interesting to establish its nature. It decays only to $K^*\bar{K}$ and it cannot be a glueball because of its large $\gamma\gamma$ width. It has been proposed as a $(u\bar{u} + d\bar{d})g$ hybrid state, a $K^*\bar{K}$ molecule or a 4-quark state [41].

14 STILL OTHER PSEUDOSCALAR RESONANCES

This section on pseudoscalars cannot terminate without mentioning the structures observed in radiative J/ψ decays to Vector Vector:

$$J/\psi \to \gamma\text{Vector Vector, where Vector}=\phi, \rho, \omega, K^*.$$

A summary of the results obtained by Mark III is shown in Fig. 22 [42]. We notice a threshold enhancement in all the spectra together with an η_c signal. The higher statistics $2\pi^+2\pi^-$ mass spectrum from DM2 shows evidence for resonant structures [43]. A possible resonance could also be present around 2.2 GeV in the $\phi\phi$ mass spectrum.

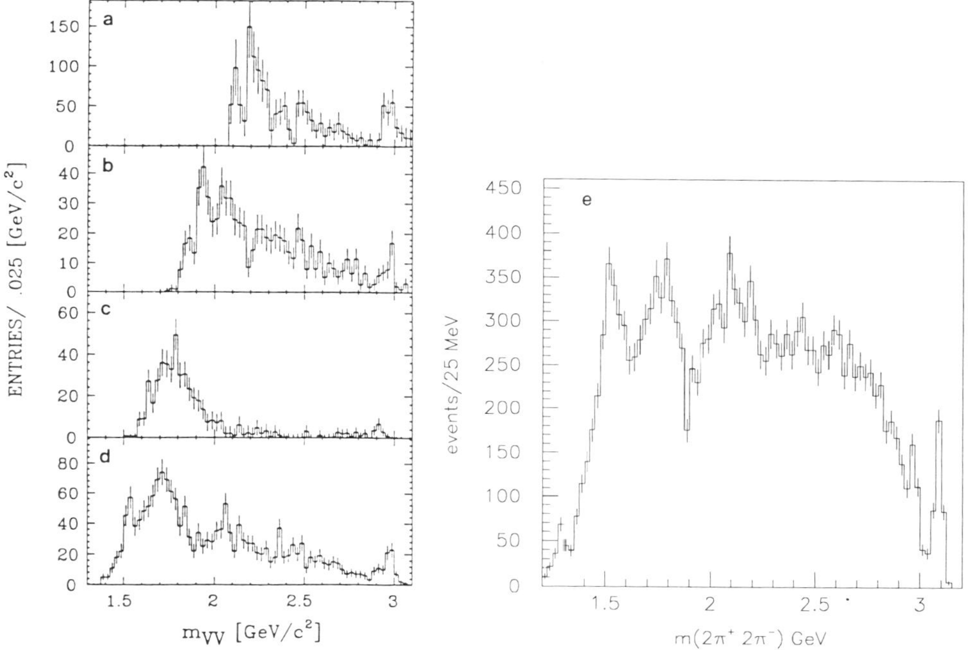

Figure 22: a) Summary of the results obtained by Mark III in the study of $J/\psi \to VV$ where $V = \phi, K^*, \omega, \rho^0$; b) $2\pi^+2\pi^-$ mass spectrum from J/ψ radiative decay (DM2).

A spin parity analysis of all these mass spectra has been performed and all the experiments find dominance for $J^{PC} = 0^{-+}$. However, the Mark III $J/\psi \to \gamma(4\pi)$ data have been re-analyzed by Bugg et al. [39], using a different approach in the spin analysis. They find evidence of complex structure of the $2\pi^+2\pi^-$ mass spectrum, with different $J^{PC} = 0^{++}, 0^{-+}, 2^{++}$ contributions.

15 THE a_1 PROLEM

One of the uncertainties related to the $J^{PC} = 1^{++}$ nonet is the still undefined properties of the I=1 member of the nonet, the $a_1(1270)$ meson. This meson has a dominant decay to $\rho(770)\pi$, i.e. $1^+ \to 1^-0^-$ where the ρ and the π have zero angular momentum. Such a combination is affected, in diffractive hadroproduction, by a $\rho\pi$ background which goes under the name of the Deck effect [44]. Since the a_1 resonance is rather broad, it is not easy in this process to distinguish resonance from background.

The result is that the 3π mass spectrum changes in shape with the four momentum transfer to the diffractive 3π system. The effect is so large that the real existence of the a_1 resonance has been questioned.

The question has been settled by two high statistics π-induced reactions which studied the reactions [45]:

$$\pi^- p \to \pi^- \pi^+ \pi^- p$$
$$\pi^- p \to \pi^+ \pi^- \pi^0 n.$$

The 3π cross section has been parametrized using three different contributions, shown in Fig. 23.

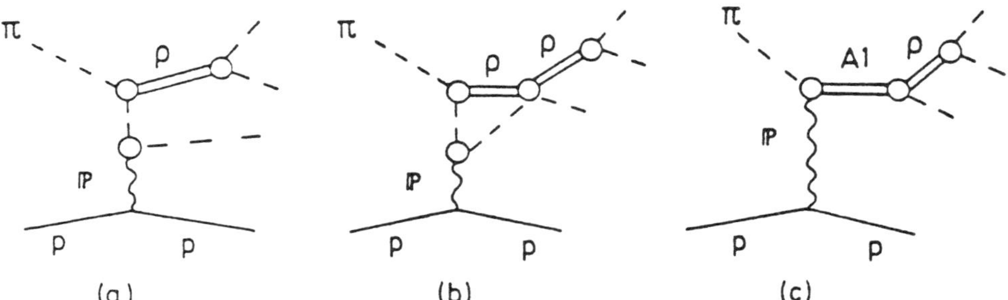

Figure 23: Diagrams contributing to the diffractive production of the 3π system in πp interactions.

- a) The incident pion dissociates into a ρ and a virtual π. The virtual π scatters diffractively into a real state. The amplitude for such a process has been written as:

$$D(M_{\pi\pi}, t_\pi, t) = \frac{A_{\pi pi}(M_{\pi\pi}, t_\pi) A_{\pi N}(S_{\pi N}, t)}{m_\pi^2 - t_\pi},$$

  where $A_{\pi\pi}$ is the l=1 $\pi\pi$ scattering amplitude through the ρ and $A_{\pi N}$ is the $\pi^- p$ elastic scattering amplitude,
- b) A rescattering term: $D' = De^{i\delta} cos\delta$ where δ is the S-wave $\rho\pi$ elastic scattering,
- c) The direct production of the a_1 resonance.

The subtraction of such a background was able to establish the existence of the resonance whose parameters are summarized in Table 4 [46]. Recent high statistics data on this problem are shown in Fig. 24.

We notice that the $a_1(1270)$ parameters are somewhat model dependent. We also notice, looking at the KEK results [48], how the resonance parameters are still a matter for debate.

Further information has been obtained from τ decays:

$$\tau^- \to \pi^- \pi^- \pi^+ \nu_\tau$$

which should proceed completely through $\tau \to a_1 \nu$, free from the Deck effect. The mass spectrum is shown in fig. 25 together with the $\pi\pi$ mass combination, which shows a large $\rho^0(770)$ production. The resulting a_1 parameters seem to be in rough agreement with what is measured in hadroproduction.

We also recall the striking result obtained in backward production of the 3π system [49] in the reaction:

$$K^- p \to \Sigma^- \pi^+ \pi^+ \pi^-.$$

Table 4: Determinations of the $a_1(1270)$ parameters. The measurements from τ decays use the Isgur et al. model [47]. The fits from different models are performed using the ARGUS data.

Experiment	m_{a_1} GeV	Γ_{a_1} GeV
ARGUS	1.211 ± 0.007	0.446 ± 0.021
DELCO	1.180 ± 0.060	0.430 ± 0.190
MARKII	1.250 ± 0.050	0.580 ± 0.100
$\pi^- p \to \pi^-\pi^-\pi^+ p$	1.280 ± 0.030	0.300 ± 0.050
$\pi^- p \to \pi^+\pi^-\pi^0 n$	1.240 ± 0.080	0.380 ± 0.100
$\pi^- p \to \pi^+\pi^-\pi^0 n$	1.122 ± 0.017	0.254 ± 0.011
$pp \to p_f(\pi^+\pi^-\pi^0)p_s$	1.208 ± 0.015	0.430 ± 0.050
Model		
Isgur et al.	1.211 ± 0.007	0.446 ± 0.021
Bowler	1.236 ± 0.006	0.450 ± 0.022
Kühn et al.	1.274 ± 0.007	0.594 ± 0.023
Ivanov et al.	1.246 ± 0.006	0.483 ± 0.021
Törnqvist	1.224	0.592

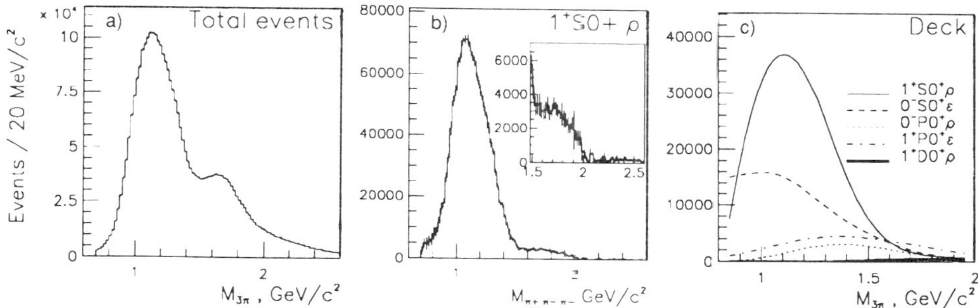

Figure 24: a) 3π effective mass distribution in πN interactions at 37 GeV/c, b) $J^P = 1^+$ contribution and c) Deck contributions. Results from VES.

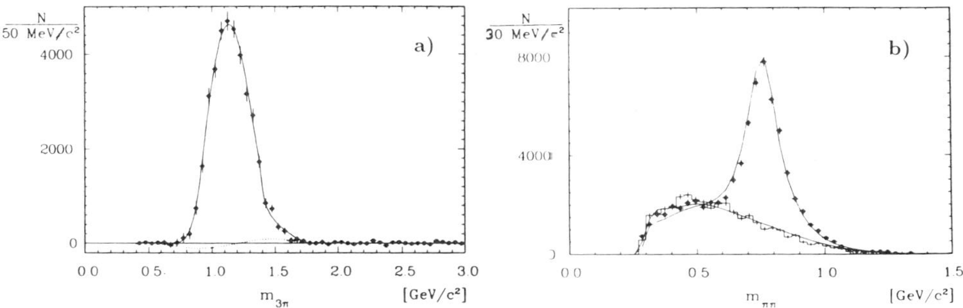

Figure 25: a) 3π mass distribution from $\tau^- \to \pi^-\pi^-\pi^+\nu_\tau$, b) $\pi^+\pi^-$ mass combination. The data are from ARGUS.

Although this reaction is free from the Deck effect, they find a very low value for the a_1 mass:

$$m = 1041 \pm 13 \quad MeV, \quad \Gamma = 230 \pm 50 \quad MeV.$$

16 THE 1^{+-} AND 2^{-+} NONETS

The status of the $J^{PC} = 1^{+-}$ and $J^{PC} = 2^{-+}$ nonets are summarized in Fig. 26. We notice the the pseudotensor nonet still has missing resonances. However, new results from $\gamma\gamma \to \eta\pi\pi$ and $\bar{p}p \to \pi(\eta\pi\pi)$ seem to find evidence for the missing η_2 members of this nonet [50].

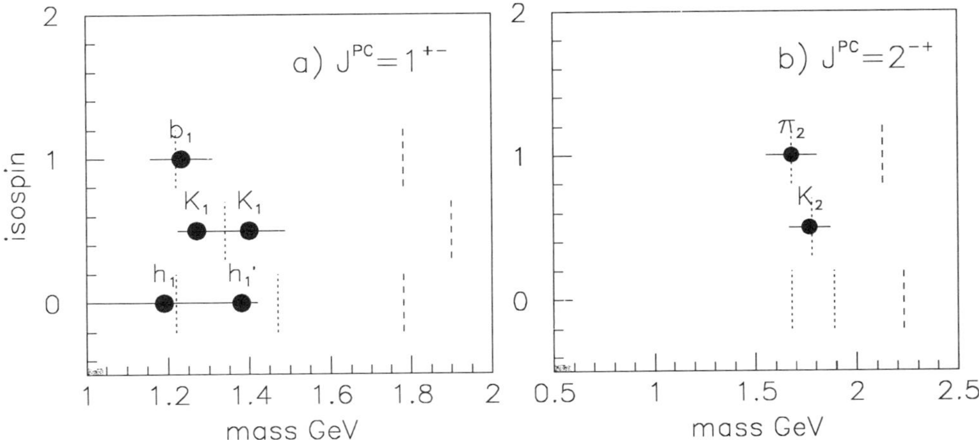

Figure 26: Comparison between measurements and expectations from the GI model for the masses. Symbols represents masses, horizontal bars their full widths. Vertical lines show the expectations from the GI model. Dotted lines: ground states, dashed: radial excitations. a) $J^{PC} = 1^{+-}$, b) $J^{PC} = 2^{-+}$.

17 CONCLUSIONS

A review of the status of pseudoscalars and axial nonets has shown the presence of states $\iota(1450)$ and $f_1(1420)$, which are difficult to accomodate in the quark model. New, clean high statistics data in different reactions are needed in order to solve the question of the possible existence of non-$\bar{q}q$ mesons in these multiplets.

18 ACKNOWLEDGMENTS

I want to thank David Bugg for the hospitality and his efforts in organizing such an interesting meeting.

REFERENCES

[1] Particle Data Group, Phys. Rev. D 50 (1994).
[2] T.H. Burnett, Physics of 0^{++} and 2^{++} mesons, these Proceedings.
[3] S. Godfrey and N. Isgur, Phys. Rev. D32 (1985) 189.
[4] A. Falvard et al., Phys. Rev. D41 (1990) 1389.
[5] D. Coffman et al., Phys. Rev. D38 (1988) 2695.
[6] J.L. Rosner, Phys. Rev. D27(1983) 1101.
[7] L. Köpke and N. Wermes, Phys. Rep. 174 (1989) 67.
[8] R.M. Baltrusaitis et al., Phys. Rev. D32(1985) 2883.
[9] H. Kolanoski, Proceedings of the IX Workshop on photon-photon collisions, San Diego, 22-26 March 1992, World Scientific, 3.
[10] see for example F. Close, Rep. Prog. Phys. 51 (1988) 833 for a general review.
[11] G.S. Bali et al. (UKQCD Collaboration), Phys. Lett. B 309 (1993) 378.
[12] D.L. Scharre et al., Phys. Lett. 97B (1980) 329.
[13] C. Edwards et al., Phys. Rev. Lett. 49 (1982) 259.
[14] J.D. Richman Ph.D. Thesis, Caltech, CALT-68-1231(1985).
[15] L.F. Landau, Dok. Akad. Nauk USSR 60 (1948) 207; C.N. Yang, Phys. Rev. 77 (1950) 242.
[16] M. Chanowitz, Phys. Rev. Lett. 46 (1981) 981.
[17] T.A. Armstrong et al., Z. Phys. C52 (1991) 389.
[18] G. Usai, Tesi di Dottorato (1994), Cagliari.
[19] S.M. Flatté, Phys. Lett. B63(1976) 224.
[20] Z. Bai et al., Phys. Rev. Lett. 65 (1990) 2507.; J. Augustin et al., Phys. Rev. D46 (1992) 1951.
[21] P. Baillon et al., Nuovo Cimento 3 (1967) 393.
[22] K. Duch et al., Z. Phys. C45 (1989) 223.
[23] A. Masoni, CERN-PPE/95-120.
[24] Z. Bar-Yam, Proceedings of Hadron91, Maryland (1991) World Scientific.
[25] C. Dionisi et al., Nucl. Phys. B169 (1980) 1.
[26] A. Birman et al., Phys. Rev. Lett. 61 (1988) 1557.
[27] N. Stanton et al., Phys. Rev. Lett. 42 (1979) 346; A. Ando et al., Phys. Rev. Lett. 57 (1986) 1296; S. Fukui et al., Phys. Lett. B267 (1991) 293.
[28] J.P. Stroot, CERN/EP/85-01, 9 January 1985.
[29] S.I. Bityukov et al., Phys. Lett. B188 (1987) 383.
[30] G.M. Beladidze et al., Phys. Lett. B313 (1993) 276.
[31] Ph. Gavillet et al., Z. Phys. C16 (1982) 119.
[32] D. Aston et al., Phys. Lett. B201 (1988) 573
[33] A. Palano, Proceedings of the IX Workshop on photon-photon collisions, San Diego, 22-26 March 1992, World Scientific, 308.
[34] T.A. Armstrong et al. Z. Phys. C52 (1991) 389.
[35] G. Gidal et al., Phys. Rev. Lett. 59 (1987) 2012.
[36] G. Gidal, Proceedings of the BNL Workshop on Glueballs, Hybrids and Exotic Hadrons, BNL, August 29 1988, AIP Conf. Proccedings.
[37] A. Seiden et al., Phys. Rev. D38 (1988) 824.
[38] D.V. Amelin et al., Phys. Lett. B356 (1995) 595.
[39] D.V. Bugg et al., Phys. Lett. B353 (1995) 378.
[40] D.O. Caldwell, Mod. Phys. Lett. A2 (1987) 771.

[41] S. Ishida et al., Progress Theor. Phys. 82 (1989) 119; R.S. Longacre, Phys. Rev. D42 (1990) 874.
[42] G. Eigen, Proceedings of the IX Workshop on photon-photon collisions, San Diego, 22-26 March 1992, World Scientific, 291.
[43] L. Stanco, LAL 87-42, August 1987.
[44] R.T. Deck, Phys. Rev. Lett. 13 (1964) 169.
[45] C. Daum et al., Nucl. Phys. B182 (1981) 269; J. Dankowych et al., Phys. Rev. Lett. 46 (1981) 580.
[46] H. Albrecht et al., DESY 92-125, September 1992.
[47] N. Isgur et al., Phys. Rev. D39 (1989) 1357.
[48] A. Ando et al., Phys. Lett. B291 (1992) 496.
[49] Ph. Gavillet et al., Phys. Lett. 68B (1977) 119.
[50] K. Karch et al., Z. Phys. C54 (1992) 33.

AN INTRODUCTION TO THE VECTOR MESONS

F. Nichitiu *
Lab. Naz. di Frascati dell'INFN, Frascati, Italy

ABSTRACT

In the present paper we survey the present status of the experimental knowledge on vector mesons built from light quarks.

1. $(q\bar{q})$ systems

Using the non relativistic quark model, most of the many hadronic states known nowadays can be described as states made of only q and $\bar{q}$. S.Godfrey and N.Isgur [1] have proposed a relativised quarkonium model which describes all the mesons from the pion to the heaviest upsilon using a unified framework.

Ignoring the heavy quarks, there are nine possible $(q\bar{q})$ combinations (quark flavour: u,d,s) which are grouped into an octet and a singlet. The mass formula for an octet is given by the Gell-Mann-Okubo formula:

$$m_{I=0}^2 = (4m_{I=1/2}^2 - m_{I=1}^2)/3. \quad (1)$$

The states with the same I, J and P (and additive quantum numbers) can mix due to SU(3) breaking, a mixing which in general is mass dependent. Because of this, it is common to group the mesons into nonets rather than octets and singlets.

If the quark flavour decomposition for a pure singlet and octet is

- \- singlet: I=0 $(u\bar{u} + d\bar{d} + s\bar{s})/\sqrt{3})$
- \- octet: I=1 $u\bar{d}; (u\bar{u} + d\bar{d})/\sqrt{2}; d\bar{u}$
- \- I=1/2 $u\bar{s}; d\bar{s}$
- \- I=1/2 $s\bar{d}; s\bar{u}$
- \- I=0 $(u\bar{u} + d\bar{d} - 2s\bar{s})/\sqrt{6}$

*On leave of absence from IFA-Bucharest, Romania

Hadron Spectroscopy and the Confinement Problem
Edited by D.V. Bugg, Plenum Press, New York, 1996

then the two I=0 physical states of a given nonet

$$|Phys-a\rangle = |8\rangle\cos(\theta) - |1\rangle\sin(\theta) \tag{2}$$

$$|Phys-b\rangle = |8\rangle\sin(\theta) + |1\rangle\cos(\theta) \tag{3}$$

can have such a mixing angle that one of the physical states can be pure or mostly $|s\bar{s}\rangle$. This angle is named 'ideal' and is given by $\tan\theta = 1/\sqrt{2}$ or $\theta = 35.3$.

The lowest vector nonet, $\rho(770)$, K*(892), $\omega(782)$ and $\phi(1020)$ is indeed ideally mixed; $\phi(1020)$ is practically a pure state of $|s\bar{s}\rangle$ and ω a pure state containing no strange quarks $|(u\bar{u} - d\bar{d})/\sqrt{2}\rangle$. It is believed that all the vector mesons are almost ideally mixed. This is an important fact which is used in phenomenological analysis of new vector meson states.

Assuming the mesons to be $q\bar{q}$ states, a pair of two 1/2 spin 'particles' (fermions) with fractional charges (u and d forming an isodoublet), the quantum numbers permitted for a meson-nonet are grouped as :

- the normal (or natural) spin-parity series where P=$(-1)^J$ and CP=+1

  0^{+-}; 1^{--}; 2^{++}; 3^{--}, or using spectroscopic notation

  $^3P_0$; $^3S_1$; $^3P_2$,

- the unnatural spin-parity series where P=$(-1)^{J+1}$

  0^{-+}; 1^{++}; 2^{-+}; 3^{++},

  1^{+-}; 2^{--}; 3^{+-}, or

  $^1S_0$; $^3P_1$; $^1D_2$,

  $^1P_1$; $^3D_2$.

The (q$\bar{q}$) ground states are (L=0):

- J^{PC}= 0^{-+}; $^1S_0$ pseudoscalar mesons, and
- J^{PC}= 1^{--}; $^3S_1$ vector mesons.

The first radial excitation of the vector meson nonet is denoted by 2^3S_1 and the first orbital excitation by 1^3D_1.

Some combinations of J, P and C are forbidden for normal $q\bar{q}$ states.The forbidden 'normal' $q\bar{q}$ states are all the mesonic states which belong to the normal spin-parity series (P=$(-1)^J$) for which C = –P, as well as the 0^{--} state. Such a quantum-mechanical selection rule for exotic J^{PC} assignments can exist only for mesons. These states are 0^{--}; 0^{+-}; 1^{-+}; 2^{+-} etc.

The present understanding of the strong interaction is that it is described by Quantum Chromo-Dynamics (QCD). This theory describes not only how quarks and antiquarks interact, but also predicts that the gluons which are the quanta of the field will themselves interact to form mesons. So, if a hadronic object (a meson) is composed only from valence gluons it is called a glueball (gluonia). The simplest colour-singlet gluonia are 2-gluon states with quantum numbers 0^{++}; 0^{-+}; 1^{-+}; 2^{-+} ...(the third is exotic, having non-($q\bar{q}$) quantum numbers).

A gluonic vector state 1^{--} requires three gluons, which may exist as bound states just as baryons are made of 3 quarks. Exotic vector meson states which are not simple $q\bar{q}$ bound states can also exist if the state is composed of a mixture of valence quarks, antiquarks and gluons ($q\bar{q}g$) (hybrids) or contain many quark states ($qq\bar{q}\bar{q}$) denoted as ($q^2\bar{q}^2$). These are multiquark states which should not be confused with simple resonant states of two mesons or with so-called molecular states. An unambiguous confirmation of such exotic states would be an important test of QCD and will give fundamental information on the behaviour of this theory in the confinement region.

2. Production of meson resonances; $e^+ e^-$ and γp.

An important part of the information on meson spectroscopy has been obtained from peripheral di-meson production where the dominant contribution comes from one pion exchange.

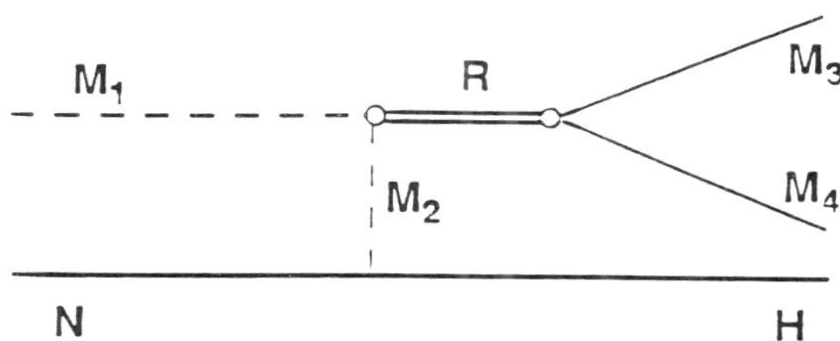

Figure 1: One Meson Exchange diagram. M_i are π or K mesons, N is the target nucleon and H is a nucleon or a hyperon.

Apart from the One-Meson-Exchange mechanism, Fig.1, basic for reactions of the type

- $\pi\pi \rightarrow \pi\pi$ or $(\bar{K}K)$
- $K\bar{K}$ $(\pi) \rightarrow K\bar{K}$ (π),

there are also other mechanisms which contribute to the production of a multi-meson system in reactions such as:

- - diffractive reactions (meso/photo production)
- - $e^+ e^-$ collisions
- - $\gamma\ \gamma$ collision
- - central production,
- - nucleon antinucleon annihilation.

The first two, photoproduction and $e^+ e^-$ annihilation, Fig. 2, are very important for vector meson spectroscopy. Because the vector mesons have the same quantum numbers as the photon, 1^{--} (I=0,1), they will be easily produced either in $e^+ e^-$ collisions or in photoproduction. A very useful description of both types of vector meson production mechanisms is given by the so called Vector Dominance Model (VDM)[2].

The VDM arose much earlier than the quark model,[3] but seems to be still compatible with it. The idea of the VDM, Fig. 3, consists in assuming that the interaction of a photon with hadrons is dominantly due to conversion of the γ into a hadronic vector meson ($J^{PC}=1^{--}$), which then interacts strongly with the hadronic target, as in any other strong collision. The Restricted VDM corresponds to the use of the low-lying vector meson states[4] (ρ, ω, ϕ); the so called Extended (or Generalized) VDM [5] uses a complete series of excited vector meson states $(\rho, \rho', \omega, \omega'$, etc). The latter is expected to provide a good description of the electromagnetic interactions for a very large range of q^2.

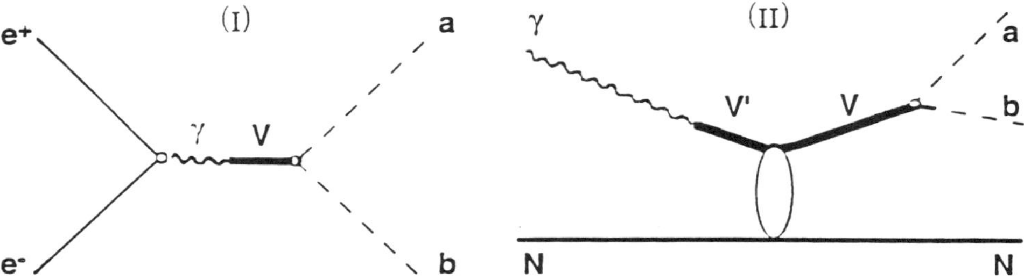

Figure 2: Vector mèson production in e^+e^- (I) and γN (II).

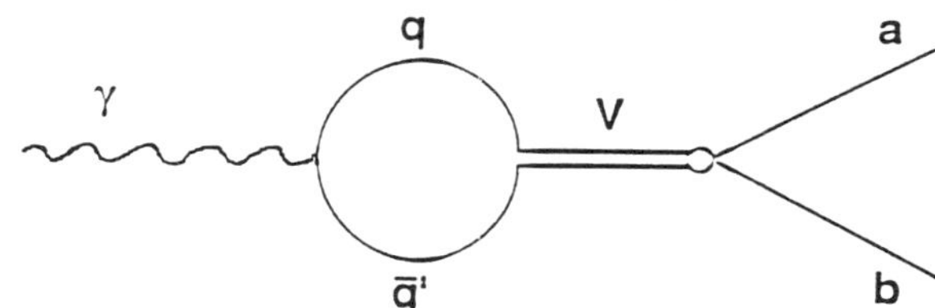

Figure 3: VDM view in quark language.

To explain the VDM in quark language, we restrict ourselves to the low energy domain, where q^2 is of the order of hadronic masses. Because the typical distance of interaction is of the same order of magnitude as the confinement radius ($R_c \sim 1/m_v \approx d \sim 1/\sqrt{q^2}$), the $q\bar{q}$ pair converted from the photon cannot be free (confinement effects will drive the $q\bar{q}$ pair to the nearby bound state with the same quantum numbers as the photon 1^{--} , therefore to vector mesons) and then the subsequent strong process takes place.

2.1. e^+e^- collisions.

When the e^+e^- center of mass energy is varied, one observes a series of peaks in the total cross section for some specific channels $e^+e^- \to$ hadrons. These peaks are due to the formation in the s channel of vector meson resonances which we interpret as quark-antiquark bound states. The general features of e^+e^- annihilation are given simply by a cross section expression calculated as a product of the usual 1/s dependence of the lepton pair production [14] and a Breit-Wigner resonance shape for the vector meson:

$$\sigma(e^+e^- \longrightarrow V) = \frac{3\pi}{s}\frac{\Gamma_{ee}\Gamma_V}{(m_V - \sqrt{s})^2 + \Gamma^2/4}. \tag{4}$$

The maximum of the cross section (σ at the peak) gives information on the leptonic width :

$$\Gamma_{ee} = \frac{m_V^2}{12\pi}\Gamma_V\sigma_{\text{peak}}, \tag{5}$$

where Γ_V is the total (hadronic) width of the vector meson. The connection between Γ_{ee} and the coupling constant is calculated by the explicit evaluation of the corresponding diagram and is given by :

$$\Gamma_{ee} = \frac{\alpha^2}{12}\left(\frac{4\pi}{g_V^2}\right)m_V; \tag{6}$$

the cross section will be given by :

$$\sigma = \frac{\pi^2}{m} \frac{(\alpha/g_V^2)}{(m_V - \sqrt{m})^2 + \Gamma^2/4}. \tag{7}$$

This semi-phenomenological picture appears to be very successful. In the framework of a potential model (an effective potential V(r) between q and $\bar{q}$), the coupling constant between photon and vector meson will be proportional to the charge of the quark and to the value of the radial bound state wave function at the origin (at r=0).

In this way a very important relation is obtained between the leptonic width of a 1^{--} meson (Γ_{ee}) and the wave function of a 1^{--} $q\bar{q}$ system which is connected with the corresponding quark structure of the vector meson

$$\Gamma_{ee} = \frac{16\pi\alpha^2}{m_V^2} Q^2 \left|\Psi_{q\bar{q}'}(0)\right|^2. \tag{8}$$

Quantitative predictions for the coupling constant between vector meson and photon depend on the choice of the potential V(r) which determines $\Psi(0)$. But there are some qualitative features which have some generality and have become important in data analysis.

In the non-relativistic limit, $\Psi(0)$ is non-zero only for S states and therefore only the $^3S_1$ and 2^3S_1 states (S radial excitation) should be coupled to e^+e^-. The orbital excitation $^3D_1$ and radial-orbital excitation 2^3D_1 will not be coupled to e^+e^-. These orbital excitations will be coupled only through relativistic corrections which involve the derivative of the wave function at the origin, or through configurations which mix with S states.

Also, because the radially excited states generally have more extended wave functions than those of the fundamental states (with n=1), the wave function at the origin, and therefore the coupling constant, are weaker, giving a smaller electronic width:

$$\Gamma_{ee}(L=0) > \Gamma_{ee}(L \neq 0) \tag{9}$$

$$\Gamma_{ee}(n>1) < \Gamma_{ee}(n=1). \tag{10}$$

A more extended wave function can be expected for multiquark states and therefore :

$$\Gamma_{ee}(\text{multiquark}) << \Gamma_{ee}(q\bar{q}). \tag{11}$$

Finally, if the $q\bar{q}$ potential is flavour independent and if we consider m_V roughly equal for all the vector mesons, an interesting rule appears :

$$\Gamma_{ee}(\rho,\omega,\phi)/Q^2 \simeq \text{constant}. \tag{12}$$

Here Q is the effective charge of the quarks involved. Because :

$$|\Psi(0)|^2/m_V^2 \simeq \text{constant}, \tag{13}$$

we can expect :

$$\Gamma_{ee} \sim Q^2. \tag{14}$$

For the fundamental vector meson nonet we have:

$$\Gamma_{ee}(\rho) = \left(\frac{1}{\sqrt{2}}\left(\frac{2}{3}+\frac{1}{3}\right)\right)^2 \qquad \frac{1}{\sqrt{2}}(u\bar{u} - d\bar{d}) \tag{15}$$

$$\Gamma_{ee}(\omega) = \left(\frac{1}{\sqrt{2}}\left(\frac{2}{3}-\frac{1}{3}\right)\right)^2 \qquad \frac{1}{\sqrt{2}}(u\bar{u}+d\bar{d}) \tag{16}$$

$$\Gamma_{ee}(\phi) = \left(-\frac{1}{3}\right)^2 \qquad (s\bar{s}) \tag{17}$$

which numerically means :

$$\begin{array}{l} \Gamma_{ee}(\rho)\,:\,\Gamma_{ee}(\omega)\,:\,\Gamma_{ee}(\phi) = 9\,:\,1\,:\,2 \\ \text{experimental} = (8.8 \pm 2.6)\,:\,1\,:\,(1.7 \pm 0.4) \end{array} . \tag{18}$$

A question is whether these ratios are correct for other vector meson nonets? The electromagnetic widths for the fundamental vector nonet taken from [7] are given below:

1^{--}	Γ_{ee}(keV)	Γ_{ee}/Q^2	$g^2/4\pi$
$\rho(0.7699)$	6.77 ± 0.32	13.5	0.6
$\omega(0.7819)$	0.60 ± 0.02	10.8	8.
$\phi(1.0194)$	1.37 ± 0.05	12.3	4.

2.2. Photoproduction.

The VDM suggests that the vector meson 'produced' (created) from the photon is scattered on the target, becomes real and decays in the hadronic final state. If the initial state and the final one of this vector meson-target scattering are identical, this part of the interaction can be considered as elastic vector-meson-nucleon scattering and has the properties of a diffractive process:

- - the cross section is nearly energy independent,
- - there is a helicity transfer in the s channel from γ to the vector meson obeying so-called s-channel helicity conservation (SCHC), which is very important in practical analysis of experimental data (Partial Wave Analysis),
- - there is an exponential behaviour of the differential cross section

$$\frac{d\sigma}{dt} \sim e^{-b|t|} \tag{19}$$

with slope $b^2 \sim R$, where R is the radius of γ N interaction.

The differential cross section for photoproduction processes is determined by the product of the coupling constant ($\gamma \to V$) and the corresponding hadronic cross section of the subprocess $VN \to VN$:

$$\frac{d\sigma}{dt}(\gamma N \longrightarrow VN) = \alpha \frac{\pi}{g_V^2}\frac{d\sigma}{dt}(VN \longrightarrow VN). \tag{20}$$

Assuming that the VN $\to$ VN scattering amplitude is purely imaginary (we are at high energies) and using the optical theorem which connects the total cross section to the imaginary part of the forward amplitude, we obtain:

$$\left.\frac{d\sigma}{dt}(\gamma N \longrightarrow VN)\right|_{t=0} = \frac{\alpha}{16b}\frac{1}{g_V^2}\sigma_{\rm tot}^2(VN), \tag{21}$$

from which is obtained an estimate of the photoproduction cross section:

$$\sigma(\gamma N \longrightarrow VN) = \frac{\alpha}{16b}\frac{1}{g_V^2}\sigma_{tot}^2(VN). \tag{22}$$

The total cross section (VN) ought to be be measured in an ideal experiment with V beams, but is in fact estimated by the vector quark content as for example by:

$$\sigma_{tot}(\Phi N) = (\sigma_{tot}^2(K^+N) + \sigma_{tot}^2(K^-N) - \sigma_{tot}^2(\Pi^-N))^{1/2} \tag{23}$$

$$\sigma_{tot}(\rho N) = \frac{1}{2}(\sigma_{tot}(\pi^+p) + \sigma_{tot}(\pi^-p)). \tag{24}$$

Due to the proportionality to the γV coupling constant in both expressions for vector meson production, in e$^+$ e$^-$ annihilation as well as in γp photoproduction the cross sections are related:

$$\sigma(e^+e^- \longrightarrow V \longrightarrow \text{hadrons}) = \frac{32\pi^3\alpha b}{m_V\sigma_{tot}^2(VN)}\sigma(\gamma N \longrightarrow VN \longrightarrow \text{hadrons}). \tag{25}$$

This formula allows therefore (in the framework of VDM) a comparison, i.e. a check on the theory; it can also be used to constrain a fit of the experimental data in order to find new states.

Some complications arise due to the fact that the 'two-step process' (vector meson scattering) is not only of the elastic type. There are so called 'off diagonal contributions' which are especially relevant for radial excitations (when the V$'$ meson is different from the V meson as in fig.2).

In the 'intermediate energy' region, e$^+$ e$^-$ annihilation can be as complicated as any other interaction process in which the final hadronic state consists of a superposition of many resonances. It seems however that, in many cases, multiparticle production can be well described by a cascade of decays (as for example in the isobar model used in $p\bar{p}$ annihilation [12]) of the type:

$$\begin{array}{llll} e^+e^- \longrightarrow \gamma \longrightarrow (V) \longrightarrow & X_1 & + & X_2 \\ & \hookrightarrow A+B & & \hookrightarrow C+D \\ & & & \hookrightarrow a+b \quad \text{etc.} \end{array} \tag{26}$$

Simple examples are :

$$\begin{array}{ll} e^+e^- \longrightarrow (\phi) \longrightarrow & \rho \quad \pi \\ & \hookrightarrow \pi\pi \end{array} \tag{27}$$

$$\begin{array}{ll} e^+e^- \longrightarrow (\rho') \longrightarrow & \rho \quad (\pi\pi)_s \\ & \hookrightarrow \pi\pi \end{array} \tag{28}$$

$$\begin{array}{ll} e^+e^- \longrightarrow (\phi') \longrightarrow & K^* \quad +\overline{K} \\ & \hookrightarrow K\pi \end{array}. \tag{29}$$

The present experimental situation in the vector meson sector can be illustrated by the 'official' view of the PDG94 [7] and is given below :

		$I=1$	$I=0$		$I=1/2$
1^3S_1	1^{--}	$\rho(770)$	$\omega(782)$	$\phi(1024)$	$K^+(892)$
2^3S_1	1^{--}	$\rho'(1450)$	$\omega'(1420)$	$\phi'(1680)$	$K'^+(1410)$
$^3D_1$	1^{--}	$\rho''(1700)$	$\omega''(1600)$	?	$K''^+(1680)$
$^3D_2$	2^{--}	?	?	?	$K_2(1820)$
$^3D_3$	3^{--}	$\rho_3(1690)$	$\omega_3(1670)$	$\phi_3(1850)$	$K_3(1780)$

3. ρ family

From the experimental point of view, the most important member of the vector meson nonet is the state with I=1 , the 'ρ family'. Besides $\rho(770)$, up to some years ago only a second ρ with mass around 1600 MeV and width 400-500 MeV ($\rho'(1600)$) was known. Now it seems to be resolved into at least two ρ - states. The early evidence for this single second ρ came from some phase-shift analyses of the $\pi\pi$ system (to which the excited ρ state or states seems to be weakly coupled),[13] as well as from the direct observation of enhancements in 4 pion spectra produced in $e^+ e^-$ collisions and in both electro and photo- production (for a list of experimental evidence of $\rho'(1600)$ see the mini-review of the PDG 84, ref [8]).

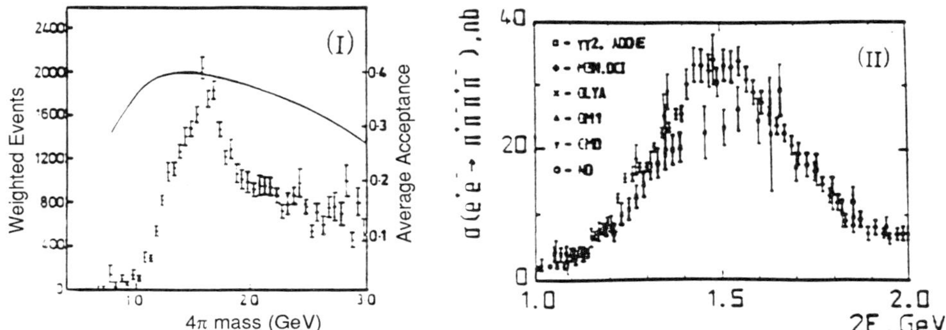

Figure 4: $2\pi^+2\pi^-$ from γp (I) and e^+e^- (II).

The first demonstration that data on e^+e^- annihilation into 2π and 4π can be understood in terms of the standard ρ and another two ρ excited states, ρ' and ρ'', appeared in 1986 in ref. [6] and was adopted by the PDG in 1988 [7]. The considerable uncertainty concerning the $\rho'(1600)$ 'bump' emerges from the complicated experimental situation, Figs.4 and 5. In reactions such as :

$$\gamma p \longrightarrow 2\pi^-2\pi^+p \tag{30}$$

$$e^+e^- \longrightarrow 2\pi^-2\pi^+ \tag{31}$$

a broad resonance peak is observed with mass and width in reasonable agreement in both reactions m=1.57 GeV Γ=0.51 GeV, but in the γp reaction over a significant non-resonant background. Fig.4 shows the experimental mass distribution for $2\pi^-2\pi^+$ from γp reactions ((I) [15]) and e^+e^- annihilations (II, data quoted in ref. [11,36]).

For the $\pi^+\pi^-2\pi^0$ final state there is a very broad peak in the e^+e^- reaction and in γp there are two clear peaks at 1.25 GeV and at 1.60 GeV. The lower part of this spectrum is completely dominated by the $\omega\pi^0$ channel through γ photoproduction of $b_1(1250)$ (a non-vector meson : $1^+(1^{+-})$ state). Fig.5 shows the experimental mass spectrum for $\pi^+\pi^-2\pi^0$ in γp reactions (I [16]) before (a) and after (b) an ω cut ; and in e^+e^- annihilations (II, data quoted in ref. [11,36]).

The mass and width observed in mass distributions and in the energy dependence of the cross sections need not be the true resonance parameters because :

- - there may be a significant background which can interfere with the resonance,
- - in a multimeson final state there can exist more than one isobar (a quasi two-body state from the isobar decomposition of the multimeson final state) which can contribute, each with its own J^P.

A complete understanding of the final state can be therefore achieved only by a spin-parity analysis where a separation of different J^P contributions is obtained. The J^P determination is usually done using the premises of the 'SCHC' production mechanism in a complex fit of experimental angular distributions for different energy (mass) intervals.

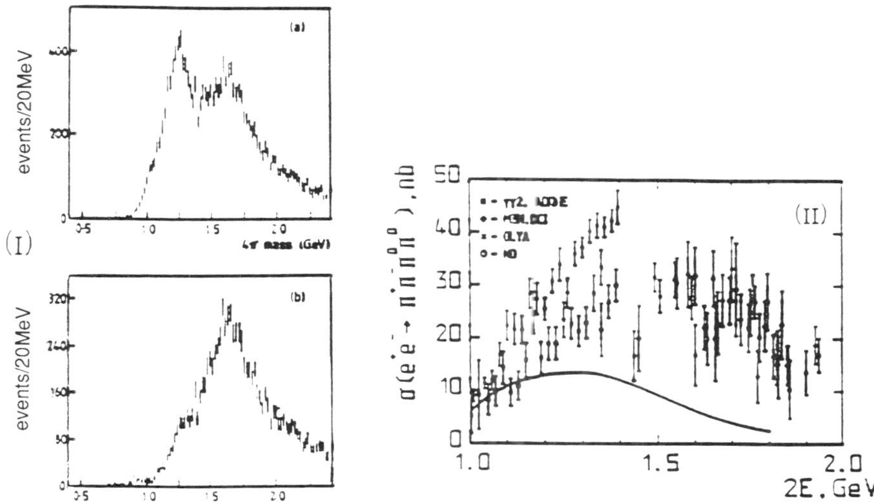

Figure 5: $\pi^+\pi^-2\pi^0$ from γp (I) and e^+e^- (II). In (I), (a) before and (b) after an ω cut.

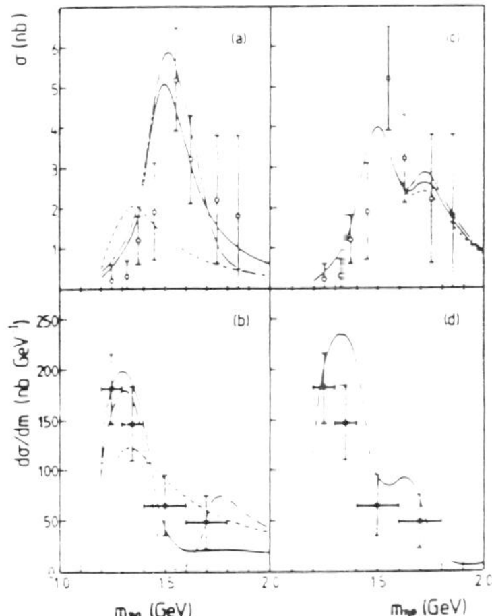

Figure 6: Fit of $e^+e^- \to \eta\rho$ (a) and (c), and $\gamma p \to \eta\rho$ (b) and (d)

The next step of the analysis will be the study of the mass dependence of each partial wave contribution (intensity of the PWA or partial wave cross section, and if it exists, the phase motion of the PW amplitude). An example of such a study is the analysis of $\eta\rho$ produced in diffractive γp reactions and in e^+e^- annihilations, Fig.6, done by Donnachie and Clegg [10] .

The mass spectra of $\eta\rho$ states with $J^P=1^-$ from diffractive photo-production and from e^+e^- annihilation appear at first sight to be completely incompatible. The apparent discrepancy was resolved through interferences between the tail of the $\rho(770)$

meson and two ρ' meson states.

This analysis points strongly to the existence of a ρ' resonance with a mass of about 1.47 GeV and a branching ratio into $\eta\rho$ of 3-4 % whose amplitude interferes with the tail of the ρ. The relative sign of the ρ' amplitude relative to that of $\rho(770)$ is different in e^+e^- annihilation and photoproduction. The contribution from a higher vector meson in the $\eta\rho$ channel seems to be only marginal.

In order to understand the structure of the equations used to fit the experimental data on the $\eta\rho$ final state produces in two different reactions (γp and e^+e^-) we recall here the specific Breit-Wigner forms of the partial wave amplitudes used in such cases. The Breit-Wigner partial wave amplitude for a simple transition shown in Fig.7 is given by:

$$T \sim \frac{\text{Vertex1} \cdot \text{Vertex2}}{\text{propagator}}, \tag{32}$$

where the vertex functions are in fact the partial widths of the resonance:

$$\text{vertex} = \Gamma_{R\to ab} \qquad \text{and} \qquad \Gamma_{R\to cd} \tag{33}$$

which can be expressed through the total hadronic width :

$$\Gamma_{R\to ab} = Br(R \to ab)\Gamma_{tot}. \tag{34}$$

When a relativistic resonance propagator is used, the vertex function becomes:

$$\text{vertex} \longrightarrow \sqrt{m\Gamma} \qquad \text{(for relativistic propagator)}. \tag{35}$$

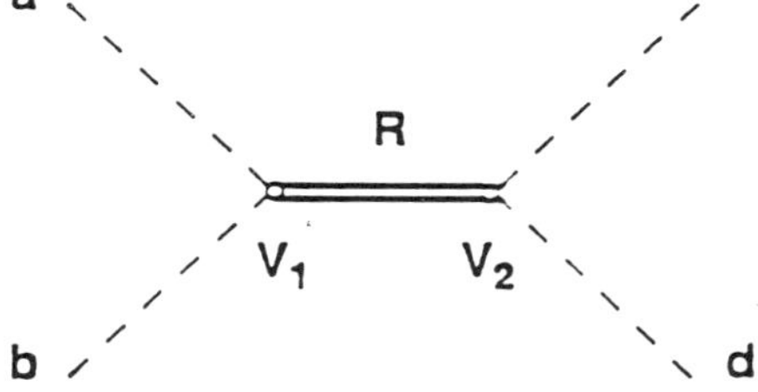

Figure 7: S channel resonance in two body scattering.

Therefore, for a reaction such as $e^+e^- \longrightarrow ab$, Fig.2 (I), the partial wave amplitude is given by:

$$T \sim \frac{\sqrt{m_V\Gamma_1} \cdot \sqrt{m_V\Gamma_2}}{m_V^2 - m^2 - im_V\Gamma} = \frac{m_V\Gamma\left[B(V \to e^+e^-)B(V \to ab)\right]^{1/2}}{m_V^2 - m^2 - im_V\Gamma}. \tag{36}$$

For reactions such as $\gamma p \longrightarrow abp$, Fig.2 (II), the partial wave amplitude will contain only one vertex function (the decay one) and therefore the Breit-Wigner form is given by :

$$T \sim \frac{\sqrt{m_V\Gamma_V \to ab}}{m_V^2 - m^2 - im_V\Gamma} = \frac{(m_V\Gamma)^{1/2}\left[B(V \to ab)\right]^{1/2}}{m_V^2 - m^2 - im_V\Gamma}. \tag{37}$$

The expression used for the analysis of the energy dependence of the $e^+e^- \longrightarrow \eta\rho$ total cross section is given by:

$$\sigma = \frac{12\pi}{m^2}\left|G_{\rho_0} + \sum \eta_i G_{\rho_i}\right|^2 \tag{38}$$

and that for the $\eta\rho$ mass distribution from the reaction $\gamma p \longrightarrow \eta\rho p$ is:

$$\frac{d\sigma}{dm} = \frac{2m}{\pi}\sigma(\gamma p \longrightarrow \rho p)\left|F_{\rho_0} + \sum_i \xi_i \alpha_i F_{\rho_i}\right|^2 \tag{39}$$

with

$$\alpha_i = \left[\frac{\sigma(\gamma p \longrightarrow \rho_i p)}{\sigma(\gamma p \longrightarrow \rho_0 p)}\right]^{1/2} \tag{40}$$

and partial wave amplitudes of the the Breit-Wigner type :

$$G_\rho = \frac{m_\rho \Gamma_\rho}{m_\rho^2 - m^2 - i m_\rho \Gamma_\rho}\left[B(\rho \longrightarrow e^+e^-)B(\rho \longrightarrow f)\right]^{1/2} \tag{41}$$

$$F_\rho = \frac{1}{m_\rho^2 - m^2 - i m_\rho \Gamma_\rho}(m_\rho \Gamma_\rho)^{1/2} B^{1/2}(\rho \longrightarrow f). \tag{42}$$

The phase factors (the relative sign of the coupling constants) observed in Fig.6 in the interference pattern for $\eta\rho$ masses below 1.5 GeV requires a change in sign of the ρ' amplitude relative to ρ between photoproduction and e$^+$e$^-$ annihilation, as was found earlier in [11]. In ref. [10] it was found that two classes of solutions are possible and they differ by the relative sign of the ρ components. These solutions are shown in Fig.6, (a and b) for the first solution and (c and d) for the second one.

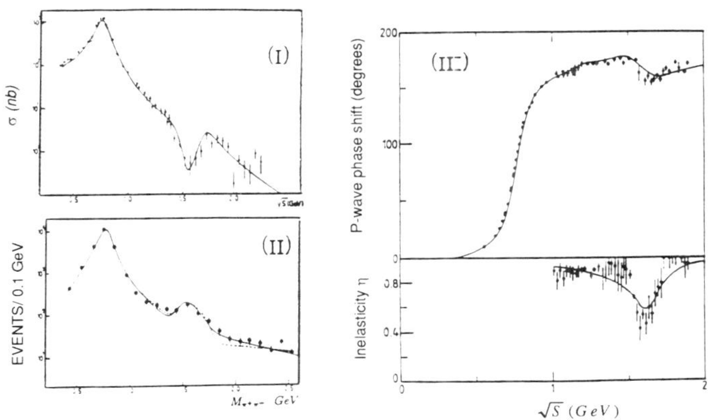

Figure 8: Fit of e$^+$e$^-$ $\rightarrow 2\pi$ (I) , γ p $\rightarrow 2\pi$ (II) and fit of $(\pi\pi)$ P wave (III)

A more complex analysis of many different channels has been done by Donnachie and Mirzaie [11] using experimental data on γp and e$^+$e$^-$ having $\pi^+\pi^-$, $2\pi^+2\pi^-$ and $\pi^+\pi^-2\pi^0$ in the final state as well as the data on the phase shift and inelasticity for $\pi^+\pi^-$ scattering with l=1 (P wave). It was found that a good fit is obtained with two vector resonances with masses M=1.465 $\pm$ 0.025 GeV, M=1.700 $\pm$ 0.025 GeV and widths Γ=0.235 $\pm$ 0.025 GeV and Γ=0.220$\pm$ 0.025 GeV respectively.The $(\pi^+\pi^-)$ branching ratio is 5 % for ρ_1 and 20 % for ρ_2. Some results of this analysis are shown in Fig.10.

As is seen in Fig.8 (I and II) in the region of m=1.5 GeV there is a dip (e$^+$e$^-$) - bump (γp) structure which is resolved here by the sign changes of the relative phases (η_i and ξ_i in eq.38 and eq.39) between photoproduction and e$^+$e$^-$ annihilation. The key feature in correlating all the data is the importance of 'off-diagonal' terms in photoproduction of states with J^P=1$^-$ and the contribution from states with $J^P \neq 1^-$.

It has been assumed and generally accepted that in applications of VDM the contributions of off-diagonal terms will be small and can be neglected. Now there is unambiguous evidence for diffraction dissociation of γ to states different of $J^P=1^-$. An example is the diffractive γ production of $b_1(1250)$ ($I^G(J^{PC}) = 1^+(1^{+-})$) which gives rise to the peak at 1.25 GeV in $\pi^+\pi^-2\pi^0$, Fig.5(I). There is also evidence for γ production [17] of $h_1(1170)$ ($0^-(1^{+-}$)) and $\rho_3(1690)$ ($1^+(3^{--})$). In the quark model these off-diagonal terms correspond to transitions of the type $1^3S_1 \rightarrow 1^1P_1$ or $1^3S_1 \rightarrow 1^3D_3$ and it is not unreasonable therefore to expect also :

- $1^3S_1 \rightarrow 2^3S_1$ for the lowest ρ radial excitation, and
- $1^3S_1 \rightarrow 1^3D_1$ for the lowest ρ orbital excitation.

The strong off-diagonal terms in photoproduction for higher ρ resonances destroys the simple VDM relations between phases and magnitudes of the amplitudes in photoproduction and e^+e^- annihilation. This effect is more evident in channels where the ρ tail is large (as in $\pi\pi$ final state where the dip-bump structure is observed). The relative phases ρ:ρ':ρ'' found in [11] are + + - in γp and + - + in e^+e^- (where there is a dip at m=1.5 GeV). Thus the sign of the $\rho'(1450)$ amplitude is negative in $e^+e^- \rightarrow \pi^+\pi^-$ and positive in $e^+e^- \rightarrow \eta\rho$. The coupling of a resonance to a final state is a signed quantity which cannot be fixed a priori and may have the same sign for different final hadronic states.

The analyses which are based on simple Breit-Wigner formulae do not satisfy the unitarity relation in the case of overlapping resonances. An analysis in which the explicit unitarised multichannel S-matrix is taken into account has been done by Hernner and Wolfson [18]. This analysis gives a good description of the ($\pi\pi$) P-wave, and also the e^+e^- annihilation into 2π and 4π up to 2 GeV. The partial wave which describes the ($\pi\pi$) system is written as a sum over all overlapping resonances:

$$f_{fi} = \sum_R \frac{m_R\Gamma_R B_{fR} C^*_{iR}}{m_R^2 - s - im_R\Gamma_R} \tag{43}$$

where B_{fR} is the decay branching ratio of the resonance R into the channel f. In the case of one resonance, C_{iR} is the usual branching ratio ($C_{iR} = B_{iR}$) and both B_{fR} and B_{iR} are real constants; the unitarity relations leads to $\sum_f B^2_{fi} = 1$. In the case of overlapping resonances it is impossible to satisfy multichannel unitarity relations :

$$Im f_{fi} = \sum_n f^*_{fn} f_{ni} \tag{44}$$

if $C_{iR} = B_{iR}$. After lengthy algebra, an expression is obtained for the matrix C_{iR} which makes the partial amplitude satisfy the multichannel unitarity relations. The result of the fit provides evidence for three ρ states : at 1.25 GeV with Γ=0.35 GeV, at 1.61 GeV with Γ=0.23 GeV and at 1.8 GeV with Γ=0.35 GeV. An interesting conclusion is that, in spite of the possibility of obtaining good enough fits with the usual sum of Breit-Wigner terms, the results with and without unitarisation are substantially different.

Another important problem in analysing experimental data concerns the ρ (ω and ϕ) tail interference; the definition of the tail is somewhat model dependent. The simple approach (used earlier[37]), taking the Breit-Wigner amplitude in the form :

$$BW \sim \frac{q^3}{m^2 - s - im\Gamma} \tag{45}$$

with Γ constant, turns out to be inadequate. The total width is in fact a sum of partial widths, each having an appropriate momentum dependence; a centrifugal barrier term, which depends on the radius of interaction R and is given by the so called Blatt-Weisskopf factor (or damping factor)[23]. This is in fact a general problem of resonance amplitude parametrization which becomes important in the case of a broad resonance, for overlapping resonances and also in the case of a resonance with a mass very near to a threshold. For the case of ρ P wave decay (l=1), the simple $q^{(2l+1)}$ momentum dependence of the partial width changes:

$$q^3 \longrightarrow q^3(1+(qR)^2)^{-1}. \tag{46}$$

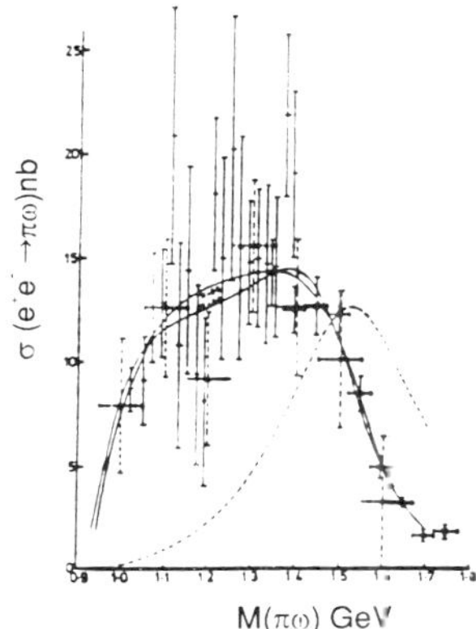

Figure 9: Typical fits of $e^+e^- \to \omega\pi$. Full curves include ρ' and a ρ tail with R=0.2, 0.8 fm; the dashed curve shows the ρ' contribution.

Using these modifications, the analysis [37] of $e^-e^- \to \omega\pi$ shows a strong correlation between the mass of the ρ' and R; the mass is largest when R is small, but compatible with previous results (a mass in the range 1.46 to 1.50 Gev and the width in the range 0.34 to 0.40 GeV). There is no need for the second ρ''; the interference found is the canonical (+ -) and the parameter R is in the range 0.0 to 0.8 fm, Fig. 9.

Information on the vector mesons can also be obtained from the analysis of the electromagnetic form factors. In the VDM framework, all the electromagnetic form factors are completely dominated by the contribution of the ρ, ω, and ϕ mesons. Neglecting the G parity violating transitions (which allow transitions of ω and ϕ into 2π), the pion electromagnetic form factor is given by the main ρ contribution:

$$F_\pi = \frac{g(\rho\gamma)g(\rho\pi\pi)}{D(\rho)} \tag{47}$$

where D(ρ) is the ρ meson propagator given by:

$$D(\rho) = s - m_\rho^2 + i\frac{m_\rho^3}{2}\Gamma_\rho\left(\frac{s-4m_\pi^2}{m_\rho^2-4m_\pi^2}\right)^{3/2} \tag{48}$$

and with pion coupling constant

$$g(\rho\gamma)g(\rho\pi\pi) = 0.705 \pm 0.089\ GeV^2. \tag{49}$$

The analysis of the pionic form factor in the region $1.34 < \sqrt{s} < 2.4$ GeV by the DM2 Collaboration [20] shows that the experimental data are not compatible with the simple VDM, even when adding the interfering contribution of an exited ρ'. The mass obtained from this last fit is $m_{\rho'} = 1.73 \pm 0.02$ GeV, $\Gamma_{\rho'} = 0.11 \pm 0.02$ GeV with a $\chi^2/N_{dof} = 43.3/13$. The fit is improved (to $\chi^2/N_{dof} = 19.9/8$) if three ρ resonances are permitted to interfere, Fig.10:

$$F_\pi = \sum_{k=0}^{2} g_k(\rho\gamma) g_k(\rho\pi\pi) \frac{1}{D_k(\rho)} e^{i\phi_k}. \tag{50}$$

Fixing the phase of the fundamental $\rho(770)$ to zero, the relative phases for the first and the second ρ recurrence are :

$$\phi_1 = 3.14 \pm 0.025\text{rad} \longrightarrow \pi \tag{51}$$

$$\phi_2 = 0.004 \pm 0.06\text{rad} \longrightarrow 0, \tag{52}$$

confirming experimentally the so called 'canonical phases' (+ - +) found before by Donnachie[11]. The masses and the widths are m=1.424 GeV , Γ=0.269 GeV and m=1.768 GeV , Γ=0.224 GeV for ρ' and ρ'' respectively.

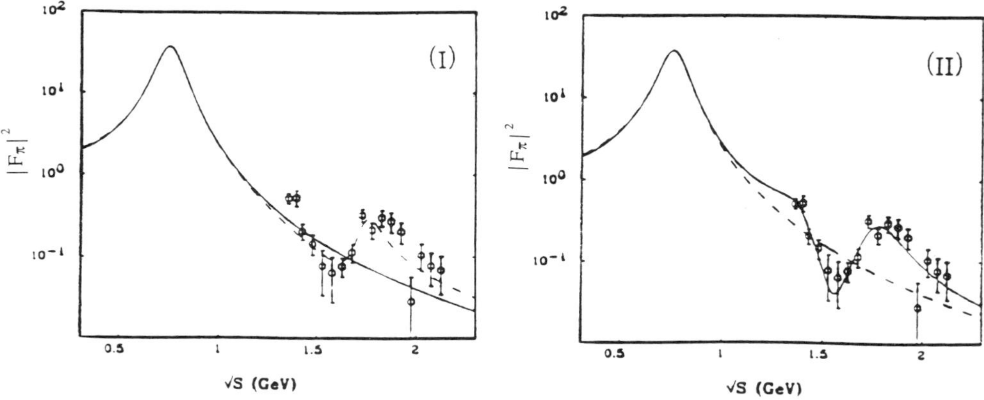

Figure 10: The fit of the pion squared form factor with a single ρ' interferring resonance (I), and with two ρ' interfering resonances (II). The tail of $\rho(770)$ is also shown.

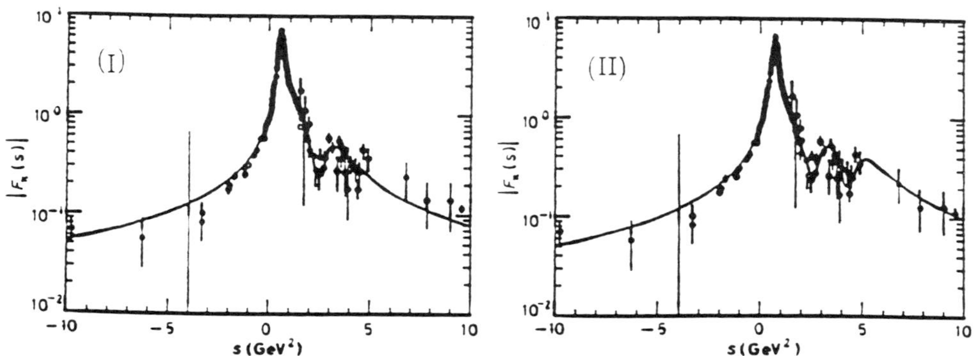

Figure 11: Fit of the modulus of the pion form factor with $\rho(770)$, $\rho'(1450)$ and $\rho''(1700)$ (I) and a fit with three excited states, as before and $\rho'''(2150)$ (II).

A more complex study of the pionic form factor (using in the fit data on space-like as well as time-like regions, but also some data on kaonic form factors) confirms the existence of two ρ excited states at 1.45 GeV and 1.70 GeV. This phenomenological analysis also shows there is a case for the existence of a third excitation of $\rho(770)$ with mass m=2.15 GeV and width Γ=0.320 GeV, Fig.11; this points to the same structure identified by Clegg and Donnachie [22] in an analysis of electron positron annihilation into 6π. This third ρ excited state has also been observed by the GAMS Collaboration in a study of the $\omega\pi^0$ system produced in the π^-p charge exchange reaction at 38 GeV/c. The OPE mechanism limits the quantum numbers of the $\omega\pi^0$ system to the following $J^{PC} = 1^{--}, 3^{--}, \ldots$ and $I^G = 1^+$. The X(2200) observed in ref. [24] is confirmed in ref.[25] to be a 1^{--} state, the $\rho(2150)$; this may be a radial-orbital excitation belonging to the 2^3D_1 nonet with mass and width m=2.140 $\pm$ 0.030 GeV, Γ = 0.320 $\pm$ 0.070 GeV. In conclusion, the sector of isovector vector mesons contains two broad ρ excited states, the first one $\rho(1450)$ with Γ=0.310 GeV belonging to the 2^3S_1 nonet and a second one $\rho(1700)$ with Γ=0.235 GeV belonging to 1^3D_1 nonet. There is also evidence for states around m=2.00 GeV and m=2.15 Gev, possible 3^3S_1 and 2^3D_1 states.

4. Isoscalar and strange vectors

4.1. Isoscalar ω states.

In this sector it is highly probable that there is a strong (ω' - ϕ') mixing, so the notation of these states as ω' or ϕ' is not so straightforward. There is also the possibility of there being isoscalar hybrid states which can also confuse the situation. The study of ω' states is made on data $e^+e^- \longrightarrow \rho\pi$ and $e^+e^- \longrightarrow \omega\pi\pi$. These data show a very different behaviour, the second one dominating at 1.6 GeV.

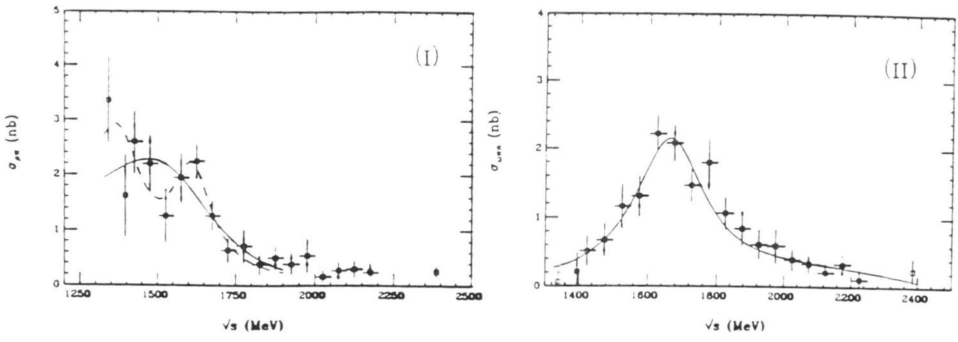

Figure 12: Cross section for $e^+e^- \longrightarrow \rho\pi$ (I) and $e^+e^- \longrightarrow \omega\pi\pi$ (II).

If one follows the conventional view of the ρ' excitations, then in fitting the data it is natural to take into account one or two new states. From the fit of the experimental data[37] on $\rho\pi$, two ω states are required with so called 'canonical phases', $\omega : \omega' : \omega''$ = (+ - +). The masses and widths of the ω excited states are m=1.44 GeV, Γ=0.24 GeV for the first and m=1.60 GeV, Γ=0.140 GeV for the second. However, the data are also compatible with only one ω' at m=1.635 GeV and Γ= 0.350 GeV. The $\omega\pi\pi$ final state exhibits only one clear resonant peak at m=1.598 GeV with Γ=0.109 GeV [19].

A more recent analysis of new experimental data[26] gives practically the same values of masses and widths as the previous analyses (in Fig.12, the continuous line in (I) is a single Breit-Wigner fit of the $\rho\pi$ final state).

Because the fits above 1.6 GeV for electron positron annihilation to $\rho\pi$ and $\omega\pi\pi$

are similar, a coupled fit of both types of data has also been performed giving m=1.662 GeV and Γ=0.280 GeV. The large cross section in $\omega\pi\pi$ suggests that this state is of ω type (mainly non strange quarks).

The first ω', seen only in the $\rho\pi$ final state is unstable due to the experimental problems (cross section normalization). The present status of the isoscalar ω type vector mesons is illustrated by the PDG[7]. The first state is $\omega(1420)$, belong to the first radial excitation (2^3S_1 nonet), decaying dominantly to $\rho\pi$; it has mass and width m=1.419 ± 0.031 GeV, Γ=0.174 ± 0.060 GeV. The second one, $\omega(1600)$, belonging to the first orbital nonet (1^3D_1), decays to $\rho\pi$ and $\omega\pi\pi$ and has the parameters: m=1.662 ± 0.013 GeV, Γ= 0.280 ± 0.024 GeV.

4.2. *Isoscalar ϕ states.*

Data used for ϕ' observations are from photoproduction reactions with two kaons in the final state:

$$\gamma p \longrightarrow K\bar{K}X \tag{53}$$

as well as from electron positron annihilation :

$$e^+e^- \longrightarrow K^+K^- \tag{54}$$

$$e^+e^- \longrightarrow K_SK_L \tag{55}$$

$$e^+e^- \longrightarrow K^*\bar{K}. \tag{56}$$

The photoproduction mechanism is found to conserve s-channel helicity and to proceed via natural-parity exchange in the t-channel[27]. From the analysis of photoproduction data, only one resonant state is found in the mass region of 1.7 GeV. The values found in two important analyses are in agreement : m=1.760 ± 0.020 GeV, Γ=0.080 ± 0.040 GeV [27] and m=1.726 ± 0.022 GeV, Γ=0.121 ± 0.047 GeV [28] .

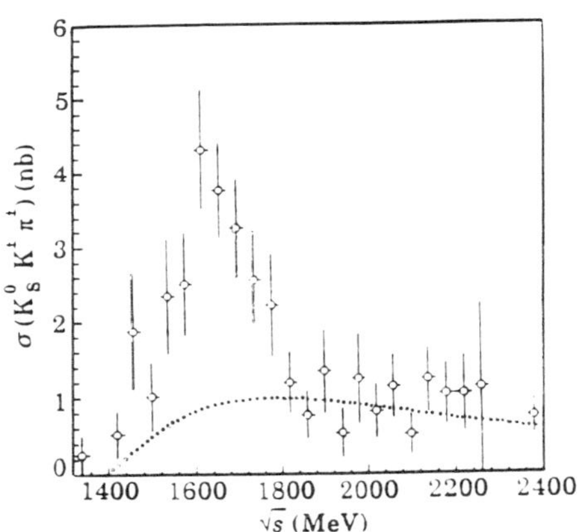

Figure 13: Cross section for $e^+e^- \longrightarrow K^0_S K^\pm\pi^\mp$.

The analysis of e$^+$e$^-$ annihilation shows also only one resonant state in $K\bar{K}$ but with parameters slightly different: m= 1.655 ± 0.017 GeV, Γ = 0.207 ± 0.045 GeV[29]. The cross section for e$^+$e$^-$ annihilation into three bodies (Eq.56) shows an enhancement around 1.650 GeV which cannot be explained simply by the contributions of only ρ, ω and ϕ (the dashed line in fig.13). A maximum likelihood fit to the Dalitz plot requires a clear $K^*(892)$K dynamics with K^{*0} dominance; this indicates an interference between a weak isovector (ρ like) and a resonant isoscalar (ϕ like) amplitude; Figs.14 (a) and (b).

The relative phase as well as the absolute isoscalar phase (assuming for ρ' m=1.57, Γ=0.51 GeV) is found to cross 90^0 in the same mass region where the peak in the cross section is observed; this supports a resonant nature for the observed signal, Figs.14 (c) and (d). The mass and width values found are m=1.657 ± 0.027 GeV and Γ = 0.146 ± 0.055 GeV. This result, in close agreement with other measurements, supports the identification of this isoscalar vector meson decaying mainly to $K^*\bar{K}$; this $\phi'(1680)$ is the first radial excitation of ϕ (2^3S_1 nonet).

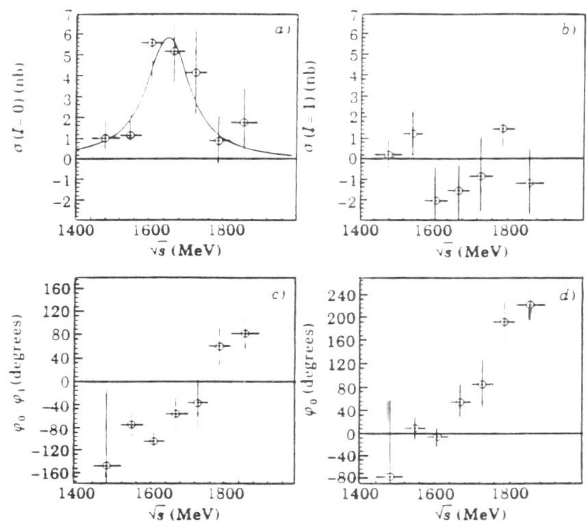

Figure 14: The Dalitz plot analysis of $K^0_S K^\pm \pi^\mp$. Isoscalar (a) and isovector (b) cross sections, relative phase between them (c) and absolute isoscalar phase (d).

4.3. Strange vectors.

The strange vectors appear to be an ideal place to study $(q\bar{q})$ spectroscopy due to the absence of glueballs and absence of isoscalar mixing. The ($K\pi$) and ($K\pi\pi$) systems can be produced in both diffractive and charge-exchange reactions. The ($K\pi$) provides information on natural spin-parity amplitudes. The ($K\pi\pi$) systems produced diffractively is dominated by unnatural spin-parity states and by natural ones in charge-exchange production.

In this sector of strange 1^- mesons, the main results come from the LASS collaboration at SLAC [31,32]. The reactions analysed are $K^-p \longrightarrow K\pi(n/p)$ and $K^-p \longrightarrow K\pi^+\pi^-n$.

The 1^- partial waves arise from the production and decay of two states. The lower in mass decays mostly to $K^*\pi$. The mass and width averaged over three measurements of three final states ($\bar{K}^0\pi^+\pi^-$, $K^-\pi^+$ and $\bar{K}^0\pi^-$) are m=1.412 ± 0.012 GeV and Γ= 0.227 ± 0.022 GeV (average given by ref.[7]).

The second state decays almost equally to $K\pi$, $K^*\pi$ and $K\rho$ and has mass and width m=1.714 ± 0.020 , Γ = 0.323 ± 0.110 GeV (as given in ref. [7]).

Fig.15 shows the $K^*\pi$ and $K\rho$ 1^- contributions from the partial wave analysis of $K\pi\pi$ production. The curves are the result of a simultaneous two resonance fit to 1^- waves and K^* in the 2^+ and 3^- waves including their relative phases. The phases are set using the predicted behaviour of the $(K^*\pi)_P$ reference wave, Fig.15 (c).

It is therefore natural to associate the first 1^- state K*(1410) with the first radial excitation of the K*(890) (the 2^3S_1 nonet) and the second one, K*(1680) with the first orbital excitation (the 1^3D_1 nonet).

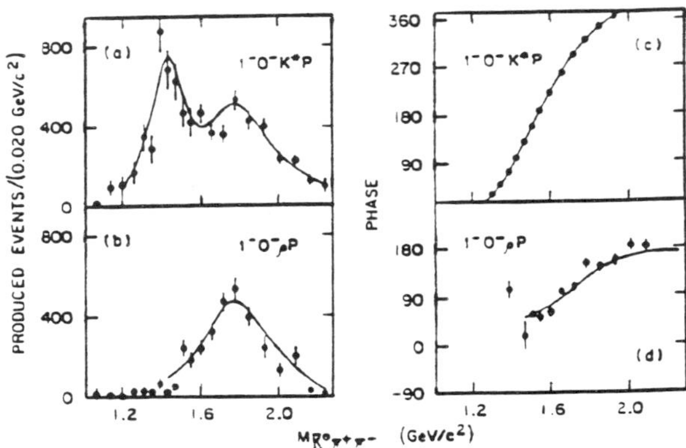

Figure 15: Fit of 1^- waves from $K^-p \longrightarrow K\pi^+\pi^-n$.

5. Exotica ?

The low mass $\rho^{'}(1300)$ has been claimed many times (in PDG 86 [9] it is mentioned as $\rho(1250)$) but never generally accepted. We reall here the analysis with unitarised amplitudes [18], some evidence for it in $\pi\pi$ spectra [34] and of J/Ψ decay in $\pi^+\pi^-\pi^0$; the recent LASS analysis [33] of $K^-p \longrightarrow \pi^+\pi^-\Lambda$ found the parameters m=1.302 ± 0.028 GeV and Γ=0.140 ± 0.048 GeV. Fig.16 shows the magnitude of the P wave and its phase motion relative to the D wave (fixed to $f_2(1270)$ parameters). The solid curve is the fit with ρ, ω-ρ interference and $\rho^{'}(1300)$.The dotted curve shows the effect of omitting $\rho^{'}(1300)$.

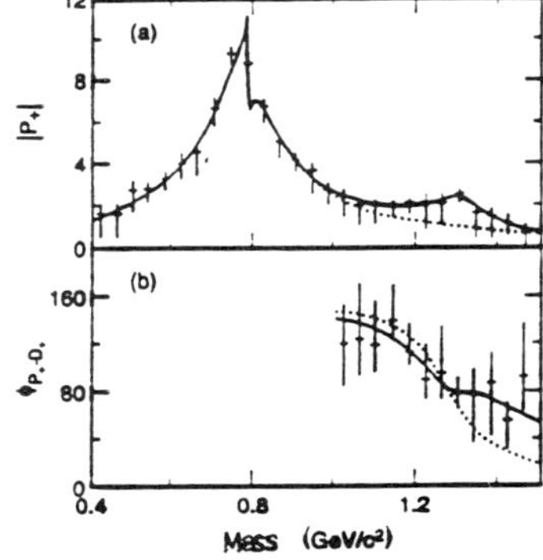

Figure 16: The magnitude and the relative phase of P wave from the reaction $K^-p \longrightarrow \pi^+\pi^-\Lambda$

The problem is therefore the number of $\rho^{'}$ states in the region 1.25 - 1.5 GeV and,if we accept two states, which is the partner of K*(1410)?

A possible explanation was given in ref. [38]. Invoking the concept of $q^2\bar{q}^2$ states coupled to the normal $q\bar{q}$ mesons, a mixing scheme is proposed in which both $\rho^{'}$ exist, at m=1.45 GeV and at m=1.270 GeV, one being built mostly of $q^2\bar{q}^2$. Anyway, if there are indeed two $\rho^{'}$ states in the mass region of 1.3 to 1.45 GeV, from ref. [39] the lower one can be an 'almost multiquark state' and from ref. [40] the heavier one can be a 'possible hybrid state'.

Another candidate for exotica in the vector meson sector can be the so-called C(1480) meson, seen for the moment only in one experiment [35] in the reaction $\pi^- p \rightarrow \phi\pi$. The mass and width are found to be m=1.480 ± 0.04 Gev and $\Gamma = 0.160 \pm 0.06$Gev with the quantum numbers $1^+(1^{--})$ (just like the ρ meson).

The coincidences of the mass of C(1480) with those of the ρ' raises the problem of whether these can be different observations of the same vector meson. The identification of a strong ρ' decay in $\omega\pi$ and the analysis of different partial widths of the 1.45 GeV resonant structure [36,37] supports the claim that the ρ' and C(1480) cannot be the same particle. So, if C(1480) is confirmed by other experiments it can be a candidate for an extra vector meson, a possible signature of exotica in this sector.

The most obvious signature for an exotic meson is in fact the existence of an object with an exotic combination of J^{PC}, for example a 'vector meson' with positive C parity (1^{-+}) (so called $\hat{\rho}$ states).

The simplest system where exotic J^{PC} states can be sought is $\eta\pi$ in odd angular momentum between η and π. Such states were claimed around m=1.400 GeV in $\eta\pi$ invariant mass from the analysis of the reaction $\pi^- p \rightarrow \pi^0 \eta n$ at 100 Gev/c (m=1.406 ± 0.020 Gev, $\Gamma = 0.180 \pm 0.030$ GeV) [41] and of reaction $\pi^- p \rightarrow \pi^- \eta p$ at 6.3/c Gev/c (m=1.323 ± 0.005 Gev, $\Gamma = 0.143 \pm 0.016$ GeV)[42]. From the partial wave analysis, in spite of possible ambiguities [43,44] (eight solutions for a S,P,D analysis), both experiments found a clear signal of a well known $a_2(1320)$ (D wave), an enhancement of the P wave and an almost flat D-P relative phase, Fig.17 [42], (which means that both waves are resonant or both non-resonant).

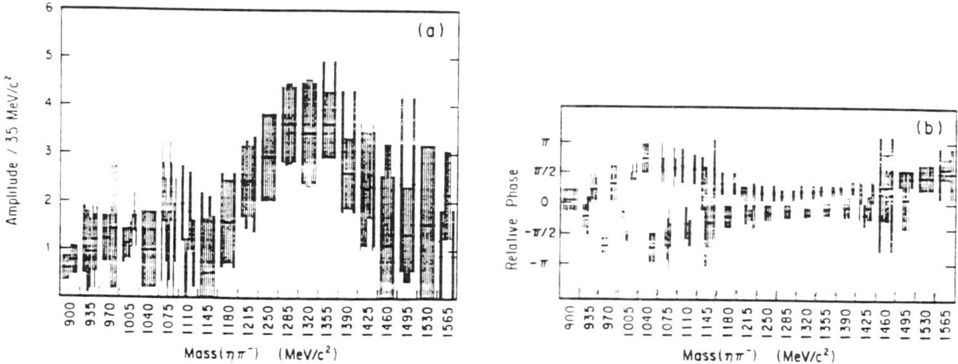

Figure 17: Distribution of amplitudes of the P wave (a) and relative D-P phase (b).Eight solutions are plotted adjacently.

In a recent study [45] of the reaction $\pi^- p \rightarrow f_1(1285)\pi^- p$ at 18 GeV/c (with $f_1(1285) \rightarrow K^+ \bar{K}^0 \pi^-$), an exotic signal was also found in the $f_1(1285)\pi$ invariant mass, a broad 1^{-+} structure ($f_1\pi$ in S wave) at 1.6-2.2 GeV.

The limited statistics and the existence of ambiguities in the partial wave analysis implies that all these observations need urgently confirmation.

6. Discussion

The first simple linear dependence in m^2 which correlates the mass and the spin of the resonances was the Regge trajectory :

$$m^2 = aJ + b_1. \tag{57}$$

This relation is suitable for examining the systematics of the isovector states because they do not suffer from hidden strangeness admixture and have bigger production cross sections than others. For the natural spin-parity series , the leading trajectory (leading means that the corresponding states have the lowest known m^2 for each spin J) is given by the parameters [25]: a= 1.16± 0.04 GeV^2 and b=-0.56 ± 0.05 GeV^2, Fig.18.

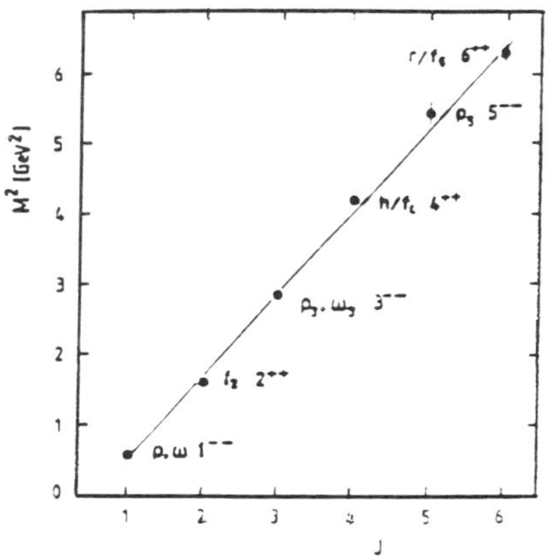

Figure 18: Regge trajectory for normal spin-parity series.

For the radial excitations we can expect a similar trajectory , with the same slope but with a displacement in b ($b_n > b_1$). Indeed, the Veneziano [46] formula gives a relation for b_n :

$$b_n = an + b_0 \tag{58}$$

where 'a', to first order is the same as before. Therefore the Veneziano-Regge formula useful also for excited states has the general form [47]:

$$m^2 = an + b(I^G, J^{PC}) \tag{59}$$

which means a linearity in excitation number. Such a dependence is shown for the vector mesons in Fig.19[24] in which, as a curiosity, we indicate by an arrow the position of $\rho'(1.300)$ at m=1.32±0.04 GeV.

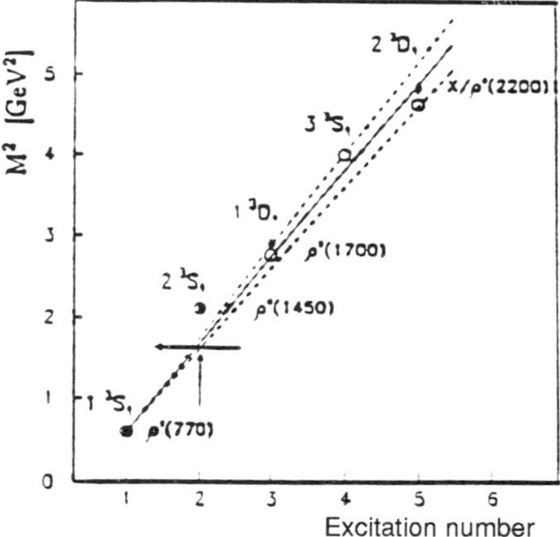

Figure 19: Excitation trajectory for vector mesons. Dots are experimental values and open circles are Godfrey-Isgur predictions.

The present status of the vector mesons composed by light quarks can by summarized in the table shown below, where for completeness we add also higher Jstates.

J^{PC}		$I=1$	$I=0$		$I=1/2$
1^3S_1	1^{--}	$\rho(770)$	$\omega(782)$	$\phi(1024)$	$K^*(892)$
2^3S_1	1^{--}	ρ' (1.30) / (1.45)	$\omega'(1420)$	$\phi'(1680)$	$K'^*(1410)$
3^3S_1	1^{--}	$1.8-2.15$	?	?	?
1^3D_1	1^{--}	$\rho''(1700)$	$\omega''(1600)$	?	$K''^*(1680)$
2^3D_1	1^{--}	2.20	?	?	?
$^3D_3$	3^{--}	$\rho_3(1690)$	$\omega_3(1670)$	$\phi_3(1850)$	$K_3(1780)$
$^3G_5$	5^{--}	2.35	?	?	$K_5(2380)$
1^{-+}			$\hat{\rho}(1400)$	$\hat{\rho}(2000)$	

The first order problem in this sector is the correct filling of the nonets in order to be sure on possible extra states (exotica) which can exist near the 'normal' ($q\bar{q}$) states. The second order problem concerns fitting different characteristics of the mesons such as the partial widths with more or less complex models in order to obtain more physical information on $q\bar{q}$ in the confinement region.

5. Acknowledgements

The author wishes to thank Prof. David Bugg for organizing this Summer School. He also wishes to thank his co-workers from the Obelix collaboration for helpful discussions.

6. References

1. S. Godfrey and N. Isgur, Phys. Rev. D32 (1985) 189.
2. T.H.Bauer et al., Rev. Mod. Phys. 50 (1978) 261.
3. Y. Nambu, Phys. Rev. 106 (1957) 1366 ; W.R.Frazer and J.Fulco Phys. Rev. Lett. 2 (1959) 365, Phys. Rev. 117 (1959) 1603; J.J.Sakurai, Ann. Phys. 11 (1960) 1.
4. D. Schildknecht, Springer Tracts Mod. Phys. 63 (1972) 57.
5. J.J. Sakurai and D. Schildknecht, Phys. Lett. B40 (1972) 121; B41,489; B42,216; F.M.Renard, Proc. 7 th Renc. Moriond, ed. Tran Thanh Van.
6. C. Erkal and M.G. Olsson Z. Phys. C31 (1986) 615.
7. Particle Data Group, Phys. Rev. D50 (1994) 1173.
8. Particle Data Group, Rev. Mod. Phys. 56 No 2 part II (1984).
9. Particle Data Group, Phys. Lett. B170 (1986) 1.
10. A. Donnachie and A.B. Clegg, Z. Phys. C34 (1987) 257.
11. A. Donnachie and H. Mirzaie, Z. Phys. C33 (1987) 407.
12. F. Nichitiu, Models and Mathematical tools in meson spectroscopy. inv. talk, LEAP94 Sept 1994 Bled, Slovenia.
13. P wave resonant are in some phase-shift solutions in the analyses : A.D.Martin and M.R.Pennington, Ann. Phys. (NY) 114 (1978) 1 ; H.Becker et al., Nucl. Phys. B151 (1979) 46 ; M.J.Corden et al., Nucl. Phys. B157 (1979) 250 , and no P resonance in all four solutions of : B. Hyams et al., Nucl. Phys. B64 (1973) 134: B100 (1975) 205.
14. F.M.Renard, Basics of electron positron collisions,1981,Editions Frontiers, Gif sur Yvette, France.
15. D. Aston et al., Nulc. Phys. B189 (1981) 15.
16. M. Atkinson et al., Z.Phys. C26 (1985) 499.

17. M. Atkinson et al., Nucl. Phys. B231 (1984) 15; B243 (1984) 1; Z. Phys. C30 (1986) 531.
18. V.K. Henner and D.N. Wolfson. Conf. Hadron 93. Como. Italy, June 1993, Nuovo Cim. A107 (1994) 2511.
19. A. Donnachie and A.B. Clegg, Z. Phys. C42 (1989) 663.
20. D.Bisello et al., Phys. Lett. B220 (1989) 321.
21. M.E.Biagini et al., Nuovo Cim. A104 (1991) 363.
22. A.B.Clegg and A.Donnachie, Z.Phys. C45 (1990) 677.
23. J.M.Blatt and V.F. Weisskopf, Theoretical Nuclear Physics. New York, J. Wiley and Sons 1952, p.361; F.v. Hippel and C.Quigg, Phys. Rev. D5 (1972) 624.
24. D. Alde et al., Z. Phys. C54 (1992) 553.
25. D. Alde et al., Z. Phys. C66 (1995) 379.
26. A. Antonelli et al., Z. Phys. C56 (1992) 15.
27. M. Atkinson et al., Z.Phys. C27 (1985) 233.
28. Busenitz, Phys. Rev. D40 (1989) 1.
29. D. Bisello et al., Z. Phys. C39 (1988) 13.
30. D. Bisello et al., Z. Phys. C52 (1991) 227.
31. D. Aston et al., Nucl. Phys. B292 (1987) 693, B296 (1988) 493 ; P.F. Bird. Ph.D. Thesis. SLAC Report 332 (1988).
32. D. Aston et al., SLAC-PUB-5721 (1991).
33. D. Aston et al., SLAC-PUB-5657 (1991).
34. L.P. Chen and W. Dunwoodie (mark III) SLAC-PUB-5674 (1991) , Int. Conf. Hadron 91,Univ. of Maryland at College Park. World Scientific 1992, pag.100.
35. S.I. Bityukov et al., Phys. Lett. B188 (1987) 383.
36. A.B. Clegg and A. Donnachie, Z. Phys. C40 (1988) 313.
37. A. Donnachie and A.B. Clegg, Z. Phys. C51 (1991) 689.
38. Yu.S. Kalashnikova, Workshop 'DAFNE', Frascati, April 1991, pag. 415.
39. A. Donnachie and Yu.S. Kalashnikova, Z. Phys. C59 (1993) 621.
40. F.E. Close and P.R.Page, preprint,Univ. of Oxford, OUTP-95-29P (1995).
41. D. Alde et al., Phys. Lett. B205 (1988) 397.
42. H. Aoyagi et al., Phys. Lett. B314 (1993) 246.
43. F. Nichitiu. Phase Shift Analysis in Physics of the Nuclear Interactions, Ed. Acad. Romania (Bucharest) (1980) ; Ed. Mir (Moskow) (Russian) 1983.
44. S.A. Sadovsky , preprint IHEP 91-75 (1991).
45. J.H.Lee et al., Phys. Lett. B323 (1994) 227.
46. C.Veneziano Nuovo Cim. 57 (1968) 190.
47. D.S. Peaslee, Conf. Hadron 89, Ajaccio,Corsica (France), sept. 1989, Editions Frontiers,Gif sur Yvette,p255.

HEAVY FLAVOUR SPECTROSCOPY

Robert K. Kutschke

Department of Physics
University of California
Santa Barbara, CA, 92106
USA

1 Introduction

During the first week of this institute many people expressed some wonder that four lectures had been allotted to discuss the spectroscopy of hadrons with heavy flavour. Surely there are simply not enough such hadrons to warrant a full four hours! A glance at figures 1 and 2, however, should reassure any doubters. Figure 1a) shows the measured mass spectrum of all of the known charmed mesons, while part b) shows the same for of all of the known charmed baryons and figure 2 shows the same for all known bottom mesons. The quark content of the various states named in these figures is summarized in Tables 1 and 2. At this time the only established bottom baryon is the Λ_b, so there is no corresponding figure for the bottom baryons.[1] Because most of the known states are charmed hadrons, this review will mostly cover charm spectroscopy, with a few references to bottom spectroscopy. Finally, whenever these lectures discuss a particular state or reaction, the charge conjugate state or reaction is usually implied. The exceptions will occur when discussing charm production from Σ^- beams or $\overline{\nu}$ beams.

These lectures are organized as follows. They will begin with a short section on naming conventions followed by an overview of the decay modes in which the various excited states are seen. This will also serve as a guide to where the most interesting parts of the main discussion will be. The main section will be a survey of the theory, with frequent asides to compare with data. After that will come a discussion of experimental

[1] Over the summer several LEP experiments have shown evidence for several new bottom hadrons, however the author has not yet had time to digest this new information and it will not be presented here

Table 1: Quark content of the ground state charmed and bottom mesons. The nomenclature for excited states is summarized in table 6. Because it is redundant, the charge superscript on the strange mesons is often dropped.

Quark Content	*Name*	*Quark Content*	*Name*
$c\overline{u}$	D^0	$\overline{b}u$	B^+
$c\overline{d}$	D^+	$\overline{b}d$	B^0
$c\overline{s}$	D_s^+	$\overline{b}s$	B_s^0
		$\overline{b}c$	B_c^+
$\overline{c}u$	$\overline{D}^0$	$b\overline{u}$	B^-
$\overline{c}d$	D^-	$b\overline{d}$	$\overline{B}^0$
$\overline{c}s$	D_s^-	$b\overline{s}$	$\overline{B}_s^0$
		$b\overline{c}$	B_c^-

techniques and, finally, a presentation of some experimental results which did not find a home in the main section. Because the field is now over 20 years old, many of the elementary results have long been "assumed knowledge" and are rarely discussed, even in modern textbooks. These proceedings provide an opportunity to remind people of many of these results, which often boil down to nothing more than conservation of angular momentum or elementary quantum mechanics.

2 Survey of Names and Decays

2.1 Taxonomy

The subset of heavy flavour mesons considered in these lectures will be those that consist of either a charmed quark, c, or a bottom quark, b, bound together with a light anti-quark, $\overline{q}$, where $q = u, d, s$. That is, states such as $b\overline{c}$, for which there are no experimental data, will not be discussed. When it is not important to distinguish between charm and bottom mesons the notation $Q\overline{q}$ will be used, where Q denotes a heavy quark. On occasion the notation q will refer only to the u and d quarks, but not to s; such instances will be noted clearly.

The naming scheme for the hadrons, described in the Review of Particle Properties[6], allows one to determine the quark content of a heavy flavour meson directly from its name. This correspondence between meson name and quark content is summarized, for the ground state mesons, in Table 1. The unadorned symbol D is often used when it is not important to make distinctions between the D^+ and the D^0, while an undifferentiated anti-charm meson is denoted as $\overline{D}$. Similarly, for the symbols B and $\overline{B}$.

Notice the convention that a $c\overline{q}$ system is a particle, D, whereas a $b\overline{q}$ system is an anti-particle, $\overline{B}$. This sign convention was chosen for a variety of reasons, one of which is that all flavours carried by a charged meson have the same sign as the charge of the meson. For example, the D_s^+ has both charm and strangeness of +1. A second reason is that all particles have semileptonic decays to $e^+\nu_e$ whereas anti-particles have semileptonic decays to $e^-\overline{\nu_e}$.

The Particle Data Group (PDG) naming scheme for hadrons also allows one to extract the quark content of a baryon from its name. This correspondence between

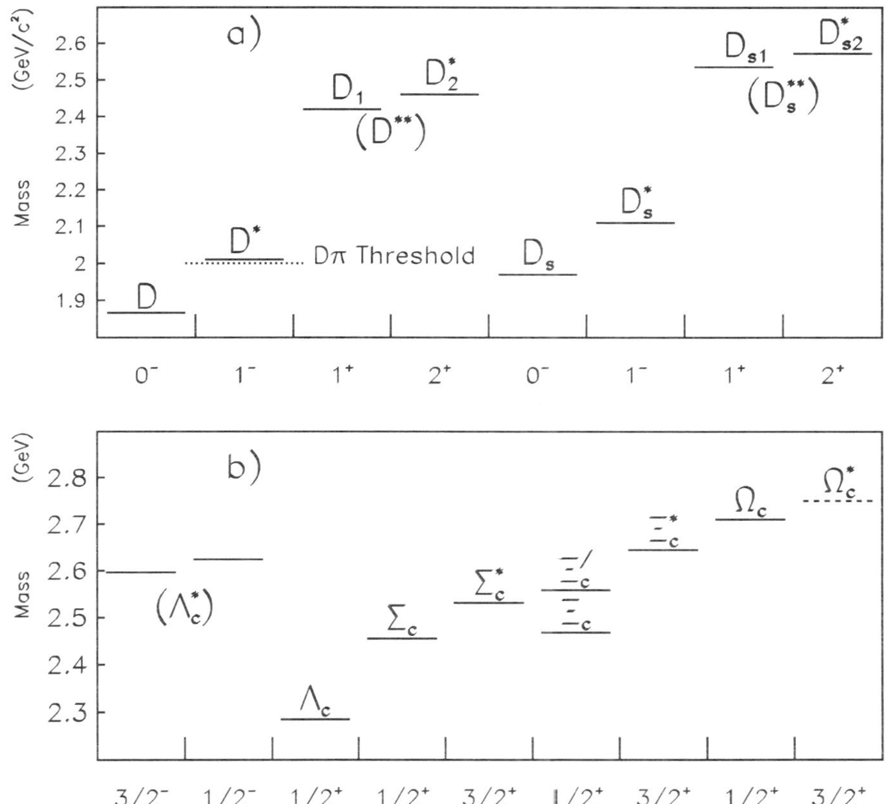

Figure 1: Measured mass spectrum of the known charmed a) mesons and b) baryons. The horizontal axis indicates the spin and parity, J^P, of the state. In most cases, the J^P is a theoretically motivated assignment, not the result of a measurement. There is one undiscovered member in the spatial ground state of the singly charmed baryons, the Ω_c^*; the predicted mass for this state is indicated with a dashed line. The Ξ_c^* and the Σ_c^* multiplets are incomplete, with only the Σ_c^{*++} and Ξ_c^{*0} having been observed. The names given in brackets are not official nomenclature but are frequently used. The formal names of the two excited Λ_c states shown to the left of part b) are the $\Lambda_c(2593)$ and the $\Lambda_c(2625)$. In part a) the $D\pi$ threshold is shown as a dotted line. The data are taken from [1](D_s^*), [2](D^{**}, D_s^{**}), [3](Ξ_c^*), [4](Ω_c), [5](Λ_c^*), and the remaining data are taken from [6]. The errors on the masses are small compared to the line thickness.

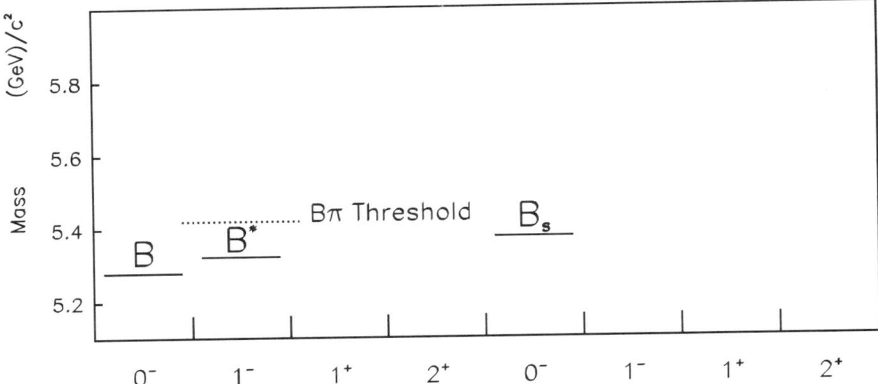

Figure 2: Mass spectrum of the known mesons which contain a b quark. The vertical scale is the same as that used in Figure 1. Note that, unlike in the charm sector, the first excitation lies below the threshold for decay to $B\pi$. No corresponding figure has been made for the b baryon sector because only one state is known, the Λ_b, at a mass of $5641. \pm 50.$ MeV/c^2. The data are taken from [6].

Table 2: Quark content of the ground state charmed and bottom baryons. The column headed I denotes the isospin of the baryon. The nomenclature of excited states, which are denoted by adding a variety of superscripts to the base name, will be discussed in sections 3.2.1 and 3.2.2. A cross-reference between these names and a naming convention used in some older papers is found in table 3.

Quarks	I	Nomenclature	Quarks	I	Nomenclature
cud	0	Λ_c	bud	0	Λ_b
cuu, cud, cdd	1	$\Sigma_c^{++}, \Sigma_c^{+}, \Sigma_c^{0}$	buu, bud, bdd	1	$\Sigma_b^{+}, \Sigma_b^{0}, \Sigma_b^{-}$
csu, csd	$\frac{1}{2}$	Ξ_c^{+}, Ξ_c^{0}	bsu, bsd	$\frac{1}{2}$	Ξ_b^{0}, Ξ_b^{-}
css	0	Ω_c	bss	0	Ω_b
ccu, ccd	$\frac{1}{2}$	$\Xi_{cc}^{++}, \Xi_{cc}^{+}$	bbu, bbd	$\frac{1}{2}$	$\Xi_{bb}^{0}, \Xi_{bb}^{-}$
ccs	0	Ω_{cc}	bbs	0	Ω_{bb}
ccc	0	Ω_{ccc}	bbb	0	Ω_{bbb}
			bcs	0	Ω_{bc}
			bcc	0	Ω_{bcc}
			bbc	0	Ω_{bbc}

Table 3: Correspondence between Particle Data Group baryon names and a scheme used in some older papers [7].

PDG Name	*Old Name*	*PDG Name*	*Old Name*	*PDG Name*	*Old Name*
Λ_c	C_0	Σ_c	C_1	Ξ_c	A
Ξ'_c	S	Ξ^*_c	S^*	Ω_c	T
Ω^*_c	T^*	Ξ_{cc}	X_u, X_d	Ω_{cc}	X_s

the name of a baryon and its quark content is summarized in Table 2. Following the example of mesons, the symbol, Σ_c for example, is used to denote an unspecified member of the Σ_c multiplet and the symbol $\overline{\Sigma}_c$ is used to denote an unspecified member of the anti-baryon multiplet.

Recall that the distinction between the Λ and the Σ^0, both of which have quark content *sud*, is that the Λ has isospin $I = 0$ whereas the Σ is the $I_3 = 0$ member of an $I = 1$ multiplet. All other differences between these two baryons, such as the spin of the *ud* diquark, are derived from this fundamental difference. This same distinction carries over into the charmed sector.

2.2 Survey of Decays

This section summarizes the allowed transitions among the states in figures 1 and 2. The following states only decay weakly and their decays will not be discussed in these lectures: D^0, D^+, D_s, B^0, B^-, B_s, Λ_c, Ξ_c and Ω_c. As will be discussed in section 3.2.1, it is interesting that the Σ_c is absent from this list. The following states are below threshold to decay via π emission and have been observed to decay electromagnetically: B^{*-}, B^{*0} and Ξ'_c. It is expected that the unobserved Ω^*_c will also only decay electromagnetically.

While the D^*_s is above threshold to decay to $D_s\pi^0$, that decay violates isospin and it was believed that the D^*_s only decayed electromagnetically. However, because isospin is not an exact symmetry the decay $D^*_s \to D_s\pi^0$ is merely suppressed, not forbidden, and it has recently been observed.

Because the D^* states are so close to the $D\pi$ threshold, their strong decay widths are severely limited by phase space and the electromagnetic decay $D^{*0} \to D^0\gamma$ has a rather large branching fraction, about 36%. For many years it was thought that $B(D^{*+} \to D^+\gamma)$ was approximately 15% however it is now known to be less than 4.2% at 90% C.L. In this case the theorists insisted that they had to be right and that the experiments were wrong. They were right, and their simple argument is sketched in section 3.4. The story of the strong decays of the D^* mesons is dominated by their small Q values, which are listed in table 4.

The remaining states decay only strongly. The D^{**} states decay to $D\pi$ and $D^*\pi$, while the D^{**}_s states decay to DK and D^*K. The Σ_c and Σ^*_c both decay to $\Lambda_c\pi$, while the Ξ^*_c decays to $\Xi_c\pi$. This leaves the Λ^*_c states, which decay to $\Lambda_c\pi\pi$, but not to $\Lambda_c\pi$. As is discussed in section 3.2.2, the absence of the $\Lambda_c\pi$ decay mode suggests the classification of these states as Λ_c excitations, not Σ_c excitations.

Table 4: The D^* branching ratios from [8]. For the strong decays, the Q value of the decay is also given. The first row of the table corresponds to the upper limit, $B(D^{*+} \to D^+\gamma) \leq 4.2\%$ @ 90% C.L.

Decay	*Branching Fraction* (%)	$Q(MeV)$
$D^{*+} \to D^+\gamma$	$1.1 \pm 1.4 \pm 1.6$	
$D^{*+} \to D^+\pi^0$	$30.8 \pm 0.4 \pm 0.8$	5.66 ± 0.09
$D^{*+} \to D^0\pi^+$	$68.1 \pm 1.0 \pm 1.3$	5.85 ± 0.05
$D^{*0} \to D^0\gamma$	$36.4 \pm 2.3 \pm 3.3$	
$D^{*0} \to D^0\pi^0$	$63.6 \pm 2.3 \pm 3.3$	7.14 ± 0.07
$D^{*0} \to D^+\pi^-$	*kinematically forbidden*	

3 Theory, with Many Experimental Asides

3.1 Heavy Flavour Mesons

While it is generally believed that QCD is the correct fundamental theory of the strong interactions there are, as yet, no practical means to produce full QCD calculations of hadron masses and their decay widths. In order to get around this problem a variety of different approximate methods have been tried, among them lattice calculations, QCD sum rules, the Bag Model and potential models. The last of these has proved to be the most useful in understanding the mass spectrum and the decay patterns of heavy flavour hadrons. Moreover many of its predictions can be expressed in simple closed forms, rather than in vast tables of numbers. Therefore these proceedings will discuss the data in terms of potential models. The simplest of these is the Non-relativistic Quark Potential Model (NRQPM), for which useful reviews can be found in [9] and [10].

Shortly after the discovery of the J/ψ the properties of the charmed mesons and baryons were predicted by the use of a potential model [11]. While the details of modern potential models have evolved a great deal, the essential features remain unchanged. The most important features are that the inter-quark potential has two distinct regimes: at short distances the potential is dominated by a one gluon exchange term, proportional to $1/r$, while, at large distances it is dominated by a confining potential which is proportional to r and which comes from an effective scalar exchange. Here r is the inter-quark separation. Another important feature is that, except for mass dependences, the inter-quark potential is independent of quark flavour. Reference [12] provides a review of what has been learned about the inter-quark potential by studies of $Q\overline{Q}$ mesons.

While the above features have not been derived formally from QCD, they are certainly motivated by QCD. In an asymptotically free theory, such as QCD, the short distance behaviour should be dominated by lowest order perturbation theory, that is, by one gluon exchange. The linear behaviour at large r can be seen by considering a thought experiment in which a $q\overline{q}$ pair are held apart by an external force. Because gluons carry colour charge, they are attracted to each other and the colour field lines, instead of looking like an electromagnetic dipole field, lie in a narrow cylinder along the line which joins the $q\overline{q}$ pair. It is said that a colour flux tube forms between the pair. Lattice calculations suggest that the tension per unit length in this flux tube is a constant [13][14]. Therefore the energy stored in the flux tube is proportional to the separation of the charges. Finally, because the underlying gauge theory, QCD, has flavour independent couplings, the effective theory does too.

The strongest experimental evidence that the long distance part of the potential is due to an effective scalar exchange comes from the recent observation of the h_c. The mass of this meson is within about 1 MeV of the spin weighted center-of-mass of the χ_c multiplet. In a model with effective scalar exchange at long distances this state should be degenerate with the spin weighted center-of-mass of the χ_c multiplet. On the other hand, an effective vector exchange would give rise to a splitting on the order of several tens of MeV [15].

The various potential models which have been advanced all follow roughly the above picture. Most differences fall into two categories: one is the limiting behaviour at large r, which is exactly linear or approximately linear. The second class of differences concern the details of the shape at intermediate values of r, a range in which the potential is not well constrained by the available data. Most of the data lies either at large r (π and ρ) or at small r (Υ). It is expected that the excited charm states will provide more data in this region.

The most complete potential model, one which describes the mass spectrum and decays of all hadrons from the π to the Υ, is that developed by Godfrey and Isgur [16] for mesons and by Capstick and Isgur for baryons[17]. They start with a non-relativistic potential model and add some of the consequences of relativity by hand. Their method changes the parameterization of the wavefunctions and the details of some of the decays. Because they do not perform a full relativistic treatment their model is said to be "relativized", rather than relativistic. The following will sketch their model for mesons using a notation which is a mixture of that from [16] and that from [18]. The model for baryons will be discussed in section 3.2

The Hamiltonian of the Godfrey-Isgur model is written in terms of the inter-quark separation, $\vec{r} = \vec{r}_1 - \vec{r}_2$, and its conjugate momentum, $\vec{p} = \vec{p}_1 - \vec{p}_2$. Here the labels 1 and 2 refer to the quark and anti-quark which make up the meson. Other notation that is needed are the masses and spins of the quarks, m_i and $\vec{S}_i$, the orbital angular momentum between the quarks $\vec{L}$, the total spin of the quarks, $\vec{S} = \vec{S}_1 + \vec{S}_2$, and the total angular momentum of the meson, $\vec{J} = \vec{S}_1 + \vec{S}_2 + \vec{L}$. Finally, $r = |\vec{r}|$ and $p = |\vec{p}|$. The Hamiltonian has the form,

$$H = H_0 + H_c + H_{BF}, \tag{1}$$

where the subscripts denote the free Hamiltonian (H_0), the confinement part of the Hamiltonian (H_c) and a spin dependent part which is reminiscent of the Breit-Fermi Hamiltonian of atomic and nuclear physics (H_{BF}). Written in full, these pieces are,

$$\begin{aligned} H_0 &= m_1 + \frac{p^2}{2m_1} + m_2 + \frac{p^2}{2m_2} \\ H_c &= -\frac{4}{3}\frac{\alpha_s}{r} + a + br \\ H_{BF} &= c_1 \vec{L}\cdot\vec{S}_1 + c_2 \vec{L}\cdot\vec{S}_2 + c_T\left(3\vec{S}_1\cdot\hat{r}\vec{S}_2\cdot\hat{r} - \vec{S}_1\cdot\vec{S}_2\right) + c_H \vec{S}_1\cdot\vec{S}_2, \end{aligned} \tag{2}$$

where,

$$\begin{aligned} c_i &= \left(\frac{1}{m_i^2} + \frac{1}{m_1 m_2}\right)\frac{4}{3}\frac{\alpha_s}{r^3} - \frac{1}{2m_i^2}\frac{b}{r} \\ c_T &= \frac{1}{m_1 m_2}\frac{4}{3}\frac{\alpha_s}{r^3} \\ c_H &= \frac{8\pi}{3}\frac{1}{m_1 m_2}\frac{4}{3}\frac{\alpha_s}{r}\delta^3(\vec{r}). \end{aligned} \tag{3}$$

Table 5: Quantum numbers J^{PC} for quark model meson states of specific ${}^{2S+1}L_J$.

J	PC			
	$-+$	$+-$	$--$	$++$
4	${}^1G_4$		${}^3G_4$	${}^3F_4$, ${}^3H_4$
3		${}^1F_3$	${}^3D_3$, ${}^3G_3$	${}^3F_3$
2	${}^1D_2$		${}^3D_2$	${}^3P_2$, ${}^3F_2$
1		${}^1P_1$	${}^3S_1$, ${}^3D_1$	${}^3P_1$
0	${}^1S_0$			${}^3P_0$

In the above there is one crucial sign, which is a prediction of the model — the negative sign between the α_s and the b contributions to the spin-orbit coupling. While each of these contributions is large compared to the spin-spin coupling, the two contributions cancel almost exactly. In the light baryon sector, this cancelation explains the unexpected success of the original Isgur-Karl model[19], in which the spin-orbit terms were simply ignored!

It is important to point out that, in the relativized quark potential model, most of the parameters were determined from the light hadrons, charmonium and bottomonium. This model, therefore, makes actual predictions for most of heavy flavour spectroscopy and decays.

The last part of the model is to specify the constituent masses of the quarks. Typical values used in non-relativistic models are $m_u = m_d = 350$ MeV, $m_s = 550$ MeV, $m_c = 1500$ MeV and $m_b = 5000$ MeV. In the relativized model of Godfrey and Isgur the masses used are, $(m_u + m_d)/2 = 220$ MeV, $m_s = 419$ MeV, $m_c = 1628$ MeV and $m_b = 4977$ MeV. These two models also have slightly different values for a and b.

3.1.1 A Reminder about $Q\overline{Q}$ and $q\overline{q}$

First, consider the predictions of this Hamiltonian for the mass spectra of the $c\bar{c}$ and $b\bar{b}$ systems. Because of angular momentum conservation and parity conservation, the quantum numbers J^P must be good quantum numbers of the eigenstates of this Hamiltonian. Further, the constituents turn into each other under charge conjugation, C. Also, QCD conserves C and I. Therefore the complete set of observable quantum numbers for the eigenstates of this system is IJ^{PC}. Because P and C are good quantum numbers, S must also be a good quantum number. Within the context of quark potential models one should also consider the quantum number L. Even though L is not a good quantum number, it is useful to start in a basis with definite L and to treat mixing between states of the same J^{PC} but different L as a perturbation. States with definite L and S are denoted by the spectroscopic notation ${}^{2S+1}L_J$ and they have the well known properties, $P = (-1)^{L+1}$ and $C = (-1)^{L+S}$. A list of these states, sorted by J^{PC}, is summarized in Table 5. When two spectroscopic states give rise to the same J^{PC}, the physical states will be mixtures of these two states. Therefore the physical states with, for example, $J^{PC} = 1^{--}$, will be a mixture of ${}^3S_1$ and ${}^3D_1$. Also, all states of a given J^P have the same S — this is just the statement that S is a good quantum number.

Recall that the specific example being discussed is the mass spectra of charmonium and bottomonium. In these systems the multiplets of given L are separated from each other by several hundred MeV while the states within the multiplets have widths of, at most, a few MeV. Recall that perturbation theory suppresses the mixing of such states with large energy difference denominators. The lowest lying 1^{--} meson will, therefore, be almost pure $^3S_1$, with only a small $^3D_1$ component. This example illustrates that, although L is not formally a good quantum number, there are cases in which it is effectively a good one. It is therefore meaningful to speak of things such as S-wave mesons and P-wave mesons, which have $L = 0$ and $L = 1$, respectively. On the other hand this will not work for all cases. It may happen that, for example, a radial excitation of the $^3S_1$ configuration lies close enough to the $^3D_1$ that the physical states will have significant mixing.

In systems containing only u and d quarks, C is no longer (except for the neutral states) a good quantum number but G parity is good. The above arguments can all be repeated with G replacing C and most of the same results are obtained. The only significant difference is that, because some of the states are broad, more care must be taken when discussing the mixing between the two spectroscopic states which contribute to a given J^{PC}.

3.1.2 The Quantum numbers of $Q\overline{q}$

In order to apply quark potential models to hadrons containing heavy flavours one must first consider how $Q\overline{q}$ mesons differ from the $Q\overline{Q}$ and $q\overline{q}$ mesons. The most important difference is that, because neither C nor G are good quantum numbers, S is no longer a good quantum number. An exception is that for mesons with $L = 0$, $S = J$ and, because J is a good quantum number, so is S.

On the other hand, the limit of the potential model Hamiltonian as $m_Q \to \infty$ is particularly simple,

$$\begin{aligned} H &= m_Q + m_q + \frac{p^2}{2m_q} - \frac{4}{3}\frac{\alpha_s}{r} + a + br \\ &\quad + \frac{1}{m_q^2}\left(\frac{4}{3}\frac{\alpha_s}{r^3} - \frac{1}{2}\frac{b}{r}\right)\vec{L}\cdot\vec{S}_q. \end{aligned} \tag{4}$$

This Hamiltonian has a new good quantum number which is present only in this limit and not in the general Hamiltonian. The new quantum number is the total angular momentum of the light quark,

$$\vec{j}_\ell = \vec{S}_q + \vec{L}. \tag{5}$$

As will be discussed at the end of this section, j_ℓ is usually known as the total angular momentum of the "light degrees of freedom" and it has a direct analogy with the total angular momentum of an electron in a hydrogen atom. It is because of this that heavy-light mesons are often referred to as the "hydrogen atom of QCD"[18].

For mesons with $L = 0$, $S = J$, and working in the $m_Q \to \infty$ limit leads to the same spectroscopy as does the full Hamiltonian, a $^1S_0$ state and a $^3S_1$ state. When L is non-zero, however, there is a question of how to couple the three angular momenta in the problem. For example, one could couple the angular momenta as $\vec{J} = (\vec{S_Q} + \vec{S_q}) + \vec{L}$ or as $\vec{J} = \vec{S_Q} + (\vec{S_q} + \vec{L})$. While the final physics must not depend on the order of coupling, a poor choice can lead to a large increase in the amount of algebra required to reach the final answer. In the case that S is a good quantum number, the physical states will be obtained directly by coupling the angular momenta as $\vec{J} = (\vec{S_Q} + \vec{S_q}) + \vec{L}$. On the other hand, in the limit that $m_Q \to \infty$, the physical states will be obtained

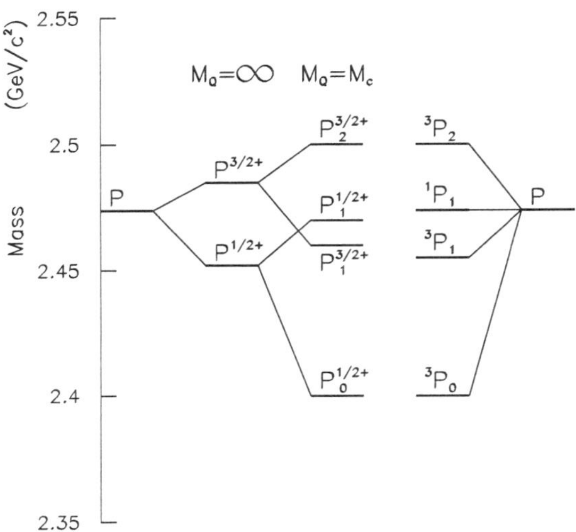

Figure 3: Splittings within the P-wave charm multiplet, following the predictions of [20]. See the discussion in the text for details.

directly by coupling the angular momenta as $\vec{J} = \vec{S_Q} + (\vec{S_q} + \vec{L})$. When neither case is appropriate one must do perturbation theory about one of these choices.

Figure 3 shows an example of the different coupling schemes for the $c\bar{q}$ mesons in the case that $L = 1$. The spin weighted center-of-mass of the P-wave mesons is shown at the left of the figure and is denoted by the label P. In the limit that $m_Q \to \infty$, this multiplet splits into a doublet in which the members have $j_\ell = 3/2$ and $j_\ell = 1/2$. These states are denoted by the spectroscopic notation $L^{j_\ell P}$. Notice that the spin weighted center-of-mass is preserved by this splitting. When one sets $m_Q = m_c$, each of these states splits into a doublet, leading to a final pattern of a doublet of doublets. These states are labeled using a spectroscopic notation $L_J^{j_\ell P}$. For comparison, the right half of the figure shows what the splitting would be like if the pattern of the charmonium splitting were reproduced. For this figure, the ${}^3P_2 -{}^3 P_0$ splitting was fixed to that found on the left half of the figure. While the differences between the mass spectra on the left and right sides of this figure are not dramatic, it will be seen in section 3.3 that the differences in the decay patterns are.

There is one feature missing from the discussion presented above. Because the charmed quark mass is not infinite, only large, j_ℓ is not a good quantum number and $P_1^{3/2+}$ state can mix with the $P_1^{1/2+}$ state, either through the spin dependent parts of the Hamiltonian or through their common decay channels.

The masses used in drawing this figure were obtained from [20], who predict that $M(P_1^{1/2+}) > M(P_1^{3/2+})$. Other authors, however, predict that the opposite will be true. That is, there is no agreement among theorists as to whether or not the doublets cross over. While this calculation does not include decay channel mixing, it does include all other effects, including the finite value of m_c. Most other authors present their results in either the $m_c \to \infty$ limit or in the ${}^3P_1$-${}^1P_1$ limit and do not make an attempt to calculate the deviations from the limit. This is one of the reasons that these proceedings focus on the model of [20, 16,17].

Candidates for the first charmed [21] and charmed-strange [22] mesons were first

Table 6: Correspondence between quark content, angular momentum quantum numbers and meson name for the known charmed mesons.

Quark Content	Notation	S − wave		P − wave	
	J^P	0^-	1^-	2^+	1^+
	$^{2S+1}L_J$	$^1S_0$	$^3S_1$		
	$L_J^{j_\ell}$	$S_0^{\frac{1}{2}}$	$S_1^{\frac{1}{2}}$	$P_2^{\frac{3}{2}}$	$P_1^{\frac{3}{2}}$
$c\overline{u}$		D^0	D^{*0}	D_2^{*0}	D_1^0
$c\overline{d}$		D^+	D^{*+}	D_2^{*+}	D_1^+
$c\overline{s}$		D_s^+	D_s^{*+}	D_{s2}^{*+}	D_{s1}^+

observed by ARGUS. Since then, all of the $c\overline{q}$ states and $c\overline{s}$ states with $j_\ell = 3/2$ have been observed and the predictions presented here are in reasonable agreement with the experimental results. Most of the modern potential models predict the intra-multiplet splittings to within about 10 MeV of the measured values and the inter-multiplet splittings to within about 40 MeV. An extensive review article on P-wave charmed mesons has been accepted for publication [2] and some of the world averages computed in that review will be discussed in section 5.3.

Now that the necessary spectroscopic notation has been defined, it is possible to complete the discussion on meson naming conventions which was begun in table 1. Table 6 continues that table to include some of the excited charmed states. Following the Particle Data Group conventions, the vector mesons are denoted by adding a superscript * to the base name. All of the remaining excited states have a subscript denoting the spin of the meson. Those excited states from the natural J^P series, $0^+, 1^-, 2^+ \ldots$, also have a superscript *. Those from the unnatural J^P series, $0^-, 1^+, 2^- \ldots$, do not have a superscript *. Any remaining confusion is separated by adding the value of the mass to the name. For example, the $P_1^{3/2+}$ $c\overline{u}$ state is well established, has a mass of about 2420 MeV, and is known as the $D_1(2420)^0$. When discovered, its $P_1^{1/2+}$ partner will be known as the $D_1(M)^0$, where M is the mass of the state. Because j_ℓ is only approximately a good quantum number, the Particle Data Group naming scheme does not include that information.

The Particle Data Group [6] refers to a state called the $D_J(2440)^+$, where J indicates that the spin is unknown. This state was seen using a partial reconstruction technique [23] which was only able to obtain large errors on the mass. The state lies within one standard deviation of the mass of the $D_1(2420)^0$, as determined by the same experiment, and it has always been understood to be the isospin partner of the $D_1(2420)^0$. Now that the $D_1(2420)^+$ has been firmly established [24], at a mass of $2425 \pm 2 \pm 2$ MeV, the $D_J(2440)^+$ should be renamed the $D_1(2420)^+$.

Finally, figure 4 shows the predicted invariant mass spectrum from [16] and [20] for many states in the charmed non-strange sector. Because the separation from the F-wave states is about 500 MeV, the lowest lying P-wave states should have very little F wave admixture. In this model the overall mass scale in this sector is about 20 MeV too high but the model otherwise describes the data well.

The above discussion is the first exposure in these lectures to a simple but important idea which, over the past 6 or 7 years, has transformed the theory of hadrons with heavy flavour, Heavy Quark Effective Theory (HQET)[25, 26]. In the limit that $m_Q \to \infty$, QCD has an exact symmetry that the mass and the spin of the heavy quark completely decouple from all dynamics. The preceding discussion shows how this occurs in a

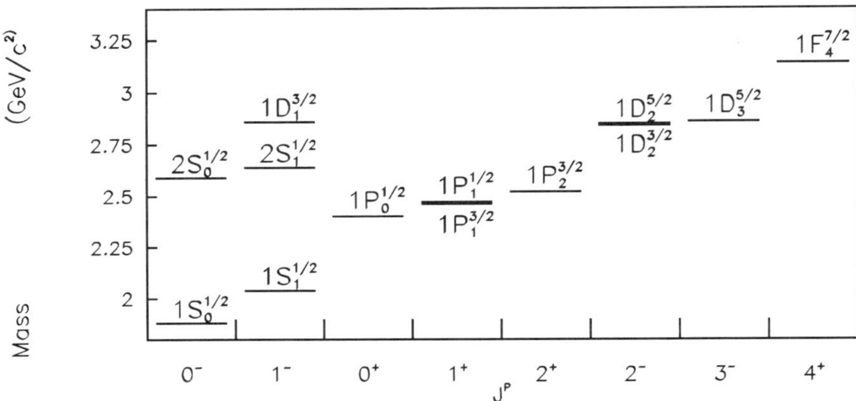

Figure 4: Mass spectrum of charmed meson predicted by the model of Godfrey and Isgur [16], [20].

potential model but the result is also true in the full theory. Motivated by this idea, one says that as $m_Q \to \infty$, the quantum numbers of a $Q\overline{q}$ meson are just those of the "light degrees of freedom". In a constituent quark model the light degrees of freedom are simply those of the light quark. In full QCD the light degrees of freedom are the sum of a current light quark, along with its associated virtual gluons and $q\overline{q}$ pairs. A third example concerns the decay of an excited charm meson to $D\pi$. In this case the light degrees of freedom of the final state are the sum of the light degrees of freedom in the D, plus all of the degrees of freedom in the π, plus the orbital angular momentum between the D and the π; that is, the light degrees of freedom include everything except the heavy quark. In such a decay both J and j_ℓ must be conserved, which leads to new selection rules.

For completeness, a list of other authors who have made predictions about the masses of P-wave charmed mesons includes, [27],[28],[29],[30], [31],[32]. While the earlier papers are sometimes far off the mark, most authors get the splittings close to correct but sometimes have a large shift in the center-of-mass of the multiplet.

3.2 Heavy Flavour Baryons

In this section, the theory of heavy flavour baryons will be discussed in terms of potential models. As for the charmed mesons, other descriptions of these states exist but potential models have been successful and their results can often be presented in closed form. In this section most of the discussion follows that of [17, 33, 34]. In this model the Hamiltonian of a baryon is,

$$H = \sum_{i=1}^{3}(m_i + \frac{p_i^2}{2m_i}) + \sum_{i<j}(H_c^{ij} + H_{BF}^{ij}), \tag{6}$$

where i and j run over the valence quarks inside the baryon. Except for one change, H_c^{ij} and H_{BF}^{ij} are simply the 2-body confining and Breit-Fermi terms from the meson Hamiltonian, equation 2. The exception is that colour factors of 4/3 must be replaced by colour factors of 2/3. This arises from the difference in the strength of the interaction between two quarks compared to that between a quark and an anti-quark. At this point

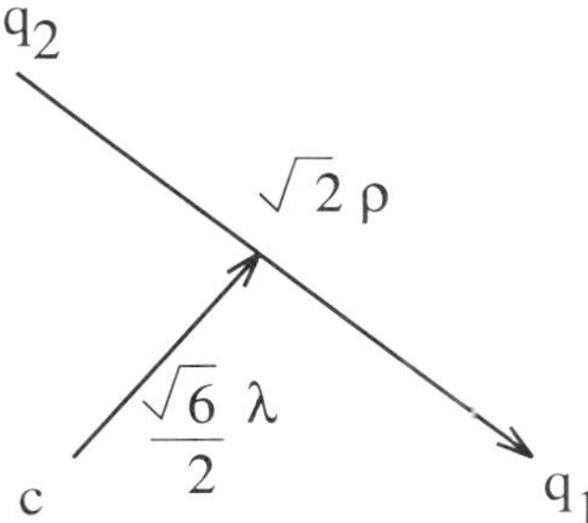

Figure 5: Definition of the ρ and λ variables used in the separation of variables for a charmed baryon, cq_1q_2.

a change of variables is made from the three inter-quark separations to the center-of-mass motion plus two internal coordinates. For the css baryons and the cqq baryons with $q \neq s$, these internal coordinates are defined as,

$$\vec{\rho} = \frac{\vec{r}_1 - \vec{r}_2}{\sqrt{2}} \qquad \vec{\lambda} = \frac{\vec{r}_1 + \vec{r}_2 - 2\vec{r}_c}{\sqrt{6}}. \tag{7}$$

These are illustrated in figure 5. The reduced masses associated with each of these degrees of freedom are,

$$m_\rho = m_q \qquad m_\lambda = \frac{3m_q m_c}{2m_q + m_c}. \tag{8}$$

There is also a conjugate momentum associated with each of these variables. Finally, for the csq baryons the change of variables is a little more complicated and is described in [33].

The quark model also provides a physically intuitive way to build up a picture of the mass spectrum of the Hamiltonian. The one important rule is that the wavefunctions must be antisymmetric under the interchange of identical quarks. Among other things, this will provide a principle which explains the correct order in which to couple the spins of the three quarks.

In the quark model, the wavefunctions for the baryons being considered here can be written as simple products of five parts,

$$\psi \quad = \quad \psi_C \, \chi \, \psi_L \, \psi_F \, \psi_R \tag{9}$$

where ψ_C denotes the colour part of the wavefunction, χ denotes the spin wavefunction, ψ_L denotes the angular parts of the spatial wave function, ψ_I denotes the flavour (isospin) wavefunction, and where ψ_R denotes the radial part of the spatial wave function. Each of these pieces will either be symmetric or antisymmetric under the interchange of a pair of quarks. The colour wavefunction is totally antisymmetric under the interchange of a pair of quarks. The ρ angular parts of the spatial wavefunction have an interchange symmetry of $(-1)^{L_\rho}$ under the interchange of quarks 1 and 2, and the λ angular parts of the spatial wavefunction are symmetric under the interchange of quarks 1 and 2. The radial parts of the wavefunction are symmetric under the interchange of quarks. A pair of quarks can either have $S = 0$, which is antisymmetric, or $S = 1$, which is symmetric. The flavour parts will be discussed below. The above

Table 7: Allowed values of S for the qq diquark inside the Λ_c or the Σ_c. For example, the negative sign for the Λ_c/ψ_I entry indicates that the isospin part of the diquark wavefunction of the Λ_c is antisymmetric under the interchange of flavour labels.

State	ψ_C	ψ_L	ψ_I	χ
Λ_c	$-$	$+$	$-$	$-\,(S=0)$
Σ_c	$-$	$+$	$+$	$+\,(S=1)$

breakdown also holds true for the corresponding strange, non-charmed mesons, which will be discussed for purposes of comparison. Not all states in the full theory of all baryons exhibit this factorization and one must worry about cases when, for example, the product $\chi\psi_L$ has a definite interchange symmetry but the individual pieces do not. For the states under consideration in these lectures, however, all of the wavefunctions can be factorized into pieces with definite symmetry.

3.2.1 S-wave Charmed Baryons

The first state to consider is the spatial ground state of the Λ_c, a cud state in which the ud diquark has, by definition, I=0. In the limit that isospin is an exact symmetry, the u and d quarks must be considered identical and the overall wavefunction must be antisymmetric under their interchange. The overall antisymmetry is built up from the pieces of the wavefunction as follows. A ud pair in an $I = 0$ state is antisymmetric under interchange of the flavour indices. The Λ_c is in its spatial ground state so both the ρ and λ degrees of freedom have $L = 0$ and both are symmetric. As always the colour wavefunction is antisymmetric. In order for the overall wavefunction to be antisymmetric under the interchange of the u and d quarks, they must be in a spin 0 state. This is summarized in the first line of table 7. The final step is to combine the spin of the charmed quark with the spin of the diquark. This gives the result that the Λ_c has a total spin[2] of $J = 1/2$. The parity of a baryon in its spatial ground state is positive.

When one applies the same logic to the Σ_c states, which have $I = 1$, one finds that the ud diquark must have $S = 1$. This implies that the spatial ground state of the Σ_c system is a doublet, the members of which have $J^P = 1/2^+$ (Σ_c) and $J^P = 3/2^+$ (Σ_c^*).

In the strange baryon sector, precisely the same logic yields the same general features for the strange, non-charmed baryons, the Λ, Σ and Σ^*. The one important difference is that the splitting of $M(\Sigma_c^*) - M(\Sigma_c)$ should be smaller than the splitting of the $M(\Sigma^*) - M(\Sigma)$. This is a hyperfine splitting which arises from the colour magnetic moment of the c or s quark. Because $m_s/m_c \simeq 1/3$ in the NRQPM, the splitting is expected to be about 3 times smaller in the charm sector than it is in the strange sector.

The above arguments are summarized in the upper part of figure 6. The left hand side shows the measured mass spectrum of the sqq baryons, while the right hand side shows that of the cqq baryons. The vertical scales on the two sides have different offsets, which have been chosen so that the Λ and Λ_c are at the same level on the page. Otherwise the vertical scale is the same on the two sides. Indeed, the vertical scale is the same, excepting offsets, for all parts of this figure. For both the sqq and the cqq systems, the spin weighted center-of-mass (*com*) of the Σ states is also shown.

[2] The notation S will be used for the spin of the light diquark while J will be used for the spin of the baryon.

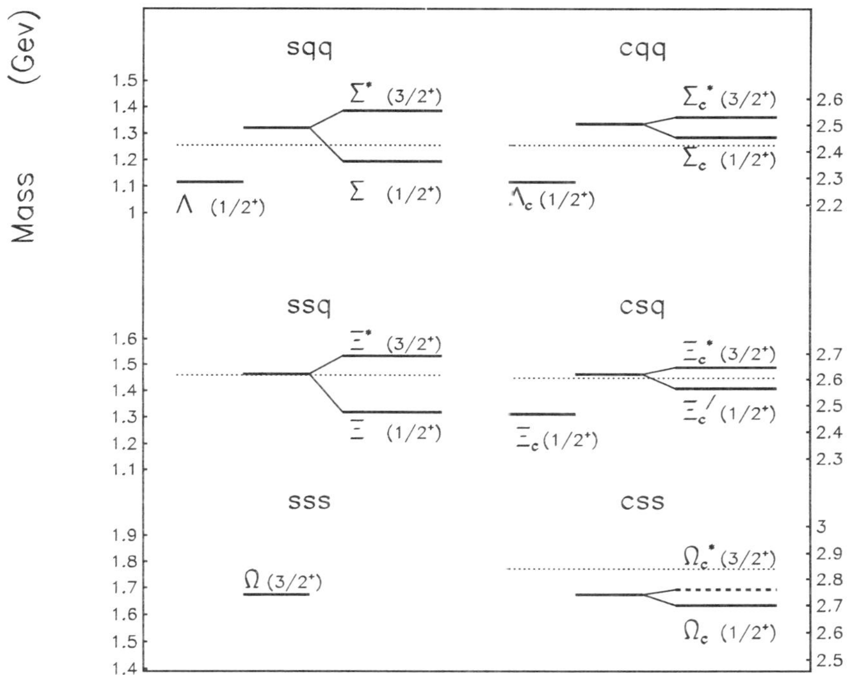

Figure 6: Measured masses of the spatial ground state strange and charmed baryons. The quantities in brackets are the J^P of each state. The unlabeled lines represent the spin weighted center-of-mass of a multiplet. For example, in the upper left part, the unlabeled line is the center-of-mass of the Σ multiplet. In each part of the figure, the dotted line indicates the threshold for decay via π emission to the lowest lying state in that part. In all cases the measurement errors are small compared to the line thickness. The dashed line shows a theoretical prediction for the mass of the unobserved Ω_c^*. The data shown here come from the same sources as that shown in figure 1.

In this figure the smaller hyperfine splitting in the charm sector is evident,

$$\frac{M(\Sigma^*) - M(\Sigma)}{M(\Sigma_c^*) - M(\Sigma_c)} = 2.5 \pm 0.2, \tag{10}$$

which confirms that prediction above is in the right ball park. In order to calculate this ratio some assumptions had to be made about the missing Σ_c^{*+} and Σ_c^{*0}. They were simply ignored, and the mass of the Σ_c^{*++} was used as the center-of-mass of the isospin multiplet. This seems reasonable as the error on the mass of the Σ_c^{*++} is much larger than the expected isospin splitting within the multiplet. Also, the splittings $M(\Sigma_c(com)) - M(\Lambda_c)$ and $M(\Sigma(com)) - M(\Lambda)$ differ by only a small amount, 15 ± 5 MeV. That the difference is small is naturally understood in terms of potential models. When considering the splitting between the Λ and the $\Sigma(com)$, the only thing which contributes is the spin of the light diquark; that is the c or s quark is strictly a spectator. Ultimately this arises from isospin symmetry, which drove the choice of the separation of variables of the Hamiltonian. These observations confirm that it makes sense to talk of these baryons as a diquark, with specific internal properties, interacting with the third quark, s or c. The usual sound-bite is that the quark models are working with "right" degrees of freedom.

Finally, the figure provides a clear answer to a question which often arises when people first encounter charmed baryons. While the decay $\Sigma \to \Lambda\pi$ is forbidden by kinematics, the decay $\Sigma_c \to \Lambda_c\pi$ is allowed. People have asked, "If splittings are supposed to be smaller for heavier mass quarks, why is it not so in this example". The answer is: only the hyperfine splittings involving the c quark are expected to behave as m_s/m_c. The splitting in question, $M(\Sigma) - M(\Lambda)$, is not a hyperfine splitting. Rather it is composed of two pieces, one of which is a hyperfine splitting, and which does behave as expected. This same reasoning predicts that the mass splitting $M(\Sigma_b^*) - M(\Sigma_b)$ should be on the order of 25 or 30 MeV and that,

$$M(\Sigma_b(com)) - M(\Lambda_b) \simeq M(\Sigma_c(com)) - M(\Lambda_c). \tag{11}$$

The next sector to consider is the Ξ_c baryons. Because Ξ_c baryons, csu and csd, contain no identical quarks, there is no requirement that the wavefunction be antisymmetric under interchange of any of the constituents. While the identical quark symmetries do not provide an insight into the correct way to couple the three spins, HQET does: because of its small colour magnetic moment, the c quark should be added last. For the same reasons as were discussed in the $Q\overline{q}$ meson section, the total spin, S, of the sq diquark is not a good quantum number. One can, however, start work in a basis of definite S and add perturbations later. When the charmed quark is added to an $S = 0$ diquark it gives the Ξ_c which has $J^P = 1/2^+$. When the charmed quark is added to an $S = 1$ diquark it gives a doublet, the Ξ_c', with $J^P = 1/2^+$, and the Ξ_c^*, with $J^P = 3/2^+$. Finally, in order to find the physical states, the model must allow the two $J^P = 1/2^+$ states to mix. The measured mass spectrum for these states is shown in the right middle section of figure 6. Because of the mixing between the two $J^P = 1/2^+$ states, the center-of-mass of the Ξ^*-Ξ_c' system does not have a simple interpretation but it has been used to set the scale between the left and right halves of the figure.

Because the corresponding strange baryons, ssu and ssd, do contain identical quarks, the antisymmetry rules apply. The ss pair is symmetric under interchange of flavour labels, is in its spatial ground state and has an antisymmetric colour wavefunction. Therefore it must have $S = 1$ and one obtains the familiar baryons, the Ξ with $J^P = 1/2^+$

and the Ξ^* with $J^P = 3/2^+$. This is shown in the left middle section of figure 6. The scales have been set so that the center-of-mass of the Ξ^*-Ξ multiplet is at the same level as the center-of-mass of the Ξ_c^*-Ξ_c' multiplet.

In this figure the smaller hyperfine splitting in the charm sector is again evident. However, because the symmetries of the wave functions are different, there is no simple formula which gives the hyperfine splitting in the charm sector in terms of that in the strange sector.

The last sector in the spatial ground state is the Ω_c sector, css. Here, again, the identical quark symmetries can be used and they require that the ss pair have $S = 1$. This gives rise to two states, the Ω_c, with $J^P = 1/2^+$, and the Ω_c^*, with $J^P = 3/2^+$. The Ω_c has recently been observed [4] and its mass is plotted in the lower right part of figure 6. The Ω_c^* has not yet been observed and it is indicated by the dashed line in the figure. To compute the position of this line, the predicted Ω_c^*-Ω_c mass splitting was taken from [35] and added to the measured Ω_c mass. It can be seen from this figure that one should expect to see the Ω_c^* only via its decay to $\Omega_c\gamma$, which will be difficult to observe.

The corresponding state in the strange sector is the Ω, sss. This must have a wavefunction which is totally antisymmetric under the interchange of any pair of quarks. Because the flavour and space parts are totally symmetric, and the colour part is totally antisymmetric, the spin part must be totally symmetric. This can only be accomplished if the three quarks have a total spin of 3/2. Thus the Ω has $J^P = 1/2^+$ and its measured mass is drawn in the lower left part of figure 6 at the same level as the center-of-mass of the Ω_c multiplet.

3.2.2 P-wave Λ_c and Σ_c Baryons

Now consider the allowed quantum numbers when one unit of orbital angular momentum is added to a Λ_c or to a Σ_c. This can be added either between the two light quarks or between the light diquark and the charmed quark. Following the nomenclature of the last section, these are known, respectively, as ρ and λ excitations. The mass shift associated with the one unit of angular momentum is given by $\Delta m_\rho = L^2/2m_\rho R_\rho^2$ and $\Delta m_\lambda = L^2/2m_\lambda R_\lambda^2$, where the quantity mR^2 is a moment of inertia. To a first approximation the mean radius of both the ρ and λ excitations should be set by the confinement radius. Therefore the mass splittings should be dominated by the differences in the reduced mass. Because $m_\lambda > m_\rho$, one expects that the λ angular excitations will be the lowest lying spatial excitations of these baryons.

First consider adding one unit of orbital angular momentum to the λ excitation of the Λ_c. Because the ρ excitation is still in an S-wave, the arguments of the previous section imply that the ud diquark must have $S = 0$. This couples with the λ orbital angular momentum to give $j_\ell = 1$. Here the j_ℓ notation defined in the meson sector has been used with the same meaning as before; it is the total angular momentum of everything except the charmed quark. Finally, j_ℓ is split by its hyperfine interaction with the charmed quark to give two states with $J^P = 3/2^-$ and $1/2^-$. This development is shown in the middle of figure 7a). To give an idea of the scale of the mass splittings, the figure also shows the S-wave Λ_c and part b) shows the S-wave Σ_c and Σ_c^*.

When one unit of orbital angular momentum is added to the ρ excitation of the Λ_c, the reasoning of the previous section implies that the ud diquark must have $S = 1$. This couples with the angular momentum to produce $j_\ell = 0, 1, 2$. Each of these is then split by its hyperfine interaction with the charmed quark, to give the five states shown at the top of figure 7a).

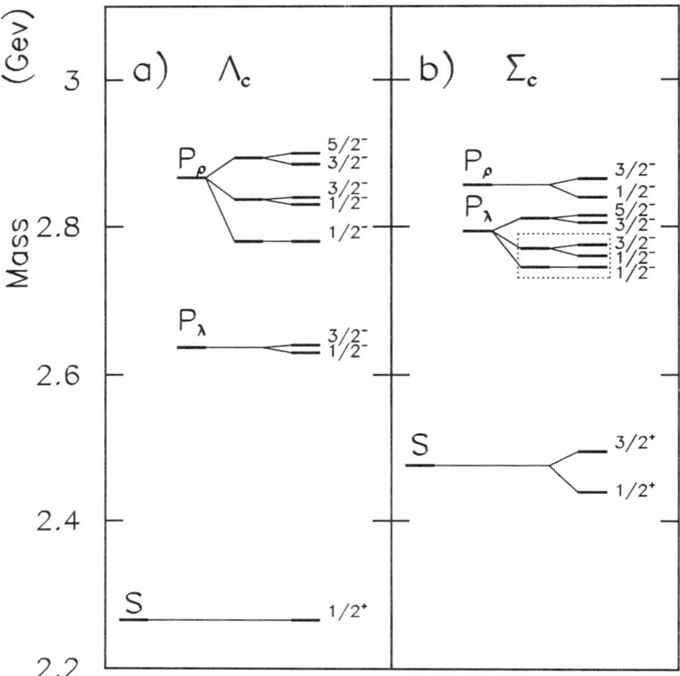

Figure 7: Predicted mass spectrum of S and P wave charmed baryons, from [17]. The region within the dotted lines is drawn with an expanded vertical scale; all of those three states are predicted to lie within about 10 MeV of where the upper one is drawn.

When one unit of orbital angular momentum is added to the λ excitation of the Σ_c, the reasoning of the previous section implies that the ud diquark must have $S = 1$. This then couples with the orbital angular momentum to give $j_\ell = 0, 1, 2$. Each of these is then split by its hyperfine interaction with the charmed quark, to give the five states shown in the middle of figure 7b).

Finally, when one unit of orbital angular momentum is added to the ρ excitation of the Σ_c, the reasoning of the previous section implies that the ud diquark must have $S = 0$, which gives $j_\ell = 1$. This is then split by its hyperfine interaction with the charmed quark, to give the doublet shown at the top of figure 7b).

This figure shows that the simple argument that the λ excitations should be at a lower mass than the ρ excitations is borne out in the complete theory, which allows different mean radii associated with each excitation. However the argument only just works in the case of the Σ_c. Notice too that the degeneracy of the ρ and λ excitations switches when one changes from I=0 to I=1.

The lowest lying Λ_c excitations stand well away from the confusion which exists a few hundred MeV above them. Candidates for these states have been observed [5][36][37] at masses of 2592.6 ± 1.3 MeV for the $1/2^-$ state and 2627.3 ± 0.8 for the $3/2^-$ state. These states are believed to be Λ_c excitations, and not Σ_c excitations, because they are observed in the final state $\Lambda_c\pi^+\pi^-$, but not in the final state $\Lambda_c\pi^0$ [5]. The decay $\Lambda_c^* \to \Lambda_c\pi^0$ is forbidden by isospin, but the decay of an excited Σ_c to that same final state is allowed. If the $\Lambda_c\pi^0$ final state were allowed then phase space requires that, for states so close to the $\Lambda_c\pi\pi$ threshold, the $\Lambda_c\pi$ final state should dominate. Reference [5] also presents a Dalitz plot of the $\Lambda\pi\pi$ final state, a candidate explanation for the relative widths of each state, and a candidate explanation for the absence of a $\Sigma_c\pi$ component in the decay of the higher mass state. Finally, the hyperfine splitting between these two $j_\ell = 1$, P-wave, charmed baryon states has been measured to be $34.7 \pm 0.5 \pm 1.2$. This is identical to the splitting observed within the $j_\ell = 3/2$ multiplets of the P-wave charmed and charmed-strange mesons. While one naively expects these splittings to be similar, it appears to be an accident that they are identical.

The relativized quark model [17] overestimates the masses of these two states by about 30 MeV. The model also appears to underestimate substantially the hyperfine splitting, but that may be partly due to the precision with which masses are quoted in their paper. The analogs of these states in the strange sector are the $\Lambda(1520)$ with $J^P = 3/2^-$, and the $\Lambda(1405)$, with $J^P = 1/2^-$. Again one sees the effect of the ratio $m_c/m_s \simeq 3$ on the hyperfine splitting,

$$\frac{M(\Lambda(1520)) - M(\Lambda(1405))}{M(\Lambda_c(2625)) - M(\Lambda_c(2593))} = 3.6 \pm 0.2. \tag{12}$$

3.3 Strong Decays

The favoured model of the strong decays of hadrons is known as the $^3P_0$ quark pair creation model (QPCM) and was developed about 1970 by Micu[38] and by Le Yaouanc and collaborators [39, 40, 41]. One of its first successes was to explain the S/D ratio in the decay $b_1(1235) \to \omega\pi$. The basic idea is illustrated by the quark line diagram shown in figure 8. In this model the initial state consists of the $q\bar{q}$ pair, $a\bar{b}$. The strong interaction then creates a new $q\bar{q}$ pair, $c\bar{d}$, from the vacuum and the final state mesons are made by rearranging the four quarks such that the initial constituents, a and $\bar{b}$, end up in different final state mesons. The model is that, relative to the initial

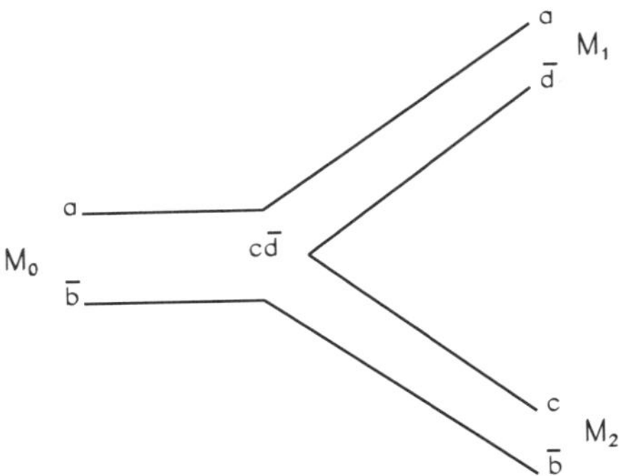

Figure 8: Quark line diagram for the decay of a meson in the $^3P_0$ quark pair creation model.

state meson, the created pair must have the quantum numbers of the vacuum. That is the created pair must have $IJ^{PC} = 00^{++}$ and there must be no momentum or orbital angular momentum between the two pairs. Consulting table 5 one can see that the only spectroscopic state with $J^{PC} = 0^{++}$ is $^3P_0$, hence the name of the model. Moreover the created pair must be in a colour singlet and it must have a flavour wavefunction of $(u\bar{u} + d\bar{d} + s\bar{s})/\sqrt{3}$. That is, there is no explicit flavour dependence but some such dependence may appear via quark mass and phase space effects. The model does not specify the dynamics by which QCD creates the pair. The model has an obvious extension to baryon decay. Finally, the most powerful part of this model is that it has only one free parameter, the overall pair creation strength, which is the same for all strong decays of mesons and baryons.

Within the quark model, one computes transition amplitudes by sandwiching this pair creation amplitude between the initial and final state wavefunctions. What follows, in the general case, is the mother of all angular momentum recoupling problems. However, so long as two or three of the mesons are S-wave mesons, the algebra is reasonable. In this formalism, the partial wave amplitudes for the decay are overlaps of radial wavefunctions and one can covert these to helicity amplitudes using the Jacob Wick identity. An example of the algebra of the full problem, in all of its generality, can be found in [42]. The final answer ends up as a product of four $9j$ symbols.

One aspect of the model which was glossed over above is that it specifies that the created pair creation amplitude is uniform throughout the universe! There have been attempts to modify this by requiring the pair creation to take place inside a flux tube joining the two quarks [43]. However this variation did not produce significantly different results.

3.3.1 Selection Rules

In order to define some notation, this section will provide a brief review of angular momentum and parity selection rules. Consider the strong decay $A \to BC$ where the intrinsic parities of the particles are denoted by η_A, η_B, η_C and their intrinsic spins are denoted by J_A, J_B, J_C. To describe the final state in such a decay one also needs to know the allowed values of the orbital angular momentum, L, between the two final state hadrons. In order for the decay to proceed in a partial wave with a given L, that

Table 8: Allowed partial waves for decays of P-wave $c\overline{q}$ mesons. No HQET selection rules have been applied.

Decay	*Allowed Partial Waves*
$D_2^* \to D\pi$	D
$D_2^* \to D^*\pi$	D
$D_1 \to D\pi$	*forbidden*
$D_1 \to D^*\pi$	S, D
$D_0^* \to D\pi$	S
$D_0^* \to D^*\pi$	*forbidden*

L must satisfy

$$\vec{J_A} = \vec{L} + \vec{J_B} + \vec{J_C} \quad (13)$$

$$\eta_A = \eta_B \eta_C (-1)^L. \quad (14)$$

Here the symbol L has been used to denote the angular momentum between the final state mesons, not the angular momentum between the quarks in one of the mesons. It should be clear from the context which is meant in any future part of the discussion — in one case a phrase "P-wave mesons" will be used and in the other case a phrase like "D-wave decays" or "S-wave transitions" will be used.

Consider, for example, the decays of the P-wave D mesons. Start by looking at figure 8 with an initial state of $c\overline{q}$, where $q = u, d$. Now, consider what sort of final states can be made by the various combinations of quarks: $D\pi$, $D^*\pi$, D_sK, D_s^*K, $D\eta$, D_ρ, D_sK^*, $D(\pi\pi)_{non-res}$ and so on. Next, remember that these states are found about 500 MeV above the center-of-mass of the S-wave multiplet. Therefore, the only decays which are certain to be important are $D\pi$ and $D^*\pi$. Most other modes are either kinematically forbidden or are strongly suppressed by phase space. Some authors [44], however, have suggested that the decay $D\rho$ may contribute significantly via the low mass tail of the ρ. It has also been suggested the non-resonant $\pi\pi$ transitions may be found; however there are few examples in particle physics in which a non-resonant $\pi\pi$ transition competes favourably with a fully allowed resonant transition, such as $D\pi$ or $D^*\pi$. The decays to $D\pi$ and $D^*\pi$ decays are well established but there are no experimental limits on the presence of $D\rho$, $D^*\rho$ or non-resonant $D\pi\pi$. Finally, table 8 shows, for each of the established decays, the partial waves which are allowed by the selection rules 13 and 14.

The same exercise for the P-wave charmed-strange mesons leads to the conclusion that the allowed final states are DK and D^*K. The K^* is certainly too narrow to contribute any significant rate through its low mass tail. The experimentally established decays are indeed to DK and D^*K. Notice that the final state $D_s\pi^0$ is both OZI forbidden and isospin forbidden, while the final state $D_s\pi\pi$ is OZI forbidden.

3.3.2 Partial Widths

Shortly after the first P-wave charmed meson was discovered, Rosner [18] published a model of the level structure for the P-wave charm system. This model also made estimates for the partial widths of the decays listed in table 8. That derivation will be sketched below. He first wrote down the states in the ${}^3P_1 - {}^1P_1$ basis, and computed the mass matrix, which has the following non-diagonal part,

$$M \begin{bmatrix} {}^3P_1 \\ \\ {}^1P_1 \end{bmatrix} = \begin{bmatrix} \overline{M} + \frac{1}{2}(c_T - c_1 - c_2) & \frac{1}{\sqrt{2}}(c_2 - c_1) \\ \\ \frac{1}{\sqrt{2}}(c_2 - c_1) & \overline{M} - \frac{3}{4}c_H \end{bmatrix} \begin{bmatrix} {}^3P_1 \\ \\ {}^1P_1 \end{bmatrix}. \tag{15}$$

Here $\overline{M}$ is the center-of-mass of the multiplet and the various c coefficients are defined by equation 3. The eigenstates of this mass matrix are found by mixing the two $J^P = 1^+$ states into the combination which diagonalizes the mass matrix,

$$\begin{aligned} |\alpha> &= \cos\phi|^3P_1> + \sin\phi|^1P_1> \\ |\beta> &= -\sin\phi|^3P_1> + \cos\phi|^1P_1>, \end{aligned} \tag{16}$$

where the mixing angle, ϕ, is specified by,

$$\tan\phi = \frac{c_1 - c_2}{\sqrt{2}(\overline{M} - M_\alpha)} = \frac{\sqrt{2}(M_\beta - \overline{M})}{c_1 - c_2}. \tag{17}$$

In the limit that $m_c \to \infty$ the mixing angle becomes ϕ_0, which is given by $\tan\phi_0 = 1/\sqrt{2}$. Finally, the reduced partial widths of each state, $\tilde{\Gamma}$, are given by,

$$\begin{aligned} \tilde{\Gamma}(D_2^* \to D^*\pi) &= \frac{3}{10}D^2 \\ \tilde{\Gamma}(D_2^* \to D\pi) &= \frac{1}{5}D^2 \\ \tilde{\Gamma}(D_1(\alpha) \to D^*\pi) &= \frac{1}{2}\left(S^2\cos^2(\phi - \phi_0) + D^2\sin^2(\phi - \phi_0)\right) \\ \tilde{\Gamma}(D_1(\beta) \to D^*\pi) &= \frac{1}{2}\left(S^2\sin^2(\phi - \phi_0) + D^2\cos^2(\phi - \phi_0)\right) \\ \tilde{\Gamma}(D_0^* \to D\pi) &= \frac{1}{2}S^2. \end{aligned} \tag{18}$$

Here a "reduced" width means the phase space factors have been set equal for all states. The quantities S and D are partial wave amplitudes which, in the quark model, are just overlaps of radial wave functions. In potential models all of the P-wave charmed mesons have the same radial wave function and all of the S-wave charmed mesons have the same radial wavefunction. Therefore D is just a single number which is the same wherever it appears in equation 18; it is not an expression which takes on different values for different decays. Similarly for S. In this model, indeed in all models, the width of the D_2^* is on the order of 20 MeV and the width of the D_0^* is on the order of 100 to 400 MeV. Therefore, predictions for the widths of the two D_1 states are sensitive to the predicted mixing angle.

In the limit that $m_c \to \infty$, these partial widths exhibit two interesting properties. First, the $D_1(\alpha)$ state decays in a pure S-wave while the $D_1(\beta)$ state decays in a pure D-wave. Second, the reduced total width of the D_2^* is equal to that of the $D_1(\beta)$ state and the reduced total width of the D_0^* is the same as that of the $D_1(\alpha)$ state. In physics zeros and ones are rarely accidents but, instead, are usually the results of symmetries. In this case, the above observations are the result of Heavy Quark Symmetry (HQS). Now the question arises, can one use HQS to get the results directly, without the need for the intermediate machinery of a potential model?

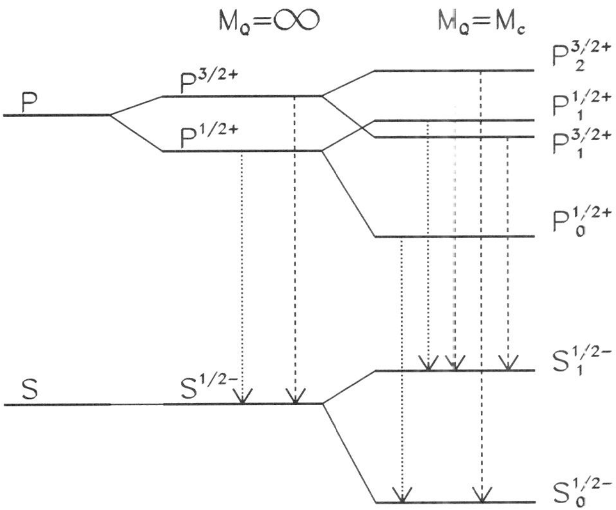

Figure 9: HQET Selection rules for the decay of P-wave charmed mesons. The S-wave transitions are indicated by dotted lines and the D-wave transitions are indicated by dashed lines. This figure is not to scale.

3.3.3 HQET Selection Rules

In the limit $m_c \to \infty$, one considers the four P-wave charmed mesons to be only two distinct states, one with $j_\ell^P = 1/2^+$ and one with $3/2^+$. Similarly the two S-wave charmed mesons are considered to be a single state with $j_\ell^P = 1/2^-$. So there are only two independent transitions, $P^{3/2+} \to S^{1/2-}\pi$ and $P^{1/2+} \to S^{1/2+}\pi$[45]. The first of these has the j_ℓ^P structure, $3/2^+ \to 1/2^-\ 0^-$, which can only proceed in a D-wave. Similarly, the second decay has the j_ℓ^P structure $1/2^+ \to 1/2^-\ 0^-$, which can only proceed in an S-wave. These rules hold true for each $P_J^{j_\ell P}$ component of each $P^{j_\ell P}$ state, not just for the overall multiplet. For example, $P_1^{3/2+} \to D^*\pi$ may only proceed in a D-wave. This picture also explains why the total width of the $P_2^{3/2+}$ is the same as that of the $P_1^{3/2+}$ — because, in the $m_c \to \infty$ limit, they are identical states.

The above discussion is illustrated in figure 9. The left hand part of that figure shows the centers of mass of the S-wave and P-wave multiplets. The middle section shows the mass spectrum and transitions in the $m_c \to \infty$ limit; the dotted line shows the S-wave transition and the dashed line shows the D-wave transition. The right hand part of the figure shows the spectrum and transitions when the heavy quark has the mass of the charmed quark. In that case, the rules derived above remain approximately true (although the figure shows them as exactly true). The rules will not be exactly true because the two $J^P = 1^+$ states will mix with each other. To summarize, this short discussion has reproduced, without any mathematics, the results obtained in the previous section. One can use this to identify Rosner's α state as the $P_1^{1/2+}$ and his β state as the $P_1^{3/2+}$.

One of the questions also addressed by this formalism is how to predict the ratio, $\Gamma(D_2^* \to D\pi)/\Gamma(D_2^* \to D^*\pi)$. Both of these branching fractions are proportional to some symmetry coefficient, which may only depend on the angular momenta in the problem, multiplied by a common overall strength. So predicting the ratio reduces to

Table 9: Predicted decay rates for strong decays of P-wave charmed mesons, in MeV. The results from [20] are for a) their pseudoscalar emission model, b) their equal radii model and c) their effective radii model.

Decay	Partial Wave	Γ(MeV) Ref [20]a	Ref [20]b	Ref [20]c	Ref [44]
$D_2^* \to D^*\pi$	D	19	17	14	9
$D_2^* \to D\pi$	D	44	35	23	16
$D_2^* \to D^*\rho$					3
$D_1(j_\ell = 1/2) \to D^*\pi$	S	250	270	1100	
	D	0.53	0.48	0.34	
$D_1(j_\ell = 3/2) \to D^*\pi$	S	6.0	6.7	26	0
	D	20	17	12	11
$D_1(j_\ell = 3/2) \to D\rho$	D				6
$D_0^* \to D\pi$	S	290	170	990	
$D_{s2}^* \to D^*K$	D	1.0	1.4	1.0	< 1
$D_{s2}^* \to DK$	D	20	26	15	7
$D_{s1}(j_\ell = 1/2) \to D^*K$	S	140	300	860	
	D	$\simeq 0$	$\simeq 0$	$\simeq 0$	
$D_{s1}(j_\ell = 3/2) \to D^*K$	S	0.31	0.64	1.82	
	D	0.05	0.07	0.04	< 1
$D_{s0}^* \to DK$	S	310	270	990	

calculating the symmetry coefficients. The most straightforward discussion of this can be found in [46], which identifies the symmetry coefficient as nothing more than a $6-j$ symbol.

However HQET itself does not predict one important thing — the size of the violation of the strict HQET limit. For example, in the HQET limit $\Gamma(P_1^{1/2+})$ is on the order of 100 to 400 MeV, while $\Gamma(P_1^{3/2+})$ is on the order of 20 MeV. So even a small violation of the HQS limit may cause enough mixing between the 1^+ states to make a profound difference in the observed decay patterns. Using only HQET, one can make only a limited statement about the S-wave partial width of the $P_1^{3/2+}$: it is small compared with typical S-wave partial widths, such as $a_1 \to \rho\pi$. It says nothing about the relationship between the S-wave and D-wave partial widths. Nevertheless the following misconception about HQET is often repeated: in the decay $P_1^{3/2+} \to D^*\pi$, HQET predicts that the D-wave amplitude will dominate over the S-wave amplitude. While all models do indeed predict this, HQET itself does not. At first glance, the data appears to confirm this prediction, but, as will be discussed in section 5.3, the data actually provides a surprisingly poor constraint on the amount of S-wave allowed in this decay.

In order to make predictions of the amount of 1^+ mixing, and therefore of the amount of S-wave in the decay of the $P_1^{3/2+}$, one must resort to models. At present the only model to include this effect is that of Godfrey and Kokoski [20]. Their predictions for the D and D_s sectors can be found in table 9. Their model does indeed predict that the $P^{3/2+}$ will decay mostly via a D wave. However something interesting happens in the D_{s1}: because the decay is so close to threshold, the p^5 phase space factor heavily suppresses the D-wave amplitude. It is so strongly suppressed that the HQET suppressed S-wave amplitude dominates. This conclusion is also reached by another group [47]. They do not calculate the amount of mixing but instead show that for a reasonable range of assumptions about the mixing angle, the S-wave decay dominates. Again, the data does not provide a strong constraint on the relative amounts of D-wave and S-wave.

The final picture of the charmed and charmed-strange P-wave mesons is quite simple. There should be six states with $j_\ell^P = 3/2^+$, which have widths on the order of 1 to 20 MeV, and six states with $j_\ell^P = 1/2^+$, which have widths on the order of 100 to 400 MeV. All six $3/2^+$ states have been observed.

3.4 Electromagnetic Decays

Both of the decays $D_s^* \to D_s\gamma$ and $D^* \to D\gamma$ are magnetic dipole (M1) decays, which proceed by one quark flipping its spin and emitting a photon. The matrix element is a coherent sum over the spin flip amplitudes for the two quarks [10],

$$A(D^* \to D\gamma) \quad \propto \quad \langle D|\,(\mu_Q\vec{\sigma}_Q + \mu_{\overline{q}}\vec{\sigma}_{\overline{q}}) \cdot \vec{\epsilon}|D^*\rangle\,. \tag{19}$$

Here $\vec{\epsilon}$ is the polarization vector of the emitted photon, quantized along the quantization axis of the initial D^*; that is, it is not the helicity of the photon. Also, $\vec{\sigma}_i$ is a Pauli matrix which operates on the spins of constituent i, and $\mu_i \propto e_i/m_i$ is the magnetic moment of the constituent. In a potential model the wave functions of the D and D^* are identical except for their spin parts and the above amplitude can be computed using the identity,

$$\vec{\sigma}\cdot\vec{\epsilon} \quad = \quad -\frac{1}{2}(\sigma_+\epsilon_- + \sigma_-\epsilon_+) + \sigma_z\epsilon_z, \tag{20}$$

where $\sigma_\pm = \mp(\sigma_x \pm i\sigma_y)$ are the spin raising and lowering operators. For example, when the D^* is in a state with a magnetic quantum number of $+1$, its spin wavefunction is $|\uparrow\uparrow\rangle$, where the order is $|Q\overline{q}\rangle$. The spin wavefunction of the D meson is $\left|(\uparrow\downarrow - \downarrow\uparrow)/\sqrt{2}\right\rangle$. For a D^* in this state, only the σ_- parts of the amplitude are non-zero, and the decay rate is,

$$\begin{aligned}\Gamma(D^* \to D\gamma) \quad &\propto \quad \left|\left\langle \frac{1}{\sqrt{2}}(\uparrow\downarrow - \downarrow\uparrow)\,|(\mu_Q\sigma_{Q-} + \mu_{\overline{q}}\sigma_{\overline{q}-})|\uparrow\uparrow\right\rangle\right|^2 \\ &\propto \quad |(-\mu_Q + \mu_{\overline{q}})|^2 \\ &\propto \quad |(\mu_Q + \mu_q)|^2\,.\end{aligned} \tag{21}$$

This last form is the usual way in which the transition amplitude is reported, which can be a little misleading since it is expressed in terms of the charge of q, not in terms of the charge of the $\overline{q}$. The other components of the σ operators describe the transition for the other magnetic substates of the D^* and they lead to the same final expression.

This exercise can be summarized as follows. For the radiative decays of the D^{*+} and D_s^{*+}, there is destructive interference between the amplitudes for photon emission from the two quarks. On the other hand, for the radiative decay of the D^{*0} the interference is constructive. Using typical potential model values of $m_u = m_d = 300$ MeV and $m_c = 1500$ MeV leads to,

$$R_\gamma^{+0} = \frac{\Gamma(D^{*+} \to D^+\gamma)}{\Gamma(D^{*0} \to D^0\gamma)} \quad \simeq \quad 0.05. \tag{22}$$

Because of angular momentum and parity conservation, the photon must be emitted in a P-wave. Therefore the correction for differences in phase space multiplies the above

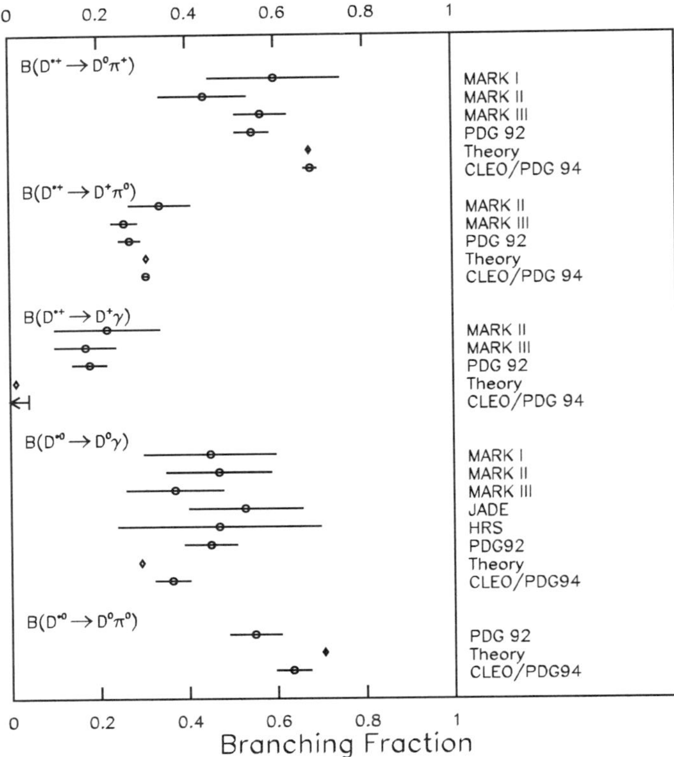

Figure 10: History of Measurements of the D^* branching fractions.

ratio by a factor of $(p_+/p_0)^3$. Here p_+ is the momentum of the photon, in the rest frame of the D^{*+}. Similarly for p_0. This correction amounts to only a few percent.

The history of the measurements of the D^* branching fractions is summarized in figure 10. In each section, the upper data points show the early measurements of the various branching fractions. The average of these points, as reported by the Particle Data Group in 1992 [48], is shown next. The diamonds without error bars show the results of a typical theoretical calculation [18]. The agreement between the 1992 world average values and the theory is poor and the biggest problem lies in the value of $B(D^{*+} \to D^+\gamma)$.

The essence of the disagreement can be seen as follows. Consider the two ratios,

$$R_\gamma^0 = \frac{B(D^{*0} \to D^0\gamma)}{B(D^{*0} \to D^0\pi^0)} \tag{23}$$

$$R_\gamma^+ = \frac{B(D^{*+} \to D^+\gamma)}{B(D^{*+} \to D^+\pi^0)}. \tag{24}$$

One can measure these two ratios and then relate the denominators using isospin and a correction for differences in phase space. This allows one to extract a value for R_γ^{+0} which is in disagreement with equation 22. If the 1992 data were correct then, either equation 22 is horribly wrong or the phase space correction is horribly wrong. No one believed that either of these calculations was seriously wrong and, for many years, there was a question mark hanging over the data.

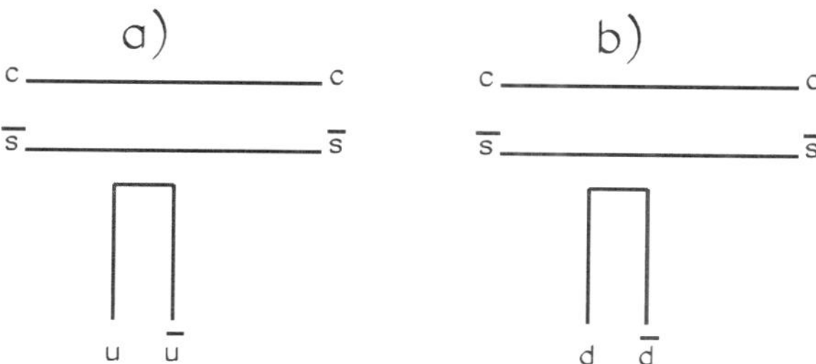

Figure 11: Isospin conservation, at the quark level, in the decay $D_s^* \to D_s\pi^0$.

In 1992 the question was resolved in favour of the theory by new CLEO measurements[8] of R_γ^0 and R_γ^+. These measurements are shown as the final data points in each section of figure 10. One might argue that the D^{*+} points agree with the theory too well, and, in fact, they do. The measurement made by CLEO was only of the two ratios R_γ^0 and R_γ^+. In order to turn these into the data points shown in the figure, three constraints were needed,

$$B(D^{*+} \to D^+\pi^0) + B(D^{*+} \to D^0\pi^+) + B(D^{*+} \to D^+\gamma) = 1 \quad (25)$$

$$B(D^{*0} \to D^0\pi^0) + B(D^{*0} \to D^0\gamma) = 1 \quad (26)$$

$$\frac{B(D^{*+} \to D^0\pi^+)}{B(D^{*+} \to D^+\pi^0)} = 2.21 \pm 0.07. \quad (27)$$

The last constraint is the consequence of isospin symmetry corrected for the differences in phase space. Its small error was made possible by improved measurements of the $D^* - D$ mass splittings [49]. Because this same constraint is also built into any theory, if $B(D^{*+} \to D^+\gamma)$ is small, then the experiment can only reproduce the theoretical predictions for the other two D^{*+} modes.

The above results are not correctly reported in the 1994 Review of Particle Properties [6]. The Particle Data Group decided that the CLEO results were of sufficient quality that it no longer included the previous results in their overall fit. Their fitting procedure includes the constraints in equations 25 and 26. When they fed the CLEO branching fractions into their fit, these two constraints were applied a second time. This incorrectly reduced the errors quoted in the Review of Particle Properties. The central values are correctly reported. The correct errors are given in table 4.

3.5 $D_s^* \to D_s\pi^0$ and Isospin Conservation

The strong decay $D_s^* \to D_s\pi^0$ provides an opportunity to see how isospin conservation works at the quark level. Figure 11 a) shows one quark line diagram through which the decay $D_s^* \to D_s\pi^0$ can proceed. While this diagram is OZI suppressed it is not identically zero and one can imagine that it might compete with a radiative decay. However there is a second diagram, shown in Figure 11 b) which can also contribute to the decay. Because the strong interactions do not see any difference between the u and d quarks, the two diagrams in Figure 11 have identical amplitudes. They will constructively interfere for final states of the form $D_s(u\bar{u} + d\bar{d})$, such as $D_s\eta$, but they will destructively interfere for final states of the form $D_s(u\bar{u} - d\bar{d})$, such as $D_s\pi^0$. In the limit that isospin is an exact symmetry this cancelation is exact.

Cho and Wise [50] recently proposed a mechanism through which the decay $D_s^* \to D_s\pi^0$ might proceed. In their model, the D_s^* decay proceeds via a virtual η which mixes into a π^0 via an isospin violating strong interaction proportional to $(m_u - m_d)(m_s - (m_u + m_d)/2)$. The decay to the virtual η occurs through the $s\bar{s}$ component of the η and so it is an OZI allowed decay. They note that the D_s is a likely place to observe such a small effect because its total width is small. Not only is its dominant decay an electromagnetic one, but even that decay is suppressed by the partial cancellation of the magnetic moments of the two quarks. Recently CLEO has presented evidence for this decay [1] and obtains,

$$\frac{\Gamma(D_s^* \to D_s\pi^0)}{\Gamma(D_s^* \to D_s\gamma)} = 0.062^{+0.020}_{-0.018} \pm 0.022. \tag{28}$$

Because the Q value of this decay is small, the resolution on the mass difference $M(D_s^*) - M(D_s)$ is also small, and the mass difference can be precisely measured, even with a small number of events. The 15 events observed in this mode contribute a precision about equal to that of several hundred events in the final state $D_s\gamma$. The 1994 Particle Data Group value for this mass difference was 141.6 ± 1.8 MeV. Since then CLEO has published a value of $144.22 \pm 0.47 \pm 0.37$ MeV using $D_s\gamma$ and a value of $143.76 \pm 0.39 \pm 0.40$ MeV using $D_s\pi^0$. Because these measurements are independent they can be averaged, which gives a mass difference of 143.97 ± 0.41.

The observation of this new decay mode provides the first significant constraints on the J^P of the D_s^*. A state which decays by a parity conserving interaction into two pseudoscalars must have natural J^P. Because the D_s^* also decays to $D_s\gamma$ it cannot have a spin of zero. This leaves the possibilities, $1^-, 2^+, 3^- \ldots$, which includes the quark model expectation of 1^-.

4 Survey of Experimental Techniques

While a great variety of experiments have contributed to the study of the ground state D^+ and D^0 mesons, most of the results in the spectroscopy and decays of the heavier charmed states has come from five experiments, ARGUS, CLEO, Fermilab experiments E691 and E687, and from CERN WA89. These five experiments form three distinct classes.

One class includes ARGUS [51] and CLEO [52], universal 4π magnetic spectrometers, which operate at e^+e^- rings with a center-of-mass energy of about 10 GeV. ARGUS is located at the DORIS II storage ring at DESY and took data for about 10 years, starting in 1982. The collaboration is still producing physics results with this data set. CLEO, which has now operated for more than 10 years, is located at CESR at Cornell University. The detector is continuing to take data and, with an approved upgrade program, will do so well into the future. One advantage enjoyed by these experiments is that the charm production cross-section is approximately 40% of the total hadronic cross-section. Moreover, because the beam energy is relatively low, the multiplicity of charm events is also reasonably low. Typically only a few additional pions accompany each charmed hadron. Historically the two detectors competed on a reasonably equal footing, with the edge in luminosity going to CLEO and the edge in detector quality going to ARGUS. However, the CLEO detector has continued to evolve while the ARGUS data taking has come to an end. The main differences between ARGUS and the present version of CLEO are the magnetic field of the analyzing magnet, the electromagnetic calorimetry, and the luminosity delivered by the storage ring. CLEO runs with a magnetic field of 1.5 T compared with 0.8 T for ARGUS. Therefore CLEO has better momentum resolution at the cost of less efficiency at low p_T. The ARGUS calorimetry is a lead-scintillator sandwich design and was a major advance in its time; in particular, it was placed inside the coil of the analyzing magnet. In the late 1980's CLEO upgraded their calorimetry to CsI crystals, also placed inside the analyzing magnet. This crystal calorimeter obtains energy resolutions about 4 times better than does ARGUS. Indeed this calorimeter is unique among the magnetic spectrometer experiments contributing to the study of heavy flavour spectroscopy. Finally the biggest advantage enjoyed by CLEO, over all of its competitors, is that the CESR storage ring continues to set world records for both instantaneous luminosity and for integrated luminosity delivered over a year.

The second class of experiments includes FNAL experiments E691 and E687, which use a high energy photon beam striking a fixed Be target. Both of these experiments have finished data taking and E687 is still producing results. The advantage that these experiments enjoy is a larger production cross-section for charm. However this cross-section is only a tiny part of the total hadronic cross-section, which complicates triggering, data acquisition, and data reduction. On the other hand, the background level is still much lower than is found in fixed target experiments with hadron beams. Therefore, using a high energy photon beam is an excellent operating point in the space of signal cross-section and background level. The tool which makes these experiments work is their silicon microstrip vertex detectors (SMD). For example, in the E691 apparatus, the mean decay lengths of D^+ and D^0 mesons are, respectively, 8.5 mm and 3.3 mm. For comparison, the resolution on the decay length is typically 0.440 mm. Because most of the backgrounds have a decay length of zero, this provides a powerful tool to reject random combinations of tracks which happen to lie in the mass region of interest. Historically, the inclusive D^0 and D^+ yields from these detectors have been comparable to those from the e^+e^- machines. However, because of their SMDs, the fixed target experiments have a much better signal to background ratio for those D^0 and D^+ modes in which all of the daughters are charged.

The final class of experiments includes just CERN WA89 which uses a Σ^- beam incident on a fixed target. The goal of this experiment is to investigate hadron production mechanisms in which the charmed particle shares a spectator quark with the beam particle. It is hoped that these processes will lead to an enhancement in the production rate of charmed baryons, in particular that of charmed-strange baryons. This experiment also relies heavily on its silicon microstrip vertex detectors. Another important feature of the detector is its RICH system used for particle identification.

Many other groups have contributed to heavy flavour spectroscopy but their contributions have, in these lectures, been buried in the references to the Review of Particle Properties. For example, the LEP experiments and CDF have contributed the measurement of the mass of the B_s. The only measurement of any Σ_c^* state comes from a Serpukov heavy liquid bubble chamber group. An ITEP bubble chamber group has made contributions to the study of D_s mesons and their excited states. The CUSB group made observations of the B^*. Other groups have made contributions which have since been superseded. The Mark I, II and III groups made important early contributions to the study of D^* mesons and numerous fixed target experiments provided measurements of the weakly decaying charmed baryons.

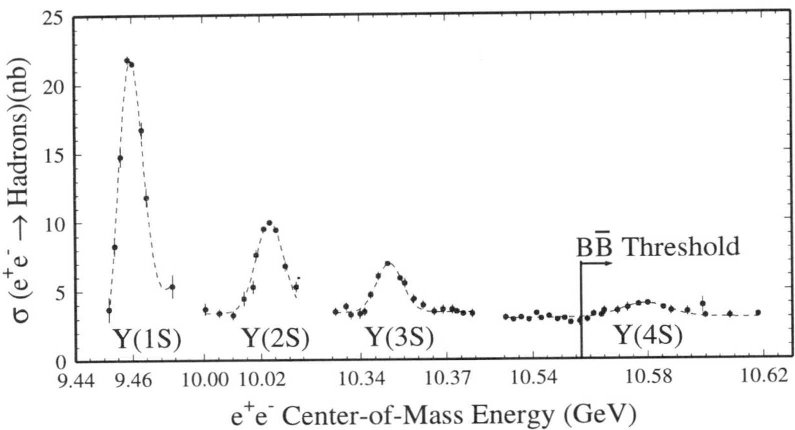

Figure 12: Visible hadronic cross section as a function of center-of-mass energy for e^+e^- collisions in the neighbourhood of the Υ resonances. Notice the gaps in the horizontal scale.

4.1 Continuum Charm Production in e^+e^- Machines

The main physics interest of the ARGUS and CLEO collaborations is the study of the decays of B mesons and, therefore, their e^+e^- colliders usually operate at the energy of the $\Upsilon(4S)$ resonance. Figure 12 shows the measured total hadronic cross-section in the energy range that CLEO operates. Notice the broken horizontal scale. The three narrow structures, at energies of about 9.46, 10.02 and 10.35 GeV, are, respectively, the $\Upsilon(1S)$, $\Upsilon(2S)$ and $\Upsilon(3S)$ resonances. These are understood to be $b\bar{b}$ bound states which are kinematically forbidden to decay into two B mesons. Their dominant decay mode is via OZI suppressed strong decays, which explains their narrow width. There is no significant charm production from these resonances. The figure also shows the threshold for $B\overline{B}$ production and the low, broad resonance immediately above this is the $\Upsilon(4S)$. This is also understood as a $b\bar{b}$ bound state and its greater width is explained by the $B\overline{B}$ final state being kinematically allowed. Indeed the $\Upsilon(4S)$ is believed to decay almost 100% of the time to $B\overline{B}$.

The baseline level of the cross-section in figure 12 comes from the reaction, $e^+e^- \rightarrow q\bar{q}$, where $q = u, d, s, c$, followed by the hadronization of the $q\bar{q}$ pair. About 40% of the time the initial $q\bar{q}$ pair is a $c\bar{c}$ pair. The hadronization process typically produces two collimated jets of hadrons but it cannot create any new charmed quarks [53]. Therefore all charmed hadrons produced by this mechanism contain one of the initial $q\bar{q}$ pair. Charmed hadrons produced in this fashion are said to come from "continuum production". One of the properties of continuum production is a pronounced leading particle effect: a final state hadron which contains one of the primary $q\bar{q}$ has a much higher average momentum than other particles produced in the hadronization process. The jargon used in these experiments is that the hadronization processes is known as fragmentation, the momentum spectrum is known as the fragmentation function, and the leading particle effect is called "having a hard fragmentation function". Moreover the excited charm states tend to have a harder fragmentation function than do the ground states. Usually one discusses fragmentation functions in terms of the scaled momentum,

$x_p = p/\sqrt{E_{beam}^2 - m^2}$, where p and m are the momentum and mass of the candidate and where E_{beam} is the beam energy. If one requires that $x_p > 0.5$, one will accept between about 60% and 85% of the charmed particles and will reduce the combinatoric background by a factor of at least 15 to 20.

There is a second, copious source of charmed hadrons in the events collected by CLEO and ARGUS. When a b quark decays, about 99% of the time it produces a c quark and a W boson. Understanding the remaining rare decay modes is one of the main challenges facing both CLEO today and B-factories in the future. A charmed quark produced via $b \to cW$ will hadronize as a charmed hadron and, perhaps, some additional light hadrons. Most of the time the W has sufficient mass to decay into either a $u\bar{d}$ or a $c\bar{s}$ pair but phase space considerations ensure that it usually decays into a $u\bar{d}$ pair. The net result is that for every B meson decay, about 1.15 charmed hadrons are produced. The precise value of this number is currently the subject of a great deal of experimental and theoretical effort. However the momentum spectrum of charmed particles from B decay is soft; the kinematics of the two step decay chain, ensures that they must have $x_p < 0.5$. Therefore, this source of charm has a poor signal to background ratio and is usually not used in studies of charm spectroscopy at ARGUS and CLEO.

4.2 The D^* Trick

One of the important tools available to people working in charm spectroscopy is the so called "D^* trick", which exploits the small Q value in the decays $D^* \to D\pi$. A list of Q values for the three different decays is given in table 4. The first consequence of the small Q value is that there is little phase space available for combinatoric background. A second consequence is that the resolution on the mass difference $\Delta m = M(D\pi) - M(D)$ is narrow. In the CLEO detector, for example, the resolution is less than 1 MeV (σ) for a broad range of D^* momenta. The combination of these two properties ensures that D^* signals have an excellent signal to background ratio. This, in turn, results in relatively good signal to background ratios for objects which decay into D^* mesons. This trick is especially important at the e^+e^- machines which have much poorer D signals than do the fixed target experiments.

While a proper explanation of the excellent resolution on Δm is beyond the scope of these lectures, a few parts of the story stand on their own. The bottom line is that, when measuring Δm, one is measuring the Q value and adding it to the mass of the pion. Therefore, even a 20% error on the Q value results in a resolution on Δm on the order of 1 MeV. Another piece of the puzzle is to see that most of the errors contributing to the error on the D^* mass also contribute to the error on the D mass. These errors largely cancel in Δm.

The following, somewhat simplistic, picture might also shed some light on the subject. First, one can then write,

$$\begin{aligned} \Delta m &= \frac{m_{D\pi}^2 - m_D^2}{m_{D\pi} + m_D} \\ &= \frac{m_\pi^2 + 2E_D E_\pi - 2p_D p_\pi \cos\theta}{m_{D\pi} + m_D}, \end{aligned} \tag{29}$$

where θ is the opening angle between the D and the π. For a D^* with momentum above, say, 1 GeV, one can make the approximation that the Q value is zero, everywhere except in the $\cos\theta$ term. In this case, the Lorentz boost equations from the D^* frame to the lab frame are,

$$E_D = \gamma m_D, \quad p_D = \beta\gamma m_D, \quad E_\pi = \gamma m_\pi, \quad p_\pi = \beta\gamma m_\pi, \tag{30}$$

where $\beta = p_{D^*}/E_{D^*}$, $\gamma = E_{D^*}/M_{D^*}$, and where everything is measured in the lab frame. The next step is to use these approximations, make a small angle approximation for $\cos\theta$ and drop terms of order m_π/m_D. This gives,

$$\Delta m \;=\; m_\pi + \frac{p_\pi^2}{2m_\pi}\theta^2. \tag{31}$$

In this picture one can see that most of the uncertainties associated with the D have cancelled out; all that is left is the contribution of the D to the error on θ. Now, consider the specific example of the decay chain, $D^{*+} \to D^0\pi^+_{slow}$, followed by $D^0 \to K^-\pi^+$, as observed in the CLEO detector. The pion from the D^* decay is labeled π^+_{slow} because the kinematics of the decay constrain it to have a momentum below about 0.450 GeV. On the other hand, the two daughters of the D^0 typically have momenta in the range of 1.0 to 2.0 GeV. Because of their higher momentum, the D^0 daughters suffer much less multiple scattering and do not make a significant contribution to the error on the opening angle. Therefore the error on the opening angle is dominated by the multiple scattering component of the error on the direction of the slow π, which is on the order of 5 to 10 mr. For typical D^* candidates the opening angle is on the order of 50 mr. The error on the momentum of the slow π is on the order of a few percent. Therefore the error on the mass difference is dominated by the error on the opening angle, which is itself dominated by the error on the direction of the slow π^+.

The above is an illustration of a general rule, that, when the lab opening angle between two systems is very small, then the error on the invariant mass of the combined system is dominated by the error on the opening angle. Conversely, when the lab opening angle is close to 180°, then the error on the invariant mass of the combined system is dominated by the error on the magnitude of the momentum of each of system.

While the D^* trick works well in all detectors for the decay $D^{*+} \to D^0\pi^+$, the CLEO calorimetry is good enough that it also works for the decays $D^* \to D\pi^0$.

4.3 Benchmark Modes

The excited charmed mesons all eventually decay strongly or electromagnetically to one of the D^0, D^+ or D_s^+ mesons. Therefore any measurement of one of the excited states requires first reconstructing one of these three mesons. While these mesons have hundreds of decay modes, only two or three are used regularly. These are known as the benchmark modes and they are characterized by both a relatively large value of $\epsilon \cdot B$ and a relatively good signal to background ratio. Here ϵ is the overall efficiency to reconstruct the mode, including fiducial volume effects, and B is the branching fraction to the benchmark mode. The benchmark modes for the charmed mesons and for some of the charmed baryons are summarized in table 10.

The general quality of some of the benchmark modes for CLEO can be seen in figures 13 and 14. For these figures, the particle identification criteria are as described in [55]. In the ARGUS detector the benchmark modes have a slightly poorer signal to background ratio, which is explained by the lower magnetic field. In the fixed target detectors the signal to background ratio for the D^+ is improved by about a factor of 15

Decay	BranchingFraction(%)
$D^0 \rightarrow K^-\pi^+$	4.01 ± 0.14
$D^0 \rightarrow K^-\pi^+\pi + \pi^-$	8.1 ± 0.5
$D^0 \rightarrow K^-\pi^+\pi^0$	13.8 ± 1.0
$D^0 \rightarrow K^0_S\pi^+\pi^-$	2.6 ± 0.3
$D^+ \rightarrow K^-\pi^+\pi^+$	9.1 ± 0.6
$D_s \rightarrow \phi\pi$	3.5 ± 0.4
$\Lambda_c \rightarrow pK^-\pi^+$	4.4 ± 0.6
$\Xi_c^+ \rightarrow \Xi^-\pi^+\pi^+$	2.1 ± 0.9

Table 10: Benchmark decay modes for some charmed hadrons. The scale of the branching fractions for the D_s, Λ_c and Ξ_c are poorly determined. All of the branching fractions are taken from the Review of Particle Properties [6], except that for the Ξ_c^+, which is taken from [54].

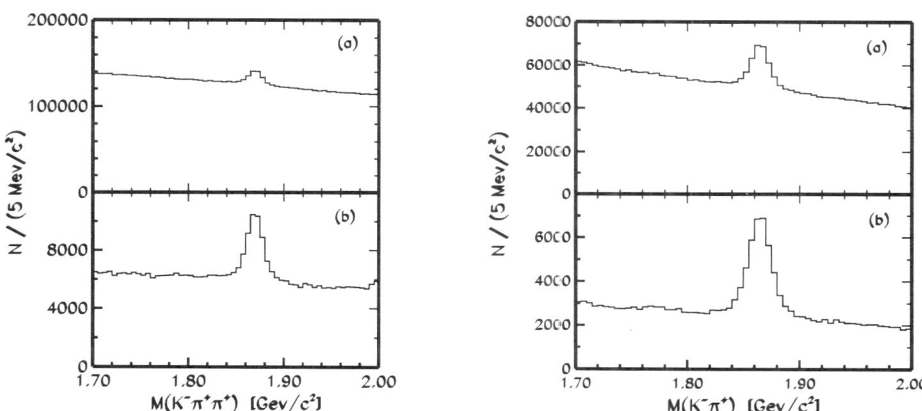

Figure 13: Benchmark D^+ and D^0 modes from CLEO. The left hand figure shows invariant mass spectra of $K^-\pi^+\pi^+$ combinations while the right hand figures shows spectra for $K^-\pi^+$ combinations. In both cases a) $x_p > 0.$ and b) $x_p > 0.6$.

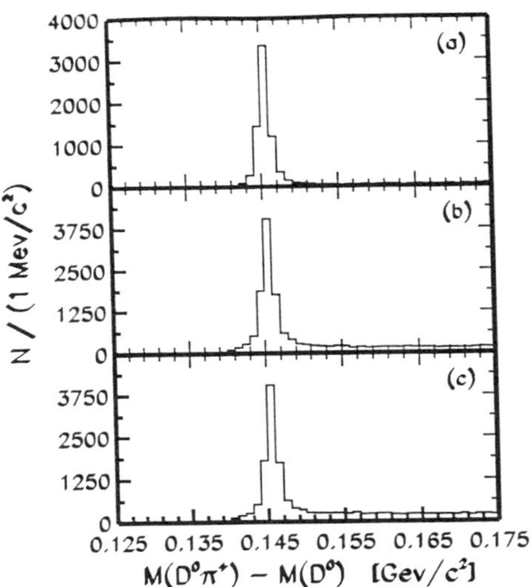

Figure 14: Benchmark $D^{*+} \to D^0\pi^+$ modes from CLEO; a) $D^0 \to K^-\pi^+$, b) $D^0 \to K^-\pi^+\pi^0$ and c) $D^0 \to K^-\pi^+\pi^+\pi^-$. All histograms are for $x_p > 0.6$.

while that for the D^0 is improved by about a factor of 2. The dramatic improvement for the D^+ is the result of its long lifetime which ensures a high efficiency for clean separation between the charm production and decay vertices. In each detector, the signal to background for the D_s benchmark modes is about the same as that for the D^0 mode in the same detector. At CLEO and ARGUS the benchmark Λ_c modes also have a signal to background ratio similar to that of the D^0 mode. At the fixed target experiments, the signal to background ratio for the D^* benchmark modes is slightly poorer than is shown here.

5 More Experimental Results

5.1 Vector-Pseudoscalar Mass Splittings

In potential models the mass splitting between the vector and pseudoscalar mesons is due to the colour hyperfine interaction and the value of the splitting is given by,

$$\Delta M(^3S_1 - ^1S_0) \quad \propto \quad \frac{1}{m_Q m_q} |\psi(0)|^2 , \tag{32}$$

where $\psi(0)$ is the value of the wavefunction at the origin. The experimental data for several such splittings are summarized in Table 11. In the light hadron sector this formula worked well if one ignored the factor of $|\psi(0)|^2$ [10] and, therefore, the first predictions in the heavy flavour sector were made ignoring possible variations in $|\psi(0)|^2$. These predictions were that the splitting $M(D_s^*) - M(D_s)$ should be on the order of 100 MeV and early, but erroneous, experimental data seemed to confirm that [56]. However it is now firmly established that the correct splitting in the D_s sector is within a few MeV of that in the D sector. In the relativized quark model [16] this is understood to be a numerical accident — the change in the wavefunction at the origin almost exactly compensates for the change in the mass of the light quark. On the other hand, HQS suggests that $|\psi(0)|^2$ is the same in both the D and B sectors. This is borne out by the data.

Table 11: Vector-Pseudoscalar splittings for various heavy-light meson systems [6, 1]. The entries for the $I = 1/2$ states give the average of the two members of the multiplet. The third column presents the predictions of an early model. The reasons behind the successes and failures of that model are now understood.

$Q\bar{q}$	*States*	*Experiment* (MeV)	$[M(D^*) - M(D)]\frac{m_c m_u}{m_Q m_q}$ (MeV)
$s\bar{u}$ $s\bar{d}$	$K^*(892) - K$	398.1 ± 0.3	385
$c\bar{u}$ $c\bar{d}$	$D^*(2010) - D$	141.4 ± 0.08	*Input*
$b\bar{u}$ $b\bar{d}$	$B^*(5325) - B$	46.0 ± 0.06	42
$c\bar{s}$	$D_s^*(2110) - D_s$	143.97 ± 0.41	90

Table 12: Mass differences of mesons which differ by the substitution of an s quark for a d quark. Most of the mass splittings are from [6], except for those of the P-wave mesons which are taken from [2]. The D_s^* mass also contains some information from [58, 1].

State	$c\bar{s}$	$c\bar{d}$	$\Delta m(MeV)$
$^1S_0$	D_s^+	D^+	99.1 ± 0.6
$^3S_1$	D_s^{*+}	D^{*+}	$103.3 \pm 1.$
$P_1^{3/2}$	D_{s1}^+	D_1^+	$114 \pm 3.$
$P_2^{3/2}$	D_{s2}^{*+}	D_2^{*+}	$110 \pm 4.$
	$\bar{b}s$	$\bar{b}d$	
$^1S_0$	B_s^0	B^0	$96 \pm 6.$

The story of the first correct observation of the D_s^* is an example of an experimental *tour de force* [57]. Recall that the dominant decay mode of this state is $D_s^* \to D_s\gamma$. This discovery was made using the ARGUS detector which had, for its time, the most advanced electromagnetic calorimetry ever installed in a 4π magnetic spectrometer. In order to suppress backgrounds, a high minimum energy cut was placed on the photon. However, this distorts the lower edge of the line shape and shifts the observed mass upwards. After a lot of work on the simulation of the response of the calorimeter, a correction of -12 ± 4 MeV was applied to the raw measured mass, giving a $D_s^* - D_s$ mass splitting of $144 \pm 9 \pm 7$ MeV. This compares well with the presently accepted value given in table 11.

5.2 Substitution of an s Quark for a d Quark

In the light hadron sector a rough rule holds that substituting a strange quark for a light quark, while keeping everything else fixed, adds approximately 150 MeV to the mass of a hadron. Table 12 shows the corresponding comparison for the heavy flavour sector. Clearly the same effect is present, although the magnitude of the shift is somewhat less than in the light hadrons. The HQS prediction that the splitting $M(B_s) - M(B^0)$ should be the same as the splitting $M(D_s) - M(D^+)$ is upheld at the level of the present errors.

Table 13: Mass splittings with the $j_\ell = 3/2$ P-wave charm multiplets.

States	*Splitting(MeV)*
$M(D_2^{*+}) - M(D_1^+)$	$34 \pm 5\ MeV$
$M(D_2^{*0}) - M(D_1^0)$	$37 \pm 3\ MeV$
$M(D_{s2}^{*+}) - M(D_{s1}^+)$	$38 \pm 2\ MeV$

There are two splittings in the charmed baryon sector where one can make the same comparison, $M(\Xi_c^{*+}) - M(\Sigma_c^{*+})$ and $M(\Xi_c^{*0}) - M(\Sigma_c^{*0})$. At first thought, these comparisons cannot be made properly because only the Σ_c^{*++} and the Ξ_c^{*0} have been established. However the experimental uncertainty on the mass of the Σ_c^{*++} is much larger than the expected isospin mass splittings within the Σ_c^* multiplet. So one can add to the list of comparisons, $M(\Xi_c^0) - M(\Sigma_c^*) = 113 \pm 5\ MeV$, which falls within the pattern seen in the meson sector.

Because of the mixing between the Ξ_c and the Ξ_c' one does not expect the same simple picture for $M(\Xi_c') - M(\Sigma_c)$ and $M(\Xi_c) - M(\Lambda_c)$. In fact one could use these splittings to learn about the magnitude of the mixing.

5.3 P-wave Charmed Meson Summary

As was discussed in section 3.1.2, there should be six P-wave charmed and charmed-strange mesons with $j_\ell = 3/2$. All of these have now been found and their masses have been well measured. They have been identified as the $j_\ell^P = 3/2^+$ mesons on the basis of the angular distributions in their decay and because their widths are on the scale of a few MeV to a few tens MeV, not on the scale of a few hundred MeV. More complete discussions of the J^P and j_ℓ assignments can be found in [55],[24],[59],[60]. A good summary of the status of the J^P assignments is that, while they are not rigorous, neither are they subject to any serious challenge [2].

Recently a review has been prepared which summarizes the current state of knowledge about the P-wave charmed and charmed strange mesons [2]. The authors have gone back to the original papers and corrected the results for improvements to the contributing measurements. For example, the reported masses have been corrected to incorporate new values of the D^* mass. Also world averages have been calculated using the methods specified by the Particle Data Group, including their method of calculating scale factors for errors. In general the experiments agree with each other well, the one problem being that there is some disagreement between ARGUS and CLEO on the total cross-sections and on the fragmentation functions.

The world average masses and widths computed in [2] have been summarized in figures 15, 16 and 17, and in table 13. Here the agreement of the data between experiments is fine. However, in the non-strange sector, the world average width measurements still have fairly large errors. These errors are large because of several effects in the backgrounds. First, because the signals sit on the top of a broad bump in the background, the measured width is strongly correlated with the background shape. Second, the background shapes are not well known and they contain many regions which are contaminated by feed-down from and reflections of other P-wave charmed mesons. Taken together these effects cause a poor determination of the widths. Moreover there are few immediate prospects for improvement. For example, most of the background in the $D^{*+}\pi^-$ final state is due to real D^{*+} candidates combined with fragmentation

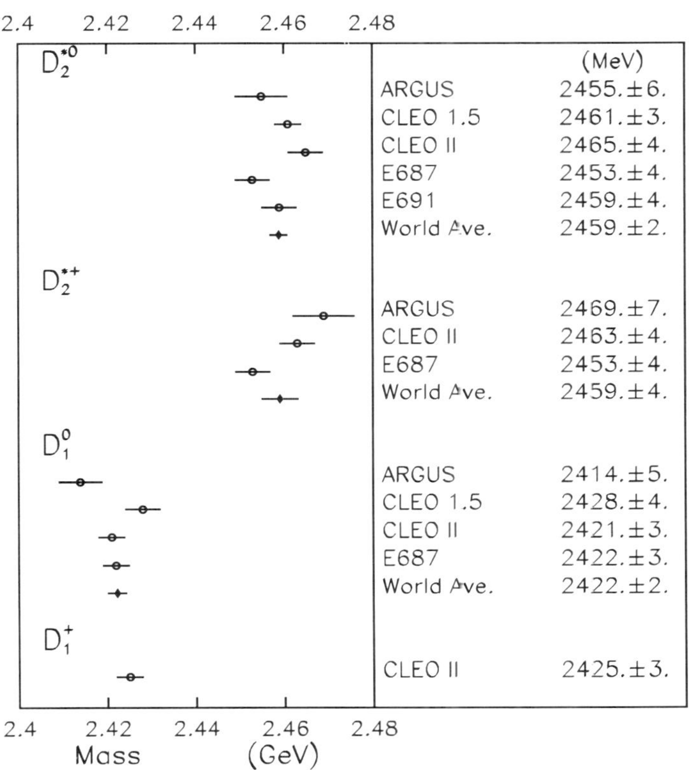

Figure 15: Masses of P-wave charmed mesons, from [2].

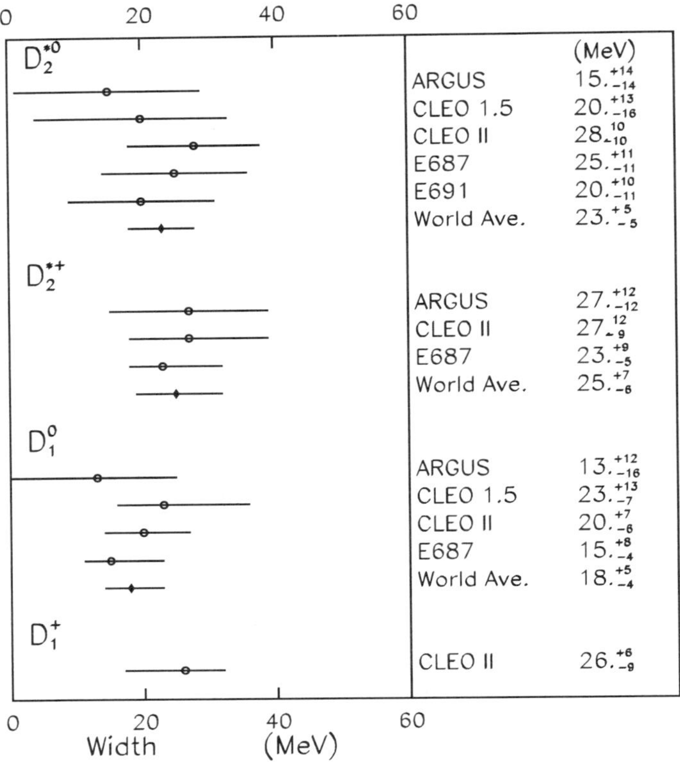

Figure 16: Widths of P-wave charmed mesons, from [2].

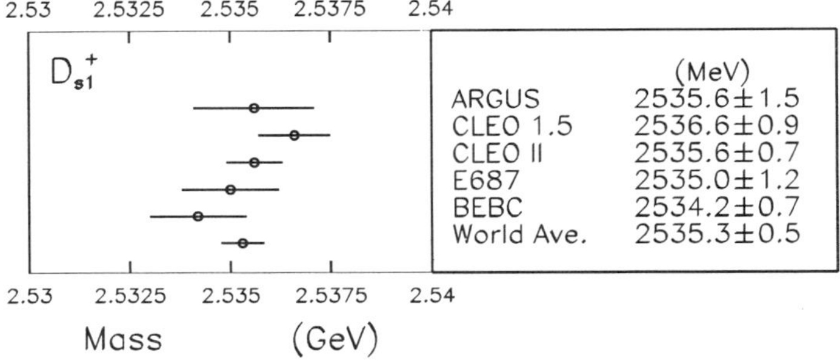

Figure 17: Measurements of the mass of the D_{s1}^+ meson, from [2]. The state has a width of $\Gamma \leq 2.3$ MeV @ 90% C.L.. The $J^P = 2^+$ partner has a mass of 2573. $\pm$ 2 MeV and a width of 16 ± 2 MeV.

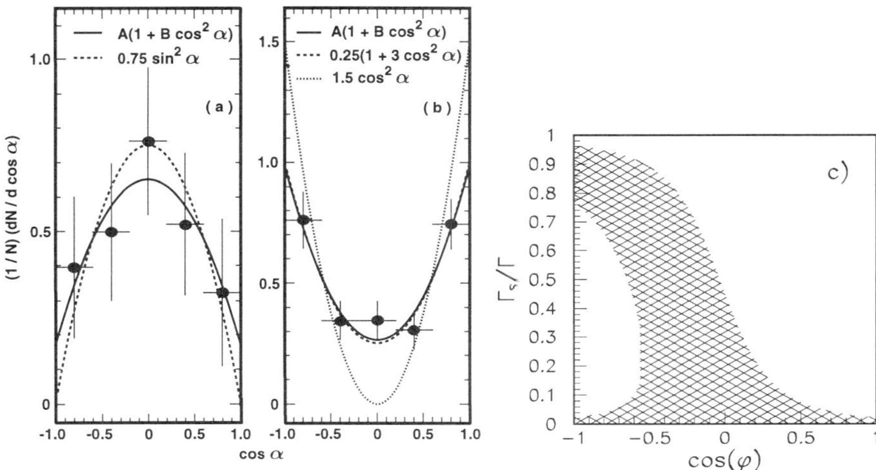

Figure 18: The measured distributions of $\cos\alpha$ for the decays a) $D_2^*(2460)^0 \to D^{*+}\pi^-$ and b) $D_1(2420)^0 \to D^{*+}\pi^-$. The superimposed curves are described and interpreted in the paper from which the data are taken [?]. For the decay of the D_1^0, part c) shows the allowed region in plane of Γ_S/Γ vs φ. The notation is defined in the text.

pions. That is, the background is from physics and cannot be removed by upgraded detectors.

One of the interesting distributions in the decay of the P-wave charmed mesons is that of the helicity angle. In the decay chain, $D_1 \to D^*\pi_1$, $D^* \to D\pi_2$, the helicity angle, α, is defined as the angle between the direction of π_2 and the direction of π_1, when both are measured in the rest frame of the D^*. It is defined similarly for the decay $D_2^* \to D^*\pi$. Figure 18shows measured distributions of $\cos\alpha$ for the decays of the D_2^* and the D_1 [55]. Assuming that the D_2 does indeed have $J^P = 2^+$, the shape of the distribution in part a) must be proportional to $\sin^2\alpha$ and the data are consistent with that shape.

Assuming that the D_1 does indeed have $J^P = 1^+$, the shape of the distribution in part b) depends on the partial wave amplitudes, S and D, in the decay of the D_1. In the limit that the decay takes place in a pure D-wave, the expected shape of the distribution is $(1 - 3\cos^2\alpha)$, and the data are consistent with that shape. However this is not sufficient evidence to conclude that the ratio $|S|/|D|$ is small — one must consider the interference between the two partial waves. Following [55], the general form of the distribution is written in terms of the variables R and φ, defined by, $\Gamma_S = |S|^2$, $\Gamma_D = |D|^2$, $\Gamma = \Gamma_S + \Gamma_D$, $R = \Gamma_S/\Gamma$ and $\varphi = \arg(S^*D)$. In terms of these variables, the shape of the angular distribution is,

$$\frac{dN}{d\cos\alpha} \propto R + (1-R)\frac{1+3\cos^2\alpha}{2} + \sqrt{2R(1-R)}\cos\varphi\left(1 - 3\cos^2\alpha\right). \quad (33)$$

Clearly many different values of R and φ can lead to a $\cos\alpha$ distribution with the observed shape. Figure 18c) shows the region in the R-φ plane which, at the 90% C. L., is allowed by the data in figure 18b). Therefore R is actually quite poorly constrained by the data.

At present the theoretical calculations of the partial wave amplitudes have been done in the Born approximation, in which the S and D are both real. If one believes that this is correct then R is constrained to be quite small. On the other hand, if one allows a relative phase between the amplitudes, then the theory gives no guidance about how to constrain R. Conversely, table 9 shows that the prediction of R is quite uncertain, which means that figure 18 cannot be used to constrain ϕ.

There are two speculations which have been advanced about figure 18 parts b) and c). The first speculation is an Occam's razor approach. If there is a substantial S-wave amplitude, then it must be an accident that the shape in b) agrees so well with the limit that there is no S-wave. People who do not believe in accidents can then argue that the S-wave component must be small. The second speculation was advanced by Falk and Peskin [61]. They start with equation 18 and introduce phase space factors of p^{2L+1}. Next, they recognize that, for small mixing angles, $\sin(\phi - \phi_0) = (\phi - \phi_0)$ while $\cos(\phi - \phi_0) = 1$. Therefore, to first order in the mixing angle, these equations predict, with no free parameters, the D-wave partial width of the D_1 in terms of the total width of the D_2^*. Using their method and $\Gamma(D_2^0) = 23 \pm 5$ MeV [2] gives the prediction $\Gamma_D(D_1) = 6.2 \pm 1.4$ MeV, which is about 3 standard deviations from the measured total width of the D_1, 18^{+5}_{-4} MeV [2]. Taken at face value, this suggests that the S-wave partial width of the D_1 is larger than its D-wave partial width.

It is formally straightforward to resolve the question about the S-wave to D-wave ratio in these decays. One needs to consider three angles. The first angle is α and the other two are θ and χ, defined below. In the above decay chain, θ is the angle between the direction of the π_1 and the direction of the lab frame, when both are measured in the rest frame of the D_1. And χ is the angle between the production and decay planes of the D^*, measured in the rest frame of either the D^* or the D_1. By looking at various projections of the distributions in these angles one can disentangle R, φ and ρ_{00}. This last variable is the fraction of D_1 mesons with helicity zero, as measured in the lab frame. It may be possible to obtain a significant result with the existing data and it will certainly be possible when several times more data are available.

6 The future

Over the next 10 years, there will be many advances in heavy flavour spectroscopy, particularly in charm spectroscopy. For example, many of the existing CLEO results used only 40% to 60% of their final data sample. Updates and new results will be forthcoming as the final data set is passed through the main track-reconstruction program. Similarly, WA89 and E687 have not yet published final results on all of their data. In October 1995, CLEO will begin taking a new data set, for which the innermost tracking system will be replaced by a silicon microstrip detector. This represents a substantial upgrade, particularly for decays involving the long-lived D^+.

In the longer term, three B-factories are under construction, BaBar at SLAC, Belle at KEK and CLEO III at Cornell. Because these experiments will focus on weak decays of B mesons produced in the decays of the $\Upsilon(4S)$, they will not contribute to b hadron spectroscopy. However, in the next 10 years, each of these experiments will reconstruct more than 20 times as many charmed hadrons as have been reconstructed to date. Moreover these decays will be observed using detectors which are better than the existing ones. Not only will the existing states be measured more precisely but many of the missing states will be observed. For example, the $j_\ell = 1/2$ P-wave charmed mesons, which are expected to be very broad, will be observed in the Dalitz plots of the decays $B \to D\pi\pi$ and $B \to D^*\pi\pi$. There is also some hope for seeing the broad D_s

P-wave states by looking in at the Dalitz plots for $B \to D^{(*)}\overline{D}^{(*)}K$. There are excellent prospects for finding the missing singly charmed S-wave baryons. Some of the P-wave charmed baryons are expected to have widths on the order of tens of MeV and these should also be observable. The subjects which show little promise are to observe states with more than one unit of orbital angular momentum between the quarks or states with more than one heavy flavour, such as $b\bar{c}$ or ccq

In a few years there will be another run of the Fermilab fixed target program, with excellent prospects for new contributions to charm spectroscopy. Any future runs of the CERN fixed target program are also likely to produce new results.

The near term future of bottom hadron spectroscopy also looks good. Preliminary results from LEP show indications of signals for the P-wave $b\bar{q}$ and $b\bar{s}$ mesons and there is some indication of the Σ_b and the Σ_b^*. Also the hadron collider groups are working hard to observe the P-wave B mesons to use as flavour tags in $B\overline{B}$ mixing and CP violation studies. Almost certainly the signals are there, and, with the help of silicon microstrip detectors, there is some hope of seeing them. On the down side, the LEP program will likely be cut short before accumulating enough luminosity to make much progress in the bottom baryon sector. Finally, the longer term status depends on how well the current program at the hadron colliders succeeds. There are plenty of bottom hadrons around, but they are usually accompanied by large backgrounds.

7 Acknowledgments

I would like to thank David Bugg for inviting me to speak at this institute, and for putting up with my tardiness in preparing these proceedings. He and the other organizers provided us all with the wonderful experience of discussing interesting physics, at a comfortable pace and in pleasant surroundings. The other speakers did a wonderful job of opening my eyes to several areas of physics which I have not had the opportunity to follow over the years. Finally, I would also like to thank the students for their questions which help to keep the field exciting.

References

[1] CLEO II Collab., J. Gronberg *et al.*, Preprint CLNS 95/1346,CLEO 95-10,EPS0177, CLEO (unpublished).

[2] J. Bartelt and S. Shukla, Ann. Rev. Nucl. Part. Sci. **45**, 83 (1995).

[3] CLEO II Collab., P. Avery *et al.*, Preprint CLNS 95/1352,CLEO 95-14,EPS0172, CLEO (unpublished).

[4] E. Chudakov, talk at the Conference on Heavy Quarks at Fixed Targets, Virgina, 1994.

[5] CLEO II Collab., K. W. Edwards *et al.*, Phys Rev. Lett. **74**, 3331 (1995).

[6] Particle Data Group, Phys. Rev. **D50**, 1173 (1994).

[7] M. K. Gaillard, B. W. Lee, and J. L. Rosner, Rev. Mod. Phys. **47**, 277 (1975).

[8] CLEO II Collab., F. Butler *et al.*, Phys. Rev. Lett. **69**, 2041 (1992).

[9] N. Isgur, in *The New Aspects of Subnuclear Physics; Proceedings of the XIV International School of Subnuclear Physics*, edited by A. Zichichi (Plenum, New York, 1978).

[10] F. E. Close, *An Introduction to Quarks and Partons* (Academic Press, New York, 1979).

[11] A. Rújula, H.Georgi, and S. Glashow, Phys. Rev. **D12**, 147 (1975).

[12] W. Kwong, J. L. Rosner, and C. Quigg, Ann. Rev. Nucl. Part. Sci. **37**, 325 (1987).

[13] K. Wilson, Phys. Rev. **D10**, 2445 (1974).

[14] J. Kogut and L. Susskind, Phys. Rev. **D9**, 3501 (1974).

[15] T. Barnes, private communication.

[16] S. Godfrey and N. Isgur, Phys. Rev. **D32**, 189 (1985).

[17] S. Capstick and N. Isgur, Phys. Rev. **D34**, 2809 (1986).

[18] J. Rosner, Comm. Nucl. Part. Phys. **16**, 109 (1986).

[19] N. Isgur and G. Karl, Phys. Rev. **D18**, 4187 (1978).

[20] S. Godfrey and R. Kokoski, Phys. Rev. **D43**, 1679 (1991).

[21] ARGUS Collab., H. Albrecht *et al.*, Phys. Rev. Lett. **56**, 549 (1986).

[22] ARGUS Collab., H. Albrecht *et al.*, Phys. Lett. **230B**, 162 (1989).

[23] E691 Collab., J.C. Anjos *et al.*, Phys. Rev. Lett. **62**, 1717 (1989).

[24] CLEO II Collab., T. Bergfeld *et al.*, Phys. Lett. **340B**, 194 (1994).

[25] N. Isgur and M. Wise, Phys. Lett. **232B**, 113 (1989).

[26] N. Isgur and M. Wise, Phys. Lett. **237B**, 527 (1990).

[27] R. Barbieri *et al.*, Nucl. Phys. **B105**, 125 (1976).

[28] D. Pignon and C. Piketty, Phys. Lett. **81B**, 334 (1979).

[29] E. Eichten *et al.*, Phys. Rev. **D21**, 203 (1980).

[30] J. Morishita *et al.*, Phys. Rev. **D37**, 159 (1988).

[31] V. Gupta and R. Kögerler, Zeit. Phys. **C41**, 277 (1988).

[32] A. Kaidalov and A. Nogteva, Sov. J. Nucl. Phys. **47**, 321 (1988).

[33] K. Maltman and N. Isgur, Phys. Rev. **D22**, 1701 (1980).

[34] L. A. Copley, N. Isgur, and G. Karl, Phys. Rev. **D20**, 768 (1979).

[35] J. M. Richard and P. Taxil, Phys. Lett. **128B**, 453 (1983).

[36] ARGUS Collab., H. Albrecht *et al.*, PL **317B**, 227 (1993).

[37] E687 Collab., P. L. Frabetti *et al.*, Phys. Rev. Lett. **72**, 961 (1994).

[38] L. Micu, Nucl. Phys. **B10**, 521 (1969).

[39] A. L. Yaouanc *et al.*, Phys. Rev. **D8**, 2223 (1973).

[40] A. L. Yaouanc *et al.*, Phys. Rev. **D9**, 1415 (1974).

[41] A. L. Yaouanc *et al.*, Phys. Rev. **D8**, 2223 (1973).

[42] J. P. Ader *et al.*, Zeit. Phys. **C5**, 85 (1980), note: Equation (12) has L_A and S_A interchanged.

[43] R. Kokoski and N. Isgur, Phys. Rev. **D35**, 907 (1987).

[44] E. J. Eichten, C. T. Hill, and C. Quigg, Phys. Rev. Lett. **71**, 4116 (1993).

[45] N. Isgur and M. Wise, Phys. Rev. **D66**, 1130 (1991).

[46] A. F. Falk and M. Luke, PL **292B**, 119 (1992).

[47] P. Cho and S. P. Trivedi, PR **D50**, 381 (1994).

[48] Particle Data Group, Phys. Rev. **D45**, S1 (1992).

[49] CLEO II Collab., D. Bortoletto *et al.*, Phys. Rev. Lett. **69**, 2046 (1992).

[50] P. Cho and M. B. Wise, PR **D49**, 6228 (1994).

[51] ARGUS Collab., H. Albrecht *et al.*, Nucl. Instr. Methods **A275**, 1 (1989).

[52] CLEO II Collab., Y. Kubota *et al.*, Nucl. Instr. Methods **A320**, 66 (1992).

[53] B. Andersson *et al.*, Phys. Rep. **97**, 31 (1983).

[54] CLEO II Collab., J. Alexander *et al.*, Phys. Rev. Lett. **74**, 3113 (1995), the number quoted in the text comes from an erratum which has been submitted to Phys. Rev. Lett.

[55] CLEO II Collab., P. Avery *et al.*, Phys. Lett. **331B**, 236 (1994).

[56] R. Brandelik *et al.*, PL **70B**, 132 (1977).

[57] ARGUS Collab., H. Albrecht *et al.*, PL **146B**, 111 (1984).

[58] CLEO II Collab., D. N. Brown *et al.*, Phys. Rev. **D50**, 1884 (1994).

[59] CLEO II Collab., Y. Kubota *et al.*, Phys. Rev. Lett. **72**, 1972 (1994).

[60] CLEO II Collab., J. P. Alexander *et al.*, Phys. Lett. **303B**, 377 (1993).

[61] A. F. Falk and M. E. Peskin, PR **D49**, 3320 (1994).

DECAY AND PRODUCTION OF FLUX–TUBE EXCITATIONS IN MESONS

Philip R. Page

Theoretical Physics, University of Oxford
1 Keble Road, Oxford OX1 3DW, U.K.

The organization is as follows. We start by developing flux–tube decay and production dynamics in an *a priori* fashion, and then briefly indicate its foundation in QCD. We then specialize to the flux–tube model and indicate various selection rules. We proceed to highlight the experimental candidates and their signatures at current and future facilities.

FLUX–TUBE DYNAMICS

The chromo-electric flux–lines emanating from a coloured quark and absorbed by a neighbouring antiquark is known to confine the system. This is not achievable when flux–lines spread out equally in space around a single charge, as is the case in classical electromagnetism. The opposite extreme where the flux-lines between the quark and antiquark are compactified into a *flux–tube*, however, means that the attraction force is constant as the interquark distance is traversed, and the energy of dissociation hence infinite : exactly what we need for the experimental datum of *confinement*. We hence adopt the idea of a chromo-electric flux–tube connecting a quark and antiquark.

Having understood that *mesons* can be regarded as $Q\bar{Q}$ systems connected by a flux–tube, it is natural to consider quantum *excitations* of these systems, which we shall call *hybrids*.

In order to simplify the discussion as a first orientation we would like to conceptually separate the dynamics of the "quark" and "flux" components of the system. This is called the *adiabaticity* assumption, and is valid for large quark masses. As the quarks move, the flux–tube is considered to re-assemble itself spontaneously; and hybrids are $Q\bar{Q}$ systems with an excited flux–tube. There would in principle be an infinite tower of excitations, but we shall specialize to the energetically lowest lying one in this lecture. The flux–tube can possess angular momentum Λ around the $Q\bar{Q}$–axis

r_A, which is conserved as the flux–tube must *a priori* be invariant under rotations around the $Q\bar{Q}$–axis. This system hence has *three* quantum numbers : the usual angular momentum L with its projection M *as well as* Λ. The only mathematical function of the direction variables θ, ϕ defining the direction of the $Q\bar{Q}$–axis, which is an eigenfunction of the above three conserved operators, is the Wigner rotation function $\mathcal{D}^L_{M\Lambda}$. It then accordingly gives the angular dependence of the wave function of the system. The quarks move adiabatically in an effective potential generated by the flux–tube dynamics, and hence obeys a Schrödinger equation which determines the dependence on $Q\bar{Q}$ separation of the wave function of the system.

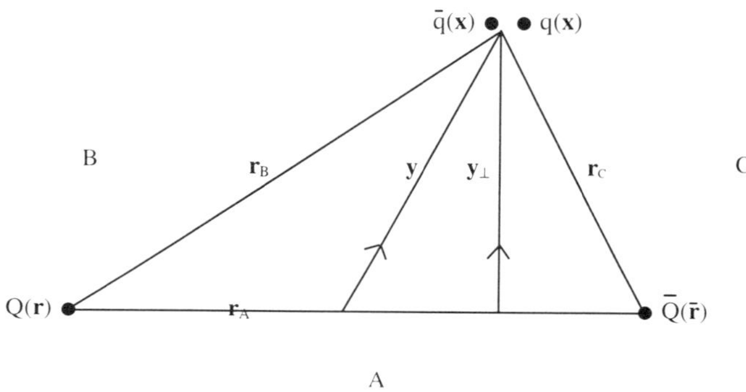

Figure 1: The topology of flux–tube decay and production

We now consider how these systems might decay. For a two–body decay it would be natural to assume the creation of a $q\bar{q}$ pair. If we constrain ourselves to a *local* theory, the $q\bar{q}$ pair must be created at *one* point y (see figure 1). This point is also called the point of "flux–tube breaking" because a very thin tube of flux for the incoming state A would "break" into two similar flux–tubes for the outgoing states B and C. In general, however, the flux–tube would be expected to have a definite *thickness*. According to the adiabaticity assumption, the initial flux–tube would have to re–assemble into the flux–tubes of the two final states. This would have a certain re–arrangement amplitude, which we call the *flux–tube overlap* $\gamma(r_A, y)$.

The $q\bar{q}$ pair creation operator is naïvely not expected to add angular momentum to the system, since we would expect it to be rotationally invariant and derivable from a Lorentz invariant operator. Thus $J_{q\bar{q}} = 0$. The simplest such operator is $\bar{\psi}\psi = \psi^+\gamma_0\psi$ which can be expanded to give an operator of the form $\boldsymbol{\sigma} \cdot p$, noting that γ_0 is off–diagonal and $\boldsymbol{\sigma}$ are Pauli matrices. Here the $q\bar{q}$ pair has spin $S_{q\bar{q}} = 1$ (due to $\boldsymbol{\sigma}$) and orbital angular momentum $L_{q\bar{q}} = 1$ (due to p). The decay process is hence called "$^3P_0$ pair creation". Here $\vec{J}_{q\bar{q}} = \vec{S}_{q\bar{q}} + \vec{L}_{q\bar{q}}$; and we use the condensed notation $^{2S+1}L_J$.

$^3P_0$ pair creation can arise from the strong coupling expansion of the Hamiltonian formulation of lattice gauge theory [1] or from the instantaneous colour Coulomb interaction [2] and is phenomenologically surprisingly successful [3]. We adopt it.

By the conservation of spin (since the decay operator creates a $S_{q\bar{q}} = 1$ pair) we obtain the following selection rule :

(A) Decays of net spin $S = 0$ states to two $S = 0$ states are forbidden.

The $^3P_0$ decay amplitude can be formulated most intuitively in the momentum frame with the $q\bar{q}$–pair being created with equal but opposite momenta k_3, k_4 as

$$\int d^3k_1\, d^3k_2\, d^3k_3\, d^3k_4\, \delta^3(k_1 + k_2 - p_A)\, \delta^3(k_1 + k_3 - p_B)\, \delta^3(k_2 + k_4 - p_C)\, \delta^3(k_3 + k_4)$$
$$\times\, \boldsymbol{\sigma} \cdot (k_3 - k_4)\, \psi_A(\frac{k_2 - k_1}{2})\, \psi_B^*(\frac{k_3 - k_1}{2})\, \psi_C^*(\frac{k_2 - k_4}{2}) \qquad (1)$$

where k_1, k_2 represent the momenta of the incoming quarks and care has been taken to constrain the constituent momenta k_i to equal the meson momenta $p_{A,B,C}$. The factor $p = k_3 - k_4$ is the relative momentum of the $q\bar{q}$ pair. The expression in Eq. 1 can be shown by change of variables [4],[5][Appendix A.1] to equal (for a flux–tube overlap $\gamma(r_A, y) = 1$)

$$(2\pi)^3\delta^3(p_B + p_C) \int d^3r_A\, d^3y\, \psi_A(r_A) \exp(\frac{i}{2}p_B \cdot r_A)\, \gamma(r_A, y)$$
$$\times\, \boldsymbol{\sigma} \cdot (2i\boldsymbol{\nabla}_{r_A} + p_B)\, \psi_B^*(\frac{r_A}{2} + y)\psi_C^*(\frac{r_A}{2} - y) \qquad (2)$$

in position space, where we define the Fourier transformed wave functions $\psi_{A,B,C}(r) = (2\pi)^{-3} \int d^3k\, \exp(ik \cdot r)\psi_{A,B,C}(k)$; and restrict $p_A = 0$. We shall now generalize to an arbitrary function $\gamma(r_A, y)$.

We have thus discussed the generic formulation of decay dynamics in an adiabatic and non–relativistic context with $^3P_0$ pair creation. For more information on the flux–tube overlap we specialize to a specific model.

THE FLUX–TUBE MODEL

A detailed form of the flux–tube overlap has successfully been obtained in the Isgur-Paton non–relativistic flux–tube model of QCD [6], which is almost the *only* non–perturbative context in which decays and production of flux–tube excitations have been discussed in the literature. We shall briefly state two important properties without deriving them.

1. For *meson decays to two mesons* the overlap reduces to a constant for infinitely thick flux–tubes, yielding the old and phenomenologically successful [3] $^3P_0$–model amplitude of Eq. 1. Flux–tube model predictions with flux–tubes of finite thickness

differ insignificantly [1, 5] from the $^3P_0$–model, effectively disguising the presence of the flux–tube : this explains the success of quark–model ideas for conventional mesons.

2. For *hybrid decays to two mesons* the overlap is odd under the transformation $y \rightarrow -y$, and hence pair creation on the incoming $Q\bar{Q}$–axis is *forbidden*.

The second property has important experimental consequences. If the final states are identical S–wave states, the wave functions carry no angular momentum indices and are thus equal : $\psi_B = \psi_C$. The function in Eq. 2 inside the integral is then odd under $y \rightarrow -y$ (due to the oddness of $\gamma(r_A, y)$), yielding a vanishing decay amplitude on y–integration. Thus the decay amplitude is proportional to the difference of the wave functions of the outgoing mesons. This difference is usually small [7] leading to a selection rule :

(B) Decays of hybrids to two S–wave mesons are suppressed.

This selection rule can heuristically be understood as follows. Heavy quark lattice gauge theory (HQLGT) [8] suggests that the lowest lying hybrids have $\Lambda = 1$, while mesons have $\Lambda = 0$. There is thus no available angular momentum amongst the outgoing mesons to absorb the $\Lambda = 1$ of the incoming hybrid.

Rule B has the immediate corollary

(C) Decays to one $L = 1$ and one $L = 0$ meson are the preferred decay channel for hybrids.

This is due to significant amounts of phase space often available and easier detection relative to even higher L final states.

In the case of *production*, where an exchanged π, ρ or ω is involved, it is possible that the strength of production could be significant at least to the extent that the exchanged off-mass-shell state may have different structure to the incident on-shell beam particle. This is because rule B is invalidated, making this method of production important. Hence :

(D) Meson exchange can be a significant production mechanism for hybrid mesons.

A specific example is the *photoproduction* of hybrids where the interaction $\gamma +$ *meson* $\rightarrow$ *hybrid* offers unique opportunities. The photon γ can be regarded as a linear combination of ρ, ω, ϕ mesons (with $J^{PC} = 1^{--}$) by vector meson dominance. The photon can be produced directly or in e^+e^-–annihilation.

To complete the discussion of the flux–tube model, we mention the J^{PC} quantum numbers of the *lowest lying* hybrid mesons, which are the same as in HQLGT [8]. They may be divided into two classes :

(a) $0^{-+}, 1^{+-}, 1^{--}, 1^{++}, 2^{-+}$ (deemed "conventional" in that they can also be shared by standard $q\bar{q}$ states), and

(b) $0^{+-}, 1^{-+}, 2^{+-}$ (deemed "exotic").

In the hybrid 1^{--} and 1^{++}, $S = 0$; all other J^{PC} have $S = 1$.

THE HYBRID HUNTER'S GUIDE

A central theme in this section will be to apply the preceding rules to a case study. We shall see that even when hybrid and conventional mesons have the *same* J^{PC} quantum numbers, they may be distinguished. The essential reason is that although superficially identical in their overall quantum numbers, the two states have different internal structures. We illustrate this with particular reference to the vector meson "$\rho_1(1460)$" [9], usually denoted as $\rho(1450)$, whose decays typify those of 1^{--} hybrid dynamics, making it a strong hybrid candidate.

(i) For $J^{PC} = 1^{--}$, rule A distinguishes between conventional vector mesons which are $^3S_1$ or $^3D_1$ states and hybrid vector mesons where the $q\bar{q}$ have $S_{q\bar{q}} = 0$. This implies that in the decays of hybrid ρ, the channel πh_1 is forbidden whereas πa_1 is allowed and that πb_1 is analogously suppressed for hybrid ω decays; this is quite opposite to the case of $^3L_1$ conventional mesons where the πa_1 channel is relatively suppressed and πh_1 or πb_1 are allowed [1]. The extensive analysis of data in Ref. [9] revealed the clear presence of $\rho(1450)$ with a strong πa_1 mode but no sign of πh_1, in accord with the hybrid hypothesis.

(ii) Applying rules B and C, we observe that in decays of hybrid $\rho \to 4\pi$ the πa_1 content is predicted to be dominant and the $\rho\rho$ to be absent. The analysis of Ref. [9] again finds such a pattern for $\rho(1450)$.

(iii) Rule B no longer operates if the internal structure or size of the two $L = 0$ states differ, as noted before. Thus, for example, decays to $\pi + \rho$, $\pi + \omega$ or $K + K^*$ may be significant in some cases [7]. Though still suppressed relative to the dominant pathway, (ii) above, this too is the case for $\rho(1450)$.

(iv) Couplings to $\rho\omega$ or $\rho\pi$ could be considerable when the ρ is effectively replaced by a photon and the ω or π is exchanged in photoproduction, illustrating rule D. We hence would advocate searching for the lightest 1^{--} hybrids in diffractive photoproduction : $\gamma(p) \to \pi a_1(p) \to 4\pi(p)$.

LIGHT HYBRID CANDIDATES

1^{--}

Continuing the previous section, we predict [7] for a vector hybrid $\rho(1450)$ the widths (in MeV)

$$\pi a_1 : \pi a_2 : \pi h_1 : \rho\rho : \pi\omega : \pi\pi \;=\; 140 : \sim 0 : 0 : 0 : 5 - 10 : 0 \tag{3}$$

These are very different from the predictions of radial or $^3D_1$ decays of quarkonia [1]. In particular the suppression of πh_1 relative to πa_1 alluded to in the previous section, is a crucial test of the hybrid initial state. It is therefore interesting that the detailed analyses of experimental data in Ref. [9] comment on the apparently anomalous decays that they find for the 1^{--} state $\rho(1450)$, in particular the suppression of πh_1 relative to πa_1 and the dominance of the latter over the $\pi\omega$:

$$\pi a_1 : \pi h_1 + \rho\rho : \pi\omega : \pi\pi \;=\; 190 : 0 - 39 : 50 - 80 : 17 - 25 \tag{4}$$

It is noticeable that the $\pi\pi$ decay also is strongly suppressed though non–zero; if this is substantiated it could indicate either a deviation from the flux–tube model or in addition some mixing between hybrid and radial 1^{--} mesons in this region. The latter could also rather naturally explain the enhancement of the $\pi\omega$ channel as well as the repulsion of the eigenstate to low mass.

Ref. [9] also finds evidence for $\omega(1440)$ with no visible decays into πb_1 which is in significant contrast to the expectations for conventional $Q\bar{Q}$ ($^3S_1$ or $^3D_1$) initial states.

In the hybrid interpretation, this suppression is natural and is the isoscalar analogue of the πh_1 selection rule alluded to above.

If the $\rho(1450)$ has signposted the existence of the 1^{--} hybrid nonet, then we need to establish which of the other seven multiplets should also be visible. States whose couplings are predicted to be strong, with highly visible decay channels and moderate widths relative to the $\rho(1450)$ candidate, *must* be seen if hybrids are to be established. Conversely, channels where no signals are seen should be those whose signals are predicted by the flux–tube model to be weak. These criteria *do* appear to be realised in the data, as shall now be discussed.

For conventional J^{PC} hybrids at $\sim$ 2 GeV made from u, d flavoured quarks the $1^{+-}, 1^{++}$ states are over 500 MeV wide [7] in both I=0,1 states; by contrast the $0^{-+}, 2^{-+}$ and the 1^{--} are predicted to be potentially accessible. We now discuss the former two states.

0^{-+}

The VES Collaboration sees an enigmatic and clear 0^{-+} signal in diffractive π production [10]. They study the channels $\pi^- N \to \pi^-\pi^+\pi^- N$; $\pi^- K^+ K^- N$; $\omega\pi^-\pi^0 N$; $\eta\eta\pi^- N$ and see a resonant signal of mass 1770 MeV and width 190 MeV in the classic $(L = 0) + (L = 1)$ $\bar{Q}q$ channels $\pi^- + f_0(1400)$; $K^- + K_0^*$ with no corresponding strong signal in the allowed $L = 0$ two body channels $\pi + \rho$; $K + K^*$. The width and large couplings to kaons (in the dominant [10] decay channels $f_0(980)\pi^-$ and $a_0^-\eta$) are both surprising if this were the second radial excitation of the pion (the first radial excitation is seen as a broad enhancement in accord with expectations). Furthermore, the apparent preference for decay into $(L = 0) + (L = 1)$ mesons at the expense of $L = 0$ pairs is qualitatively in accord with expectations for hybrids.

Our quantitative estimates [7] on the relative importance of available channels further suppport this identification. We find widths to $\pi f_0(1300)$ of $\sim$ 170 MeV and to πf_2 of $5 - 10$ MeV. The KK_0^* channel is kinematically suppressed though probably non–zero due to the $\sim$ 300 MeV width of the $K_0^*(1430)$. The decay to $L = 0$ pairs, which is naively expected to be suppressed, turns out to be potentially significant, $\pi\rho \sim 30$ MeV. This is compatible with the experimental limit [10]

$$\frac{0^{-+} \to \pi^-\rho^0}{0^{-+} \to \pi^- f_0(1300)} < 0.07 \tag{5}$$

The $KK^*(890)$ channel, by contrast, is expected to be a mere $\sim$ 5 MeV, which is consistent with the observed order of magnitude suppression observed in Ref. [10]

$$\frac{0^{-+} \to K^- K^*}{0^{-+} \to (K^- K^+ \pi)_S} < 0.1 \tag{6}$$

In addition it is found [7] that the total width is in accordance with expectations. Important tests are now that there should be a measureable coupling to the $\pi\rho$ channel with only a small πf_2 or KK^* contribution.

Of special interest is the observation of decays into $f_0(1500)\pi$, since $f_0(1500)$ is a strong glueball candidate. If sustained, this would indicate the presence of a "gluon–rich" decay mode.

The 0^{-+} may be prominent in low energy photoproduction where π exchange is important but its $\omega\gamma$ coupling is suppressed by a selection rule [7] for hybrids which will disfavour the 0^{-+} photoproduction at higher energies where ω–exchange is more dominant.

$\underline{2^{-+}}$

The prediction [7] for the 2^{-+} VES partner to the 0^{-+} is that the πf_2 channel should dominate significantly over the πf_0 channel. This is a problem if one wishes to identify the isovector 2^{-+} seen at ~ 2.2 GeV at VES as the hybrid partner of the 0^{-+}. The putative signal is claimed in $\pi f_0(1300)$ whereas no πf_2 nor $\pi f_0(980)$ are reported. The properties and existence of this state are less clearcut experimentally. If $\pi f_0(1300) \geq \pi f_2(1270)$ is sustained for this state, it is not a hybrid.

Historically the ACCMOR Collaboration [11] noted a 2^{-+} structure around 1.8 GeV, coupled to πf_2, and too near to the $\pi_2(1670)$ for these to be the 1^1D_2 and 2^1D_2 (radial excitation) of conventional quarkonium. This structure is tantalisingly similar to sightings of a possible 2^{-+} (or even 1^{-+}) at 1.77 GeV, width 100-200 MeV in photoproduction via π exchange [12] and coupled to $\pi\rho$ and πf_2. An earlier low energy photoproduction experiment [13] also shows a clear structure at 1.7 - 1.9 GeV though its quantum numbers are not identified; we note that hybrid $2^{-+}, 1^{-+}$ are both favourably photoproduced [7] *via* π exchange or ω exchange. If these various experiments are heralding activity in the 2^{-+} isovector wave, a search for the πb_1 decay channel becomes pivotal. This follows once again from rule A : this prevents the decay of ${}^1D_2(\pi_2) \to b_1\pi$ whereas this channel is allowed and potentially significant for a hybrid π_2 [7].

There are also indications of a doubling of states in the $I = 0$ $\eta\pi^0\pi^0$ channel where the Crystal Barrel at LEAR [14] finds both $\eta_2(1645)$ (which is probably the partner of $\pi_2(1650)$) and also a candidate $\eta_2(1875)$ decaying into $f_2\eta$ but not $a_2\pi$ (unlikely to be $s\bar{s}$). Our calculation [4, 7] predicts the widths (in MeV)

$$\pi a_2 : \eta f_2 : K^*K \;=\; 160 : \sim 20 : 1 \qquad (7)$$

so that the total width is consistent with the 200 MeV observed; but the πa_2 channel should now be sought.

$\underline{1^{-+}}$

The most obvious signature for a hybrid meson is the appearance of a isospin 1 state with an exotic combination for J^{PC}. It was noted in Ref. [15] and confirmed in Ref. [7] that the 0^{+-} width is predicted to be over 1500 MeV thereby rendering the state effectively invisible. The 2^{+-} is also predicted to be very broad and hard to see if its mass is ≥ 1.9 GeV. The best opportunity for isolating exotic hybrids appears to be in the 1^{-+} wave where the AGS at Brookhaven [13] may have indications for such an isospin 1 state whose mass and decay characteristics are in line with expectations. The search was motivated by rule C and concentrated on the decay channel $\pi + f_1$, which is where the candidate has been sighted. The experiment sees a broad structure in the mass region $1.6-2.2$ GeV which is suggestive of being a composite of two objects at 1.7 and 2.0 GeV. It is the latter that appears to have a resonant phase. The current data suffers from low statistics, making its existence in need of independent corroboration.

Our expectations [7] for widths are (in MeV)

$$\begin{aligned} \pi f_1 : \pi b_1 : \eta a_1 : \pi\eta(1295) &= 60 : 170 : 30 : 30 \\ \pi\rho : \eta\pi : \eta'\pi : \rho\eta &= 5-20 : 0-10 : 0-10 : 0-10 \end{aligned} \qquad (8)$$

We note the possible presence of $\pi\rho$ or even $\pi\eta$ decays that are not negligible relative to the signal channel πf_1, due to significant final state wave function differences softening rule B. This may be important in view of a puzzle, commented upon in Ref. [16], that the production mechanism appeared not to be as expected given the anticipated hybrid dynamics. Instead of b_1–exchange, leading to the $\pi + b_1(1235)$ coupling according to rule C, significant $\pi + \rho$ coupling may be responsible. In view of Eq. 8 above, it is clear that the latter coupling may be significant on the scale of the πf_1 signal; the final state decays into $\pi + \rho$ should therefore also be investigated experimentally.

If the current BNL data are sustained, a problem emerges for decays to $\eta_u(1295)\pi$. Estimates [4] for the branching ratios to the final state $(K^+\bar{K}^0\pi^-)\pi^-$ selected by the experiment, seem to indicate a larger signal in $\eta_u(1295)\pi$ (~ 2 MeV) than in $f_1(1285)\pi$ (~ 1.5 MeV), contrary to what was observed.

CHARMONIUM HYBRIDS

The reason for interest in *charmonium* hybrids "H_c" derives from the expectation that their masses are better defined than is the case for their light quark counterparts enabling firmer predictions and, due to the smaller amount of phase space available in the corresponding decay channels, their widths are smaller.

Above the $D^{**}D$ theshold hybrids decay preferentially into $D^{**}D$ [17] according to rule C. For H_c below the $D^{**}D$ threshold the flux–tube model predicts [4] very small widths by rule B (decays into DD, D^* D and D^* D^* are almost forbidden). H_c is predicted to exist at 4.1 - 4.2 GeV [3, 8], above the DD, D^* D, D^* D^* thresholds, but below the $D^{**}D$ thesholds. The dominant decays may hence be through mixing between H_c and conventional $c\bar{c}$. The simplest explanation is that these states are roughly 50 : 50 mixtures of the $c\bar{c}$ state $\psi(3S)$ existing in this mass region [18] and an "inert" H_c state. A mixing of 50% can arise naturally if there is mass degeneracy between two "primitive" states. The possibility of such a mixing between the primitive $3S$ $c\bar{c}$ at 4100 MeV and a H_c state, close by in mass, has recently been suggested [18]. If such degeneracy occurs one immediately expects that the physical eigenstates will tend to be

$$\psi_\pm \simeq \frac{1}{\sqrt{2}}(\psi(3S) \pm H_c)$$

In such an eventuality both the dominant production in e^+e^- annihilation and the prominent hadron decays will be driven by the $\psi(3S)$ component, leading to intimate relationships [18] between the properties of the two eigenstates which are then identified as $\psi(4160) \equiv \psi_+$ and $\psi(4040) \equiv \psi_-$.

These predictions may be tested at SPEAR or at a future Tau-Charm Factory. Beijing e^+e^- annihilation experiments may also soon have high enough luminosity to produce $J^{PC} = 1^{--}$ states above the DD threshold.

Bottomonium hybrids : Bottomonium hybrids are expected to be more difficult to produce than charmonium hybrids. Detection in e^+e^- annihilation of a $J^{PC} = 1^{--}$ hybrid above the BB threshold at the SLAC B-factory could be a possibility, though.

Acknowledgement : I am grateful for an invitation from the director Prof. D. Bugg to participate in a most pleasant, stimulating and timely summer school; and thank NATO for funding as well as the organizers for creating such an excellent atmosphere.

Further reading (overviews) :

1. T. Barnes, "Signatures for Hybrids", ORNL-CCIP-93-14, RAL-93-069, in *Conf. on Exclusive Reactions at High Momentum Transfers*, Elba (1993).

2. N. Isgur, "Hadron Spectroscopy", in *Hadrons and Hadronic Matter*, ed. D. Vautherin *et al.*, Plenum Press (1990).

3. J. Paton, *Nucl. Phys.* **A446** (1985) 419; *ibid.* **A508** (1990) 377.

References

[1] R. Kokoski and N. Isgur, *Phys. Rev.* **D35** (1987) 907.

[2] T. Barnes, *Proc. of HADRON95*, Manchester (1995).

[3] T. Barnes, F.E. Close and E.S. Swanson, RAL-94-106, hep-ph/9501405.

[4] P.R. Page, *D.Phil. Thesis*, Univ. of Oxford (1995).

[5] P.R. Page, *Nucl. Phys.* **B446** (1995) 189.

[6] N. Dowrick, J. Paton and S. Perantonis, *J. Phys.* **G13** (1987) 423; S.J. Perantonis, *D. Phil. Thesis*, Univ. of Oxford (1987).

[7] F.E. Close and P.R. Page, *Nucl. Phys.* **B443** (1995) 233; *ibid.*, *Phys. Rev.* **D52** (1995) 1706.

[8] C. Michael *et al.*, *Nucl. Phys.* **B347** (1990) 854; *Phys. Lett.* **B129** (1983) 351.

[9] A. Donnachie and Yu. Kalashnikova, *Z. Phys.* **C59** (1993) 621; A.B. Clegg and A. Donnachie, *Z. Phys.* **C62** (1994) 455.

[10] VES Collaboration, A. Zaitsev, *Proc. of* 27th *Int. Conf. on High Energy Physics*, Glasgow, (1994) p. 1409; VES Collaboration, D Ryabchikov, *Proc. of HADRON95*, Manchester (1995).

[11] C. Daum *et al.* (ACCMOR Collaboration), *Nucl. Phys.* **B182** (1981) 269.

[12] D. Aston *et al.*, *Nucl. Phys.* **B189** (1981) 15;
G. Condo *et al.*, *Phys. Rev.* **D43** (1991) 2787.

[13] Y. Eisenberg *et al.* Phys. Rev. Lett. **23** 1322 (1969).

[14] A. Cooper, Ph.D. Thesis, Queen Mary and Westfield College, London University, 1994 (unpublished); Chrystal Barrel Collaboration, C. Pinder, *Proc. of HADRON95*, Manchester (1995); *ibid.* T. Degener.

[15] N. Isgur, R. Kokoski and J. Paton, *Phys. Rev. Lett.* **54** (1985) 869.

[16] J.H. Lee *et al., Phys. Lett.* **B323** (1994) 227.

[17] P.R. Page, hep-ph/9410323, *Proc. of "Quark Confinement and the Hadron Spectrum"*, 20-24 June 1994, Como, Italy.

[18] F.E. Close and P.R. Page, "Do ψ (4040), ψ (4160) signal hybrid charmonium?", OUTP-95-13P, RAL-95-035.

HADRONIC PHYSICS FROM THE LATTICE

Chris Michael

DAMTP, University of Liverpool, Liverpool, L69 3BX, UK

Abstract

We present the lattice gauge theory approach to evaluating non-perturbative hadronic interactions from first principles. We discuss applications to glueballs, inter-quark potentials, the running coupling constant, the light hadron spectrum and the pseudoscalar decay constant f_P.

1 Introduction

The theory of the strong interactions is accurately provided by Quantum Chromodynamics (QCD). The theory is defined in terms of elementary components: quarks and gluons. The only free parameters are the quark mass values (apart from an overall energy scale imposed by the need to regulate the theory). This formulation is essentially the unique candidate for the theory of the strong interactions. The only feasible way to describe a departure from QCD would be in terms of quark and gluon substructure. At least at the energy scales up to which it is tested, there is no evidence for such substructure.

QCD provides a big challenge to theoretical physicists. It is defined in terms of quarks and gluons but the physical particles are composites: the mesons and baryons. Any complete description must then yield these bound states: this requires a non-perturbative approach. One can see the limitations of a perturbative approach by considering the vacuum: this will be approximated in perturbation theory as basically empty with rare quark or gluon loop fluctuations. Such a description will allow quarks and gluons to propagate essentially freely which is not the case experimentally. The true (non-perturbative) vacuum can be better thought of as a disordered medium with whirlpools of colour on different scales. Such a non-perturbative treatment then has the possibility to explain why quarks and gluons do not propagate (ie quark confinement).

The main contender for a non-perturbative description of QCD is the lattice gauge

Hadron Spectroscopy and the Confinement Problem
Edited by D.V. Bugg, Plenum Press, New York, 1996

theory approach. Following the ideas of Wilson, space-time is replaced by a discrete grid (the lattice) but gauge invariance is retained exactly. Using also periodic boundary conditions in space and time, the system will have a finite number of degrees of freedom: the gluon and quark fields at the lattice sites. Actually gauge invariance implies that the gluon field lies on the links of the lattice (they correspond to the mechanism that allows colour orientation at different sites to be compared). This finite number of degrees of freedom implies that the theory is a quantum many-body problem rather than a field theory.

The step which makes this many-body problem tractable is to consider Euclidean time. This is perhaps the step which is the most difficult to assess. Formally the quantities to be evaluated can be expressed as Green functions: vacuum expectation values of products of fields. The existence and properties of these Green functions under a Wick rotation to Euclidean time are widely used in perturbative treatments. A non-perturbative determination of the Green functions in the Euclidean time case will lead directly to quantities of relevance in the physical Minkowski time case (eg masses and matrix elements).

With the Euclidean time approach, the formulation of QCD (consider, for example, the functional integral over the gauge fields) is converted into a multiple integral which is well defined mathematically. For a lattice of L^4 sites with a colour gauge group of SU(3), this would be a $8 \times 4 \times L^4$ dimensional integral (8 from the colour group manifold, $4L^4$ from the gauge fields on each link). For any reasonable value of L, this is a very high dimension indeed. Simpson's rule is not the way forward! The standard approach is to use a Monte Carlo approximation to the integrand. This is implemented in an "importance sampling" version so that a stochastic estimate of the integral is made from a finite number of samples (called configurations) of equal weight. The construction of efficient algorithms to achieve this is a topic in itself. Here I will concentrate on the analysis of the outcome, assuming that such configurations have been generated.

So what we have at our disposal is a set of samples of the vacuum. It is then straightforward to evaluate the average of various products of fields over these samples - this gives the Green functions by definition. The Green function can then be continued from Euclidean to Minkowski time (in most cases this is trivial) and compared to experiment.

The validation of the lattice approach calls for a series of checks that everything is under control.

- the lattice spacing should be small enough (discretisation errors)
- the lattice must be big enough in space and time (finite size errors)
- the statistical errors must be under control
- Green functions must be extracted with no contamination (eg a ground state mass could be contaminated with a piece coming from an excited state)
- both the quark contribution to the vacuum (sea quarks) and the quark constituents of hadrons (valence quarks) are usually treated by using larger mass values than the experimental ones and then extrapolating. This extrapolation must be treated accurately.
- a particularly severe approximation is to treat the sea-quarks as of infinite mass. This corresponds to neglecting quark loops in the vacuum and is computationally

very much faster. This is known as the "quenched approximation".

The most subtle of these is the discretisation error. In order to extract the continuum limit of the lattice, one must show that the physical results will not change if the lattice spacing is decreased further. This is subtle because the lattice spacing is not known directly - in effect, it is measured. The lattice simulation is undertaken at a value of a parameter conventionally called β. In the limit of small coupling g^2, where perturbation theory applies, $\beta = 6/g^2$. Thus large β corresponds to small g^2. Now, perturbatively, the coupling g^2 corresponds to the lattice spacing a as

$$g^2 \approx \frac{1.0}{b_0 \log a^{-2}} \quad \text{where} \quad b_0 = \frac{11 - \frac{2}{3}N_f}{16\pi^2}$$

for N_f flavours of quarks. For the case of interest, $N_f \leq 3$, this corresponds to small values of g^2 at small distance scale a - as expected from asymptotic freedom.

The perturbative argument is appropriate to the study of results at large β (small g^2). We will find that lattice simulation of QCD uses values of $\beta \approx 6$ and hence bare couplings corresponding to $\alpha = g^2/(4\pi) \approx 0.08$. In the pioneering years of lattice work, this was thought to be a sufficiently small number that the perturbation series would converge rapidly. One of the major advances, in recent years, has been the realisation that the bare lattice coupling (our g^2 above) is a very poor expansion parameter and the perturbation series in the bare coupling does not converge well at the β values of interest. The theoretical explanation for this poor convergence is that the lattice Lagrangian differs from the continuum Lagrangian and allows extra interactions. These include tadpole diagrams [11] which have the property that they sum up to give a contribution that involves high order terms in the perturbation series. The way to avoid this problem with tadpole terms is to use a perturbation series in terms of a renormalised coupling - rather than the bare lattice coupling. I return to this topic when discussing the lattice determination of α_S later.

This change of attitude to the method of determining a from β has had considerable implications for lattice predictions, as I now explain, since we wish to work in a region of lattice spacing a where perturbation theory in the bare coupling is not precise. Because of this, in practice, a is determined from the non-perturbative lattice results themselves. Thus if the energy of some particle is measured on the lattice, it will be available as the dimensionless combination $\hat{E}$. From a value for E in physical units, then a can be determined since, on dimensional grounds, $\hat{E} = Ea$. Furthermore, by increasing β, the change in the observable $\hat{E}$ gives information about the change in a since E is fixed - assuming it is the physical mass. This should allow a calibration of β in terms of a to be established.

This procedure is overly optimistic, however. The discretised lattice theory is different from the continuum theory on scales of the order of the lattice spacing a. For the Wilson action formulation of gauge theory, this implies that the continuum energy E_c is related to the lattice observable $\hat{E}(\beta) = Ea$ as

$$\frac{\hat{E}(\beta)}{a} = E_c + \mathcal{O}(a^2)$$

A direct consequence is that the ratio of two energies (of different particles, for example) will have discretisation errors of order a^2.

Thus, to cope with discretisation errors, the procedure required is to evaluate dimensionless ratios of quantities of physical interest at a range of values of the lattice spacing a and then extrapolate the ratio to the continuum limit ($a \to 0$).

Note that when the fermionic terms are included, the discretisation error is of order a for the Wilson fermionic action. By adding further terms in the fermionic action, the error can be reduced - to order αa for the SW-clover fermion formulation.

2 Glueball Masses

I choose to illustrate the workings of the lattice method by describing the determination of the glueball spectrum. Of course, glueballs are only defined unambiguously in the quenched approximation - where quark loops in the vacuum are ignored. In this approximation, glueballs are stable and do not mix with quark - antiquark mesons. This approximation is very easy to implement in lattice studies: the full gluonic action is used but no quark terms are included. This corresponds to a full non-perturbative treatment of the gluonic degrees of freedom in the vacuum. Such a treatment goes much further than models such as the bag model.

The glueball mass can be measured on a lattice through evaluating the correlation $C(t)$ of two closed colour loops (called Wilson loops) at separation t lattice spacings. Formally

$$C(t) = <0|G(0)G^{\dagger}(t)|0> = \sum_{i=0} c_i < g_i(0)|g_i(t) > c_i^* = \sum_{i=0} |c_i|^2 e^{-\hat{m}_i t}$$

where G represents the closed colour loop which can be thought of as creating a glueball state g_i from the vacuum. Summing over a complete set of such glueball states (strictly these are eigenstates of the lattice transfer matrix where $\exp -\hat{m}_i$ is the lattice eigenvalue corresponding to a step of one lattice spacing in time) then yields the above expression. As $t \to \infty$, the lightest glueball mass will dominate. This can be expressed as

$$\hat{m}_0 = \lim_{t\to\infty} \hat{m}_{eff}(t) \quad \text{where} \quad \hat{m}_{eff}(t) = \log(\frac{C(t-1)}{C(t)})$$

Note that since for the excited states $\hat{m}_i > \hat{m}_0$, then $\hat{m}_{eff}(t) > \hat{m}_{eff}(t+1) > \hat{m}_0$. This implies that the effective mass, defined above, is an upper bound on the ground state mass. In practice, sophisticated methods are used to choose loops G such that the correlation $C(t)$ is dominated by the ground state glueball (ie to ensure $|c_0| >> |c_i|$). By using a several different loops, a variational method can be used to achieve this effectively. These techniques are needed to obtain accurate estimates of $\hat{m}_0$ from modest values of t since the signal to noise decreases as t is increased. Even so, it is worth keeping in mind that upper limits on the ground state mass are obtained in principle.

The method also needs to be tuned to take account of the many glueballs: with different J^{PC} and different momenta. On the lattice the Lorentz symmetry is reduced to that of a hypercube. Non-zero momentum sates can be created (momentum is discrete in units of $2\pi/L$ where L is the lattice spatial size). The usual relationship between energy and momentum is found for sufficiently small lattice spacing. Here we shall concentrate on the simplest case of zero momentum (obtained by summing the correlations over the whole spatial volume).

For a state at rest, the rotational symmetry becomes a cubic symmetry. The lattice states (the g_i above) will transform under irreducible representations of this cubic symmetry group (called O_h). These irreducible representations can be linked to the representations of the full rotation group SU(2). Thus, for example, the five spin components of a $J^{PC} = 2^{++}$ state should be appear as the two-dimensional E^{++} and the three-dimensional T_2^{++} representations on the lattice, with degenerate masses. This degeneracy requirement then provides a test for the restoration of rotational invariance - which is expected to occur at sufficiently small lattice spacing.

The results of lattice measurements [1, 2, 3, 4] of the 0^{++} and 2^{++} states are shown in fig 1. The restoration of rotational invariance is shown by the degeneracy of the two O_h representations that make up the 2^{++} state. Fig 1 shows the dimensionless combination of the lattice glueball mass $\hat{m}_0$ to a lattice quantity $\hat{r}_0$. We will return to describe the lattice determination of $\hat{r}_0$ in more detail - here it suffices to accept it as a well measured quantity on the lattice that can be used to calibrate the lattice spacing and so explore the continuum limit. The quantity plotted, $\hat{m}_0\hat{r}_0$, is expected to be equal to the product of continuum quantities $m_0 r_0$ up to corrections of order a^2. This is indeed seen to be the case. The extrapolation to the continuum limit ($a \longrightarrow 0$) can then be made with confidence.

The value of r_0 in physical units is about 0.5 fm and we will adopt a scale equivalent to $r_0^{-1} = 0.372$ GeV. This information yields lattice predictions for the glueball masses of around 1.6 GeV and 2.2 GeV for the 0^{++} and 2^{-+} glueballs respectively.

We return briefly to the independence of the results on the volume of the lattice. In the early days of glueball mass determination, it was expected that a spatial size L should satisfy $\hat{m}(0^{++})L > 1$ and, hence, that values of $\hat{m}(0^{++})L$ of 1 to 4 would suffice. A careful lattice study [12] showed that $\hat{m}(0^{++})L > 8$ was required to obtain rotational invariance and a result independent of L. The results collected in fig 1 all satisfy this latter inequality so can be regarded as the infinite volume determination.

The predictions for the other J^{PC} states are that they lie higher in mass and the present state of knowledge is summarised in fig 2. Remember that the lattice results are strictly upper limits. For the J^{PC} values not shown, these upper limits are too weak to be of use.

Since quark - antiquark mesons can only have certain J^{PC} values, it is of special interest to look for glueball with J^{PC} values not allowed for such mesons: $0^{--}, 0^{+-}, 1^{-+}, 2^{+-}$, etc. Such spin-exotic states, often called "oddballs", would not mix directly with quark - antiquark mesons. This would make them a very clear experimental signal of the underlying glue dynamics. Various glueball models (bag models, flux tube models, QCD sum-rule inspired models,..) gave different predictions for the presence of such oddballs (eg. 1^{-+}) at relatively low masses. The lattice mass spectra clarify these uncertainties but, unfortunately for experimentalists, do not indicate any low-lying oddball candidates. The lightest candidate is from the T_2^{+-} spin combination. Such a state could correspond to an 2^{+-} oddball. Another interpretation is also possible, however, namely that a non-exotic 3^{+-} state is responsible (this choice of interpretation can be resolved in principle by finding the degenerate 5 or 7 states of a $J = 2$ or 3 meson). The overall conclusion at present is that there is no evidence for any oddballs of mass less than 3 GeV.

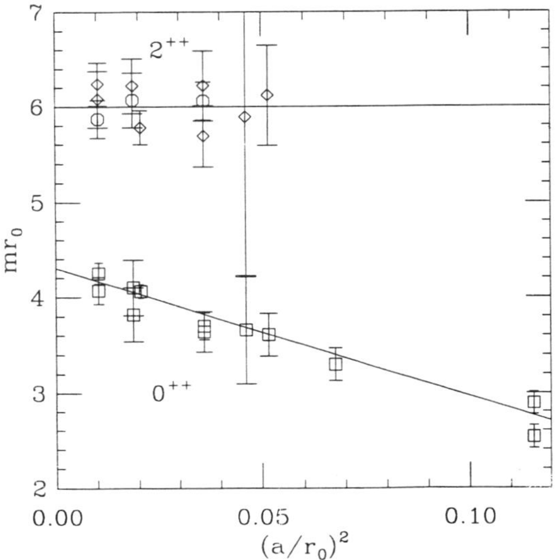

Figure 1: The value of mass of the $J^{PC} = 0^{++}$ and 2^{++} glueball states from refs[1, 2, 3, 4] in units of r_0. The T_2 and E representations are shown by octagons and diamonds respectively. The straight line shows a fit describing the approach to the continuum limit as $a \to 0$.

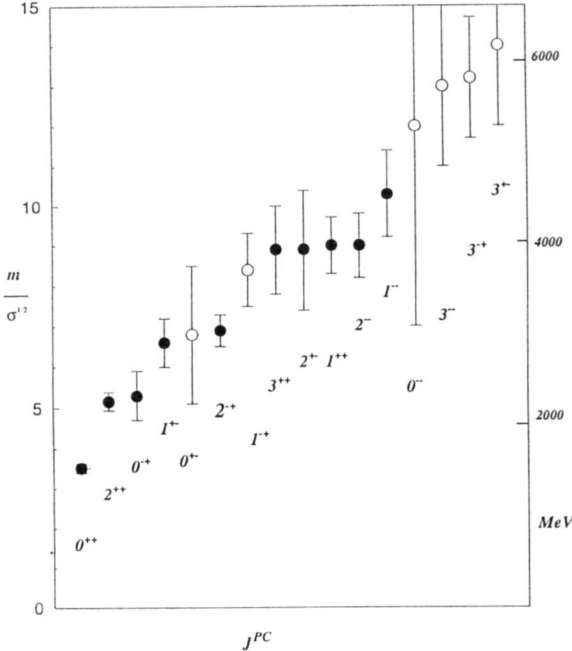

Figure 2: The mass of the glueball states with quantum numbers J^{PC} from ref[3]. The scale is set by $\sqrt{\sigma} \approx 0.44$ GeV which yields the right hand scale in GeV. The solid points represent mass determinations whereas the open points are upper limits.

Glueballs are defined in the quenched approximation and, hence, they do not decay into mesons since that would require quark - antiquark creation. It is, nevertheless, still possible to estimate the strength of the matrix element between a glueball and a pair of mesons within the quenched approximation. For the glueball to be a relatively narrow state, this matrix element must be small. A very preliminary attempt has been made to estimate the size of the coupling of the 0^{++} glueball to two pseudoscalar mesons [5]. A relatively small value is found. Further work needs to be done to investigate this in more detail, in particular to study the mixing between the glueball and 0^{++} mesons since this mixing may be an important factor in the decay process.

Another lattice study will become feasible soon. This is to study the glueball spectrum in full QCD vacua with sea quarks of mass m_D. For large m_D, the result is just the quenched result described above. For m_D equal to the experimental light quark masses, the results should just reproduce the experimental meson spectrum - with the resultant uncertainty between glueball interpretations and other interpretations. The lattice enables these uncertainties to be resolved in principle: one obtains the spectrum for a range of values of m_D between these limiting cases, so mapping glueball states at large m_D to the experimental spectrum at light m_D

3 Potentials between quarks

A very straightforward quantity to determine from lattice simulation is the interquark potential in the limit of very heavy quarks (static limit). This potential is of direct physical interest because solving the Schrödinger equation in such a potential provides a good approximation to the Υ spectrum. It is also relevant to exploring both confinement and asymptotic freedom on a lattice.

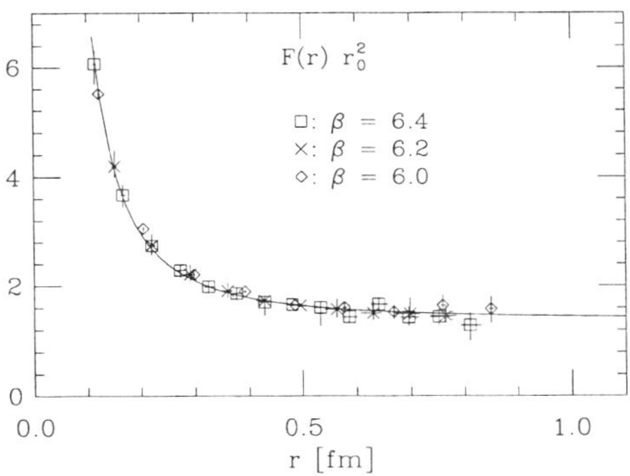

Figure 3: The force between static quarks at separation r as measured on lattices with different lattice spacing a (with the β-values shown).

The basic route to the static potential is to evaluate the average $W(\hat{R},\hat{T})$ in the vacuum samples of a rectangular closed loop of colour flux (a Wilson loop of size $\hat{R}\times\hat{T}$). here the lattice quantities $\hat{R}$ and $\hat{T}$ are related to the physical distances R and T by $\hat{R}=R/a$, etc where a is the lattice spacing which is not known explicitly. Then it can be shown that the required static potential in lattice units is given by

$$\hat{V}(\hat{R}) = \lim_{\hat{T}\to\infty} \log \frac{W(\hat{R},\hat{T}-1)}{W(\hat{R},\hat{T})}$$

The limit of large T is need to separate the required potential from excited potentials. This limit can be made tractable in practice by using more complicated loops than the simple rectangular loop described above.

A summary of results [6] for the potential at large R is shown in fig 3. The result that the force dV/dR tends to a constant at large R (and thus $V(R)$ continues to rise as R increases) is a manifestation of the confinement of heavy quarks (in the quenched approximation). The force appears to approach a constant at large R. A simple parametrisation is traditional in this field:

$$V(R) = V_0 - \frac{e}{R} + KR$$

where K (sometimes written σ) is the string tension. The term e/R is referred to as the Coulombic part in analogy to the electromagnetic case. The equivalent relationship in terms of quantities defined on a lattice is

$$\hat{V}(\hat{R}) = \hat{V}_0 - \frac{e}{\hat{R}} + \hat{K}\hat{R}$$

where the lattice string tension $\hat{K} = Ka^2$.

Since the string tension is given by the slope of $V(R)$ against R as $R \to \infty$, this implies that some error will arise in determining K coming from the extrapolation of lattice data at finite R. A practical resolution is to define a value of R where the potential takes a certain form. The convention is to use r_0 where

$$\hat{R}^2 \frac{d\hat{V}(\hat{R})}{d\hat{R}}\bigg|_{\hat{r}_0} = 1.65$$

Thus $\hat{r}_0$ can be determined by interpolation in $\hat{R}$ rather than extrapolation. In practice, this means that $\hat{r}_0$ is very accurately determined by lattice measurements and so is a useful quantity to use to set the scale since $\hat{r}_0 = r_0/a$. With the simple parametrisation above, we have $r_0^2 = (1.65 - e)/K$ so r_0 is closely related to the string tension since $e \approx 0.25$. The string tension is usually taken from experiment as $\sqrt{K} = 0.44$GeV where the value comes from $c\bar{c}$ and $b\bar{b}$ spectroscopy and from the light meson spectrum interpreted as excitations of a relativistic string. Similar analyses also imply that $r_0 \approx 0.5$ fm. Here we use $r_0^{-1} = 0.372$ GeV to be specific. Since we shall be describing quenched lattice results, the energy scale set from different physical quantities will not necessarily agree (since *experiment* has full QCD not the quenched vacuum) and so a systematic error must be applied to any such choice of scale. This, for instance, must be kept in mind when taking glueball mass values from the lattice.

The lattice potential $V(R)$ can be used to determine the spectrum of $b\bar{b}$ mesons by solving Schrödinger's equation since the motion is reasonably approximated as non-relativistic. The lattice result is similar to the experimental Υ spectrum. The main difference is that the Coulombic part (e) is effectively too small (0.25 rather than 0.4). This produces [14] a ratio of mass differences $(1P-1S)/(2S-1S)$ of 0.71 to be compared with the experimental ratio of 0.78. This difference is understandable as a consequence of the Coulombic force at short distances which would be increased by $33/(33-2N_f)$ in perturbation theory in full QCD compared to quenched QCD. We will return to discuss this.

Another feature of the lattice determination of potentials is that the energy of static quarks at separation R with an excited gluonic field can be determined. This enables predictions to be made for hybrid mesons (eg. $bG\bar{b}$ where G stands for the gluonic excitation). Such mesons can have J^{PC} values not allowed to $b\bar{b}$ states and this allows them to be explored experimentally. The current situation [15, 14] is that such states, as predicted by the lattice, will be at high masses and hard to isolate experimentally.

At small R, the static potential can be used, in principle, to study the running coupling constant. Small R corresponds to large momentum and thus the coupling should decrease at small R. Thus the Coulombic coefficient e introduced above should actually decrease logarithmically as R decreases. Perturbation theory can be used to determine this behaviour of the potential at small R.

In the continuum the potential between static quarks is known perturbatively to two loops in terms of the scale $\Lambda_{\overline{MS}}$. For $SU(3)$ colour, the continuum force is given by [7]

$$\frac{dV}{dR} = \frac{4}{3}\frac{\alpha(R)}{R^2} \tag{1}$$

with the effective coupling $\alpha(R)$ given by

$$\frac{1}{4\pi[b_0 \log(R\Lambda_R)^{-2} + (b_1/b_0)\log\log(R\Lambda_R)^{-2}]} \tag{2}$$

where $b_0 = 11/16\pi^2$ and $b_1 = 102\ b_0^2/121$ are the usual coefficients in the perturbative expression for the β-function, neglecting quark loops in the vacuum. Here $\Lambda_R = 1.048\Lambda_{\overline{MS}}$.

On a lattice the force can be estimated by a finite difference and one can extract the running coupling constant by using [8]

$$\alpha(\frac{\hat{R}_1+\hat{R}_2}{2}) = \frac{3}{4}\hat{R}_1\hat{R}_2\frac{\hat{V}(\hat{R}_1)-\hat{V}(\hat{R}_2)}{\hat{R}_1-\hat{R}_2} \tag{3}$$

where the error in using a finite difference is negligible in practice.

This is plotted in fig 4 versus $R\sqrt{K}$: this combination is dimensionless and so can be determined from lattice results since $R\sqrt{K} = \hat{R}\sqrt{\hat{K}}$, where $\hat{K}$ is taken from the fit to $\hat{V}(\hat{R})$. The interpretation of α as defined above as an effective running coupling constant is only justified at small R where the perturbative expression dominates. Also shown are the two-loop perturbative results for $\alpha(R)$ for different values of Λ_R.

The figure clearly shows a *running* coupling constant. Moreover the result is consistent with the expected perturbative dependence on R at small R. There are systematic errors, however. At larger R, the perturbative two-loop expression will not be an accurate estimate of the measured potentials, while at smaller R, the lattice artefact corrections (which arise because $R/a \approx 1$) are relatively big. Setting the scale using $\sqrt{K} = 0.44$ GeV implies $1/a(\beta = 6.5) = 4.13$ GeV, so $R < 4a(6.5)$ corresponds to values of $1/R > 1$ GeV. This R-region is expected to be adequately described by perturbation theory.

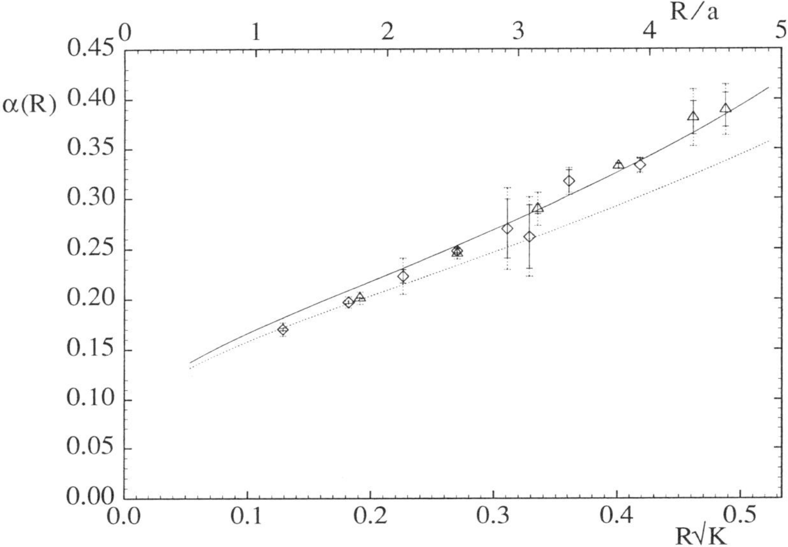

Figure 4: The effective running coupling constant $\alpha(R)$ obtained from the force between static quarks at separation R from ref[9]. The scale is set by the string tension K. Data are at $\beta = 6.5$ (diamonds) and at $\beta = 6.2$ (triangles). The dotted error bars represent an estimate of the systematic error due to lattice artefact correction. The curves are the two-loop perturbative expression with $a(6.5)\Lambda_R = 0.060$ (dotted) and 0.070 (continuous).

This determination from the interquark force of the coupling α allows us to compare the result with the bare lattice coupling determined from $\beta = 6/g^2$. At $\beta = 6.5$, $\alpha_{bare} = g^2/4\pi = 0.073$. the values of α shown in fig 4 are much larger. The effective coupling constant is thus almost twice the bare coupling. This is quite acceptable in a renormalisable field theory. The message is that the bare coupling should be disregarded - it is not a good expansion parameter. The measured α, however, proves to be a reasonable expansion parameter in the sense that the first few terms of the perturbation series converge. This successful calibration of perturbation theory on a lattice is important in practice. For instance, when matrix elements are measured on a lattice they have finite correction factors (usually called Z) to relate them to continuum matrix elements. These Z factors are evaluated perturbatively - so an accurate continuum prediction needs trustworthy perturbative calculations.

4 Full QCD

So far we have discussed the glueball spectrum, interquark potentials and α_S in the quenched approximation. This corresponds to treating the sea quarks as of infinite mass (so they don't contribute to the vacuum). To make direct comparison with experiment, it is necessary to estimate the corrections from these dynamical quark loops in the vacuum.

The strategy is to use a finite sea-quark mass but still a value larger than the empirical light quark mass. The reason is computational: the algorithms become very inefficient as the sea-quark mass is reduced. The target is to study the effects as the sea-quark mass is reduced and then extrapolate to the physical value. The present situation, in broad terms, is that there is no significant change as the sea-quark mass is reduced. This could be because there are no corrections to the quenched approximation. Alternatively, the corrections may only turn on at a much lower quark mass than has been explored so far.

Let us try to make this argument a little more quantitative. For heavy sea-quarks of mass m, their contribution will be approximately proportional to $e^{-2m/E}$ where E is a typical hadronic energy scale (a few hundred MeV). Thus the quark loop contributions will be negligible for $m >> E$, which corresponds to the quenched approximation. As $m \approx E$, the effects will turn on in a non-linear way.

The computational overhead of full QCD on a lattice is so large because the quark loops effectively introduce a long range interaction. The quark interaction in the Lagrangian is quadratic and so can be integrated out analytically. This leaves an effective Lagrangian for the gluonic fields which couples together the fields at all sites. This implies that, in a Monte Carlo method, a change in gluon field at one site involves the evaluation of the interaction with all other sites. In practice, one makes small changes at all sites in parallel, but this still amounts to inverting a large sparse matrix for each update. This is computationally slow.

As the sea-quark mass becomes small, one would expect to need a larger lattice size to hold the quarks. For heavy quarks, the effective range of the quark loops in the vacuum will be of order $1/m$. Thus the quenched approximation corresponds to $m \to \infty$ and a local interaction. For light quarks of a few MeV mass, the range will not be $1/m$, because quarks are confined. The lattice studies that have been made suggest that spatial sizes of order twice those adequate for the quenched approximation are needed for full QCD. This also implies considerable computational commitment.

A popular indication of how close a full QCD study is to experiment is to ask whether the ρ meson can decay to two pions. Since the decay is P-wave, it needs non-zero momentum. On a lattice spatial momentum is quantised in units of $2\pi/L$. Thus we need $m_V > 2m_P + 2\pi/L$ for the decay channel to be allowed energetically. At present this criterion is rarely satisfied in quenched studies, let alone in the more computationally demanding case of full QCD.

The conclusion of current full QCD lattice calculations is that the expected sea-quark effects are not yet fully present. The main effect observed in full QCD calculations is that the lattice parameter β which multiples the gluonic interaction term in the Lagrangian is shifted. Apart from this renormalisation of β, there is little sign of any other statistically significant non-perturbative effect.

Consider the changes to be expected for the inter-quark potential when the full QCD vacuum is used:

- At small separation R, the quark loops will increase the size of the effective coupling α compared to the pure gluonic case. This effect can be estimated in perturbation theory and the change at lowest order will be from 1/33 to 1/(33-2N_f).
- At large separation R, the potential energy will saturate at a value corresponding to two 'heavy-quark mesons'. In other words, the flux tube between the static quarks will break by the creation of a $q\bar{q}$ pair from the vacuum.

Current lattice simulation [13] shows some evidence for the former effect but no statistically significant signal for the latter.

If one assumes that these lattice simulations are an approximation to the true full QCD vacuum, then one can use them to estimate the full QCD running coupling from lattice studies. A summary [10] of present lattice results is that

$$\alpha_{\overline{MS}}(M_Z) = 0.112(7)$$

This conclusion will be reinforced when the full QCD lattice results reproduce more of the features of the experimental spectrum.

5 Quenched hadron masses

Since the full QCD simulation is not feasible at present, it is worthwhile to explore the hadron spectrum in the quenched approximation. This amounts to allowing quarks to propagate in the gluonic vacuum. Computationally this can be studied by solving the lattice Dirac equation for the quarks. Since the gluonic vacuum is full of rich structure, this is a computationally intensive problem: it amounts to inverting a large sparse matrix. Indeed it proves necessary to compute the quark propagation for a range of valence quark masses larger than the physical light quark masses and then extrapolate. Because of this, the statistical precision of such calculations is still somewhat limited. Nevertheless, different groups using different methods agree on the main results. For a recent review see ref[10].

Let us first discuss the ρ meson. In fig 5, the mass is compared [16] to a lattice scale (here $\Sigma = Ka^2$ where K is the string tension). The figure shows that different treatments of fermions on the lattice (Wilson, Clover and staggered) with different discretisation errors (a, $\alpha_S a$ and a^2 respectively) are in agreement with a common continuum limit. Moreover, using the usual convention that the string tension is $\sqrt{K} =$ 0.44 GeV, the continuum value of $m_\rho/\sqrt{K}$ of 1.80(5) yields $m_\rho = 0.79(2)$ GeV which is consistent with the experimental value of 0.77 GeV.

Studies have been made of the mesons and baryons which are composed of light and strange quarks- see ref[10]. The surprise is that the quenched approximation seems to reproduce these experimental values quite well. This may reflect the relatively large errors that are still present in the lattice determinations. It may also reflect the fact that the hadronic dynamics has a similar energy scale in each case so that the quenched approximation makes similar errors - which cancel in mass ratios.

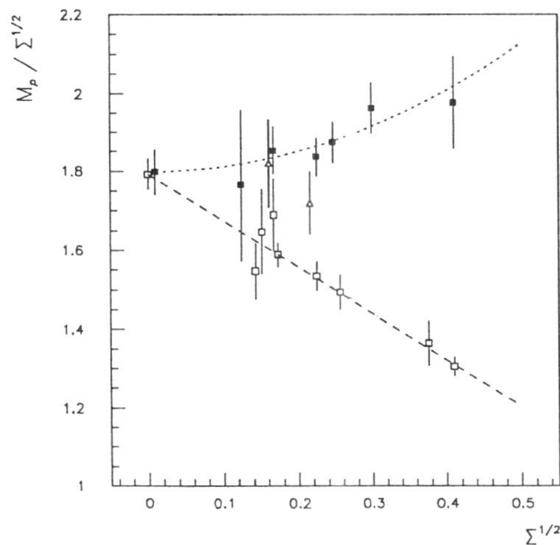

Figure 5: The mass of the ρ meson as a dimensionless ratio to the lattice string tension $\Sigma = Ka^2$. The continuum limit corresponds to $a \to 0$ at the left hand side. The filled squares are with staggered fermions, the open squares from Wilson fermions and the triangles are from Clover fermions. The lines show the discretisation errors which behave as a^2, a and $\alpha_s a$ respectively.

6 Matrix elements eg f_P

One of the advantages of lattice QCD is that it is a method to calculate hadronic matrix elements from first principles. Consider as an example the weak decay of a pseudoscalar meson P. The weak axial current will couple to the quarks in the meson. This current will be local. Thus the required quantity will relate the quark current to the hadronic state P. This is the pseudoscalar decay constant f_P which is defined by the matrix element of the divergence of the axial current. For a pseudoscalar state of zero spatial momentum,

$$< 0|q_\mu j^A_\mu|P >= \frac{f_P}{\sqrt{2m_P}}$$

where j^A is the axial current which in terms of quark fields is $\bar{q}\gamma_5\gamma_\mu q$. For the pion, this identity is the partially conserved axial current (PCAC) relation and f_π is the coupling of the pion to the weak current (and hence is relevant to the $\mu + \nu$ decay mode of the pion). Since the axial current is represented by local quark fields, f_P gives a relationship between the hadronic state ($|P>$) and the quark sub-structure (j^A).

For a pseudoscalar meson with a heavy quark, such as the B meson, this same relationship is needed. Because the relevant weak decay is not currently observable (branching ratio too small), a lattice calculation is needed to determine f_B. The picture, for a heavy quark, is clarified by the heavy quark effective theory (HQET) which treats the heavy quark as slow moving. In the static limit, the lattice calculation needed is easily visualised. A straight line of colour flux of length T in the lattice time direction represents the propagation of the heavy quark. By combining this with the propagator from one end to the other of a light quark in the lattice gluonic background field, one has a gauge invariant quantity which can be measured on a lattice (see fig 6). At each end, one joins the heavy quark to the light quark by a local axial current ($\gamma_5\gamma_4$

for a state at rest). Then the observed correlation $C(t)$ is proportional to $f_P^2 e^{-E_P T}$ for large T so allowing f_P and E_P do be determined in principle. In practice more sophisticated methods are used to improve the lattice measurement signal. Here E_P has the interpretation of the mass difference of the B meson and the b quark - although this difference is not directly useful since a non-perturbative definition of the quark mass requires careful discussion.

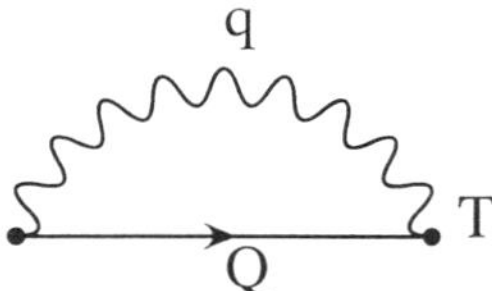

Figure 6: The diagram representing a static quark Q propagating in the time direction and a light quark q which creates a gauge invariant quantity appropriate for studying f_B.

As well as using static quarks, lattice studies have been performed using propagating heavy quarks. Combining the results from both studies [16] yields $f_B = 180(50)$ MeV. This value is needed as an ingredient in using experimental data to fix the CKM weak matrix elements. This is turn has implications for the experimental feasibility of CP violation through studies of $\bar{B}B$ mixing.

7 Outlook

Lattice techniques can extract reliable continuum properties from QCD. At present, the computational power available combined with the best algorithms suffices to give accurate results for many quantities in the quenched approximation. The future is to establish accurate values for more subtle quantities in the quenched approximation (eg. weak matrix elements of strange particles) and to establish the validity of the quenched approximation by full QCD calculations.

We need to reach the stage where an experimentalist saying ‘as calculated in QCD’ is assumed to be speaking of non-perturbative lattice calculations rather than perturbative estimates only.

REFERENCES

[1] P. De Forcrand, et al., Phys. Lett. 152B (1985) 107

[2] C. Michael and M. Teper, Nucl. Phys. B314 (1989) 347

[3] UKQCD collaboration, G. Bali, K. Schilling, A. Hulsebos, A. C. Irving, C. Michael and P. Stephenson, Phys. Lett. B309 (1993) 378-84.

[4] H. Chen, J. Sexton, A. Vaccarino, and D. Weingarten, Nucl. Phys. B (Proc. Suppl.) 34 (1994) 357-359.

[5] J. Sexton, A. Vaccarino, and D. Weingarten, Nucl. Phys. B (Proc. Suppl.) 42 (1995) 279.

[6] G.S. Bali and K. Schilling, Phys. Rev. D47 (1993) 661; H. Wittig (UKQCD collaboration) Nucl. Phys. B (Proc. Suppl) 42 (1995) 288.

[7] A. Billoire, Phys. Lett. 104B (1981) 472.
[8] C. Michael, Phys. Lett. 283B (1992) 103.
[9] UKQCD collaboration, A. Hulsebos et al., Phys. Lett. B294 (1992) 385.
[10] C. Michael, Nucl. Phys. B (Proc. Suppl.) 42 (1995) 147-61.
[11] G. P. Lepage and P. B. Mackenzie, Phys. Rev. D48 (1993) 2250.
[12] C. Michael, G. A. Tickle and M. Teper, Phys. Lett. B 207 (1988) 313.
[13] U. M. Heller et al.,Phys. Lett. B335 (1994) 71.
[14] S. Perantonis and C. Michael, Nucl. Phys. B347 (1990) 854
[15] C. Michael, Proc. of Aachen Workshop, 'QCD 20 Years Later', eds H. Kastrup and P. Zerwas, World Scientific 1993, pp 505-519
[16] R. Sommer, Nucl. Phys. B (Proc. Suppl.) 42 (1995) 186.

CONSTITUENT QUARK MODELS OF BARYON STRUCTURE AND STRONG DECAYS

Simon Capstick

Supercomputer Computations Research Institute
and Department of Physics
Florida State University
Tallahassee, FL 32306-3016

ABSTRACT

An introduction is given to the nonrelativistic quark model of baryon structure and strong decays, with an eye towards understanding the results of analyses of πN elastic scattering and single-pion photoproduction. This is followed by a brief survey of methods used to incorporate relativity into the model.

THE NONRELATIVISTIC MODEL OF ISGUR AND KARL

The nonrelativistic (NR) quark model as applied to baryons owes its origins to many people. We intend to focus on the model of Isgur and Karl [1, 2] which evolved from the pioneering work of others; an extensive list of references to the origins of the model is given in Ref. [3].

First let us try to state clearly the assumptions which are implicit in this type of description of baryon physics. If, for example, one probes the proton with an electron which transfers an amount of energy and momentum which is modest compared to several hundred MeV, then it is useful to describe the proton in this 'soft' region as being made up of three constituent quarks. The light (u and d) constituent quarks have masses of roughly 200 to 350 MeV, and can be thought of as extended objects, *i.e.* 'dressed' valence quarks. Strange quarks are about 150 MeV heavier. These are not the partons which occur in deep inelastic scattering from the proton; their interactions are described in an effective model which is not QCD but which is motivated by it.

We will assume that the gluon fields affect the quark dynamics by providing a confining potential in which the quarks move, which is effectively pair-wise linear at large separation of the quarks, and at short distance one-gluon exchange provides a Coulomb potential and the important spin-dependent potential. Otherwise the effects of the gluon dynamics on the quark motion are neglected. This model will obviously have a limited applicability: to 'soft' (low-Q^2 or coarse-grid) aspects of hadron struc-

Hadron Spectroscopy and the Confinement Problem
Edited by D.V. Bugg, Plenum Press, New York, 1996

ture, and to low-mass hadrons where gluonic excitation is unlikely. Similarly, since we have not allowed for anything but three quarks in a baryon [we have neglected other Fock-space components like $qqq(q\bar{q})$], it will only be applicable to hadrons where large mass shifts from couplings to decay channels are not expected. Recent progress has been made in understanding these mass shifts [4], and the effects of including the dynamics of the glue [5].

The Interquark Hamiltonian

In the model of Isgur and Karl [1, 2], the Schrödinger equation $H\Psi = E\Psi$ for the nonrelativistic three valence-quark system is solved for baryon energies and wavefunctions. The Hamiltonian is

$$H = \sum_i \left(m_i + \frac{p_i^2}{2m_i} \right) + \sum_{i<j} \left(V^{ij} + H_{hyp}^{ij} \right), \tag{1}$$

where the spin-independent potential V^{ij} has the form $V^{ij} = C_{qqq} + br_{ij} - 2\alpha_s/3r_{ij}$, with $r_{ij} = |r_i - r_j|$. In practice, V^{ij} is written as a harmonic-oscillator potential $Kr_{ij}^2/2$ plus an anharmonicity U_{ij}, which is treated as a perturbation. The hyperfine interaction H_{hyp}^{ij} is the sum

$$H_{hyp}^{ij} = \frac{2\alpha_s}{3m_i m_j} \left\{ \frac{8\pi}{3} S_i \cdot S_j \delta^3(r_{ij}) + \frac{1}{r_{ij}^3} \left[\frac{3(S_i \cdot r_{ij})(S_j \cdot r_{ij})}{r_{ij}^2} - S_i \cdot S_j \right] \right\} \tag{2}$$

of contact and tensor terms arising from the color magnetic dipole-magnetic dipole interaction. In this model spin-orbit forces which arise from one-gluon exchange and from Thomas precession of the quark spins in the confining potential are neglected; their inclusion spoils the agreement with the spectrum, since the resulting splittings tend to be too large. The relative strengths of the Coulomb, contact and tensor terms are as determined from the Breit-Fermi limit [the expansion to $O(p^2/m^2)$] of the one-gluon exchange potential.

Note that α_s, b, and m are not fundamental, but are phenomenological parameters extracted from a fit to the spectrum of states. This means that α_s is not less than unity in the NR model. The pairwise-linear potential has a strength b which can be justified from a string picture to be approximately half of the meson string tension $\sqrt{\sigma}$, and the masses m are the model-dependent constituent quark masses.

Spin and Flavor Wavefunctions

Baryon states are written as the product of a totally antisymmetric (under the exchange group S_3) color wavefunction C_A, and a sum $\sum \psi\chi\phi$, where ψ , χ, and ϕ are the spatial, spin, and flavor wavefunctions of the quarks respectively. The spatial wavefunctions are chosen to represent S_3 in the case of baryons with equal-mass quarks, and the spin wavefunctions automatically do so. The sum is constructed so that it is symmetric under exchange of equal mass quarks, and also implicitly includes Clebsch-Gordan coefficients for coupling the quark orbital angular momentum L with the total quark spin S.

The exchange group S_3 has three kinds of representation: totally antisymmetric (A), totally symmetric (S), and a mixed symmetry pair (M^ρ and M^λ) which transform among themselves under the group elements

$$S_3 = \{1, (12), (13), (23), (123), (132)\}. \tag{3}$$

For example $(12)M^\rho = -M^\rho$, $(12)M^\lambda = M^\lambda$, and $(13)M^\rho = M^\rho/2 - \sqrt{3}M^\lambda/2$, $(13)M^\lambda = -\sqrt{3}M^\rho/2 - M^\lambda/2$. The spin wavefunctions χ are an example of such a representation, and can be found from the Clebsch-Gordan coefficients for coupling three spins-$\frac{1}{2}$,

$$\tfrac{1}{2} \otimes \tfrac{1}{2} \otimes \tfrac{1}{2} = (1 \oplus 0) \otimes \tfrac{1}{2} = (\tfrac{3}{2} \oplus [\tfrac{1}{2}]_\rho) \oplus [\tfrac{1}{2}]_\lambda \,. \tag{4}$$

The spin-$\frac{3}{2}$ wavefunction χ^S is totally symmetric under the exchange group S_3, and the two spin-$\frac{1}{2}$ wavefunctions χ^{M_λ} (from $1 \otimes \frac{1}{2}$) and χ^{M_ρ} (from $0 \otimes \frac{1}{2}$) form a mixed-symmetry pair. For example, we have

$$\begin{aligned} \chi^S_{\frac{3}{2},\frac{3}{2}} &= |\uparrow\uparrow\uparrow>, \quad \chi^{M_\rho}_{\frac{1}{2},\frac{1}{2}} = \frac{1}{\sqrt{2}}\{|\uparrow\downarrow\uparrow> - |\downarrow\uparrow\uparrow>\} \\ \chi^{M_\lambda}_{\frac{1}{2},\frac{1}{2}} &= -\frac{1}{\sqrt{6}}\{|\downarrow\uparrow\uparrow> + |\uparrow\downarrow\uparrow> - 2|\uparrow\uparrow\downarrow>\} \,, \end{aligned} \tag{5}$$

where the subscripts are the total spin and its projection, and other projections can be found by applying a lowering operator.

The construction of the flavor wavefunctions ϕ proceeds exactly analogously to that of the spin wavefunctions in the case of baryons made up exclusively of light quarks, such as the states N and Δ. The flavor wavefunctions have either mixed symmetry $(N, I = \frac{1}{2})$ or are totally symmetric $(\Delta, I = \frac{3}{2})$ and are found from SU(2)-isospin to be

$$\begin{aligned} \phi^S_{\Delta^{++}} &= uuu, \quad \phi^S_{\Delta^+} = \frac{1}{\sqrt{3}}\{uud + udu + duu\}\,, \; etc. \\ \phi^{M_\rho}_p &= \frac{1}{\sqrt{2}}\{udu - duu\}\,, \quad \phi^{M_\rho}_n = \frac{1}{\sqrt{2}}\{udd - dud\} \\ \phi^{M_\lambda}_p &= -\frac{1}{\sqrt{6}}\{duu + udu - 2\,uud\}\,, \quad \phi^{M_\lambda}_n = \frac{1}{\sqrt{6}}\{udd + dud - 2\,ddu\}\,. \end{aligned} \tag{6}$$

Note that isospin is only a good symmetry in the absence of electromagnetic interactions between the quarks, and also in the absence of the quark mass difference $m_d - m_u \simeq 4$ MeV. Since the electromagnetic interactions are quite weak on the strong interaction scale (note $\alpha_{EM}/r_p \simeq 1.6$ MeV), and the quark mass difference is only 1 to 2% of the constituent quark mass, the Hamiltonian in Eq. (1) is to a good approximation independent of the substitution of an up quark for a down quark (or, more generally, an arbitrary rotation of the u and d labels). It is this approximate symmetry which we call isospin. The study of the violations of this symmetry is an interesting subject; see Ref. [6].

For baryons which consist of two quarks with equal masses plus one quark with a different mass it is *not* advantageous to use a basis (without the common antisymmetric color wavefunction) which is symmetrized under exchange of the unequal mass quarks. This is because the mass differences $m_s - (m_u + m_d)/2$ and $m_c - (m_u + m_d)/2$, etc., are substantial compared to the average quark momentum, which is related to the confinement scale or the string tension. The result is that there are substantial differences between, for example, the spatial wavefunctions of the ground state of the nucleon (n or p) and that of the ground state Λ. Equivalently, the approximate SU(3)$_f$ symmetry of the wavefunctions under the group which expresses independence of the strong interactions under exchange of the quark flavors u, d, and s is strongly broken.

We use instead 'the uds basis', which imposes symmetry only under exchange of equal mass quarks, and adopt the flavor wavefunctions

$$\phi_{\Lambda^0} = \frac{1}{\sqrt{2}}(ud - du)s$$

$$
\begin{aligned}
\phi_{\Sigma^{+,0,-}} &= uus, \frac{1}{\sqrt{2}}(ud+du)s, dds \\
\phi_{\Xi^{0,-}} &= ssu, ssd \\
\phi_{\Omega^-} &= sss \\
\phi_{\Lambda_c^+} &= \frac{1}{\sqrt{2}}(ud-du)c \\
\phi_{\Sigma_c^{++,+,0}} &= uuc, \frac{1}{\sqrt{2}}(ud+du)c, ddc \ ,
\end{aligned}
\tag{7}
$$

and similarly for the Λ_b states. Notice that these flavor wavefunctions are all either even or odd under (12) exchange, as are the spin wavefunctions in Eq. (5). This allows us to easily build sums $\sum \psi\chi\phi$ for these states which are symmetric under exchange of the equal mass (or approximately equal for the Λ-like states) quarks 1 and 2. It may also be convenient to describe the N and Δ states using wavefunctions which are only symmetrized under exchange of *identical* quarks. For details see Ref. [3].

Spatial Wavefunctions

The spatial wavefunctions ψ are, in zeroeth order in the perturbations U and H_{hyp}, the harmonic-oscillator eigenfunctions $\psi_{NLM}(\boldsymbol{\rho}, \boldsymbol{\lambda})$, where

$$
\boldsymbol{\rho} = (r_1 - r_2)/\sqrt{2}, \quad \boldsymbol{\lambda} = \frac{1}{\sqrt{6}}(r_1 + r_2 - 2r_3). \tag{8}
$$

These Jacobi coordinates separate the Hamiltonian in Eq. (1) into two independent three-dimensional oscillators when $U = H_{hyp} = 0$. The $\psi_{NLM}(\boldsymbol{\rho}, \boldsymbol{\lambda})$ can then be conveniently written as sums of products of three-dimensional harmonic oscillator eigenstates. These oscillator wavefunctions have quantum numbers (n, l, m), where n is the number of radial nodes and $|l, m\rangle$ is the orbital angular momentum, and 0-th order (harmonic oscillator) energies $E = (N + \frac{3}{2})\omega = (2n + l + \frac{3}{2})\omega$, where $\omega^2 = 3K/m$.

For N and Δ states these sums are arranged so that the resulting ψ_{NLM} have their orbital angular momenta coupled as $L = \boldsymbol{l} \;\; + \boldsymbol{l}$, and so that the result represents the permutation group S_3. When there are unequal mass quarks the latter condition is unneccesary. The resulting combined six-dimensional oscillator state has energy $E = (N+3)\omega$, where $N = 2(n_\rho + n_\lambda) + l_\rho + l_\lambda$, and parity $(-1)^{l_\rho + l_\lambda}$.

Positive-parity ground states (members of the ground-state baryon octet) are described, in 0-th order in the perturbations, by basis states with $N = 0$. For example the $N(938)$ and $\Delta(1232)$ have, in 0-th order, the common spatial wavefunction

$$
\psi_{00}^{S} = \frac{\alpha^3}{\pi^{\frac{3}{2}}} e^{-\frac{\alpha^2}{2}(\rho^2+\lambda^2)}, \tag{9}
$$

where the notation is $\psi_{LM}^{(S_3-symmetry)}$, $\rho = |\boldsymbol{\rho}|$, $\lambda = |\boldsymbol{\lambda}|$, and where the harmonic oscillator scale constant is $\alpha = (3Km)^{\frac{1}{4}}$. For Λ and Σ states the generalization is to allow the wavefunction to have an asymmetry between the ρ- and λ-oscillators

$$
\psi_{00} = \frac{\alpha_\rho^{\frac{3}{2}} \alpha_\lambda^{\frac{3}{2}}}{\pi^{\frac{3}{2}}} e^{-(\frac{\alpha_\rho^2}{2}\rho^2 + \frac{\alpha_\lambda^2}{2}\lambda^2)}, \tag{10}
$$

where $\alpha_\rho = (3Km)^{\frac{1}{4}}$, with $m = m_1 = m_2 = (m_u + m_d)/2$, and $\alpha_\lambda = (3Km_\lambda)^{\frac{1}{4}}$, with $m_\lambda = 3mm_3/(2m + m_3) > m$. Note that the orbital degeneracy is now broken, since $\omega_\rho^2 = 3K/m > \omega_\lambda^2 = 3K/m_\lambda$.

In 0-th order, the low-lying negative-parity excited resonances ('P waves') have $N = 1$ spatial wavefunctions with either $l_\rho = 1$ or $l_\lambda = 1$. The wavefunctions for the unequal mass case are

$$\begin{aligned}
\psi^{\rho}_{1\pm1} &= \mp\frac{\alpha_\rho^{\frac{5}{2}}\alpha_\lambda^{\frac{3}{2}}}{\pi^{\frac{3}{2}}}\rho_\pm e^{-(\frac{\alpha_\rho^2}{2}\rho^2+\frac{\alpha_\lambda^2}{2}\lambda^2)}, \\
\psi^{\lambda}_{1\pm1} &= \mp\frac{\alpha_\rho^{\frac{3}{2}}\alpha_\lambda^{\frac{5}{2}}}{\pi^{\frac{3}{2}}}\lambda_\pm e^{-(\frac{\alpha_\rho^2}{2}\rho^2+\frac{\alpha_\lambda^2}{2}\lambda^2)}, \\
\psi^{\rho}_{10} &= \frac{\alpha_\rho^{\frac{5}{2}}\alpha_\lambda^{\frac{3}{2}}}{\pi^{\frac{3}{2}}}\sqrt{2}\rho_0 e^{-(\frac{\alpha_\rho^2}{2}\rho^2+\frac{\alpha_\lambda^2}{2}\lambda^2)},
\end{aligned} \tag{11}$$

and similarly for ψ^λ_{10}, where $\rho_\pm \equiv \rho_x \pm i\rho_y$, $\rho_0 \equiv \rho_z$, etc. For the equal mass case we simply use the states ψ^{M_ρ} and ψ^{M_λ} found from Eq. (11) by equating α_ρ and α_λ, where the labels M_ρ and M_λ refer to the overall exchange symmetry of the spatial wavefunctions.

The positive parity excited resonances are initially assigned $N = 2$ wavefunctions with 'radial excitations' in one of the two oscillators, or 'orbital excitations' with $l_\rho + l_\lambda = 2$ and $L = 0$, 1 or 2. In order to write these down we have to absorb some basic rules about combining two representations of S_3 (the equivalent of Clebsch-Gordan coefficients for S_3), namely the $\boldsymbol{\rho}$ and $\boldsymbol{\lambda}$-oscillator eigenfunctions. Since $\boldsymbol{\rho}$ and $\boldsymbol{\lambda}$ have M^ρ and M^λ character respectively, and we need to make second order polynomials of a given symmetry, we will need the rules for combining a pair of mixed-symmetry representations,

$$\begin{aligned}
\frac{1}{\sqrt{2}}(M^\rho M^\rho + M^\lambda M^\lambda) &= S \\
\frac{1}{\sqrt{2}}(M^\rho M^\lambda - M^\lambda M^\rho) &= A \\
\frac{1}{\sqrt{2}}(M^\rho M^\lambda + M^\lambda M^\rho) &= M^\rho \\
\frac{1}{\sqrt{2}}(M^\rho M^\rho - M^\lambda M^\lambda) &= M^\lambda.
\end{aligned} \tag{12}$$

We may now construct symmetrized spatial basis wavefunctions at the $N = 2$ level to represent the positive-parity excited states of the non-strange baryons. For example, there are three $L^P = 0^+$ states, two made from $l_\rho = 0 \otimes l_\lambda = 0$ with one radial node in either of the two oscillators,

$$\begin{aligned}
\psi^{S'}_{00} &= \sqrt{\frac{2}{3}}\frac{\alpha^5}{\pi^{\frac{3}{2}}}\frac{1}{\sqrt{2}}(\rho^2 - \frac{3}{2\alpha^2} + \lambda^2 - \frac{3}{2\alpha^2})e^{-\frac{\alpha^2}{2}(\rho^2+\lambda^2)} \\
\psi^{M^\lambda}_{00} &= \sqrt{\frac{2}{3}}\frac{\alpha^5}{\pi^{\frac{3}{2}}}\frac{1}{\sqrt{2}}(\rho^2 - \lambda^2)e^{-\frac{\alpha^2}{2}(\rho^2+\lambda^2)},
\end{aligned} \tag{13}$$

[where $\frac{3}{2} - (\alpha\rho)^2$ is the Laguerre polynomial $L_n^{l+\frac{1}{2}} = L_0^{\frac{1}{2}}$] and one from $l_\rho = 1 \otimes l_\lambda = 1$ with no radial nodes,

$$\psi^{M^\rho}_{00} = -\frac{2}{\sqrt{3}}\frac{\alpha^5}{\pi^{\frac{3}{2}}}\boldsymbol{\rho}\cdot\boldsymbol{\lambda}e^{-\frac{\alpha^2}{2}(\rho^2+\lambda^2)}. \tag{14}$$

The radial excitation $\psi^{S'}$ is S under S_3 because it has the structure $\boldsymbol{\rho}\cdot\boldsymbol{\rho}+\boldsymbol{\lambda}\cdot\boldsymbol{\lambda}$. It is straightforward to see that the $\psi_{00}^{M^\lambda}$ and $\psi_{00}^{M^\rho}$ wavefunctions also have the advertised S_3 symmetry.

We may also construct an $L^P = 1^+$ state from $l_\rho = 1 \otimes l_\lambda = 1$,

$$\psi_{11}^{A} = -\frac{\alpha^5}{\pi^{\frac{3}{2}}}(\rho_+\lambda_0 - \rho_0\lambda_+)e^{-\frac{\alpha^2}{2}(\rho^2+\lambda^2)}, \tag{15}$$

where we have written down only the top state with $|L,M\rangle = |1,1\rangle$. There are also the $L^P = 2^+$ states constructed from $l_\rho = 0 \otimes l_\lambda = 2$, $l_\rho = 2 \otimes l_\lambda = 0$, and $l_\rho = 1 \otimes l_\lambda = 1$,

$$\begin{aligned}
\psi_{22}^{S} &= \frac{1}{\sqrt{2}}\frac{\alpha^5}{\pi^{\frac{3}{2}}}(\rho_+^2 + \lambda_+^2)e^{-\frac{\alpha^2}{2}(\rho^2+\lambda^2)} \\
\psi_{22}^{M^\lambda} &= \frac{1}{\sqrt{2}}\frac{\alpha^5}{\pi^{\frac{3}{2}}}(\rho_+^2 - \lambda_+^2)e^{-\frac{\alpha^2}{2}(\rho^2+\lambda^2)} \\
\psi_{22}^{M^\rho} &= \frac{\alpha^5}{\pi^{\frac{3}{2}}}\rho_+\lambda_+e^{-\frac{\alpha^2}{2}(\rho^2+\lambda^2)},
\end{aligned} \tag{16}$$

where once again we have shown only the top states.

A Sample Basis: N and Δ States to N=2

We now know enough to form, for example, all of the fully symmetrized N=0, 1, and 2 band basis state wavefunctions for the equal mass case, i.e. the states N and Δ. We will label states by the notation [2] $|X^{2S+1}L_\pi J^P\rangle$, where X is the baryon flavor, S is the total quark spin, L the total orbital angular momentum, and J and P are the overall spin and parity. The symbol π denotes the exchange symmetry of the spatial part of the wavefunction. We will suppress the Clebsch-Gordan coefficents coupling $\vec{L}$ and $\vec{S}$, and all quark-spin M values.

Another popular notation for classification of baryon states is that of SU(6). This would be an exact symmetry of the interquark Hamiltonian if it was invariant under exchange (rotation) of the u, d, and s quarks [SU(3)$_f$], *and* independent of the spin of the quarks [SU(2)]. As outlined above, this is only very approximately true; for example (as we shall see) the ground state N and Δ are split by the contact part of the color magnetic spin-spin ('hyperfine') interaction by $\simeq 300$ MeV, which is hardly a 'hyperfine' splitting. In this SU(6) notation there is an SU(3)$_f$ octet of ground state baryons with $(J=)S=\frac{1}{2}$, i.e. (n, p, Λ^0, $\Sigma^{+,0,-}$, $\Xi^{0,-}$), giving $(2S+1)\cdot 8 = 16$ states, plus an SU(3)$_f$ decuplet of ground state baryons with $(J=)S=\frac{3}{2}$ (the four charge states of Δ, three excited Σ^* states, two excited Ξ^* states, and Ω^-), giving $(2S+1)\cdot 10 = 40$ states, for 56 states in total. These ground states all have $L^P = 0^+$, so the SU(6) multiplet to which they belong is labelled $[56, 0^+]$.

The two $[56,0^+]$ nonstrange ground states are, in 0-th order, represented by

$$\begin{aligned}
|N^2S_S\tfrac{1}{2}^+\rangle &= C_A\psi_{00}^{S}\frac{1}{\sqrt{2}}(\phi_N^\rho\chi_{\frac{1}{2}}^\rho + \phi_N^\lambda\chi_{\frac{1}{2}}^\lambda) \\
|\Delta^4S_S\tfrac{3}{2}^+\rangle &= C_A\phi_\Delta^S\psi^S\chi_{\frac{3}{2}}^S,
\end{aligned} \tag{17}$$

where we have for now explicitly included the totally antisymmetric color wavefunction. Similarly the next set of excited states occurs at $N=1$ in the harmonic oscillator,

$$|N^4P_M(\tfrac{1}{2}^-,\tfrac{3}{2}^-,\tfrac{5}{2}^-)\rangle = C_A\chi_{\frac{3}{2}}^S\frac{1}{\sqrt{2}}(\phi_N^\rho\psi_{1M}^\rho + \phi_N^\lambda\psi_{1M}^\lambda)$$

$$|N^2P_M(\tfrac{1}{2}^-,\tfrac{3}{2}^-)\rangle = C_A\frac{1}{2}\left\{\phi_N^\rho[\psi_{1M}^\rho\chi_{\frac{1}{2}}^\lambda + \psi_{1M}^\lambda\chi_{\frac{1}{2}}^\rho] + \phi_N^\lambda[\psi_{1M}^\rho\chi_{\frac{1}{2}}^\rho - \psi_{1M}^\lambda\chi_{\frac{1}{2}}^\lambda]\right\}$$

$$|\Delta^2P_M(\tfrac{1}{2}^-,\tfrac{3}{2}^-)\rangle = C_A\phi_\Delta^S\frac{1}{\sqrt{2}}(\psi_{1M}^\rho\chi_{\frac{1}{2}}^\rho + \psi_{1M}^\lambda\chi_{\frac{3}{2}}^\lambda) \qquad (18)$$

where the notation $(\frac{1}{2}^-,\frac{3}{2}^-,\frac{5}{2}^-)$ lists all of the possible J^P values from the $\vec{L}+\vec{S}$ coupling. These are members of the $[70,1^-]$ SU(6) multiplet (here the 70 is made up of two octets of $S=\frac{3}{2}$ and $S=\frac{1}{2}$ states and a decuplet of $S=\frac{1}{2}$ states as above, plus a singlet Λ state[7] with $S=\frac{1}{2}$).

It is then straightforward but rather tedious to work out the fully symmetrized wavefunctions (which therefore represent nonstrange states) which occur at the $N=2$ level in the oscillator. Note that, from Eq. (6), there are only symmetric and mixed-symmetry flavor wavefunctions in the nonstrange case. These are

$$[56',0^+]: \quad |N^2S_{S'}\tfrac{1}{2}^+\rangle = C_A\psi^{S'}\frac{1}{\sqrt{2}}(\phi_N^\rho\chi_{\frac{1}{2}}^\rho + \phi_N^\lambda\chi_{\frac{1}{2}}^\lambda)$$

$$|\Delta^4S_{S'}\tfrac{3}{2}^+\rangle = C_A\phi_\Delta^S\psi^{S'}\chi_{\frac{3}{2}}^S$$

$$[70,0^+]: \quad |N^4S_M\tfrac{3}{2}^+\rangle = C_A\chi_{\frac{3}{2}}^S\frac{1}{\sqrt{2}}(\phi_N^\rho\psi_{00}^\rho + \phi_N^\lambda\psi_{00}^\lambda)$$

$$|\Delta^2S_M\tfrac{1}{2}^+\rangle = C_A\phi_\Delta^S(\psi_{00}^\rho\chi_{\frac{1}{2}}^\rho + \psi_{00}^\lambda\chi_{\frac{1}{2}}^\lambda)$$

$$|N^2S_M\tfrac{1}{2}^+\rangle = C_A\frac{1}{2}\left\{\phi_N^\rho[\psi_{00}^\rho\chi_{\frac{1}{2}}^\lambda + \psi_{00}^\lambda\chi_{\frac{1}{2}}^\rho] + \phi_N^\lambda[\psi_{00}^\rho\chi_{\frac{1}{2}}^\rho - \psi_{00}^\lambda\chi_{\frac{1}{2}}^\lambda]\right\}$$

$$[56,2^+]: \quad |\Delta^4D_S(\tfrac{1}{2}^+,\tfrac{3}{2}^+,\tfrac{5}{2}^+,\tfrac{7}{2}^+)\rangle = C_A\phi_\Delta^S\psi_{2M}^S\chi_{\frac{3}{2}}^S,$$

$$|N^2D_S(\tfrac{3}{2}^+,\tfrac{5}{2}^+)\rangle = C_A\psi_{2M}^S\frac{1}{\sqrt{2}}(\phi_N^\rho\chi_{\frac{1}{2}}^\rho + \phi_N^\lambda\chi_{\frac{1}{2}}^\lambda)$$

$$[70,2^+]: \quad |N^4D_M(\tfrac{1}{2}^+,\tfrac{3}{2}^+,\tfrac{5}{2}^+,\tfrac{7}{2}^+)\rangle = C_A\chi_{\frac{3}{2}}^S\frac{1}{\sqrt{2}}(\phi_N^\rho\psi_{2M}^\rho + \phi_N^\lambda\psi_{2M}^\lambda)$$

$$|\Delta^2D_M(\tfrac{3}{2}^+,\tfrac{5}{2}^+)\rangle = C_A\phi_\Delta^S(\psi_{2M}^\rho\chi_{\frac{1}{2}}^\rho + \psi_{2M}^\lambda\chi_{\frac{1}{2}}^\lambda)$$

$$|N^2D_M(\tfrac{3}{2}^+,\tfrac{5}{2}^+)\rangle = C_A\frac{1}{2}\left\{\phi_N^\rho[\psi_{2M}^\rho\chi_{\frac{1}{2}}^\lambda + \psi_{2M}^\lambda\chi_{\frac{1}{2}}^\rho] + \phi_N^\lambda[\psi_{2M}^\rho\chi_{\frac{1}{2}}^\rho - \psi_{2M}^\lambda\chi_{\frac{1}{2}}^\lambda]\right\}$$

$$[20,1^+]: \quad |N^2P_A(\tfrac{1}{2}^+,\tfrac{3}{2}^+)\rangle = C_A\psi_{1M}^A\frac{1}{\sqrt{2}}(\phi_N^\rho\chi_{\frac{1}{2}}^\lambda - \phi_N^\lambda\chi_{\frac{1}{2}}^\rho). \qquad (19)$$

In the case of states with unequal mass quarks the procedure for constructing the basis proceeds in analogy to that outlined above, except we now impose the simpler requirement that the states have exchange symmetry between the two equal-mass quarks.

The Anharmonic Perturbation

The energies and compositions (wavefunctions) of the low-lying resonances are found by solving the Schrödinger equation using first-order perturbation theory in the anharmonicity $U=\sum_{i<j}U_{ij}$ and the hyperfine interaction $H_{hyp}=\sum_{i<j}H_{ij}^{hyp}$.

The anharmonicity is flavor independent and a spin-scalar, so it is SU(6) symmetric; this means that it causes no splittings within a given SU(6) multiplet [this is one of the reasons for classifying the states within SU(6)]. It does, however, break the initial degeneracy, illustrated in figure 1, within the harmonic oscillator bands. The anharmonicity is treated in the Isgur-Karl model as a diagonal perturbation on the energies of the states, and so is not allowed to cause mixing between the $N = 0$ and $N = 2$ band states. It gives splittings between the $N = 1$ band states only when the quark masses are unequal; the diagonal expectations of U in the $N = 0$ (and $N = 1$ bands for the N and Δ states) are lumped into the band energies.

Let's examine a sample calculation of the anharmonicity in the basis we have constructed, say, its value in the $[56, 0^+]$ multiplet. We have

$$\langle \phi^S_\Delta \psi^S_{00} \chi^S_{\frac{3}{2}} | \sum_{i<j} U_{ij} | \phi^S_\Delta \psi^S_{00} \chi^S_{\frac{3}{2}} \rangle = 3 \langle \psi^S_{00} | U(r_{12}) | \psi^S_{00} \rangle = 3 \frac{\alpha^3}{\pi^{\frac{3}{2}}} \int d^3\rho U(\sqrt{2}\rho) e^{-\alpha^2\rho^2} =: a, \tag{20}$$

where we have calculated with the simplest member of the $[56, 0^+]$ multiplet, and used the symmetry of the basis state to replace $\sum_{i<j} U_{ij}$ by $3U(r_{12})$, since $r_{12} = \sqrt{2}\rho$.

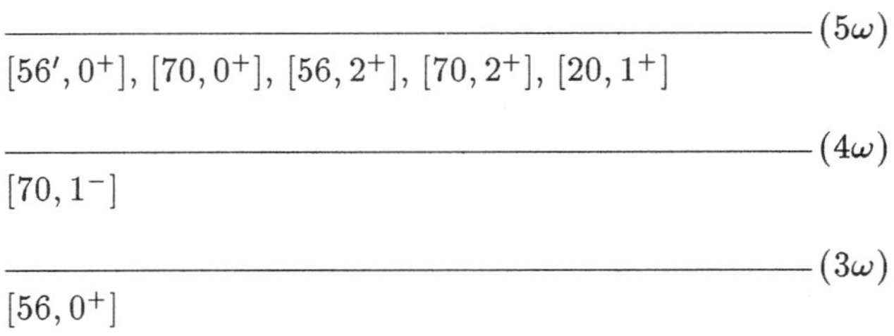

Levels	Energy
$[56', 0^+]$, $[70, 0^+]$, $[56, 2^+]$, $[70, 2^+]$, $[20, 1^+]$	(5ω)
$[70, 1^-]$	(4ω)
$[56, 0^+]$	(3ω)

Figure 1. The unperturbed oscillator levels in the SU(6) limit.

In the $[70, 1^-]$ multiplet the calculation becomes a little more involved,

$$\begin{aligned} &\langle \frac{1}{\sqrt{2}} (\phi^\rho_N \psi^\rho_{1M'} + \phi^\lambda_N \psi^\lambda_{1M'}) | \sum_{i<j} U_{ij} | \frac{1}{\sqrt{2}} (\phi^\rho_N \psi^\rho_{1M} + \phi^\lambda_N \psi^\lambda_{1M}) \rangle \\ &= \frac{3}{2} \delta_{M'M} \left\{ \langle \psi^\rho_{1M} | U(\sqrt{2}\rho) | \psi^\rho_{1M} \rangle + \langle \psi^\lambda_{1M} | U(\sqrt{2}\rho) | \psi^\lambda_{1M} \rangle \right\} \\ &= \frac{b}{3} + \frac{a}{2} \end{aligned} \tag{21}$$

where we have suppressed the spin expectation value $\langle \chi^S_{\frac{3}{2}} | \chi^S_{\frac{3}{2}} \rangle = 1$, left out some steps, and defined

$$b := 3 \frac{\alpha^5}{\pi^{\frac{3}{2}}} \int d^3\rho \rho^2 U(\sqrt{2}\rho) e^{-\alpha^2\rho^2}. \tag{22}$$

With the further definition

$$c := 3 \frac{\alpha^7}{\pi^{\frac{3}{2}}} \int d^3\rho \rho^4 U(\sqrt{2}\rho) e^{-\alpha^2\rho^2}, \tag{23}$$

it is straightforward to see that, in first order, the anharmonicity splits up the $N = 2$ band states in a pattern which is independent of the exact form of the potential U, as illustrated in figure 2, where we have defined $\Omega := \omega - a/2 + b/3$ and $\Delta := -5a/4 + 5b/3 - c/3$. Note Isgur and Karl fit Ω and Δ to the spectrum of the positive-parity excited states, using $E_0 = 1150$ MeV, $\Omega \simeq 440$ MeV, $\Delta \simeq 440$ MeV, which moves the first radial excitation $[56', 0^+]$ *below* the (unperturbed by H^{hyp}) level of the P-wave excitation $[70, 1^-]$. Note that ω is 250 MeV, which makes this first order perturbative treatment questionable (since $\Delta \simeq \Omega > \omega$).

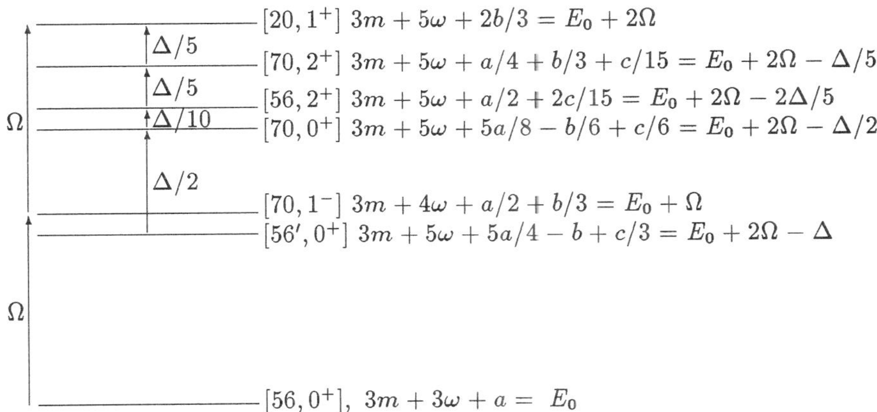

Figure 2. The anharmonic perturbation applied to the unperturbed basis.

The Hyperfine Perturbation

The hyperfine interaction is treated to first order in both the energies and wavefunctions, with the contact part of the hyperfine interaction [the first term in Eq. (2), H^{cont}] active within all bands, and the tensor interaction [the second term in Eq. (2)] active within the $N = 1$ and $N = 2$ bands. To a large degree it is the contact interaction (responsible, *e.g.*, for the $\Delta - N$ and $\Sigma - \Lambda$ splittings) and the anharmonic splitting in the $N = 2$ band which determine the coarse features of the spectrum.

Continuing with our example of the spectroscopy of the nonstrange baryons up to the $N = 2$ band, we may evaluate the contact interaction by exploiting the exchange symmetry of the wavefunctions,

$$\sum_{i<j} S_i \cdot S_j = S^2 - s_1^2 - s_2^2 - s_3^2 = S(S+1) - 3(3/4) = \begin{cases} 3/2 & S = 3/2 \\ -3/2 & S = 1/2. \end{cases} \tag{24}$$

Evaluating the spatial integral for the ground state N and Δ we find

$$\langle \psi_{00}^S | \delta^3(r_{ij}) | \psi_{00}^S \rangle = \langle \psi_{00}^S | \delta^3(r_{12}) | \psi_{00}^S \rangle = \frac{\alpha^3}{\pi^{\frac{3}{2}}} \int d^3\rho\, \delta^3(\sqrt{2}\rho) e^{-\alpha^2\rho^2} = \frac{\alpha^3}{\pi^{\frac{3}{2}}} \frac{1}{2\sqrt{2}}, \tag{25}$$

so that the contact splitting between these two $[56, 0^+]$ states is

$$\langle \Delta^4 S_S \tfrac{3}{2}^+ | H^{cont} | \Delta^4 S_S \tfrac{3}{2}^+ \rangle - \langle N^2 S_S \tfrac{1}{2}^+ | H^{cont} | N^2 S_S \tfrac{1}{2}^+ \rangle = \left(\frac{2\alpha_s}{3m_u^2} \frac{8\pi}{3} \right) \cdot 3 \cdot \left(\frac{\alpha^3}{\pi^{\frac{3}{2}}} \frac{1}{2\sqrt{2}} \right) =: \delta. \tag{26}$$

The parameter δ is fit to the spectrum, and is approximately 300 MeV.

The model is now able to predict the pattern of the contact (and tensor) interaction splittings between the various states in the $N = 1$ band of negative-parity states, as all of these splittings are proportional to the parameter δ. Since the wavefunctions have been symmetrized we may concentrate on $\langle H_{12}^{hyp} \rangle$, since $\langle H^{hyp} \rangle = \langle \sum_{i<j} H_{ij}^{hyp} \rangle = 3\langle H_{12}^{hyp} \rangle$ when evaluated in this basis. Noting that

$$S_1 \cdot S_2 = S_{12}^2 - s_1^2 - s_2^2 = S_{12}(S_{12}+1) - 2(3/4) = \begin{cases} 1/2 & S_{12} = 1 \\ -3/2 & S_{12} = 0, \end{cases} \tag{27}$$

where $S_{12} = S_1 + S_2$, we can easily see that

$$\begin{aligned}
\langle \chi^{\rho}_{\frac{1}{2}} | S_1 \cdot S_2 | \chi^{\rho}_{\frac{1}{2}} \rangle &= -\frac{3}{2} \\
\langle \chi^{\lambda}_{\frac{1}{2}} | S_1 \cdot S_2 | \chi^{\lambda}_{\frac{1}{2}} \rangle &= \langle \chi^{S}_{\frac{3}{2}} | S_1 \cdot S_2 | \chi^{S}_{\frac{3}{2}} \rangle = \frac{1}{2}.
\end{aligned} \tag{28}$$

If we put these together with the spatial matrix elements

$$\begin{aligned}
\langle \psi^{\rho}_{1M} | \delta^3(r_{12}) | \psi^{\rho}_{1M} \rangle &= 0 \\
\langle \psi^{\lambda}_{1M} | \delta^3(r_{12}) | \chi^{\lambda}_{1M} \rangle &= \frac{\alpha^3}{\pi^{\frac{3}{2}}} \frac{1}{2\sqrt{2}},
\end{aligned} \tag{29}$$

we find that, for any permissible J,

$$\begin{aligned}
\langle N^4 P_M J^- | H^{cont} | N^4 P_M J^- \rangle &= \langle \Delta^2 P_M J^- | H^{cont} | \Delta^2 P_M J^- \rangle \\
&= 3 \left(\frac{2\alpha_s}{3m_u^2} \frac{8\pi}{3} \right) \cdot \frac{1}{2} \cdot \frac{1}{2} \cdot \left(\frac{\alpha^3}{\pi^{\frac{3}{2}}} \frac{1}{2\sqrt{2}} \right) = \frac{\delta}{4} \\
\langle N^2 P_M J^- | H^{cont} | N^2 P_M J^- \rangle &= 3 \left(\frac{2\alpha_s}{3m_u^2} \frac{8\pi}{3} \right) \cdot \frac{1}{4} \cdot \left(-\frac{3}{2} + \frac{1}{2} \right) \cdot \left(\frac{\alpha^3}{\pi^{\frac{3}{2}}} \frac{1}{2\sqrt{2}} \right) = -\frac{\delta}{4}.
\end{aligned} \tag{30}$$

The resulting spectrum is illustrated in figure 3, where the unperturbed level of the nonstrange P-wave states has been taken to be around 1610 MeV.

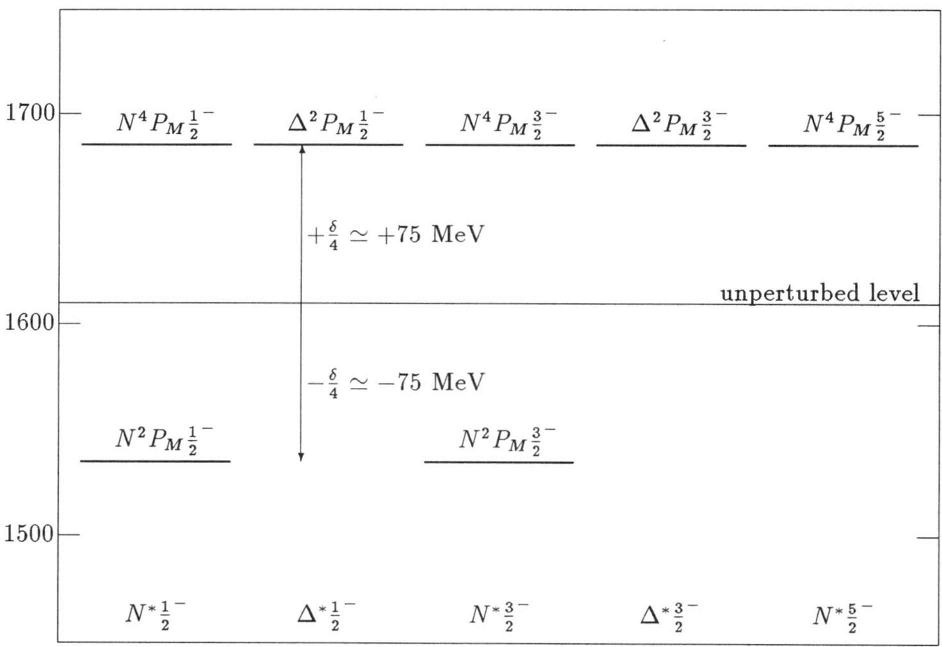

Figure 3. The contact perturbation applied to the P-wave nonstrange baryons.

The tensor part of the hyperfine interaction has generally smaller expectation values, and so is not as important to the spectroscopy as the contact interaction. It does, however, cause significant mixing in some states which are otherwise unmixed; this has important consequences for the decays of the states in question. The best techniques to use for its evaluation are based on the Wigner-Eckhart theorem for composite systems [8]. For details see Refs. [1] and [3]. Note that, as in the contact part of the hyperfine interaction, the tensor interaction is still a total angular momentum (J) scalar and an isospin scalar, so it can only mix the states in our sample spectrum with the same J but different quark spin S.

Leaving the details of the calculation to the interested reader, we simply plot in figure 4 the results of Isgur and Karl's calculation of the tensor interaction for our states. Note that the calculation of the tensor interaction for the $N^*\frac{1}{2}^-$ and $N^*\frac{3}{2}^-$ states involves diagonalizing a 2x2 matrix, and the result is that the eigenstates of the Hamitonian are admixtures of the (quark)-spin-$\frac{1}{2}$ and -$\frac{3}{2}$ basis states. Note also that the quark-spin part of the tensor interaction is a second-rank tensor, which has a zero expectation value in the purely quark-spin-$\frac{1}{2}$ states $\Delta^2P_M\frac{1}{2}^-$ and $\Delta^2P_M\frac{3}{2}^-$. For details see Ref. [9].

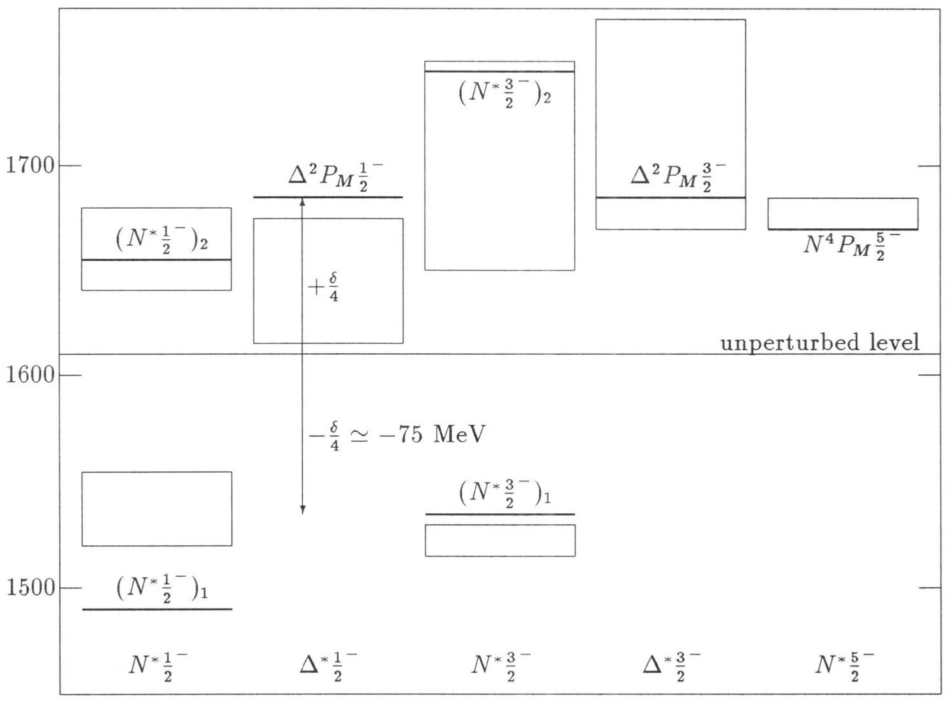

Figure 4. The hyperfine perturbation applied to the P-wave nonstrange baryons.

It is arguable whether the tensor part of the hyperfine interaction has improved the agreement between the model predictions for the masses of these states and the values extracted from various partial wave analyses, shown as boxes in Fig. 4. The boxes show the approximate range of values for the resonance mass quoted by the Particle Data Group [10]. Isgur and Karl argue [9], however, that there is evidence from analyses of decays for the mixing in the wavefunctions caused by the tensor interaction. If we write the tensor mixing between the two states $|N^2P_MJ^-\rangle$ and $|N^4P_MJ^-\rangle$ (for $J=\frac{1}{2}$

and $\frac{3}{2}$) in terms of an angle, we find

$$|(N^*\tfrac{3}{2}^-)_1\rangle = cos\theta_D|N^2P_M\tfrac{3}{2}^-\rangle - sin\theta_D|N^4P_M\tfrac{3}{2}^-\rangle, \tag{31}$$

with $\theta_D \simeq +6^o$, and

$$|(N^*\tfrac{1}{2}^-)_1\rangle = cos\theta_S|N^2P_M\tfrac{1}{2}^-\rangle - sin\theta_S|N^4P_M\tfrac{1}{2}^-\rangle, \tag{32}$$

with $\theta_D \simeq -32^o$. Empirical mixing angles were determined from an analysis of decay data [11] to be $\theta_D \simeq +10^o$ and $\theta_S \simeq -32^o$.

MODELS OF BARYON STRONG DECAYS

Models of the spectrum of baryon states are of limited usefulness without a simultaneous model of their strong decays, as otherwise it may not be possible to tell which of the predicted states corresponds to a state seen in the strong-interaction experiments such as πN elastic scattering. For example, there are more states predicted by the model in the $N = 2$ bands for N, Δ, Λ, and Σ than exist in the partial wave analyses. Strong decay models can tell us which states are likely to have substantial πN and $\bar{K}N$ couplings, and so should appear in these experiments.

Hadron Transition Models

We will consider only OZI-allowed [12] strong decays of baryons, where the quark in the final-state meson originates in the initial baryon. As a first step we could consider a model where the hadrons are treated as elementary point-like objects, as illustrated in figure 5.

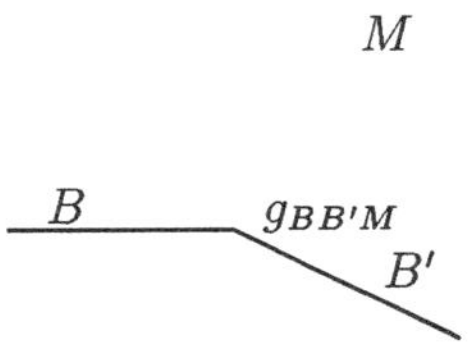

M

B $g_{BB'M}$ B'

Figure 5. The decay process $B' \rightarrow BM$ via elementary meson emission from a pointlike baryon.

This kind of model has little predictive power, since each transition is described in terms of an independent phenomenological coupling constant, $g_{BB'M}$. While the use of SU(2) or SU(3) flavor symmetry arguments can give relationships among some of these coupling constants, the overall situation is nevertheless largely unworkable.

A second class of models treats the baryons as objects with structure, but the decay takes place by the emission of an elementary meson. Such an approach is often taken in bag models; the work of Koniuk and Isgur [13] uses this approach within a potential model, using the configuration mixed wavefunctions for the baryons which result from the nonrelativistic model. Here the mesons are emitted from quark lines, as illustrated in figure 6, and the $g_{BB'M}$ coupling constants are replaced with a smaller set of $g_{qq'M}$'s. In addition, SU(2) or SU(3) flavor symmetry can be used to relate the coupling constants for mesons within a single multiplet, as well as those for different quarks.

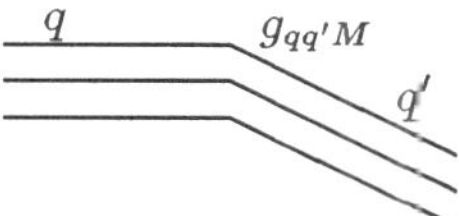

Figure 6. The decay process $B' \to BM$ via elementary meson emission from a quark.

A third class of models may be referred to broadly as pair creation models. Here both the baryons and mesons have structure, and the decay of the baryon occurs by the creation of a quark-antiquark pair somewhere in the hadronic medium. The created antiquark combines with one of the quarks from the decaying baryon to form the daughter meson, while the quark of the created pair becomes part of the daughter baryon. This is illustrated in figure 7.

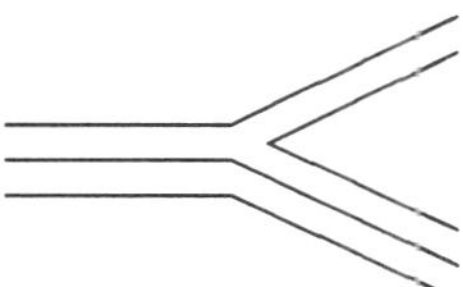

Figure 7. The OZI-allowed decay process $B' \to BM$ via quark pair creation.

There are several types of pair creation model. In the $^3P_0$ model popularized by LeYaouanc *et al.* [14], the quark-antiquark pair is created anywhere in space with the quantum numbers of the QCD vacuum, namely 0^{++}. Since the created quark and antiquark have opposite instrinsic parity and charge conjugation parity, they are placed in an L=1 relative wavefunction to restore an even P and C for the pair. In order to have $J = 0$ for the pair, they must be created with their spins aligned to give $S = 1$. This corresponds to $^3P_0$, hence the name of the model. The operator responsible for the transition is then written as

$$\begin{aligned} T &= -3\gamma \sum_{i,j} \int dp_i dp_j \delta(p_i + p_j) C_{ij} F_{ij} \\ &\times \sum_m C(1,1,m,-m|0,0)\chi^{ij}_{1m}\mathcal{Y}_{1\,-m}(p_i - p_j) b^\dagger_i(p_i) d^\dagger_j(p_j). \end{aligned} \qquad (33)$$

Here, C_{ij} and F_{ij} are the color and flavor wave functions of the created pair, both assumed to be singlet, χ^{ij} is the spin triplet wave function of the pair, and $\mathcal{Y}_{1\,-m}(p_i - p_j)$ is the vector harmonic which places the created pair in a relative p-wave. The pair creation may occur, in principle, very far away from the decaying hadron, but the greatly extended wave functions required naturally suppress such contributions to the decay amplitude. This model has been quite popular in descriptions of hadron decays, and has been applied to baryon decays [14] and meson decays [14, 15], as well as the decays of fictitious four-quark states [16]. This model has the drawback of being purely phenomenological (although there has been recent progress in understanding its origins [17], but has the advantage that the decay amplitudes are written in terms of a single parameter γ. In practice the model also involves some choices for the phase space in order to evaluate decay widths.

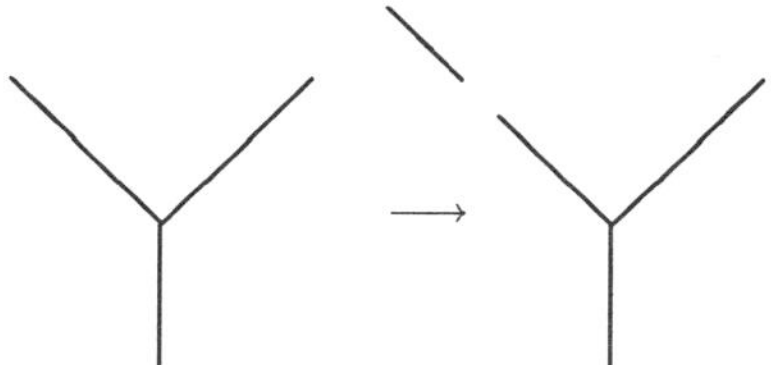

Figure 8. The decay process $B' \to BM$ via string breaking.

Other pair-creation models include the string-breaking models of [18] Dosch and Gromes, and of Alcock, Burfitt and Cottingham [19]. In these models, the lines of color flux between quarks have collapsed into a string, and the pair is created when the string breaks, as illustrated in figure 8. In the Dosch-Gromes version of this model, the created pair have the quantum numbers ${}^3P_0$, while in the Alcock *et al.* version, the quantum numbers of the created pair is ${}^3S_1$.

Several authors have used similar ideas in describing decays of hadrons in flux-tube breaking models [20]. Here, the pair still has quantum numbers ${}^3P_0$, but is constrained to be created somewhere within the flux tube connecting quarks. The 'string' breaking picture arises in the zero-width limit of the flux tube.

We have extended the ${}^3P_0$ model to describe a larger set of baryons, using the 'relativized model' wavefunctions described below and a ${}^3P_0$ model of the decays. As we shall see in the next lecture, the strong decay calculation carried out by Koniuk and Isgur [13] in the elementary-meson emission model and our ${}^3P_0$-model calculation establish that the states whose mixed wavefunctions allow them to couple to their πN and $\bar{K}N$ formation channels largely correspond, in both energy and number, with the observed states in the analyses of πN and $\bar{K}N$ scattering.

INCORPORATING RELATIVISTIC EFFECTS

As we shall see, the main features of the spectrum of the low-lying baryon resonances are quite convincingly described by the nonrelativistic model. Just as importantly, the mixing of the states caused by the hyperfine interaction is crucial to explaining their observed decays, *e.g.* the $N\eta$ decays of the $N^*\frac{1}{2}^-$ states. However, there are a number of criticisms which can be made of the model, many of which are based on the absence of a consistent relativistic treatment. We will briefly outline these criticisms, and then describe models which go beyond the limitations of the nonrelativistic model.

Criticisms of the Nonrelativistic Quark Model

In strongly bound systems such as the baryons, where $p/m \simeq 1$, the approximation of nonrelativistic kinematics and dynamics is not justified. For example, if one forms the one-gluon exchange T-matrix element *without* performing a nonrelativistic reduction, factors of m_i in Eq. 2 are replaced, roughly, with factors of $E_i = \sqrt{p_i^2 + m_i^2}$. In a potential model picture there should be 'kinematic' smearing of the interquark coordinate r_{ij} with a characteristic size given by the Compton wavelength of the quark $1/m_q$.

Neglecting the scale dependence of a cut-off field theory (QCD) has resulted in non-fundamental values of parameters like the quark mass, the string tension (implicit in the size of the anharmonic perturbations) and the strong coupling $\alpha_s \simeq 2$. A consistent theory with constituent quarks should give those quarks a commensurate size, which will smear out the interactions between the quarks. The string tension should be consistent with meson spectroscopy, and the relation between the anharmonicity and the meson string tension should be explored. If there are genuine three-body forces in baryons they are neglected in the NR model. The neglect of spin-orbit interactions in the Hamiltonian is also inconsistent, independent of our choice of *ansatz* for the short-distance and confining physics. There is some evidence in the *observed* spectrum for spin-orbit splittings, *e.g.* that between the states $\Delta^*\frac{1}{2}^-(1620)$ and $\Delta^*\frac{3}{2}^-(1700)$.

The model also uses a first-order perturbative evaluation of large perturbations. The contact term is, unless the above smearing is implemented, formally infinite. As we have seen, the size of the first-order anharmonic splitting of the $N = 2$ band must be larger than the 0-th order harmonic splitting, to get the lightest $N = 2$ band nucleon [identified with the Roper resonance $N(1440)$] below the P-wave non-strange states. This calls into question the usefulness of first order perturbation theory. It also means that the wavefunctions of states like the Roper resonance should have a large anharmonic *mixing* with the ground states. Some of these flaws of the nonrelativistic model are inessential, and can be corrected. In the following we will describe a model which puts the ideas of many authors together in an attempt to correct as many of these deficiencies as possible.

The Relativized Quark Model

Modern calculations [21] have gone beyond Isgur and Karl's model for the spectrum and wavefunctions of baryons. As an example we consider the relativized quark model, which was first applied to meson spectroscopy by Godfrey and Isgur [22], and later adapted to the baryon system [3]. Once again the Schrödinger equation is solved in a Fock space made up of valence quarks. The Hamiltonian is now given by $H = \sum_i \sqrt{p_i^2 + m_i^2} + V$, where V is a relative-position and -momentum dependent potential which tends, in the nonrelativistic limit (*not* taken here) to $V_{string} + V_{Coul} + V_{hyp} + V_{so(cm)} + V_{so(Tp)}$. Here V_{string} is the potential generated by adding the lengths of the gauge-invariant (Y-shaped) string configuration, and multiplying by the meson string tension $\sqrt{\sigma}$. The string is assumed to adjust infinitely quickly to the motion of the quarks so that it is always in its minimum length configuration; this generates a three-body adiabatic potential for the quarks [23] which includes genuine three-body forces.

The Coulomb, hyperfine, color-magnetic and Thomas-precession spin-orbit potentials [1] are as they were in the nonrelativistic model except that: (1) the inter-quark coordinate r_{ij} is smeared out over mass-dependent distances, as suggested by relativistic kinematics and QCD; (2) the momentum dependence away from the $p/m \to 0$ limit is parametrized, as suggested by relativistic dynamics. In practice (1) is brought about by convoluting the potentials with a smearing function $\rho_{ij}(r_{ij}) = \sigma_{ij}^3 e^{-\sigma_{ij}^2 r_{ij}^2}/\pi^{\frac{3}{2}}$, where the σ_{ij} are chosen to smear the inter-quark coordinate over distances of $O(1/M_Q)$ for Q heavy, and approximately 0.1 fm for light quarks; (2) is brought about by introducing factors which replace m_i in the nonrelativistic model by, roughly, E_i. For example the contact part of H_{hyp}^{ij} becomes

$$V_{cont}^{ij} = \left(\frac{m_i m_j}{E_i E_j}\right)^{\frac{1}{2}+\epsilon_{cont}} \frac{8\pi}{3}\alpha_s(r_{ij})\frac{2}{3}\frac{S_i \cdot S_j}{m_i m_j}\left[\frac{\sigma_{ij}^3}{\pi^{\frac{3}{2}}}e^{-\sigma_{ij}^2 r_{ij}^2}\right]\left(\frac{m_i m_j}{E_i E_j}\right)^{\frac{1}{2}+\epsilon_{cont}}, \qquad (34)$$

where ϵ_{cont} is a constant parameter, and $\alpha_s(r_{ij})$ is a running-coupling constant which runs according to the lowest-order QCD formula, saturating to 0.6 at $Q^2 = 0$.

The energies and wavefunctions of all the baryons are then solved for by expanding the states in a large harmonic oscillator basis. Wavefunctions are expanded to $N \leq 7$ ($N \leq 8$ for $J^P = \frac{1}{2}^+$). The Hamiltonian matrix is diagonalised in this basis, and energy eigenvalues are then crudely minimized in α, the harmonic-oscillator size parameter. To avoid construction of symmetrized wavefunctions in this large basis, we have not explicitly symmetrized the non-strange wavefunctions in u and d quarks, but have allowed the strong Hamiltonian (which cannot distinguish u and d quarks) to sort the basis into N and Δ states.

The result is that the largely successful phenomenology of the nonrelativistic model is realized with a new understanding. The relativized model shows some improvements, and some deterioration relative to the nonrelativistic model, but is much more tightly constrained. The same set of parameters fit *all* mesons and baryons, except the string tension $\sqrt{\sigma}$ which in our best fit was reduced about 15% from the meson value. One can conclude that the nonrelativistic approximation is not essential. Spin-orbit interactions in this model are small, for several reasons; we are able to use an α_s smaller by a factor of 3, and retain the same contact interaction splittings, by virtue of our non-perturbative evaluation of the smeared contact interaction. There is also, as expected [1], a partial cancellation of the color-magnetic and Thomas-precession spin-orbit terms, and we have used our freedom to choose $\epsilon_{cont} < \epsilon_{so}$ to suppress the spin-orbit interactions relative to the hyperfine terms.

SUMMARY

We have seen how a simple nonrelativistic model of baryons based on constituent quark degrees of freedom is constructed, and how this simple model largely accounts for the structure of the nonstrange baryon spectrum. We have also described how the resulting wavefunctions may be used in a calculation of the strong decays of these states. The relativized model has been introduced as an example of how to incorporate relativistic effects in a more realistic picture. We will return to a critical comparison between this model and the results of the analyses of πN and $\bar{K}N$ elastic scattering in the next lecture, where we will see that there are several problematic states which are not well understood, and where we suggest what could be done, both theoretically and experimentally, to understand them.

In much the same way that the interquark Hamiltonian can be evaluated in this harmonic-oscillator basis, one can evaluate the strength of electromagnetic transitions between baryon states such as photoproduction from the n or p, $\gamma N \rightarrow X$, with X a baryon resonance (the same amplitudes will determine the widths for baryon resonances to decay via emission of a photon). This evaluation will be outlined in the next lecture, and the resulting photocouplings will be compared to those extracted from pion photoproduction data.

ACKNOWLEDGEMENTS

This work was supported in part by the U.S. Department of Energy through Contract No. DE-FG05-86ER40273, and by the Florida State University Supercomputer Computations Research Institute which is partially funded by the Department of Energy through Contract No. DE-FC05-85ER250000.

References

[1] N. Isgur and G. Karl, *Phys. Lett.* **72B**, 109 (1977); **74B** (78) 353; Phys. Rev. **D18**, 4187 (1978).

[2] N. Isgur and G. Karl, Phys. Rev. **D19**, 2653 (1979).

[3] S. Capstick and N. Isgur, Phys. Rev. **D34**, 2809 (1986).

[4] P. Geiger and N. Isgur, Phys. Rev. **D41**, 1595 (1990); **D44**, 799 (1991).

[5] N. Isgur and J. Paton, Phys. Lett. **124B**, 247 (1983); Phys. Rev. **D31**, 2910 (1985); J. Merlin and J. Paton, J. Phys. G **11**, 439 (1985).

[6] N. Isgur, Phys. Rev. **D21**, 779 (1980); S. Capstick, Phys. Rev. **D36**, 2800 (1987); R.E. Cutkosky and P. Geiger, Phys. Rev. **D48**, 1315 (1993); V. Dmitrasinovic and S.J. Pollock, University of Colorado preprint UC/NPL-1116 (hep-ph/9504414).

[7] If we impose $SU(6)$, then we need a symmetrized wavefunction for this Λ state; since the flavor wavefunction may be totally antisymmetric if it contains three distinguishable symbols u, d, and s, this is
$|\Lambda^2 P_M(\frac{1}{2}^-, \frac{3}{2}^-)\rangle = C_A \phi^A_\Lambda(\psi^\rho_{1M}\chi^\lambda_{\frac{1}{2}} - \psi^\lambda_{1M}\chi^\rho_{\frac{1}{2}})$.

[8] *Angular Momentum*, by D.M. Brink and G.R. Satchler, Clarendon Press, Oxford (1962).

[9] N. Isgur and G. Karl, Phys. Lett. **72B**, 109 (1977).

[10] The Particle Data Group, L. Montanet et al., Phys. Rev. **D50**, 1173 (1994).

[11] A.J.G. Hey, P.J. Litchfield, and R.J. Cashmore, Nucl. Phys. **B95**, 516 (1975). Faiman and Plane found similar results in their $SU(6)_W$ analysis, see D. Faiman and D.E. Plane, Nucl. Phys. **B50**, 379 (1972).

[12] S. Okubo, Phys. Lett. **5**, 165 (1963); G. Zweig, CERN reports TH-401, TH-402 (1964,1965), in *Proceedings of the International School of Physics, 'Ettore Majorana'*, Erice, Italy, 1964, A. Zichichi ed., Academic Press, New York, 1965; J. Iizuka, Prog. Theor. Phys. Supp. 37 -38 (1966) 21.

[13] R. Koniuk and N. Isgur, Phys. Rev. **D21**, 1868 (1980).

[14] See for example A. Le Yaouanc, L. Oliver, O. Pène and J. C. Raynal, *Hadron Transitions In The Quark Model*, Gordon and Breach, 1988. This work contains many references to hadron decay models. Other references to hadron transition models are: A. Le Yaouanc, L. Oliver, O. Pène and J. C. Raynal, Phys. Rev. **D8**, 2223 (1973); **D9**, 1415 (1974); **D11**, 680 (1975); **D11**, 1272 (1975); E. Eichten, K. Gottfried, T. Kinoshita, K. Lane and T. M. Yan, Phys. Rev. **D17**, 3090 (1978).

[15] See for example L. Micu, Nucl. Phys. **B10**, 521 (1969); A. Le Yaouanc, L. Oliver, O. Pène and J. C. Raynal, Phys. Lett. **71B**, 397 (1977).

[16] M. Barbour and J. P. Gilchrist, Zeit. Phys. **C7**, 225 (1981); Zeit. Phys. **C8**, 282(E) (1981); J. P. Ader, B. Bonnier and S. Sood, Nuovo Cim. **68**, 1 (1982); Phys. Lett. **84B**, 488 (1979); W. Roberts, B. Silvestre-Brac and C. Gignoux, Phys. Rev. **D41**, 182 (1990).

[17] T. Barnes, private communication.

[18] H. Dosch and D. Gromes, Phys. Rev. **D33**, 1378 (1986).

[19] J. W. Alcock, M. J. Burfitt and W. N. Cottingham, Zeit. Phys. **C25**, 161 (1984).

[20] N. Isgur and J. Paton, in Ref.[5]; R. Kokoski and N. Isgur, Phys. Rev. **D 35**, 907 (1987); S. Kumano and V. R. Pandharipande, Phys. Rev. **D38**, 146 (1988); Fl. Stancu and P. Stassart, Phys. Rev. **D38**, 233 (1988); **D39**, 343 (1989); **D41**, 916 (1990); **D42**, 1521 (1990); P.R. Page, these proceedings.

[21] D.P. Stanley and D. Robson, Phys. Rev. Lett. **45**, 235 (1980); C.P. Forsyth and R.E. Cutkosky, Phys. Rev. Lett. **46**, 576 (1981); Z. Phys. **C18**, 219 (1983); J. Carlson, J. Kogut and V. R. Pandharipande, Phys. Rev. **D27**, 233 (1983); **D28**, 2809 (1983); R. Sartor and Fl. Stancu, Phys. Rev. **D31**, 128 (1985); **D33**, 727 (1986).

[22] S. Godfrey and N. Isgur, Phys. Rev. **D32**, 1 (1985)89.

[23] H.G. Dosch and V. Müller, Nucl. Phys. **B116**, 470 (1976); R.E. Cutkosky and R.E. Hendrick, Phys. Rev. **D16**, 786 (1977); **D16**, 793 (1977); J. Carlson, J.B. Kogut, and V.R. Pandharipande Phys. Rev. **D27**, 233 (1983); **D28**, 2807 (1983).

A COMPARISON WITH EXPERIMENTAL RESULTS AND OUTSTANDING ISSUES IN BARYON PHYSICS

Simon Capstick

Supercomputer Computations Research Institute
and Department of Physics
Florida State University
Tallahassee, FL 32306

ABSTRACT

A comparison is made between the results for baryon masses and decay widths from the analyses of $N\pi$ elastic scattering experiments, and a constituent quark model calculation of these quantities. This is also carried out for pion photoproduction, and for formation and production of strange baryons. This comparison is used to focus on the outstanding issues in baryon spectroscopy, with an eye towards how the new generation of experimental facilities can best address them.

$N\pi$ ELASTIC SCATTERING

Figure 1 shows a set of Born diagrams which can be used to describe $N\pi$ elastic scattering. According to the principle of duality, an equivalent description of this process is to add up amplitudes for all the s- and u-channel diagrams, with the nucleon and all of its excited states (with $I = \frac{1}{2}$ or $\frac{3}{2}$) in the intermediate state, or to add the amplitudes of all t-channel diagrams, with a set of mesons in the intermediate state.

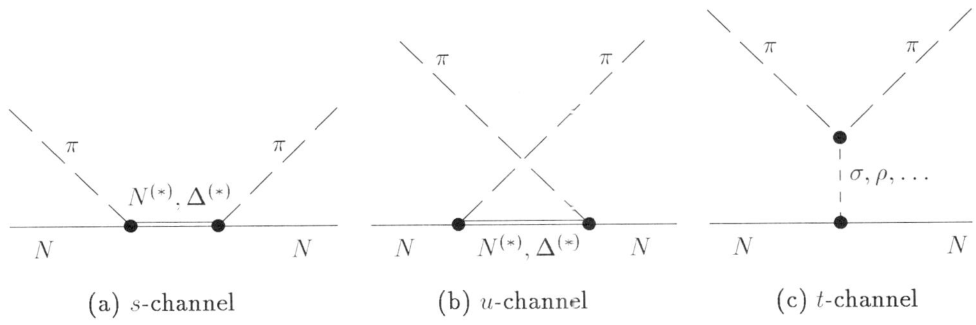

Figure 1. Born diagrams for $N\pi$ elastic scattering.

A prescription for inclusion of higher-mass resonances [1] which couple to $N\pi$ (which may be unobserved) is to use Regge-pole phenomenology to give the masses and vertex strengths for each pole term. Note that the t-channel description cannot account for the bumps in the cross-section in the resonance region, but only for the average behavior there. It is used to describe the cross section at high energy, where the cross section is strongly forward peaked, and where a description in terms of a large number of baryon resonances in the s and u channels would be highly inefficient. (It is also a necessary ingredient at very low energy, where σ exchange, for example, is needed to account for the scattering cross section). This is matched on to the description in terms of baryon resonances at the upper end of the resonance region. Likewise, in the resonance region a description in terms of s- plus u-channel diagrams is most efficient, although all analyses need to include smooth background terms in order to fit the data.

Note that we cannot extract the relative phases of the $N\pi$ coupling amplitudes for the resonances from $N\pi$ elastic scattering experiments, as the contribution to the sum from each baryon is proportional to the modulus squared of its amplitude. This is important when comparing the photocoupling amplitudes extracted from pion photoproduction data to model calculations.

In order to understand the results of the partial-wave analyses of πN elastic scattering, we need to examine the notation for πN resonances used in the Particle Data Group (PDG) [2] and elsewhere. Referring to figure 1(a), we wish to determine the quantum numbers of an intermediate s-channel baryon for a given relative angular momentum of the final π-N pair. Since the nucleon has $I = \frac{1}{2}$ and the pion has $I = 1$, the intermediate state has either $I = \frac{1}{2}$ (a nucleon or N^* resonance) or $I = \frac{3}{2}$ (a Δ or Δ^*). Since the pion has $J^P = 0^-$, and the nucleon has $J^P = \frac{1}{2}^+$, then if the final-state pion and nucleon are in a relative S-wave, the intermediate state has negative parity and can only have $J = \frac{1}{2}$; the two possibilities are labeled S_{11}, for the N^* with $J^P = \frac{1}{2}^-$, or S_{31} for the Δ^* with $J^P = \frac{1}{2}^-$ (the notation is $L_{2I\,2J}$). Similarly if the final-state pion and nucleon are in a relative P-wave, then we may have $J^P = \frac{1}{2}^+, \frac{3}{2}^+$ and either isospin, leading to the N^* states P_{11}, P_{13} and the Δ^* states P_{31} and P_{33}. There will, in general, be many resonances of a given type $L_{2I\,2J}$, with the density of states increasing at higher energies.

The Ground States N and Δ

The structure of the spectrum is to a large extent determined by the contact part of the hyperfine interaction, which is usually *fit* to the $\Delta - N$ mass splitting. Note that in the relativized model calculations of Refs [3] and [4] it was shown that the same contact interaction (identical in form and strength) can explain the $\rho - \pi$ splitting and the $\Delta - N$ splitting, once corrected for the difference between the color matrix elements of the one-gluon-exchange interactions in the two systems. The parameter γ of our ${}^3P_0$ model [5] is fit to the $\Delta \to N\pi$ decay width. This then fixes the size of $g_{NN\pi}$, which has been determined empirically in a recent analysis by the VPI group [6] to be $g_{NN\pi} = 12.92 \pm 0.13$, in contrast to the older accepted value of 13.4 ± 0.1 from Koch and Pietarinen [7]. Our model gives a value of 17.3, in only rough agreement with these values; Miller [8] obtains a value of 13.2 using a model similar to ours.

We will return to a discussion of D-wave mixing in the ground states and its effect on the $E2$ strength for the $\gamma N \to \Delta$ transition and G_E^n in the third lecture.

Low-Lying Negative Parity Excited States

As we have seen, the nonrelativistic model efficiently explains the rough pattern of mass splittings for the negative-parity states. All states found in the analyses are predicted to be present by the quark model, and there are no unseen model states, at least at these low energies. Figure 2 shows a comparison between the predictions of the relativized model of Ref. [4] and the range of mass values for the resonances quoted by the PDG [2]. Note the model wavefunctions are not restricted to the $N = 1$ harmonic oscillator band, but have compositions which are predominantly in this band if the anharmonic perturbations are not too severe. We have added a slight shift to the model predictions in the figure to better fit the band center of mass; in the nonrelativistic model this quantity is *fit* rather than predicted, so an error of this size is perhaps not unexpected in our model which *calculates* the band separations.

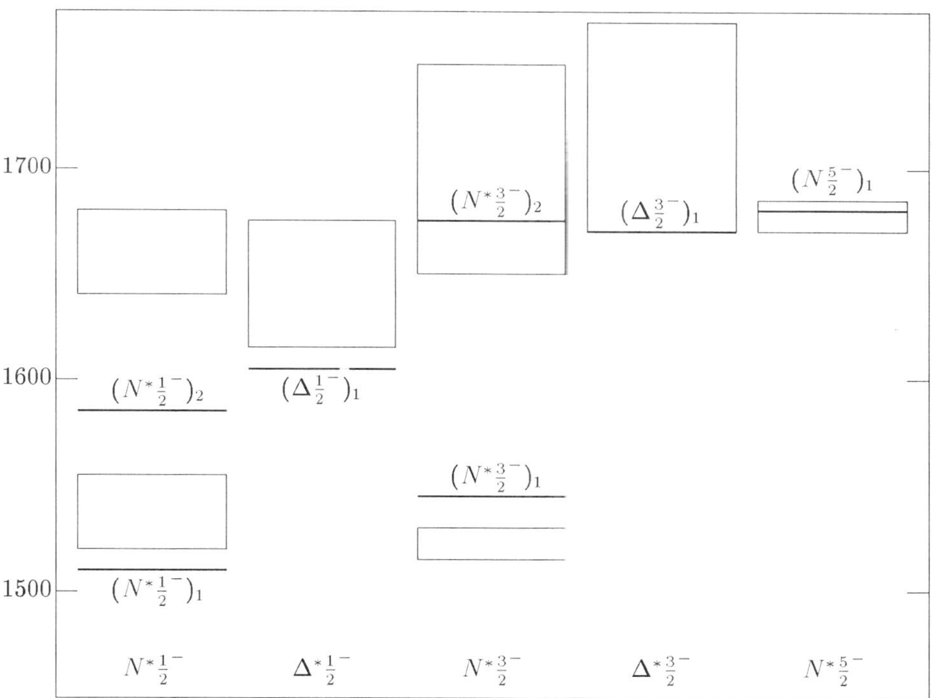

Figure 2. Non-strange negative-parity excited baryons; boxes give the range of masses quoted by the PDG [2], bars the predictions of Ref. [4], with a correction of +50 MeV added to all energies to roughly reproduce the band center of mass.

The comparison between the absolute values of the $N\pi$ decay amplitudes for these states calculated in our ${}^3P_0$ model and those extracted from the analyses is made in the lower parts of figures 3 and 4. The model is less accurate for prediction of coupling strengths than it is for masses, although it is usually able to explain the *relative* sizes of the $N\pi$ amplitudes in a given partial wave. These amplitudes depend on details of the wavefunctions, which are less certain than the energies in a variational calculation. However, the fact that our decay model has very little freedom (one parameter, and some choices for phase space) to fit the amplitudes makes this rough agreement encouraging.

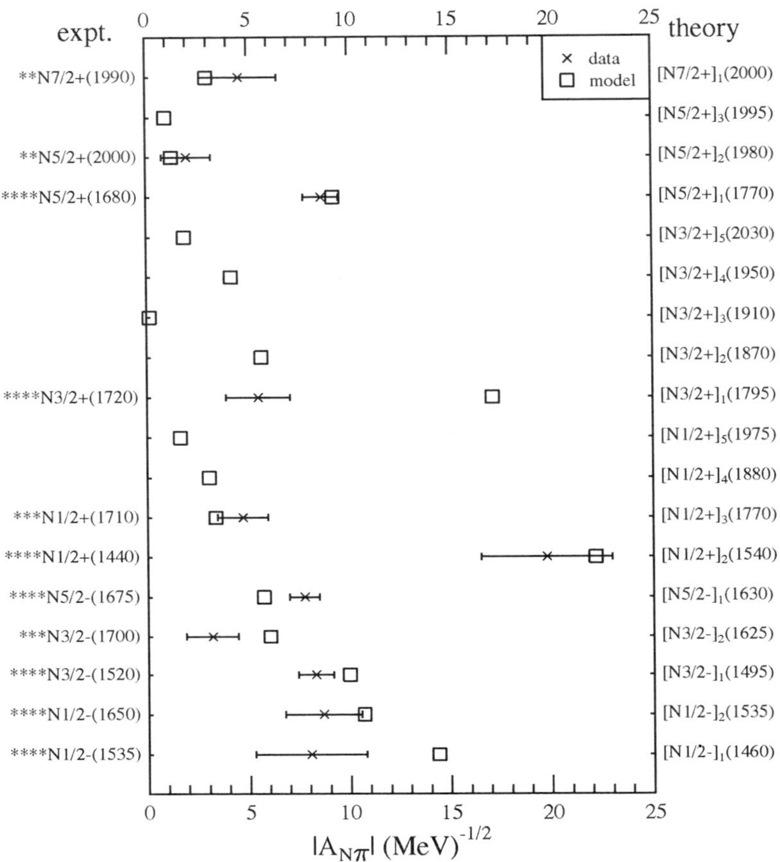

Figure 3. Absolute values of the $N\pi$ amplitudes for *all* N^* resonances in the N=1 and N=2 bands, plotted along a line parallel with the bottom axis for each state. Model states are listed along on the right axis, and states from the partial-wave [2] analyses along with their overall ratings are shown on the left axis, aligned with our model assignment. 'Missing' states are those with no experimental analogues.

The $N\eta$ decays of the two S_{11} states are an interesting test of the tensor interaction which is responsible for their mixing in the nonrelativistic model. The lightest state $N\frac{1}{2}^-(1535)$ has a strong decay width to $N\eta$, but very little phase space for the decay, while the heavier state $N\frac{1}{2}^-(1650)$ has a very small branching ratio to decay to $N\eta$. This has been explained in the nonrelativistic elementary meson emission model [9] as being due to the tensor mixing in this system, which causes a cancellation in the $N\frac{1}{2}^-(1650)$ and an enhancement in the $N\frac{1}{2}^-(1535)$. This is also the mechanism in our recent relativized $^3P_0$ model of the $N\eta$ decays [10], and can be viewed as indirect evidence for presence of the tensor interaction in baryons, as predicted by one-gluon exchange.

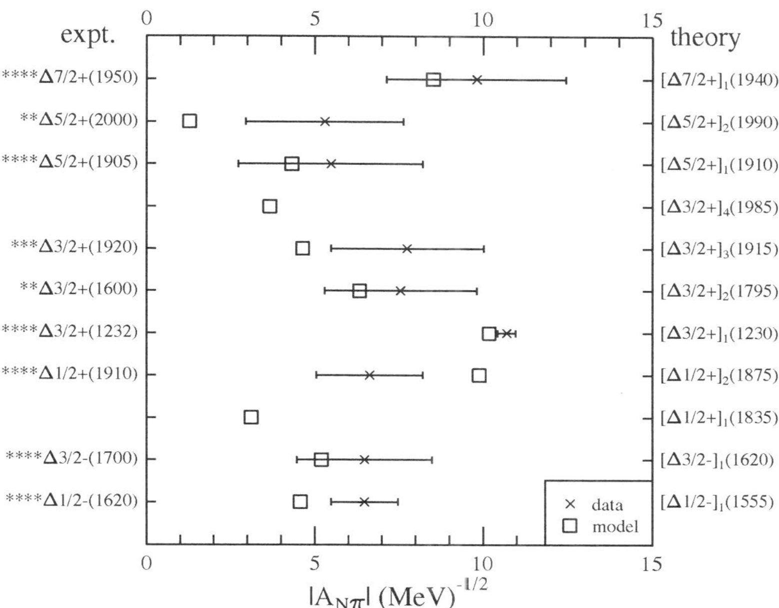

Figure 4. Absolute values of the $N\pi$ amplitudes for *all* Δ resonances in the N=0, 1, and 2 bands. Legend as in figure 3.

Interestingly, in the recent VPI analysis of Arndt *et al.* [11] the $D_{13}(1700)$ state (the second $N\frac{3}{2}^-$) is entirely missing. We will return to a discussion of this state when we discuss inelastic channels.

Positive-Parity Excited States

Figure 5 shows the comparison between the PDG quoted ranges on the masses for the low-lying positive-parity N and Δ resonances and the spectrum resulting from the relativized model. Once again we have shifted the entire spectrum a little to better fit the band center of mass. The figure only shows those states in this mass range which are predicted by our decay analysis [5] to have the dominant $N\pi$ couplings in their partial waves (see figures 3 and 4). This means that ten model states are missing in the $N\pi$ analyses; these states correspond both in energy and number to those which the model predicts to have smaller $N\pi$ couplings than other nearby states in their partial waves [9, 5].

The Roper resonance $N\frac{1}{2}^+(1440)$ is about 100 MeV too light to fit easily into the model, although this kind of error is not unusual. It has also been shown [12] that models of this kind cannot arrange for the lightest positive-parity state to lie below the negative-parity states, once we go beyond first-order perturbation theory. This discrepancy has led to controversy about the nature of the Roper resonance, including suggestions that it is a hybrid baryon [13] and that more than one state exists [14] near 1440 MeV. Recent analyses by Cutkosky and Wang [15], the VPI group of Arndt *et al.* [11], and Manley and Saleski [16] show that the scattering amplitude has a complicated pole structure in this region, but do not support the conclusion that there is more than one resonance. Refs. [15] and [16] also find a large $N\pi$ width (350 to

400 MeV) for this state, which is also true in our decay model (Fig. 4 compares our predictions for the Roper resonance with the results of these recent analyses).

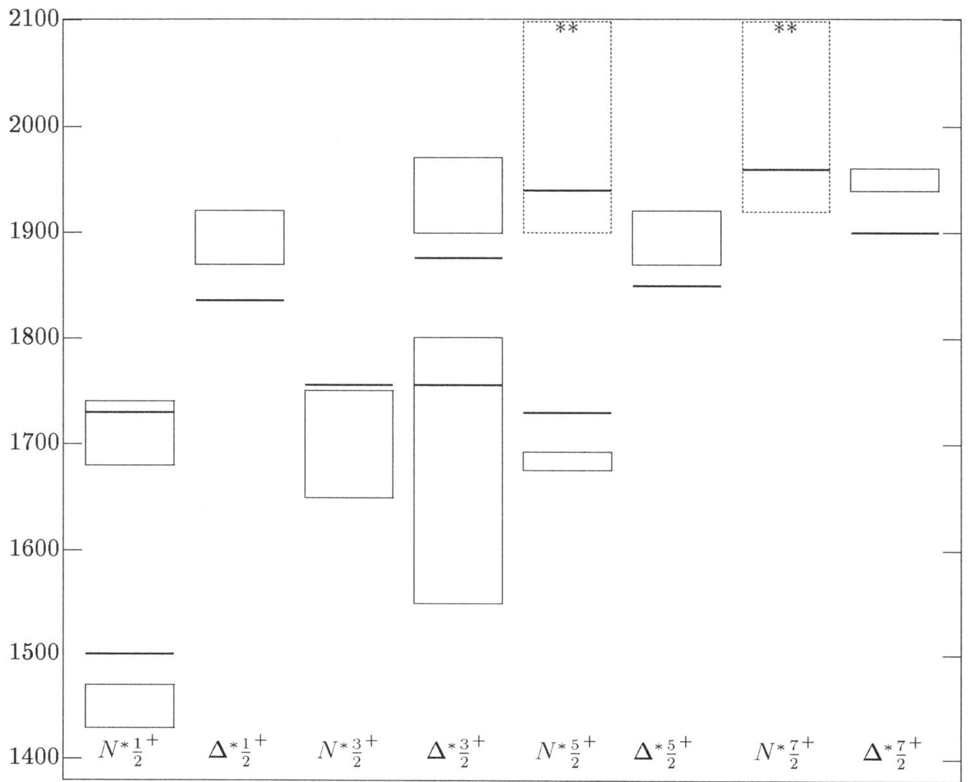

Figure 5. Non-strange positive-parity excited baryons up to $N = 2$; boxes give the range of masses quoted by the PDG [2], bars the predictions of Ref. [4], with a correction of -40 MeV added to all energies to roughly reproduce the band center of mass.

The second P_{33} state $\Delta\frac{3}{2}(1600)$ is also predicted to be too heavy if one compares with the results of the analyses based only on $N\pi$ elastic scattering experiments. The result of the Karlsruhe-Helsinki analysis [17] is 1522±13 MeV, and that of the CMU-LBL analysis [1] is 1600±50 MeV; the recent VPI analysis [11] quotes a pole position of approximately $1625 - i\,100$ MeV. However, in Manley and Saleski's coupled-channel analysis [16] described below, which also takes into account various quasi-two-body channels in the inelastic $N\pi \rightarrow N\pi\pi$ data, this state is assigned a mass of 1706 ± 10 MeV, which is within 100 MeV of our prediction for its mass. Its width is determined to be rather broad, at 430±73 MeV, coming mainly from p-wave $\Delta\pi$ (67±5%) with only a small branch (about 12%) into $N\pi$. At higher energies the model predicts many broad, overlapping states. This puts in doubt the narrow resonance approximation, implicit in models of baryons where states are treated in isolation without allowing for mass shifts and mixings caused by their decays. Hybrid baryons (those with excited glue) and states where $qqq(q\bar{q})$ configurations play an important role are likely to be present at these energies. Our initial strategy has been to make predictions with the simple three quark model at these energies (which is only possible with a large basis and an *ab initio* decay model), and to use discrepancies with the results of the partial-wave analyses to point us to this new physics.

States in Higher Bands

If we examine the lightest nonstrange baryon model states for increasing total spin J and either parity, we find that the model is able to quite accurately predict the correlation between increasing spin and mass (the Regge trajectories) seen in the elastic $N\pi$ partial wave analyses. Just as importantly, at these higher energies it is generally true that the lightest model state tends to couple to the $N\pi$ formation channel in our $^3P_0$ decay model [5]. This is a good test of the Coulomb plus linear spin-independent potential used to confine the quarks.

If one examines the lightest few negative parity N^* states which have predominantly $N = 3$ or $N = 4$ band wavefunctions, in several cases there are two almost degenerate lightest states predicted by the model, and they have comparable $N\pi$ strengths. Only one state is seen in the analyses, with a larger coupling strength. These are clear instances of the need for decay channel mixings, which can make one state lighter and couple more strongly to the formation channel than the other. Our relativized model is unable to solve the problem of the overestimate in quark model of the mass of the lightest (predominantly $N = 3$ band) $\Delta\frac{5}{2}^-$ state, about 200 MeV above the mass of the state $D_{35}(1930)$ seen in the analyses.

PION PHOTOPRODUCTION

A description of single-pion photoproduction involves a combination of an electromagnetic (EM) excitation process for an excited baryon and a strong decay (in the s and u channels), or a meson EM transition in the t channel (see figure 6). As we have described the calculation of the strong decay amplitudes in the first lecture, here we will outline how the EM excitation amplitudes are calculated in the nonrelativistic model.

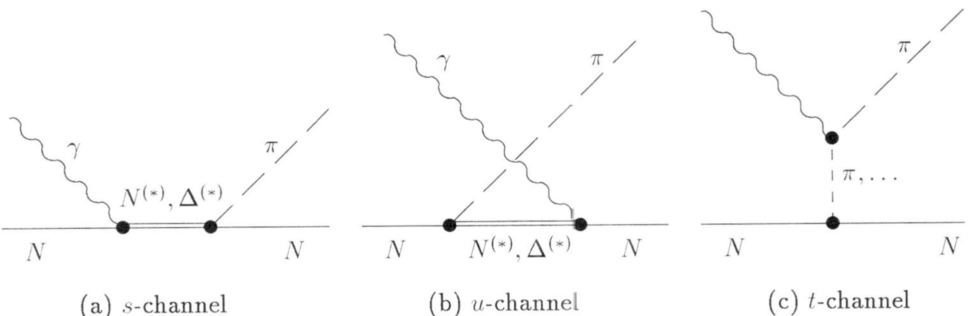

Figure 6. Born diagrams for pion photoproduction.

Evaluation of the strength of electromagnetic transitions $\gamma N \rightarrow X$ between the nucleons and excited baryon states involves finding matrix elements of an EM transition Hamitonian between states in the harmonic-oscillator basis, and is similar to the evaluation of the interquark Hamiltonian described above. The idea is to use the impulse approximation, illustrated in figure 7, which describes the target nucleon and the final resonance as made up of quarks which are free while interacting with the photon, but whose momenta, charges and spins are distributed according to the bound-state wavefunctions which result from a model of the spectrum.

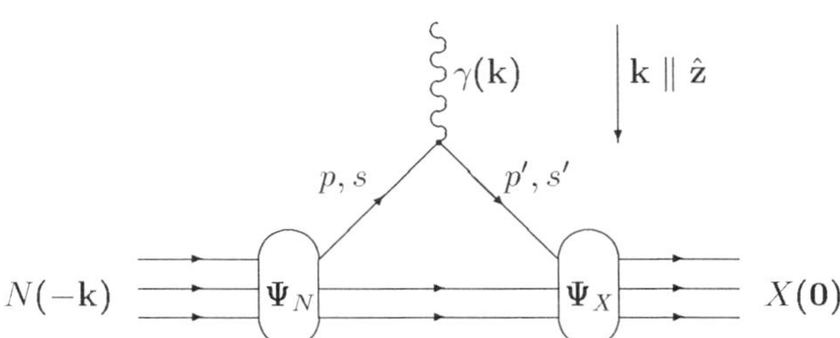

Figure 7. The impulse approximation for the electromagnetic interaction.

The EM interaction Hamiltonian is found from a nonrelativistic reduction of the quark current $e^{i(p'-p)\cdot x}\bar{u}(p', s')[-ie\gamma^{\mu}]u(p, s)$ to be

$$H^{\mathrm{em}} = \sum_i H_i^{\mathrm{em}} = -\sum_i \left\{ \frac{e_i}{2m_i} [\mathbf{p}_i \cdot \mathbf{A}(\mathbf{r}_i) + \mathbf{A}(\mathbf{r}_i) \cdot \mathbf{p}_i] + \boldsymbol{\mu}_i \cdot \boldsymbol{\nabla}_i \times \mathbf{A}(\mathbf{r}_i) \right\}, \qquad (1)$$

where $\boldsymbol{\mu}_i = e_i \boldsymbol{\sigma}_i / 2m_i$ is the magnetic moment of the i-th quark, e_i, m_i, $\boldsymbol{\sigma}_i/2$, $\mathbf{p}_i$ are its charge, (constituent) mass, spin, and momentum, and $\mathbf{A}$ is the photon field. Note that this interaction is flavor dependent, even for equal mass quarks, through its dependence on the charge e_i, so amplitudes for production from the proton and neutron contain independent information about the structure of the initial and final baryons.

There are, in general, a pair of amplitudes $A_{1/2}$ and $A_{3/2}$ associated with photoproduction from each target, which correspond to the two possibilities for aligning the spin of the photon and initial baryon in the center of momentum (c.m.) frame. This is illustrated in Figure 8 for photoproduction of a $\Delta\frac{3}{2}^+$ state; note we only need $A_{1/2}$ to describe photoproduction of baryons X with $J_X = 1/2$. These helicity amplitudes are defined in terms of helicity states by

$$A_\lambda^N = \langle X J_X; \mathbf{0}\lambda | H^{\mathrm{em}} | N\tfrac{1}{2}; -\mathbf{k}\lambda_N \rangle, \qquad (2)$$

where $\lambda_N = \lambda_\gamma - \lambda = 1 - \lambda$, and $N = p, n$. Here we specialize to the case of photoproduction of the nonstrange baryons N and Δ for which the bulk of the analysed data exists, although the generalization to the unequal mass case, *e.g.* for radiative decays of excited hyperons [18], is straightforward.

To reduce the H^{em} of Eq. (1) to an operator which acts between a nucleon and a resonance wavefunction, we first substitute the monochromatic photon field

$$\mathbf{A}_{\mathbf{k}}(\mathbf{r}_i) = \sum_\lambda \sqrt{\frac{2\pi}{k_0}} \left(\boldsymbol{\epsilon}_{\mathbf{k},\lambda} e^{i\mathbf{k}\cdot\mathbf{r}_i} a_{\mathbf{k},\lambda} + \boldsymbol{\epsilon}^*_{\mathbf{k},\lambda} e^{-i\mathbf{k}\cdot\mathbf{r}_i} a^\dagger_{\mathbf{k},\lambda} \right), \qquad (3)$$

where $k_0 = |\mathbf{k}|$ for real photons, and use the first (photon absorption) part with $\boldsymbol{\epsilon}_{\mathbf{k},+1} = -\frac{1}{\sqrt{2}}(1, i, 0)$, where we choose $\mathbf{k} \parallel \hat{\mathbf{z}}$. It is possible, when calculating the photocouplings of the N and Δ resonances, to exploit the overall symmetry of the wavefunctions (without the color wavefunction) to make the replacement $H^{\rm em} \to 3H_3^{\rm em}$. This is a convenient choice, since for the equal mass case the momentum of the third quark is just $\mathbf{p}_3 = \sqrt{2/3}\mathbf{p}_\lambda + \mathbf{P}/3$, where the momenta $\mathbf{p}_\rho$ and $\mathbf{p}_\lambda$ have equivalent definitions to their conjugate coordinates $\boldsymbol{\rho}$ and $\boldsymbol{\lambda}$, and $\mathbf{P}$ is the total momentum.

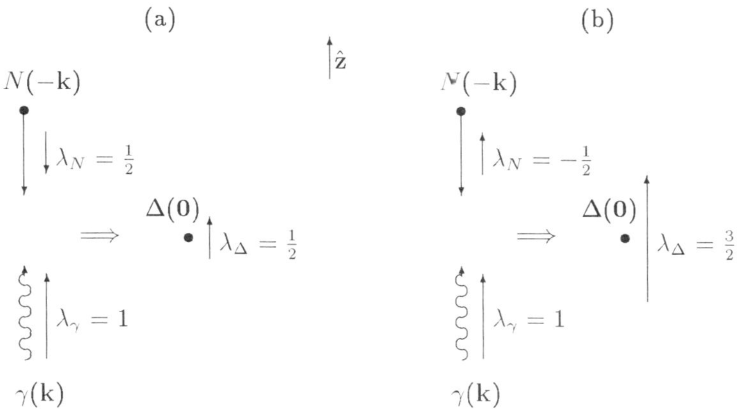

Figure 8. Momenta and helicities in the center of momentum frame for photoproduction of $\Delta\frac{3}{2}^+$; (a) defines the amplitude $A_{1/2}$, and (b) defines $A_{3/2}$.

The result of inserting plane waves for the center of mass motion of the baryons and integrating over the position coordinates is

$$A_\lambda^N = 3\langle XJ;\lambda| - \frac{e_3}{2m_u}\frac{1}{\sqrt{2}}\sqrt{\frac{2\pi}{k_0}}e^{-ik\sqrt{\frac{2}{3}}\lambda_z}\left(\sqrt{\frac{2}{3}}p_{\lambda+} - k\frac{\sigma_{3+}}{2}\right)|N\tfrac{1}{2};\lambda-1\rangle. \tag{4}$$

Here $p_{\lambda+} = p_{\lambda x} + ip_{\lambda y}$, and σ_{3+} raises the spin of the third quark. The expectation value in Eq. (4) is best carried out in momentum space, where the recoil phase factor has the effect of shifting the $\mathbf{p}_\lambda$ momentum of the final-state wavefunction by $\sqrt{2/3}\mathbf{k}$.

The helicity amplitudes for photoproduction of any excited $\Delta(I=\frac{3}{2})$ state satisfy $A_\lambda^p = A_\lambda^n$, since the expectation values of the charge operator e_3 between the flavor wavefunctions for the N and Δ satisfy

$$\begin{aligned}\langle\phi^S_{\Delta^+}|e_3|\phi^\rho_p\rangle &= \langle\phi^S_{\Delta^0}|e_3|\phi^\rho_n\rangle = 0\\ \langle\phi^S_{\Delta^+}|e_3|\phi^\lambda_p\rangle &= \langle\phi^S_{\Delta^0}|e_3|\phi^\lambda_n\rangle = -\sqrt{2}e/3,\end{aligned} \tag{5}$$

where e is the proton charge.

Koniuk and Isgur [9] calculated the amplitudes of Eq. (4) using the Isgur-Karl model [19, 20] wavefunctions, which include configuration mixing caused by the hyperfine interaction. In order to compare with the helicity amplitudes quoted in the Particle Data Group (PDG) [2], the sign of the $N\pi$ decay amplitudes must be calculated in the model. The quality of the $\gamma N \to N\pi$ data is poorer than the $N\pi$ elastic data, and so a resonance X must be present in $N\pi$ elastic in order to extract its photocoupling $A_\lambda^N(X)$. Since the sign of the coupling $A_{N\pi}(X)$ cannot be determined in $N\pi$ scattering, the PDG quote $A_\lambda^N(X)\cdot A_{N\pi}(X)/|A_{N\pi}(X)|$ for the photocoupling helicity amplitude, i.e. inclusive of the sign of $A_{N\pi}(X)$ (there are also some other conventional signs included, see Ref. [9]).

The results of this calculation, with signs calculated using the ${}^3P_0$ model of Ref. [5], are compared with the empirical photocouplings for a selection of baryon states in figures 9 and 10. For most low-lying states the agreement is quite good; for certain well determined states, such as the Roper resonance $N\frac{1}{2}^+(1440)$, there are sizeable discrepancies. This may be partially due to deficiencies in the wavefunctions and the nonrelativistic approximation, which is not as well justified when applied to some photocouplings. The photon momentum is determined by conservation of momentum in the excited baryon c.m. frame to be $k = (M_X^2 - m_N^2)/2M_X$ at resonance, and is not small compared to the average quark momentum when M_X is susbtantially larger than M_N. When the struck quark recoils against the other quarks in the baryon with a relativistic momentum, the Galilean momentum shift in Eq. (4) cannot be accurate.

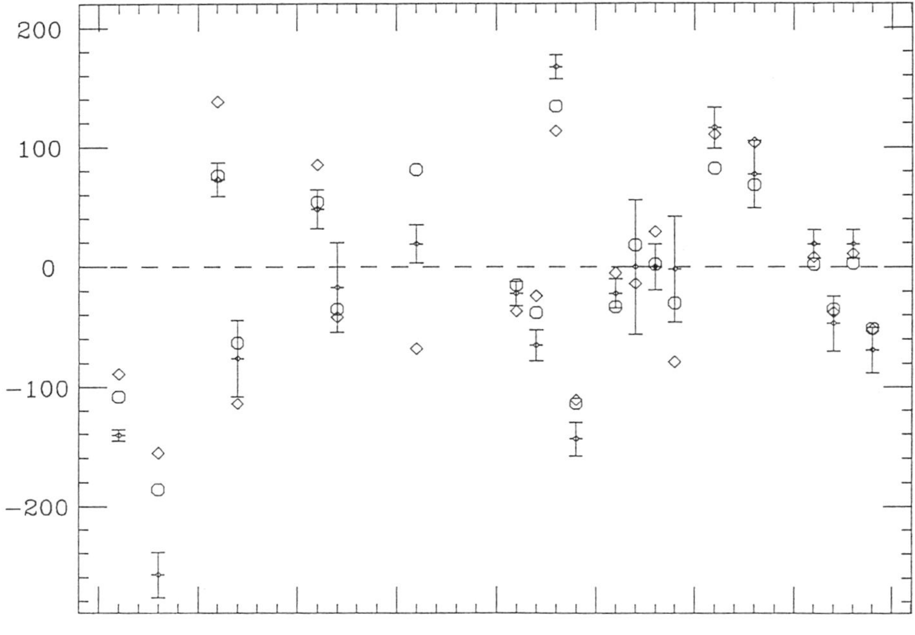

Figure 9. Breit-frame photoproduction amplitudes for $\Delta(1232)$ and the P-wave resonances, in units of 10^{-3} GeV$^{-1/2}$. Diamonds are for the nonrelativistic calculation, and circles are for the relativized calculations (for details, see Ref. [21]). Amplitudes are plotted, from left to right, in the order $A^p_{1/2}$, $A^n_{1/2}$, $A^p_{3/2}$, and $A^n_{3/2}$. Data points are taken from Ref. [22].

The nonrelativistic operator $H^{\rm em}$ adopted above is not expanded to the same order as the interquark Hamiltonian, $O(p^2/m^2)$. To this order we must add spin-orbit interactions, and Brodsky and Primack [23] have shown from the requirement of electromagnetic gauge invariance that there are also associated two-body currents (which go beyond the simple impulse approximation illustrated in Fig. 5). These corrections were put together with the configuration-mixed nonrelativistic wavefunctions in a calculation of the photocouplings by Close and Li [24], and in our calculation [25, 21] using the relativized model wavefunctions described in the first lecture. These results are also shown in figures 9 and 10. It is clear that these additional effects improve the comparison with the extracted photocouplings, although they clearly are not able to account for the model's underprediction of the amplitudes for the photoproduction of the Δ and the Roper resonance. These difficulties may be due to the neglect of the pion degree of freedom [26, 27], as these are both light states with strong couplings to $N\pi$. In our recent work [28] the Roper EM couplings (along with some others) are found to be very sensitive to further relativistic effects. Other authors [29] have taken the anomalous photocouplings of the Roper resonance to be evidence that it has a substantial amount of hybrid baryon mixed into it.

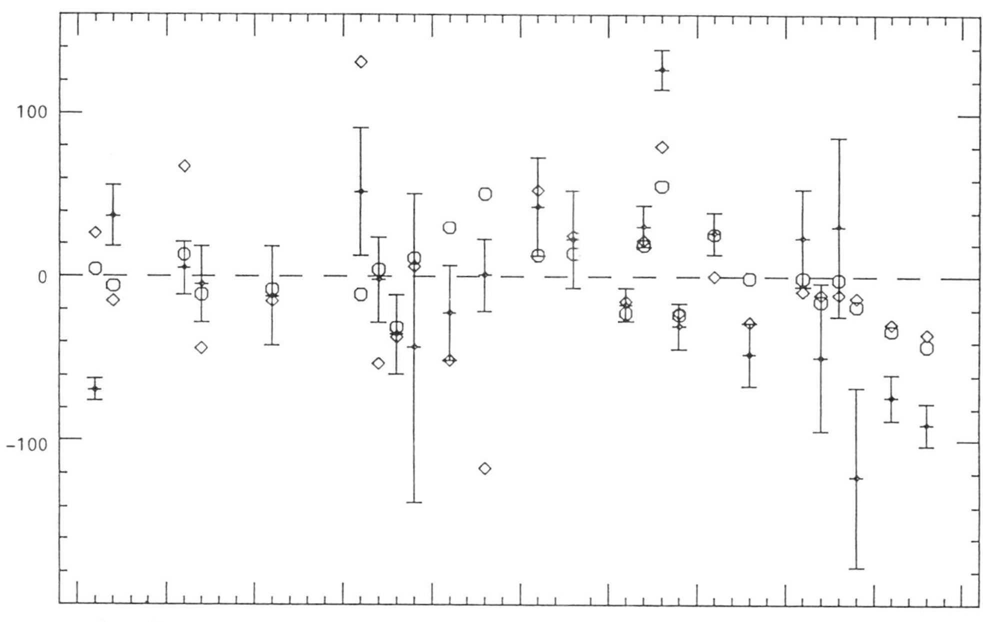

Figure 10. Breit-frame photoproduction amplitudes for low-lying (N=2 band) positive-parity excited states for which there exist data, in units of 10^{-3} GeV$^{-1/2}$. Caption as in figure 9.

The relativized model definitely improves the photocouplings of the other radial excitation of the nucleon, $N\frac{1}{2}^{+}(1710)$, and those of $\Delta\frac{3}{2}^{+}(1600)$, a radial excitation of Δ(1232), are also somewhat improved. The model is unable to account for the small photocoupling of $\Delta\frac{1}{2}^{-}(1620)$ or the (relatively precisely determined) large $A^{p}_{3/2}$ amplitude of $N\frac{5}{2}^{+}(1680)$ found in the analyses. The next generation of experiments at new facilities such as CEBAF will examine the pion photoproduction and electroproduction processes; the latter will yield the Q^2 dependence of the Roper resonance EM couplings, for example. It is also obvious that we need a model of the process $\gamma N \to N\pi$ at 1440 MeV to understand the Roper resonance EM couplings.

INELASTIC CHANNELS IN $N\pi$ TO $N\pi\pi$

The coupled channel analysis of Manley and Saleski [16] puts the $N\pi$ elastic partial wave amplitudes from the CMU-LBL and Karlsruhe-Helsinki analyses together with a set of $N\pi$ to $N\pi\pi$ isobar-model amplitudes in the range 1320-1930 MeV, to extract partial widths (with an assumed energy dependence) and the signs of the decay amplitudes relative to $N\pi$, along with the masses of the resonances. In the isobar model, the reactions are described by a coherent superposition of quasi-two-body (inelastic) channels, such as $N\pi \to \Delta\pi \to (N\pi)\pi$, $N\pi \to N\frac{1}{2}^+(1440)\pi \to (N\pi)\pi$, $N\pi \to N\rho \to N(\pi\pi)$, and $N\pi \to N\epsilon \to N(\pi\pi)$ where ϵ represents the strong s-wave isoscalar $\pi\pi$ interaction, assigned a mass and width of roughly 800 MeV. One other (unmeasured) inelastic channel is added to maintain unitarity, which is taken to be $N\eta$ below 1600 MeV ($K\Lambda$ and $N\omega$ open up at slightly higher energies).

Their approach is to impose unitarity directly on the S-matrix (this is not a K-matrix approach) by writing $S = S_R^T S_B S_R$, where $S_B = (1 + iK_B)(1 - iK_B)^{-1}$ is a multichannel background K-matrix (with no poles), so that away from poles $S_R \simeq 1$ and $S \simeq S_B$. The resonances are in the matrix $S_R = S_1^{1/2} S_2^{1/2} ... S_N^{1/2}$, where we have N resonances, and S_1 is a symmetric, unitary matrix for the first resonance found from a Breit-Wigner T-matrix, $(S_1^{1/2})^2 = 1 + 2iT_1$, etc. Near this first resonance we have $S \simeq S_1$ and $T \simeq T_1$, assuming it is isolated and that $S_B \simeq 1$.

As mentioned above, the $N\frac{3}{2}^-$ state $D_{13}(1700)$ does not appear in the analysis of $N\pi$ elastic scattering of Arndt *et al.* [11]. Manley and Saleski's analysis shows this state decaying predominantly to $\Delta\pi$, with very little coupling to $N\pi$, and so this is not a surprising result. In our $^3P_0$ model we find that this state has a smaller $N\pi$ amplitude than the $D_{13}(1520)$ [5], and our calculations of inelastic channels [10] verify that it has a large amplitude to decay via $\Delta\pi$, so that this should be its dominant decay mode. Arndt *et al.* also find little evidence for the $N\frac{1}{2}^+$ partner of the Roper, $P_{11}(1710)$ in their $N\pi$ analysis; they claim that if it is present it must have a significantly larger width than is listed in the PDG. This is borne out by Manley and Saleski, who see a substantial width of this state to decay to $\Delta\pi$. Our $^3P_0$ calculations confirm this picture; we predict this state to have roughly an 18 MeV decay width into $N\pi$, and a width of roughly 190 MeV to decay into $\Delta\pi$.

Manley and Saleski also find an $N\eta$ branch for $N\frac{1}{2}^-(1535)$ of $43 \pm 6\%$, and as discussed above find a broad (391 ± 34 MeV) Roper resonance, as well as a significantly higher mass of 1706 ± 10 MeV for the radial recurrence of the Δ, nominally $\Delta\frac{3}{2}^+(1600)$. There is also an entirely new P_{13} state, $N\frac{3}{2}^+(1880)$, which may well be a formerly missing $[70, 0^+]$ state, as it is found to have a relatively small $N\pi$ branch of $26 \pm 6\%$. Interestingly the analysis of Arndt *et al.* [11] also finds poles in this mass range. Manley and Saleski were also able to confirm many existing states and measure their inelastic branching fractions (along with the amplitudes and their signs). Among the community of those involved in partial-wave analyses for the experimental baryon resonance physics programs at new accelerators such as CEBAF, it is generally agreed that this kind of coupled channel analysis will be necessary to extract resonance parameters accurately from the simultaneous information in the many different channels which will result.

BARYONS WITH STRANGENESS

Figure 11 compares the relativized model predictions for the masses of the negative-parity excited hyperons (Λ and Σ) with the range of masses quoted by the PDG. Only those states predicted by the elementary emission model to couple to the $\bar{K}N$ formation channel (with subsequent decays to $\bar{K}N$, $\Sigma\pi$, and $\Lambda\pi$) are shown; three model states are not shown. Here the relativized model does a little worse than the nonrelativistic model, with the splittings underestimated because of a lack of freedom to fit the orbital splitting between the $l_\rho = 1$ states and the $l_\lambda = 1$ states (recall $\omega_\rho > \omega_\lambda$ in states like uus). The wavefunctions which result, however, predict a pattern of decays similar to that of the nonrelativistic wavefunctions.

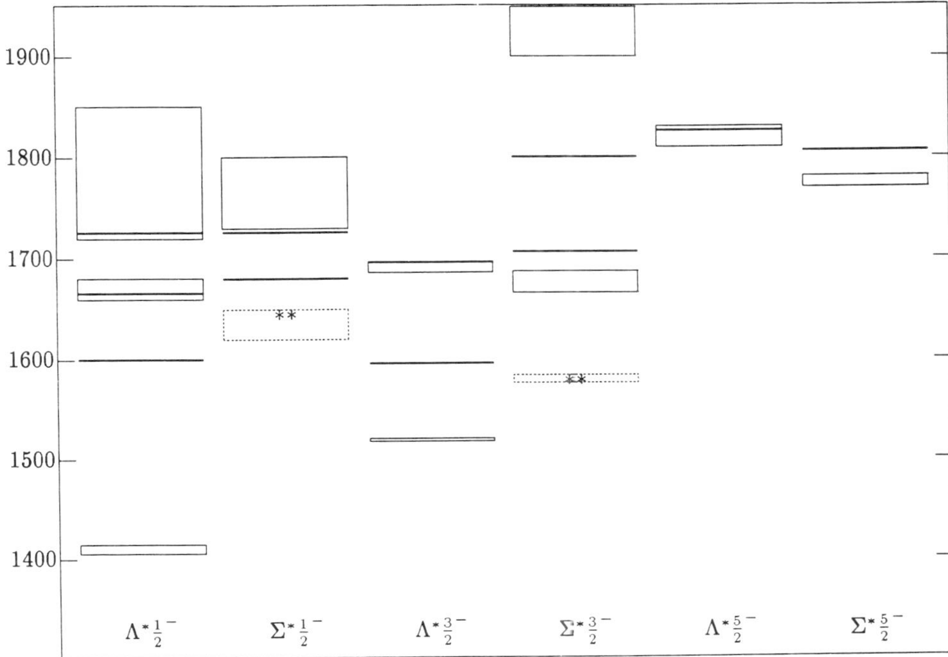

Figure 11. $S = -1$ negative-parity excited baryons; boxes give the range of masses quoted by the PDG [2], bars the predictions of Ref. [4], with a correction of +50 MeV added to all energies to roughly reproduce the band center of mass.

An obvious problem in these models is the degeneracy of the lightest $\Lambda\frac{1}{2}^-$ with its partner $\Lambda\frac{3}{2}^-$, which are $\Lambda\frac{1}{2}^-(1405)$ and $\Lambda\frac{3}{2}^-(1520)$ in the analyses. This degeneracy can be lifted by a spin-orbit interaction, but with a strength that causes unacceptably large spin-orbit splittings elsewhere in the baryon spectrum. The very strong model coupling of $\Lambda\frac{1}{2}^-(1405)$ to S-wave $\bar{K}N$ suggests that the problem here is the neglect of the $qqq(q\bar{q})$ sector of Fock space, which could cause a shift towards or below the $\bar{K}N$ threshold, essentially by decay-channel couplings. Some authors model this state as a $\bar{K}N$ bound state; this, however, leads to another problem, which is the lack of a qqq $\Lambda\frac{1}{2}^-$ state near 1520 MeV. It may be possible to distinguish between these pictures of the $\Lambda(1405)$ by measuring its radiative decays [along with those of $\Lambda(1520)$] to $\Lambda\gamma$ and $\Sigma\gamma$ [18]. There is an excellent article with references on this topic in the Particle Data Group by Dalitz [30].

Figure 12 makes a similar comparison for the positive-parity excited strange baryons (there is also the $[56, 0^+]$ ground state $\Sigma^*\frac{3}{2}^+(1385)$ which fits well in the models). Once again only states predicted to be seen in $\bar{K}N \to \bar{K}N$, $\Sigma\pi$, $\Lambda\pi$ are shown. Here the comparison is hampered by a lack of data, and there has been essentially no recent data collected on these states. Without a kaon factory it would appear that the best possibility for seeing these states is in kaon photo- and electroproduction experiments such as those at CEBAF, where $\gamma N \to \bar{K}KN$ goes through a quasi-two body channel $\bar{K}\Lambda^*$ or $\bar{K}\Sigma^*$, although the analysis of these kinds of final-state production experiments is difficult. This kind of analysis gives the information that we have on the $S = -2$ Ξ baryons. There are three Ξ states including the ground state with definite spin-parity listed by the PDG, and these fit well in our model [4]. Several other relatively certain states exist, with unknown spin-parity assignments. In the case of the Ω states only the ground state can be assigned a spin and parity. The excited Ξ and Ω states with unknown spin and parity lie in mass ranges where there are many model states predicted; see, for example, Ref. [4].

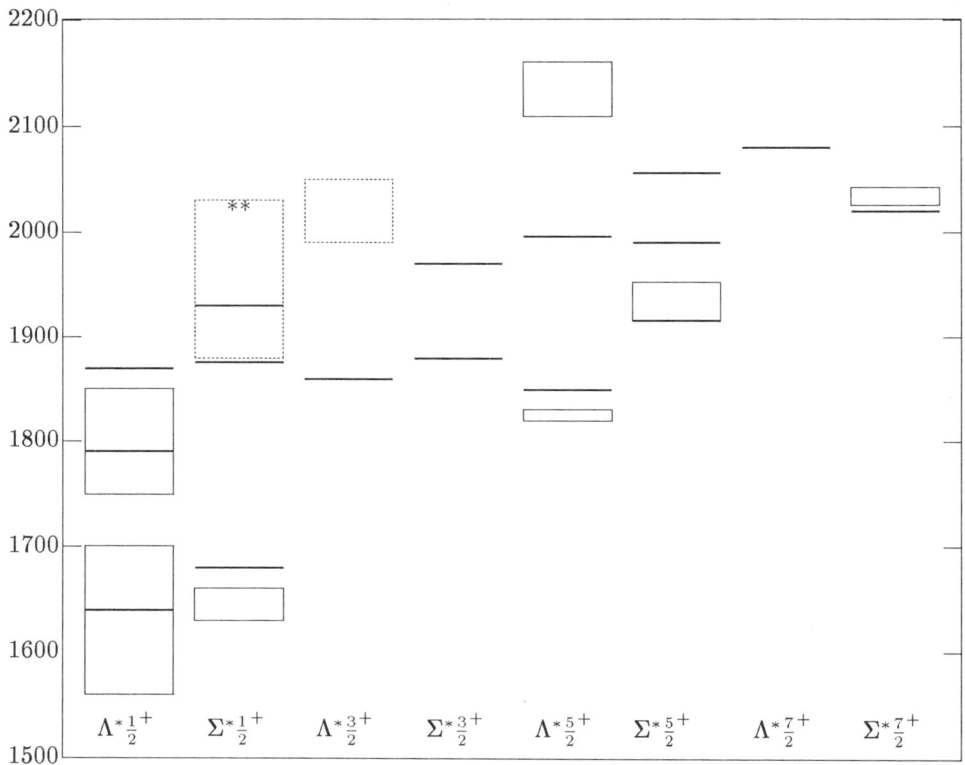

Figure 12. $S = -1$ positive-parity excited baryons up to $N = 2$; boxes give the range of masses quoted by the PDG [2], bars the predictions of Ref. [4], with a correction of -40 MeV added to all energies to roughly reproduce the band center of mass.

SUMMARY

We have seen that calculations carried out in the constituent quark model described in the first lecture are capable of explaining many features of the nonstrange baryon spectrum, as determined by $N\pi$ elastic scattering. When the wavefunctions from these calculations are combined with a simple model of the strong decays, we are able to explain why certain model states are missing in the analyses, and this decay model is also usually able to predict the dominant decay mode for states which decay into inelastic quasi-two body channels. Recent results from new analyses of the elastic and inelastic channels have shed some light on the nature of controversial states such as the Roper resonance, and have probably found a previously missing state.

The comparison between the photocouplings of these baryon states calculated in both the nonrelativistic and relativized models and those extracted from pion photoproduction data shows that relativistic corrections are important. Recent work has demonstrated the need for a relativistic treatment of the electromagnetic interaction, and possibly for the inclusion of the pion degree of freedom into such calculations. It is in this area that we expect the largest improvement in the data set in the near future, from new facilities such as CEBAF.

We have also compared model calculations of the masses and strong couplings of strange baryons to the results of analyses of $N\bar{K}$ elastic scattering and other reactions, and found reasonable agreement with the data, with some noticeable exceptions which point to deficiencies in the model. Without a kaon production facility, the best hope for improvement of the data for the $S = -1$ states may be in electromagnetic production reactions.

ACKNOWLEDGEMENTS

This work was supported in part by the U.S. Department of Energy through Contract No. DE-FG05-86ER40273, and by the Florida State University Supercomputer Computations Research Institute which is partially funded by the Department of Energy through Contract No. DE-FC05-85ER250000

References

[1] R.E. Cutkosky, C.P. Forsyth, R.E. Hendrick, and R.L. Kelly, *Phys. Rev.* **D20**, 2839 (1979); R.E. Cutkosky, *et al.*, *Phys. Rev.* **D20**, 2804 (1979).

[2] The Particle Data Group, L. Montanet *et al.*, *Phys. Rev.* **D50**, 1173 (1994).

[3] S. Godfrey and N. Isgur, *Phys. Rev.* **D32**, 189 (1985).

[4] S. Capstick and N. Isgur, *Phys. Rev.* **D34**, 2809 (1986).

[5] S. Capstick and W. Roberts, *Phys. Rev.* **D47**, 1994 (1993).

[6] R. Arndt *et al.*, *Phys. Rev. Lett.* **65**, 157 (1990).

[7] R. Koch and E. Pietarinen, *Nucl. Phys.* **A336**, 331 (1980).

[8] G. Miller, *Phys. Rev.* **C39**, 1563 (1989).

[9] R. Koniuk and N. Isgur, *Phys. Rev.* **D21**, 1868 (1980).

[10] S. Capstick and W. Roberts, *Phys. Rev.* **D49**, 4570 (1994).

[11] R. Arndt et al., *Phys. Rev.* **D43**, 2131 (1991).

[12] J.-M. Richard, *Phys. Rep.* **212**, 76 (1992).

[13] T. Barnes and F.E. Close, *Phys. Lett.* **B158**, 89 (1983); E. Golowich, E. Haqq, and G. Karl, *Phys. Rev.* **D28**, 160 (1983); F.E. Close, *Rep. Prog. Phys.* **51**, 833 (1988).

[14] R.A. Arndt, J.M. Ford, and L.D. Roper, *Phys. Rev.* **D32**, 1085 (1984).

[15] R.E. Cutkosky and S. Wang, *Phys. Rev.* **D42**, 235 (1990).

[16] D.M. Manley and E.M. Saleski, *Phys. Rev.* **D45**, 4002 (1992).

[17] R. Koch, in *Proceedings of the IVth International Conference on Baryon Resonances* (Toronto, 1980), ed. N .Isgur, p.3; G. Höhler *et al.*, *Handbook of Pion-Nucleon Scattering*, *Physics Data* 12-1 (1979); E. Pietarinen, *Nucl. Phys.* **B107**, 21 (1976).

[18] J.W. Darewych, R. Koniuk, and N. Isgur, *Phys. Rev.* **D32**, 1765 (1985).

[19] N. Isgur and G. Karl, *Phys. Lett.* **72B**, 109 (1977); **74B** (78) 353; *Phys. Rev.* **D18**, 4187 (1978).

[20] N. Isgur and G. Karl, *Phys. Rev.* **D19**, 2653 (1979).

[21] S. Capstick, *Phys. Rev.* **D46**, 2864 (1992).

[22] J.J. Hernández *et al.*, *Phys. Lett.* **B239**, 1 (1990).

[23] S.J. Brodsky and J.R. Primack, *Phys. Rev.* **174**, 2071 (1968); *Ann. Phys.* **52**, 315 (1969).

[24] F.E. Close and Zhenping Li, *Phys. Rev.* **D42**, 2194 (1990); **D42**, 2207 (1990).

[25] S. Capstick, *Phys. Rev.* **D46**, 1965 (1992).

[26] G.A. Miller, S. Théberge, and A.W. Thomas, *Phys. Lett.* **91B**, 192 (1980); *Phys. Rev.* **D22**, 2838 (1980); *ibid* **D24**, 216 (1981).

[27] T.S.H. Lee and B.C. Pearce, *Nucl. Phys.* **A530**, 532 (1991).

[28] S. Capstick and B.D. Keister, *Phys. Rev.* **D51**, 3598 (1995).

[29] Z.-P. Li, *Phys. Rev.* **D44**, 2841 (1991); Z.-P. Li, V. Burkert, and Z.-J. Li, *Phys. Rev.* **D46**, 70 (1992).

[30] R.H. Dalitz, in Ref. [2].

BARYON PHYSICS AT CEBAF

Simon Capstick

Supercomputer Computations Research Institute
and Department of Physics
Florida State University
Tallahassee, FL 32306

ABSTRACT

The accelerator and the three experimental halls at CEBAF are briefly described, and an introduction is given to the program at CEBAF to study the physics of nucleons and electromagnetically produced excited baryons. The latter will be studied mainly in Hall B with the photon tagger and CLAS detector, which are also briefly described. Experiments to look at final states involving strange quarks and to study Compton scattering from nucleons are examined. Finally, a brief summary is given of all three lectures.

OVERVIEW OF THE FACILITY

The Continuous Electron Beam Accelerator Facility (CEBAF) is located in Newport News, Virginia. The accelerator consists of a pair of parallel superconducting linear accelerators with nominal maximum acceleration energies of 400 MeV, which are used repeatedly by recirculating the beam. Electrons are injected into the first linac on one side of the racetrack shaped accelerator, pass through the accelerator into a beam pipe surrounded by bending magnets in the first recirculation arc, and then are accelerated again in the second linac. At this point the beam can be extracted with twice the linac energy, or recirculated in the second arc and combined with a pulse of electrons from the injector for a second pass through the first linac. These combined pulses are separated on the basis of their momenta before going into the recirculation arcs, as each energy requires a separate beam pipe and magnetic field strength. This repeats until beams with energies in increments of one fifth of the maximum energy are available for extraction at the end of the second linac.

The beam pulses are spaced very closely (20 cm or 2/3 ns apart), and the intensity of the pulses is variable bunch by bunch (repeating every third bunch) by a factor of 1000. This allows beams of different intensities to be delivered to each of three experimental halls with different energies, in increments of one fifth of the maximum energy, which is variable from about 2.5 GeV up to a design energy of 4 GeV by variation of the linac field gradient. Beam currents can be as high as 200 μA, and the beam emittance $\varepsilon \simeq 2\text{x}10^{-9}$ mrad and energy spread $\sigma_E/E \simeq 2.5\text{x}10^{-5}$ are quite small. The experimental program is now underway, as the accelerator has recently begun delivering beam to one of the three experimental halls (Hall C). The superconducting cavities in the linear accelerator have exceeded design specifications by about a factor of 2 in both field gradient and quality factor, which means that a maximum accelerator energy of 6 GeV is easily attainable.

The Experimental Halls

There are three experimental halls, each with complementary experimental equipment and physics programs. Hall C contains a short-orbit spectrometer and a single high-momentum spectrometer. The Hall C physics program includes experiments to examine the strange quark content of the proton, color transparency, the neutron electric form factor, and kaon electroproduction. Hall A, the largest of the halls, contains two very large high resolution spectrometers, which will detect the electron and hadron arms resulting from electron-nucleus collisions. They will be used to study nuclear structure at small inter-nucleon separations, and nuclear medium effects on bound nucleon structure, and to perform precision measurements of high-Q^2 form factors of nucleons and nuclei, and studies of the strange quark content of the proton through parity violation in electron-proton scattering.

The majority of baryon physics experiments approved so far will use the equipment in Hall B, which consists of a photon tagger and an (almost) 4π spectrometer, the CEBAF Large Acceptance Spectrometer (CLAS). The photon tagger produces a beam of bremsstrahlung photons from scattering electrons in a radiator (one electron or fewer, on average, per electron beam bunch). The energy of each photon is tagged by bending the electron which radiated it in a large magnet and detecting its deflection, which is proportional to the energy lost to the radiated photon, and time of arrival. The timing information is then used to tag an individual photon arriving at the detector with this energy. In this way experiments may be performed which use both electrons (with the radiator removed and the tagger magnet turned off) and energetic real photons to excite nucleons and small nuclei in the target.

The CLAS detector has a six-segment toroidal magnetic field surrounded by various layers of detection systems, and is roughly spherical in shape with a diameter of about 10m. Two interior layers of drift chambers lie inside the toroidal magnets, and a third large drift chamber (with 46,000 wires) lies outside the field magnets, to provide charged particle tracking. Outside the third drift chamber region are Cerenkov counters (for electron identification), scintillation counters (to provide time-of-flight information), and finally electromagnetic shower calorimeters which will provide electron, photon, and neutron detection. There is approximately 50% coverage of the azimuthal angle at 10° deflection in the lab, and 85% coverage at 90°. The drift chambers extend back further than 135°, whereas the Cerenkov counters and shower calorimeters have partial coverage to about 90°.

The detector has been designed to be multi-purpose, with large acceptance for two or more loosely correlated particles, capable of dealing with luminosities up to the order of 10^{34} $\text{cm}^{-2}\text{s}^{-1}$. The large acceptance is required because the experiments to be performed require high luminosity; in the tagged photon beams experiments the luminosity is background limited, and in those that require a polarized solid state or gas target it is constrained by limits on the beam intensity. Electronics will also be provided which will allow for the flexible selection of events to be written to tape.

THE BARYON PHYSICS EXPERIMENTAL PROGRAM

A major part of the experimental program at CEBAF has the goal of understanding baryon physics, and many of these experiments will take place in Hall B using the CLAS detector and photon tagger described above. In particular, there are experiments to examine the physics of the nucleons, and the masses and properties of excited baryon states.

These experiments will electromagnetically excite the nucleon, in two ways: by scattering electrons, which can be thought of as providing a virtual photon as the incoming projectile particle, subsequently absorbed by the nucleon; and by scattering a beam of real photons made in the tagger. In either case, if we view this process in an s-channel picture, the intermediate state can be a nucleon or a Δ in a ground or excited state. The photon carries both isospin zero or one, since it couples to the charge of the quarks through $e_q\bar{q}\gamma^\mu q$, and $e_q = \frac{1}{6} + \frac{1}{2}\tau_z$, which has both isoscalar (proportional to $\mathbf{1}$) and isovector (proportional to τ_z) components. The experiments will then look at a wide variety of final states such as the nucleon itself (in the case of electron scattering only), $N\pi$, $N\pi\pi$, $\bar{K}\Lambda$ and $\bar{K}\Sigma$ (where the Λ and Σ subsequently decay), and $N\gamma$ (real and virtual Compton scattering). In what follows we will examine the kinds of physics issues which will be addressed by each kind of experiment.

In order to use these reactions to study the physics of excited baryons, we have to work in the kinematic regime known as the resonance region. This is, roughly speaking, for W values from $N\pi$ threshold to about 2.5 or 3 GeV, and for Q^2 values where the baryon electromagnetic transition form factors have not dropped off severely, say $0 \leq Q^2 \leq 3$ GeV^2. In this region the intermediate state in the reaction can be efficiently described as a sum of overlapping baryon resonances (in both the s and u channels), and the important degrees of freedom may be thought of as constituent quarks with the glue in its adiabatic ground state, or possibly its first excited state (hybrid baryons).

A huge improvement in the data set for photoproduction is expected to emerge from CEBAF, which will allow less model-dependent approaches to analysis to extract much more precise values for the photocouplings of baryons. There are very few data in existence for pion electroproduction beyond the first resonance region (W in the vicinity of the Δ), and as a consequence the transition form factors for many resonances will be extracted with much greater precision or for the first time from CEBAF experiments.

There are also proposals to study meson physics at CEBAF. The charged pion electromagnetic form factor will be studied [1] in Hall C, where electroproduction of the pion will be studied in a kinematical regime where t-channel exchange of an off-shell pion dominates. This virtual pion is then knocked on shell by the virtual photon exchanged by the electron. Extraction of the pion form factor requires extrapolation of the results to the real-pion kinematical region. A similar Hall C experiment [2] will perform a longitudinal/transverse separation in the kaon electroproduction process $p(e, e'K^+)\Lambda$ (or Σ), once again in the kinematical regime which favors t-channel exchange of a charged kaon. In this case the extrapolation to real kaon kinematics required to extract information about the kaon form factor is more severe. There is also a Hall B experiment [3] which will study electroproduced excited light-quark mesons.

Rare radiative decays of the ϕ meson will be measured in an experiment [4] which will use the tagged photon beam incident upon a one-inch beryllium target to produce ϕ mesons whose decay products will be detected in a segmented lead glass array. A charged particle veto between the target and detector array will be used to reduce the background (electromagnetic and hadronic backgrounds are projected to be acceptable). The goal is to study neutral radiative decay modes of ϕ with branching fractions of the order of 10^{-5}, in order to explore the structure of the scalar daughter states a_0 and f_0 and to look for possible symmetry violations, *e.g.* testing C-parity in $\phi \to \omega\gamma$.

Elastic Electron-Nucleon Scattering

In the case of electron scattering one possibility is to have the nucleon itself as a final state, so that the process is elastic electron-nucleon scattering. Several experiments will be performed at CEBAF using both hydrogen and deuterium targets which are sensitive to the Q^2 dependence of the nucleon electromagnetic form factors G_E^N and G_M^N, where $N = p$ or n. Of particular interest is the electric form factor of the neutron, which is zero in the limit of SU(6) symmetry described in the first lecture. The neutron is known to have a negative charge radius from earlier experiments, but the Q^2 dependence away from small Q^2 is not well known. A negative charge radius can be explained by a simple argument [5] using the exchange symmetry of the product of the flavor and spin wavefunctions (the product of the color wavefunction and the ground state spatial wavefunction is antisymmetric) and the contact interaction. Figure 1 illustrates that $\langle r_d^2 \rangle_n > \langle r_u^2 \rangle_n$ because of a repulsive contact interaction between the two d quarks. The contact interaction between the two ud pairs must bé attractive, since the average contact interaction between the quarks is negative, as we have seen in the first lecture. This means that the d quarks must lie on the outside and the u quarks on the inside, which yields a negative charge radius.

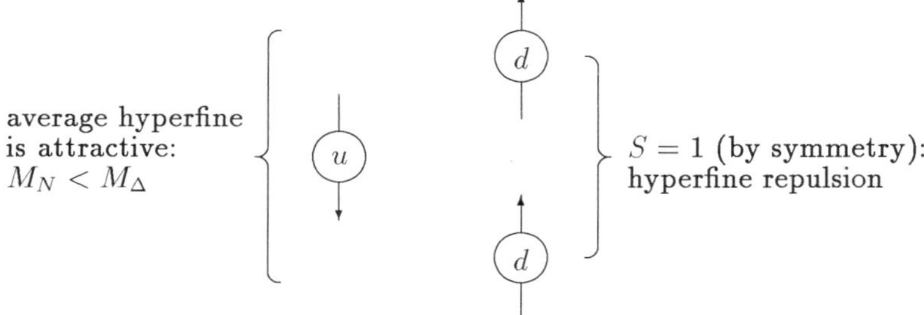

Figure 1. The hyperfine interaction in the neutron, which causes the d quarks to lie outside the u quark.

An alternative description is in terms of the pion degree of freedom. Suppose a nucleon is surrounded by a pion cloud which can also interact with the photon. Then the neutron can dissociate into $p\pi^-$ or $n\pi^0$. The dissociation into neutral particles causes no effective separation of charge, but the $p\pi^-$ pair will have the heavier proton closer to the center of the original neutron than the lighter π^-, causing a negative charge radius. In practice both mechanisms are likely to be present to some degree, and precision measurements of G_E^n will be very useful in understanding these effects.

Single and Multiple Pion Production

As mentioned in the second lecture, one of the best prospects for discovery of new baryon states, such as those missing from the analyses of pion elastic scattering, is in the electromagnetic production of multiple pions. If we accept that these states are missing because of a weak coupling to the $N\pi$ channel, then processes that form these excited baryon states electromagnetically are favorable, and some means of decay (once again employing an s-channel picture) other than $N\pi$ will also be favorable.

These multi-pion production processes can be thought of as decaying through various quasi-two-body channels involving excited baryons, such as $\Delta\pi$, $N\frac{1}{2}^+(1440)\pi$, *etc.*, and channels involving excited mesons, such as $N\rho$, $N\omega$, $N\eta$, *etc.* The CLAS detector at CEBAF is well suited to examining these multi-particle decay channels, especially with all charged particles in the final state (or a single neutral).

As an example, there is an experiment [6] to examine the photoproduction reaction $\gamma p \to p\pi^+\pi^-$ in the CLAS detector in Hall B at CEBAF. All of the final state particles are charged and so will be detected. The invariant mass of the two pions will be examined to look for intermediate baryon states in $\gamma p \to p\rho$, and similarly the invariant mass of one pion and the proton can be reconstructed to look for missing baryons in, for example, the $\Delta\pi$ channel. Models of the strong decays of these missing states predict sizeable decay widths [7, 8] into these channels. Other photoproduction experiments [9] will study single-pion photoproduction and its helicity structure, η and η' photoproduction from protons and deuterium, and photoproduction of ρ mesons from the proton with linearly polarized photons.

Δ-Region Experiments Several of the pion electroproduction experiments will look closely at the region of W close to the mass of $\Delta(1232)$ [10], and will examine the Q^2 dependence of the dominant $M1$ and small $E2$ transition multipoles. The ratio of the $E2$ multipole to the $M1$ multipole, $R = E2/M1 \equiv E1+/M1+$ is of great interest theoretically, as it is zero in the nonrelativistic model in the SU(6) limit. A non-zero ratio is caused by the tensor interaction, which mixes D-wave components into the predominantly S-wave wavefunctions of the nucleon and Δ. In the notation from the first lecture, the wavefunctions found by Isgur, Karl and Koniuk [11] for the ground states are

$$\begin{aligned} |\Delta\rangle &\simeq 0.97|\Delta^4 S_S\rangle + 0.20|\Delta^4 S_{S'}\rangle - 0.10|\Delta^4 D_S\rangle + 0.07|\Delta^2 D_M\rangle \\ |N\rangle &\simeq 0.93|N^2 S_S\rangle - 0.29|N^2 S_{S'}\rangle - 0.23|N^2 S_M\rangle - 0.04|N^4 D_M\rangle + 0.00|N^2 P_A\rangle, \end{aligned} \tag{1}$$

where the dominant components are the $[56, 0^+]$ substates, the next most important components are admixtures with radial excitations caused by the contact and anharmonic interactions, and the D-wave components are mixed in by the tensor interaction. In terms of the helicity amplitudes described in the second lecture, we have

$$E2 \equiv \frac{1}{2\sqrt{3}}(A_{\frac{3}{2}} - \sqrt{3}A_{\frac{1}{2}}), \quad M1 \equiv -\frac{1}{2\sqrt{3}}(3A_{\frac{3}{2}} + \sqrt{3}A_{\frac{1}{2}}). \tag{2}$$

In electroproduction experiments there is also a small scalar or charge multipole $C1$. Both $E2$ and $C1$ are zero in the absence of the hyperfine tensor interactions in a nonrelativistic model without pions.

The Isgur-Karl-Koniuk model [11] gives a value for the ratio $E2/M1$ at the photon point $Q^2 = 0$ of about -0.4%, with a smaller value arising from the relativized model [12]; values extracted from the current data range between -1.0% and -3.0%. The ratio is predicted by the model to remain negative and grow slowly in magnitude with Q^2. Perturbative QCD along with helicity conservation predict [13] that this ratio should approach +1 at high Q^2, and it would be very interesting to see this quantity cross zero or display a trend towards positive values at the Q^2 values accessible at CEBAF.

Other Δ region experiments will examine the $\gamma^* N \to \Delta$ transition form factor at high momentum transfer [14], and the electron asymmetry in $p(e, e'p)\pi^0$ and $p(e, e'\pi^+)n$ with $Q^2 < 2$ GeV^2 [15]. There is also an interesting experiment to study the axial-vector transition form factor $G_A^{N\Delta}$ in $ep \to e'\Delta^{++}\pi^-$ [16].

Higher Mass States The CEBAF facility is ideal for studying electroproduction of pions and other mesons, and the intermediate resonances in these reactions. These studies will provide new information on the Q^2 dependence of the transition form factors for these baryons, and can also be used to search for new states in reactions with final states other than $N\pi$. They include studies of the polarized structure functions in electroproduction of Δ and the Roper resonance [17], pion electroproduction in the second and third resonance regions [18], and two pion electroproduction [19]. It is also proposed to search for missing baryons in the electroproduction of ω mesons [20], and to look for the amplitudes for the production of $S_{11}(1535)N\frac{1}{2}^-$ and $P_{11}(1710)N\frac{1}{2}^+$ in an $ep \to e'p\eta$ experiment [21]. These two states are known to have sizeable branching ratios to decay to $N\eta$.

Strange Quarks

There is a considerable amount of effort taking place at CEBAF to study the strange quark sea in the nucleon *via* parity violating ep scattering. As mentioned above in the second lecture, experiments which involve electromagnetic production of strange hadrons are also planned at CEBAF, and should allow considerable progress in the study of excited strange baryons. Experiments will study electromagnetic production of hyperons [22], and the radiative decays of electromagnetically produced light excited hyperons [23]. As we have seen in the second lecture, the latter study should have a direct bearing on the issue of the structure of the $\Lambda(1405)\frac{1}{2}^-$ resonance. An experiment is also planned to study electroproduction of Λ, $\Lambda(1520)\frac{3}{2}^-$, and $f_0(975)$ *via* the channels K^+K^-p and $K^+\pi^-p$ [24], which once again involve only charged particles for ease of reconstruction of the intermediate state.

Real and Virtual Compton Scattering

Although the data that exist for Compton scattering from the nucleon in the resonance region [25] are limited, it appears that it may be possible to extract information about baryon resonances from such a study. Our model of this process [26] describes the intermediate state as a coherent sum of nonstrange baryon resonances, whose wavefunctions and electromagnetic couplings are given by the constituent quark model. This study has shown that considerable information about the resonances is contained in the Compton scattering amplitude at these energies, and that taken together, the resonances missing from $N\pi \to N\pi$ experiments make a sizeable contribution. Real Compton scattering is currently being studied in the $\Delta(1232)$ region with polarized initial photons at Brookhaven National Laboratory by the LEGS group, and at MAMI/Mainz. The feasibility of real Compton scattering experiments using the photon tagger in Hall B at CEBAF is also currently under investigation.

At CEBAF there will be an experiment in Hall A [27] which will study single-photon electroproduction, or 'virtual' Compton scattering (VCS), where an electron is scattered from the proton and the excited proton then emits a real photon. This can be thought of in first Born approximation as $\gamma^*p \to p\gamma$, and in an s-channel picture the intermediate state will be a superposition of ground- and excited-state nonstrange baryons. The main background for this nucleon Compton scattering is neutral pion decay of the intermediate state, followed by a subsequent colinear two-photon decay. One photon goes into the detector and appears to be a Compton photon, and the other has little lab energy and is lost in the interaction region. These processes will be discarded by using the two high-resolution spectrometers in Hall A to very precisely determine the momenta of the outgoing electron and proton, and so distinguish the missing mass of a pion from a massless photon.

Another background is from the Bethe-Heitler process, which is a simple elastic electron-proton scattering process with a photon radiated by the initial or final electron. It can be kept a small fraction of the Compton scattering amplitude by restricting the experiment to large CM-system photon scattering angle (where the direction of the virtual photon is given by q, the three-momentum transfer). This background may actually help in the analysis of the experiment, because unlike the Compton scattering amplitude it is always purely real, and is easily calculable. Predictions for threshold VCS have been published by Guichon, Liu, and Thomas [28], and we are currently extending our model of proton Compton scattering to VCS [29], along with other improvements.

SUMMARY

It is fundamental to understanding QCD that we understand baryons. Lattice gauge theory and similar *ab initio* approaches may not work for, say, the third P_{11} state $N\frac{1}{2}^+(1710)$. As we have seen, the constituent quark model has its failings, but has the distinct advantage that it is calculable and (hopefully) correctable. Theorists are currently working on the inclusion of relativity and decay-channel mixings into the model, and are concentrating especially on the electromagnetic couplings of these states. With many more data, and hard work, we may extract some exciting physics from the new experimental programs at CEBAF and elsewhere. Precise electroproduction experiments, combined with constituent quark models of the transition form factors, may allow extraction of the form factors of the constituent quarks. We may discover hybrid baryons, or 'molecular' excitations with baryon quantum numbers (baryon-meson bound states). We will certainly solve some old puzzles, such as the nature of the Roper resonance and the $\Lambda(1405)$, and the question of the existence of the 'missing' (from $N\pi \to N\pi$) positive-parity baryons.

ACKNOWLEDGEMENTS

This work was supported in part by the U.S. Department of Energy through Contract No. DE-FG05-86ER40273, and by the Florida State University Supercomputer Computations Research Institute which is partially funded by the Department of Energy through Contract No. DE-FC05-85ER250000.

References

[1] D.J. Mack *et al.*, CEBAF experiment E-93-021.

[2] O.K. Baker *et al.*, CEBAF experiment E-93-018.

[3] M.V. Kossov *et al.*, CEBAF experiment E-93-012.

[4] A.R. Dzierba, J. Napolitano, *et al.*, CEBAF experiment E-94-016.

[5] N. Isgur, G. Karl, D.W.L. Sprung, *Phys. Rev.* **D23**, 163 (1981).

[6] J. Napolitano *et al.*, CEBAF experiment E-93-033.

[7] R. Koniuk and N. Isgur, *Phys. Rev.* **D21**, 1868 (1980).

[8] S. Capstick and W. Roberts, *Phys. Rev.* **D49**, 4570 (1994).

[9] W.J. Briscoe, J. Ficenec, D. Jenkins, *et al.*, *CEBAF experiment E-94-103; D.I. Sober et al., CEBAF experiment E-91-005; B.G. Ritchie et al., CEBAF experiments E-91-008 and 94-008; P.L. Cole, J.P. Connelly, R.R. Whitney et al., CEBAF experiment E-94-109.*

[10] V. Burkert, R. Minehart, *et al.*, *CEBAF experiment E-89-037.*

[11] N. Isgur, G. Karl and R. Koniuk, *Phys. Rev. Lett.* **41**, 1269 (1978); *Phys. Rev.* **D25**, 2394 (1982).

[12] S. Capstick and G. Karl, *Phys. Rev.* **D41**, 2767 (1990); S. Capstick, *Phys. Rev.* **D46**, 2864 (1992).

[13] C.E. Carlson, *Phys. Rev.* **D34**, 2704 (1986).

[14] J. Napolitano, P. Stoler, *et al.*, *CEBAF experiment E-94-014.*

[15] V. Burkert, R. Minehart, *et al.*, *CEBAF experiment E-89-042.*

[16] L. Elouadrhiri. D. Heddle, R. Hicks, Zh. Li. *et al.*, *CEBAF experiment E-94-005.*

[17] S. Frullani, R.W. Lourie, *et al.*, *CEBAF experiment E-91-011.*

[18] V. Burkert, M. Gai, R. Minehart, *et al.*, *CEBAF experiment E-89-038.*

[19] V. Burkert, M. Ripani, *et al.*, *CEBAF experiment E-93-006.*

[20] V. Burkert, H. Funsten, D.M. Manley, B. Mecking, *et al.*, *CEBAF experiment E-91-024.*

[21] S. Dytman, K. Giovanetti, *et al.*, *CEBAF experiment E-89-039.*

[22] R. Schumacher *et al.*, *CEBAF experiment E-89-004.*

[23] G.S. Mutchler *et al.*, *CEBAF experiment E-89-024.*

[24] L. Dennis, H. Funsten, *et al.*, *CEBAF experiment E-89-043.*

[25] See Ref. [26] for a complete list of references to these experiments.

[26] S. Capstick and B.D. Keister, *Phys. Rev.* **D46**, 84 (1992); Erratum *ibid* **D46**, 4104 (1992).

[27] P.-Y. Bertin, G. Fournier *et al.*, *CEBAF experiment E-93-050.*

[28] P.A.M. Guichon, G.Q. Liu, and A.W. Thomas, DAPNIA-Saclay preprint DAPNIA-SPHN-95-02, to appear in *Nucl. Phys. A.*

[29] S. Capstick and B.D. Keister, in progress.

MODELS OF CONFINEMENT

A.V. Sarantsev

Petersburg Nuclear Physics Institute, Gatchina, Russia

1 Introduction

In all the history of science it was assumed that properties of the fundamental particles must be rather simple and all complications of nature are a consequence of the collective interaction (many body problem). In the fifties the consideration of the proton, neutron and Λ hyperon as fundamental particles interacting via pion or kaon exchanges became a popular model [1]. But already in the sixties, after the discovery of other hyperons, Gell-Mann [2] and Zweig [3] realized that hadrons can be classified on the basis of a triplet of hypothetical particles (called quarks) carrying fractional electric charges. The quark model not only explained the existing spectra of baryons and mesons but also predicted a new particle: the Ω hyperon, which was later discovered successfully. Many experiments were designed to find quarks, both using accelerators and in cosmic rays; but all failed to detect such particles. Very soon it was realized that, instead of being simple "bricks of nature", it is astonishing that these fundamental constitutents cannot be observed directly in experiments, but are confined inside mesons and baryons.

The present theory of strong interactions, QCD, takes into account this property of the quarks by introducting a new quantum number: color. It is assumed that a quark can be in one of three color states, while experimentally only colorless states can be observed. However this simply means that the problem of quark confinement is substituted by another problem: the problem of color confinement.

Nowadays we believe that QCD is the correct theory of strong interactions. It works perfectly at high energies, but the perturbation approach used in this energy region cannot be applied to the region of quark (color) confinement, where the coupling constant becomes larger then 1. One method to understand the confinement phenomenon is to develop nonperturbative QCD methods in the region of low energies. Another method is to construct a model of confinement which can describe well the experimental data and thus provide information about quark interactions. In these lectures we will consider three different models of quark confinement: the potential model, the bag model and the quark-hadron duality model.

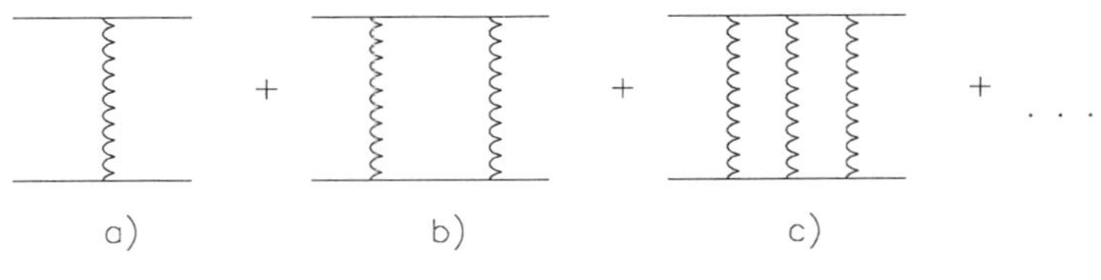

Fig.1

2 QCD at small energies

The theory of strong interactions, QCD, can be considered as a non-abelian generalization of quantum electrodynamic theory QED. In QCD the quarks interact via gluon exchange and the Lagrangian of the quark-gluon interaction has the following form:

$$\mathcal{L}_{QCD} = \bar{q} i \gamma^{\mu} D_{\mu} q - m \bar{q} q - \frac{1}{4} F^{a}_{\mu\nu} F^{a}_{\mu\nu}. \tag{1}$$

Here γ_{μ} are Dirac matrices and the quark field q is a color column of bispinors. The covariant derivative D_{μ} and gluon energy tensor $F^{a}_{\mu\nu}$ are defined as:

$$D_{\mu} = \partial_{\mu} - i g \lambda^{a} A^{a}_{\mu},$$

$$F^{a}_{\mu\nu} = \partial_{\mu} A^{a}_{\nu} - \partial_{\nu} A^{a}_{\mu} + g f_{abc} A^{b}_{\mu} A^{c}_{\nu}, \tag{2}$$

where A^{a}_{μ} is a color octet of gluon fields and the λ-matrices obey the following commutation properties:

$$[\lambda^{a}, \lambda^{b}] = -i f_{abc} \lambda^{c}. \tag{3}$$

The f_{abc} are real, totally antisymmetric structure constants. The nonabelian nature of the interaction appears through the presence of the term $f_{abc} A^{b}_{\mu} A^{c}_{\nu}$ in eq.(2) which characterizes the gluon-gluon interaction.

First let us consider the quark interaction via one gluon exchange. Diagrams for such processes are shown in fig.1 and the sum of these diagrams corresponds to the potential:

$$V(r) = -\frac{4}{3} \frac{\alpha_{s}(r)}{r}. \tag{4}$$

At intermediate distances the coupling constant $\alpha_{s}(r)$ can be taken as a constant (the direct calculation of the diagrams in fig.1 gives $\alpha_{s} = g^{2}/4\pi$) and the potential (4) is in this case the analogue of the Coulomb potential in electrodynamics.

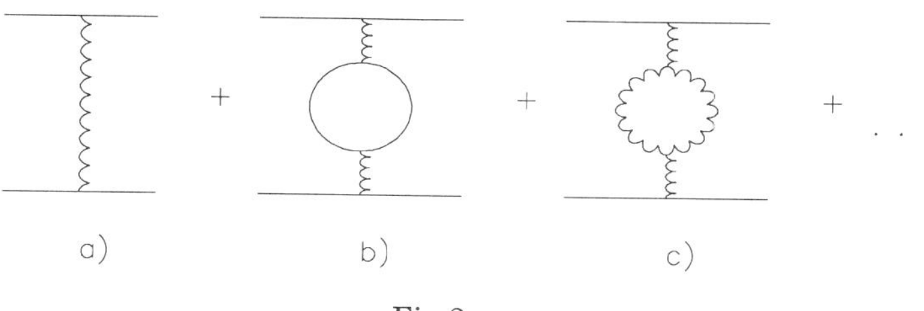

Fig.2

At small and large distances, the dependence of α_s on r plays a very important role. As in the electrodynamic case the evolution equation of the coupling constant can be calculated from the diagrams shown in fig.2 and has the following form:

$$\alpha_s(Q^2) = \frac{\alpha_s(Q_0^2)}{1 - \alpha_s(Q^2)}[\Pi(Q^2) - \Pi(Q_0^2)] \tag{5}$$

where Q is the momentum of the exchanged particle and $\Pi(Q^2)$ is its polarization operator. In QED the photon polarization operator is equal to:

$$QED: \qquad \Pi(Q^2) - \Pi(Q_0^2) = \frac{1}{3\pi} ln \frac{Q^2}{Q_0^2} \quad . \tag{6}$$

The gluon polarization operator is different from this expression because of the number of possible quark flavor states n_f and the contribution of the gluon loop diagram shown in fig.2c.

$$QCD: \qquad \Pi(Q^2) - \Pi(Q_0^2) = \frac{1}{4\pi}\left[\frac{2}{3}n_f - 11\right] ln \frac{Q^2}{Q_0^2}. \tag{7}$$

Eqs.(6) and (7) gives the following expression for the derivative of the coupling constants:

$$QED: \qquad \frac{\partial \alpha_s(Q^2)}{\partial \, ln \, Q^2} = \frac{1}{3\pi}\alpha_s^2(Q^2)$$

$$QCD: \quad \frac{\partial \alpha_s(Q^2)}{\partial \, ln \, Q^2} = -\frac{33 - 2n_f}{12\pi}\alpha_s^2 \left(1 + \frac{153 - 19n_f}{33 - 2n_f}\alpha_s^2 + \cdot\right) \tag{8}$$

It is seen that in quantum electrodynamics the derivative of α_s is always positive and the coupling constant increases with increasing Q^2 The situation in QCD is just the opposite: if the number of flavors is less than 16, the derivative (8) is negative and at large Q^2 (small distances) we face the phenomenon of asymptotic freedom: the quarks do not interact

$$\alpha_s(Q^2) \to 0 \qquad as \qquad Q^2 \to \infty \, . \tag{9}$$

Another side of this phenomenon is that at small Q^2 (large distances) the coupling constant becomes much larger than 1 and the perturbative approach does not work.

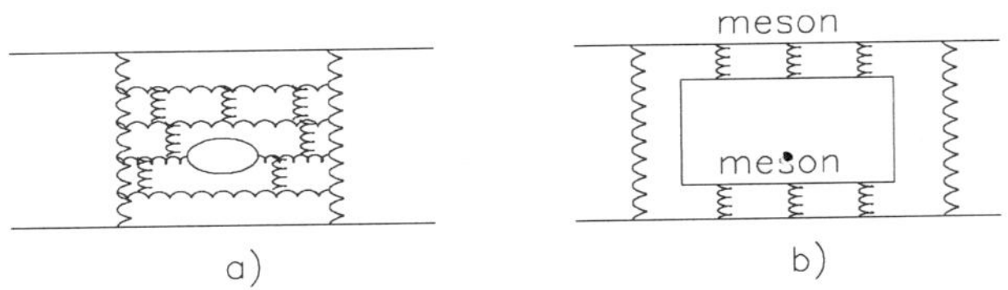

Fig.3

At small Q^2, many-particle exchanges like that shown in fig.3a start to play the main role. It is possible to assume that such kinds of interaction are responsible for quark confinement. This assumption is taken as basic in the potential model. But is it really true? Actually it is very likely that the confinement is connected with other processes: namely the process of hadron production, or the so-called infrared instability of the vacuum. The quarks can easily escape the confinement trap if their energy is large enough to produce a new $q\bar{q}$ pair. But if such processes are important for the description of hadron properties, it is relevant to assume that the diagrams shown in fig.3b will give an appreciable contribution to the quark-quark interaction and can be responsible for the creation of the confinement barrier. In these lectures, along with the well known potential and bag models we will consider an example of a model which treats the diagrams of fig.3b as central in the explanation of the confinement mechanism.

3 Potential model

Let us consider the interaction of fermions with mass m. The potential of the interaction, which can be introduced in the nonrelativistic limit, has to be a scalar function constructed from two independent momenta, for example:

$$\frac{\vec{P}}{m} = \frac{\vec{p_1} + \vec{p_1}\,'}{2m} = \frac{\vec{p_2} + \vec{p_2}\,'}{2m} \qquad \frac{\vec{q}}{m} = \frac{\vec{p_1}\,' - \vec{p_1}}{m} = -\frac{\vec{p_2}\,' - \vec{p_2}}{m}, \tag{10}$$

where p_i and p_i' are momenta of the particles in the initial and final states, spin Pauli matrices σ_1 and σ_2 of fermions, the symmetrical metric tensor δ_{ij} and totally antisymmetric tensor ϵ_{ijk}.

Taking into account that the production operator $\vec{\sigma}\vec{q}$ is a pseudoscalar and the expression

$$\chi_s'^{+} \vec{\sigma}\vec{P} \chi_s = 0 \tag{11}$$

where χ_s', χ_s are two dimentional spinors, we obtain the following structure of the potential:

$$V_{tot}(\vec{P}^2, \vec{q}^2) = V_0(\vec{P}^2, \vec{q}^2) + i\epsilon_{ijk}\frac{P_i q_j}{m^2}\,(\sigma_{1k} + \sigma_{2k})V_1(\vec{P}^2, \vec{q}^2) +$$
$$(\vec{\sigma}_1\vec{\sigma}_2)\,V_2(\vec{P}^2, \vec{q}^2) + S_{12}V_3(\vec{P}^2, \vec{q}^2), \tag{12}$$

where

$$S_{12} = 3\frac{(\vec{\sigma}_1\vec{q})(\vec{\sigma}_2\vec{q})}{m^2} - (\vec{\sigma}_1\vec{\sigma}_2)\frac{\vec{q}^{\,2}}{m^2}. \tag{13}$$

In the expression (12) we neglect all higher order terms of $\vec{q}/m$ and $\vec{P}/m$. The first term on the right side of eq.(12) corresponds to spin independent forces, responsible for the formation of the ground states; the second term to the spin-orbit forces, the third term to the spin-spin forces and the last term to the tensor interaction. Taking into account the properties of Fourier transformation:

$$\int \frac{d^3q}{(2\pi)^3}\, V(\vec{q}^{\,2})\, e^{i\vec{q}\vec{r}} = V(r)$$

$$\int \frac{d^3q}{(2\pi)^3}\, q_j V(\vec{q}^{\,2})\, e^{i\vec{q}\vec{r}} = \frac{\partial V(r)}{i\partial r_j} = \frac{r_j}{ir}V'(r), \tag{14}$$

we can rewrite the spin-orbit interaction in the usual form:

$$i\epsilon_{ijk}\frac{P_i q_j}{m^2}\,(\sigma_{1k}+\sigma_{2k})V_1(q^2) \rightarrow \epsilon_{ijk}\frac{P_i r_j}{m^2}\,(\sigma_{1k}+\sigma_{2k})\frac{V_1'(r)}{r} = (\vec{\sigma}_1+\vec{\sigma}_2)[\vec{r}\times\vec{P}]\frac{V_1'(r)}{r}. \tag{15}$$

Let us consider now what kind of interaction can provide the potential (12). First of all we have to take into account one gluon exchange, which is a vector exchange ($J^P = 1^-$) and is characterized by the operator $Q = \vec{\lambda}\gamma_\mu$. The two gluon exchange diagram shown in fig.3a can have all possible quantum numbers. Here the scalar exchange is described by the operator $Q = 1$, the pseudoscalar exchange by the operator $Q = \gamma_5$ and so on. The relativistic expression for such interaction has the following form:

$$\chi_1'^{+}\chi_2'^{+}V(\vec{P}^2,\vec{q}^{\,2})\chi_1\chi_2 = [\bar{u}(p_1')Qu(p_1)]G(q^2)[\bar{u}(p_2')Qu(p_2)] \tag{16}$$

where $G(q^2)$ is the interaction function (propagator of the effective particle) and bispinors $u(p_i)$ are equal to:

$$u(p_i) = \sqrt{\frac{1}{2E_i(E_i+m)}}\begin{pmatrix} E_i+m \\ \vec{\sigma}\vec{p}_i \end{pmatrix}\chi_i \rightarrow \frac{1}{8m^2}\begin{pmatrix} 8m^2-\vec{p}_i^{\,2} \\ 4m\vec{\sigma}\vec{p}_i \end{pmatrix}\chi_i. \tag{17}$$

Here $E_i = \sqrt{m^2+\vec{p}_i^{\,2}}$ and the last term corresponds to the transition to the nonrelativistic limit. Taking into account the normalization of the bispinors:

$$u^+(p)u(p) = \bar{u}(p)\gamma_0 u(p) = 1, \tag{18}$$

we obtain in the case of scalar exchange:

$$[\bar{u}(p_1')u(p_1)] = 1 - \frac{\vec{p}_1^{\,2}}{2m^2} + \frac{i}{4m^2}\epsilon_{ijk}p_{1i}q_j\sigma_{1k}. \tag{19}$$

The potential of the scalar interaction after performing the Fourier transform is equal to:

$$V(\vec{P}^2,\vec{r}) = G_{sca}(r)\left(1-\frac{\vec{P}^2}{m^2}\right) - \frac{\vec{\sigma}_1-\vec{\sigma}_2}{4m^2}[\vec{r}\times\vec{P}]\frac{G_{sca}'(r)}{r}. \tag{20}$$

Remember that operator $\vec{P}$ acts on the wave function but not on the potential.

The potential defined by vector exchange ($Q = \gamma_\mu$) has a much more complicated structure:

$$V(\vec{P}^2, \vec{r}) = G_{vec}(r)\left(1 + \frac{\vec{P}^2}{m^2}\right) + \frac{3(\vec{\sigma}_1 + \vec{\sigma}_2)}{4m^2}[\vec{r} \times \vec{P}]\frac{G'_{vec}(r)}{r} + \frac{\vec{\sigma}_1\vec{\sigma}_2}{6m^2}\vec{\nabla}^2 G_{vec}(r)$$
$$-\frac{S_{12}}{12}\left(G''_{vec}(r) - \frac{G'_{vec}(r)}{r}\right) + \frac{1}{4m^2}\vec{\nabla}^2 G_{vec}(r) + \frac{1}{m^2}(\vec{P}\vec{\nabla})^2 U(r) \qquad (21)$$

where

$$\vec{\nabla}^2 U(r) = -G_{vec}(r).$$

In the case of gluon exchange, the color λ matrices will give an additional factor $-\frac{4}{3}$ in the case of the quark-antiquark interaction (mesons) and a factor $-\frac{2}{3}$ in the case of quark-quark interaction (baryons). The scalar and vector exchanges are the only interactions which provide a nonvanishing contribution to the spin independent part in the nonrelativistic limit. For example the potential of the pseudoscalar interaction is equal to:

$$V(\vec{P}^2, \vec{q}^2) = \frac{\vec{\sigma}_1\vec{\sigma}_2}{12m^2}\vec{\nabla}^2 G_{pse}(r) + \frac{S_{12}}{12}\left(G''_{pse}(r) - \frac{G'_{pse}(r)}{r}\right) \qquad (22)$$

and defines the level splitting only.

Let us consider the heavy quarkonium ($c\bar{c}$ and $b\bar{b}$) states, which can be regarded as a rather good nonrelativistic system. In this case the confinement potential must have a contribution either from the vector or scalar interaction to provide the formation of the spin independent states. The wave function can be found by solving the Schrödinger equation:

$$(H_0 + H_{LS})|\Psi(\vec{r})>= E|\Psi(\vec{r})> \qquad (23)$$

where

$$H_0 = 2m + \frac{\vec{P}^2}{m} + G_{vec}(r) + G_{sca}(r),$$

$$H_{LS} = \frac{\vec{\sigma}_1 + \vec{\sigma}_2}{4m^2}[\vec{r} \times \vec{P}]\frac{3G'_{vec}(r) - G'_{sca}(r)}{r}. \qquad (24)$$

We omitted here for the sake of simplicity the spin-spin and tensor interactions. The wave function corresponding to eqs.(23)-(24) has the following form:

$$\Psi(\vec{r}) = \frac{w(r)}{r}Y_{lm}(\Theta), \qquad (25)$$

and the mass spectrum can be easily calculated. Several different forms of the vector interaction were used to calculate the spectrum of the $b\bar{b}$ and $c\bar{c}$ mesons, for example:

$$G_{vec}(r) = b\, ln\frac{r}{c} \qquad \textit{Quigg and Rosner}$$
$$G_{vec}(r) = \frac{1}{b}\left(\frac{r}{b}\right)^{0.1} + c \qquad \textit{Martin}$$
$$G_{vec}(r) = -\frac{4}{3}\frac{\alpha_s}{r} + br + c \qquad \textit{Eichten} \qquad (26)$$

where b, c are parameters of the model. All these forms describe well the charmonium and bottomonium ground states but fail to describe the splitting of the $^3P_1$ states. This splitting is given by the spin-orbit interaction and cannot be explained by vector exchange. It is necessary to introduce the scalar interaction, which has the opposite sign for this term. The best description of the spectra is obtained with the following choice of interaction:

$$G_{vec}(r) = -\frac{4}{3}\frac{\alpha_s}{r} \qquad G_{sca}(r) = br + c. \tag{27}$$

It is very important to take into account the energy dependence of the coupling constant. As a rule it is parametrized in some form to reproduce well the evolution equation (8) and has one free parameter: Λ_{QCD} (the energy where coupling constant becomes infinitely large).

The form (27) is the basic form used in the potential model. It reproduces quite well the spectrum of the heavy quarkonium states and practically does not leave any room for the introduction of other interactions. It is the main reason why such an interaction was taken to describe the spectra of the light mesons, where relativistic effects are large and in principal any interaction can be responsible for formation of the confinement barrier.

The model which most successfully describes the spectrum of light mesons as well as the heavy quarkonium states was developed by Godfrey and Isgur [4]. In this model the heavy quarkonium states were treated in a very similar way to that described above; for the light quarks a recipe was suggested to take into account the relativistic effects. The kinetic part of the Hamiltonian was taken in the form:

$$H_0 = \sqrt{\vec{p}_1^2 + m_1^2} + \sqrt{\vec{p}_2^2 + m_2^2}, \tag{28}$$

and the potential was calculated directly from (16) without transition to the nonrelativistic limit. The function $G_{vec}(q^2)$ corresponding to one gluon exchange is equal to:

$$G_{vec}(\vec{q}^2 = -Q^2) = -4\pi\frac{4}{3}\frac{\alpha_s(Q^2)}{Q^2}. \tag{29}$$

The function $G_{sca}(q^2)$ was not calculated directly but was supposed to be the Fourier transform of the expression (27).

However the direct calculation of the hadron spectrum with such a potential creates some problems: 1) it gives too singular a behavior of the potential at small distances, 2) the relativistic potential should not be a local one. To resolve these problems a semiquantative model was suggested: the potential between quarks m_i and m_j was smeared over the interaction region by some function ρ_{ij}:

$$V_{ij}(r) = \int dr'\rho_{ij}(r' - r)V_{ij}(r') \tag{30}$$

where

$$\rho_{ij} = \frac{\sigma_{ij}^3}{\pi^{3/2}}\, e^{-\sigma_{ij}^2(r'-r)}, \tag{31}$$

with

$$\sigma_{ij} = \frac{\sigma_0^2}{2}\left[1 + \left(\frac{4m_i m_j}{(m_i + m_j)^2}\right)^4\right] + s^2\left(\frac{2m_i m_j}{(m_i + m_j)^2}\right)^2. \tag{32}$$

The parameters σ_0 and s were used to fit the hadron spectra.

The energy dependence of the coupling constant α_s on Q^2 was taken in the form:

$$\alpha_s(Q^2) = 0.25\ e^{-Q^2} + 0.15\ e^{-0.1Q^2} + 0.20\ e^{-0.01Q^2} \tag{33}$$

which satisfies quite well the evolution equation (8) with $\Lambda_{QCD} = 200$ MeV.

Let us note that the model with interaction (27) does not reproduce the sector of the pseudoscalar mesons $^1S_0$; in their article Godfrey and Isgur simply suggested a parametrization of the amplitude in this sector to reconstruct the meson spectrum. The problem of this sector is related to the $U(1)$ problem of QCD and can be solved by the introduction of instanton induced forces [5].

The Godfrey-Isgur model describes rather well the meson spectrum. The best description was obtain with the following parameters: the masses of quarks $m_u = m_d = 0.220$ GeV, $m_s = 0.420$ GeV, $m_c = 1.630$ GeV, m_b=4.980 GeV; the potential $b = 0.18$ GeV^2, $c = -253$ MeV; smearing σ_0=1.80 GeV, s=1.55. The spectra of the mesons are given in Tables 1,2. The most attractive features of this model are its simplicity and the possibility to calculate a huge number of bound states.

However there are problems in the model which are clearly seen already in the scalar sector of the meson spectrum. This problem is tightly connected with the problem of calculating hadron decay properties. The potential model, which is most powerful in the calculation of hadron masses, actually has great difficulties in the calculation of other observables like decay widths or meson photoproduction processes.

4 Bag model

From an educational point of view, the bag model suggested by the MIT group [6] is a very interesting attempt to explain confinement on the basis of the QCD Lagrangian. The main idea of this approach is that the real vacuum does not tolerate the free color field. Then the color field can appear only inside some kind of bag (or better say blob) which is stabilized by the pressure balance of the internal color fields and external hadron fields. Inside the bag the physics is described by the QCD Lagrangian while outside the bag only hadron states exists.

The bag Lagrangian has the following form:

$$\mathcal{L}_{BAG} = [\mathcal{L}_{QCD} - B]\Theta(\bar{q}(x)\Gamma q(x)). \tag{34}$$

Here $\mathcal{L}_{QCD}$ is the QCD lagrangian given in eqs (1-3), and the B is the constant energy of the quark bag. The Θ function reflects the fact that color fields exist only inside the bag with boundary defined by the condition:

$$\sum_a \bar{q}_a(x)\Gamma q_a(x) = 0. \tag{35}$$

The expression $\bar{q}(x)\Gamma q(x)$ must be either scalar or pseudoscalar: it means that Γ can be either 1 or γ_5 (or their linear combination). The γ_5 case actually violates P-invariance but is important for the formulation of the chiral bag model. Using the Euler-Lagrangian equations for the quark and gluon fields:

$$\frac{\partial \mathcal{L}_{BAG}}{\partial \bar{q}(x)} - \partial_\mu \frac{\partial \mathcal{L}_{BAG}}{\partial \partial_\mu \bar{q}(x)} = 0, \qquad \frac{\partial \mathcal{L}_{BAG}}{\partial A_\nu^a(x)} - \partial_\mu \frac{\partial \mathcal{L}_{BAG}}{\partial \partial_\mu A_\nu^a(x)} = 0, \tag{36}$$

Table 1

Masses of the meson states obtained in the framework of the Godfrey-Isgur model. The experimental values are given in parentheses. All units are in GeV.

$N^{2S+1}L_J$	$u\bar{d}$, $\frac{u\bar{u}-d\bar{d}}{\sqrt{2}}$, $d\bar{u}$	$u\bar{s}$, $d\bar{s}$	$\frac{u\bar{u}+d\bar{d}}{\sqrt{2}}$, $s\bar{s}$	$\frac{u\bar{u}+d\bar{d}}{\sqrt{2}}$, $s\bar{s}$
1^1S_0	0.15 (0.14)	0.47 (0.49)	0.52 (0.55)	0.96 (0.96)
1^3S_1	0.77 (0.77)	0.90 (0.89)	0.78 (0.78)	1.02 (1.02)
1^3P_0	1.09 (–)	1.24 (1.43)	1.09 (–)	1.36 (–)
1^1P_1	1.22 (1.24)	1.34 (1.27)	1.22 (1.17)	1.47 (–)
1^3P_1	1.24 (1.26)	1.38 (1.40)	1.24 (1.28)	1.48 (1.42)
1^3P_2	1.31 (1.27)	1.43 (1.43)	1.28 (1.27)	1.53 (1.52)
1^3D_1	1.66 (1.70)	1.78 (1.68)	1.66 (1.60)	1.88 (–)
1^1D_2	1.68 (1.67)	1.78 (1.77)	1.68 (1.67)	1.89 (1.87*)
1^3D_2	1.70 (–)	1.81 (1.82)	1.70 (–)	1.91 (–)
1^3D_3	1.68 (1.69)	1.79 (1.78)	1.68 (1.67)	1.90 (1.85)
1^3F_2	2.05 (2.01)	2.15 (–)	2.05 (2.15*)	2.24 (2.30)
1^1F_3	2.03 (–)	2.12 (2.32*)	2.03 (–)	2.22 (–)
1^3F_3	2.05 (2.05*)	2.15 (2.32*)	2.05 (–)	2.23 (–)
1^3F_4	2.01 (2.04*)	2.11 (2.05)	2.01 (2.05)	2.20 (2.22*)
2^1S_0	1.30 (1.30)	1.45 (1.46*)	1.44 (1.30)	1.63 (1.44)
2^3S_1	1.45 (1.45)	1.58 (1.41)	1.46 (1.42)	1.68 (1.69)
2^3P_0	1.78 (–)	1.89 (1.95*)	1.78 (–)	1.99 (–)
2^1P_1	1.78 (–)	1.90 (1.65*)	1.78 (–)	2.01 (–)
2^3P_1	1.82 (–)	1.93 (1.65*)	1.82 (1.51)	2.03 (–)
2^3P_2	1.82 (1.81*)	1.94 (1.98*)	1.82 (1.64*)	2.04 (2.01*)
2^3D_1	2.15 (2.21*)	2.25 (–)		
2^1D_2	2.13 (2.10*)	2.23 (2.25*)		
2^3D_2	2.15 (–)	2.26 (2.25*)		
2^3D_3	2.13 (2.25*)	2.24 (–)		
3^1S_0	1.88 (1.77*)	2.02 (1.83*)		
3^3S_1	2.00 (2.15*)	2.11 (–)		

* - the resonance needs confirmation.

Table 2

Masses of the meson states obtained in the framework of the Godfrey-Isgur model. The experimental values are given in parentheses. All units are in GeV.

$N^{2S+1}L_J$	$c\bar{c}$	$c\bar{u}, c\bar{d}$	$c\bar{s}$	$b\bar{b}$
1^1S_0	2.97 (2.98)	1.88 (1.87)	1.98 (1.97)	9.40 (–)
1^3S_1	3.10 (3.10)	2.04 (2.01)	2.13 (2.11)	9.46 (9.46)
1^3P_0	3.44 (3.42)	2.40 (–)	2.48 (–)	9.85 (9.86)
1^1P_1	3.52 (–)	2.44 (2.42)	2.53 (2.53)	9.88 (–)
1^3P_1	3.51 (3.51)	2.49 (2.42)	2.57 (2.53)	9.88 (9.89)
1^3P_2	3.55 (3.56)	2.50 (1.47)	2.59 (–)	9.90 (9.92)
1^3D_1	3.82 (3.77)	2.82 (–)	2.90 (–)	10.14(–)
1^1D_2	3.84 (–)			10.15(–)
1^3D_2	3.84 (–)			10.15(–)
1^3D_3	3.85 (–)	2.83 (–)	2.92 (–)	10.16(–)
1^3F_2	4.09 (–)			10.35(–)
1^1F_3	4.09 (–)			10.35(–)
1^3F_3	4.10 (2.05)			10.35(–)
1^3F_4	4.09 (–)	3.11 (–)	3.19 (–)	10.36(–)
2^1S_0	3.62 (3.59)	2.58 (–)	2.67 (–)	9.98 (–)
2^3S_1	3.68 (3.69)	2.64 (–)	2.73 (–)	10.00(10.02)
2^3P_0	3.92 (–)			10.23(10.24)
2^1P_1	3.96 (–)			10.25(–)
2^3P_1	3.95 (–)			10.25(10.26)
2^3P_2	3.98 (–)			10.26(10.27)
2^3D_1	4.19 (4.16)			10.44(–)
2^1D_2	4.21 (–)			10.45(–)
2^3D_2	4.21 (–)			10.45(–)
2^3D_3	4.22 (–)			10.45(–)
3^1S_0	4.06 (–)			10.34(–)
3^3S_1	4.10 (4.04)			10.35(10.35)

we obtain the usual Yang-Mills equations inside the bag:

$$\gamma_\mu[i\partial_\mu + g\lambda^a A^a_\mu(x)]q(x) = mq(x),$$

$$-\partial_\mu F^a_{\mu\nu} = g[\bar{q}(x)\gamma_\nu\lambda^a q(x) - f_{abc}F^b_{\mu\nu}A^c_\mu(x)] \equiv J^a_\nu(x). \tag{37}$$

Here J^a_ν is the color current. This current is conserved:

$$\partial_\nu J^a_\nu = \partial_\nu\partial_\mu F^a_{\mu\nu} = 0 \tag{38}$$

and the integral of its zero component corresponds to the color charge of the bag:

$$Q^a = \int dV\, J^a_0(x). \tag{39}$$

The action of the partial derivative on the Θ function results in the δ function with coefficients which define the boundary conditions:

$$(\mathcal{L}_G - B)\Gamma q(x) = -\frac{i}{2}\gamma^\mu q(x)\partial_\mu(\bar{q}(x)\Gamma q(x)),$$

$$F^a_{\mu\nu}\partial_\mu(\bar{q}(x)\Gamma q(x)) = 0, \tag{40}$$

providing that Yang-Mills equations are valid on the boundary, with the result:

$$\mathcal{L}_{QCD} \to \mathcal{L}_G = -\frac{1}{4}F^a_{\mu\nu}F^a_{\mu\nu}. \tag{41}$$

The expression $\bar{q}(x)\Gamma q(x)$ is a constant on the boundary (see eq.(35)) and therefore the gradient from it can be written as:

$$\partial_\mu(\bar{q}(x)\Gamma q(x)) = n_\mu D(x), \tag{42}$$

where vector n_μ is the unit vector normal to the surface of the bag.

$$n_\mu n_\mu = -1. \tag{43}$$

Then the first boundary condition (40) has the following form:

$$(\mathcal{L}_G - B)\Gamma q(x) = -\frac{i}{2}\gamma^\mu n_\mu D(x)q(x). \tag{44}$$

Really there are two boundary conditions in this expression. If we multiply eq.(44) by $in_\mu\gamma_\mu$ and then by $(B - \mathcal{L}_G)\Gamma$ we obtain:

$$(\mathcal{L}_G - B) = \frac{D(x)}{2}. \tag{45}$$

Thus we have three boundary conditions: two for the quark fields and one for the gluon field:

$$\frac{1}{2}\partial_\mu(\bar{q}(x)\Gamma q(x)) = n_\mu(B - \mathcal{L}_G), \tag{46}$$

$$i\gamma^\mu n_\mu q_a(x) = \Gamma q_a(x), \tag{47}$$

$$n_\mu F^a_{\mu\nu} = 0. \tag{48}$$

The boundary condition (46) is called the pressure balance equation. It means that the pressure of the quarks and gluon fields must be equal to the pressure of the external vacuum on the bag. The second condition gives an important restriction on the color flow through the boundary. Using the conjugated expression:

$$\bar{q}^b(x)\Gamma = -\bar{q}^b(x) i\gamma_\mu n_\mu \tag{49}$$

we have on the boundary:

$$\bar{q}^b(x)\Gamma q_a(x) = 0 \qquad \bar{q}^b(x) i\gamma_\mu n_\mu q_a(x) = 0. \tag{50}$$

It means that there are no color field transitions through the boundary. Let us remember that we started from the condition (35) which was much less restricted than condition (50).

Let us express the energy-momentum tensor of the gluon field through the color electric and magnetic fields:

$$F^a_{0i} = E^a_i \qquad F^a_{ij} = \epsilon_{ijk} B^a_k \qquad i,j,k = 1,2,3. \tag{51}$$

Then the boundary condition (48) gives for the electric field:

$$\vec{n}\vec{E}^a = 0. \tag{52}$$

For the magnetic field in the instantaneous rest frame, where $n_\mu = (0, -\vec{n})$, the boundary condition has the following form:

$$n_i \epsilon_{ijk} B_k = [\vec{n} \times \vec{B}] = 0. \tag{53}$$

Using Yang-Mills equation (37) we obtain:

$$\vec{\nabla}\vec{E}^a = J^a_0(x) \tag{54}$$

and thus:

$$Q^a = \int_V d^3x J^a_0(x) = \int_V d^3x \vec{\nabla}\vec{E}^a = \vec{n}\vec{E}^a|_S = 0. \tag{55}$$

That means that the color charge inside the bag is equal to zero.

Unfortunately the calculation of the hadron spectrum cannot be performed directly from the boundary conditions (46-48) in the general case. Below we will consider two special cases when the result can be obtained with some additional assumptions.

In the case of the heavy quarks it is relevant to assume that quarks move slowly compared with the gluons and therefore can be considered as static sources of the gluon field. Then only the zero component of the color current is non-zero:

$$J^a_0(x) = g[\lambda^a \delta^3(x - x_1) - \lambda^a \delta^3(x - x_2)]. \tag{56}$$

The quark field near the boundary is not important and the boundary condition (49) transforms into:

$$B = \mathcal{L}_G = \frac{1}{2}\sum_a (E^2_a(x) - B^2_a(x)). \tag{57}$$

The condition (47) plays no role and the condition (48) is:

$$\vec{n}\vec{E}_a = 0 \qquad [\vec{n} \times \vec{B}_{\cdot}^{\cdot} = 0. \tag{58}$$

The solution in the case of static sources is well known in QED:

$$A^a_\mu = (\Phi_a(x), \vec{0}) \qquad \vec{E}_a = -\vec{\nabla}\Phi_a \qquad \vec{B} = \vec{0}, \tag{59}$$

where $\Phi(x)$ is the potential of the gluon field. The situation in QCD is more complicated: for example, the magnetic field is not equal to zero but has some contribution of order $O(g^3)$. But assuming the smallness of the coupling constant inside the bag we can consider (59) as a good approximation.

From eqs.(54) and (59) we have:

$$\triangle\Phi_a(x) = -J^a_0(x). \tag{60}$$

Assuming axial symmetry along the axis between $r_1 - r_2$ the solution of this equation is:

$$\Phi(r) = \frac{g'}{4\pi|\vec{r} - \vec{r}_1|} - \frac{g'}{4\pi|\vec{r} - \vec{r}_2|} + \sum_l c_l \left(\frac{r}{R}\right)^{2l} P_{2l}(cos\Theta). \tag{61}$$

It is usually enough to take into account the first five terms from the sum of eq.(61) to satisfy the boundary conditions. In the case of large separation, the bag transforms into a cylinder. The condition (52) shows that the electric field hardly changes along the cylinder cross section; using Gauss's theorem that production of the electric field over the cross section is equal to the charge of the system we obtain:

$$E_a 2\pi R_c^2 = g_a, \tag{62}$$

where R_c is a cylinder radius; from the first boundary condition we obtain:

$$\sum_a (E_a 4\pi^2 R_c^2)^2 = 2B\pi^2 R_c^4 = \sum_a g_a^2 = g'^2. \tag{63}$$

It gives the following cylinder radius:

$$R_c = \left(\frac{g'^2}{2\pi^2 B}\right)^{1/4}. \tag{64}$$

The sum of the gluon and volume energy is equal to:

$$E_g + E_V = \int_V \frac{1}{2}\sum_a (\vec{E}_a)^2 d^3x + BV = 2B\pi R_c^2 r + BV. \tag{65}$$

To obtain the potential of the quark-quark interaction, the Coulomb potential corresponding to the two first terms of eq.(61) must be added:

$$V = -\frac{g'^2}{4\pi r} + kr + V_{Q\bar{Q}}, \tag{66}$$

where

$$k = (2g'^2 B)^{1/2} \qquad V_{Q\bar{Q}} = BV.$$

In the case of small quark separations, the sphere is a good approximation for the quarks and only the first term from the sum in eq.(61) need be taken into account. In this case the equation (60) under boundary condition (58) can be solved exactly and as a result we obtain:

$$V = -\frac{g'^2}{4\pi r} + \sqrt{\frac{2}{3}}kr. \tag{67}$$

We see that the coefficients of the linear part of the potential are different for the case of large and small quark separations. It is not clear what must be done to join these regions. One of the possible approaches is to use the potential in the form:

$$V(r) = -\frac{4}{3}\frac{\alpha_s}{r} + \sqrt{\frac{2}{3}}kr \qquad r \leq R_c$$

$$V(r) = -\frac{4}{3}\frac{\alpha_s}{r} + kr + \left(\sqrt{\frac{2}{3}} - 1\right)kR_c \qquad r \geq R_c. \tag{68}$$

As in the potential model approach, the dependence of the coupling constant α_s on r must be taken into account.

The potential (68) reproduces rather well the spectrum of the ground states; however, being a pure vector interaction, it fails to describe the spin-orbit splitting in heavy quarkonium.

Let us consider now the bag with light quarks. In this case rather different assumptions are made:

1) The spherical bag is specified first and quarks are put inside it,

2) the effects of the gluon field are neglected (at least in the first approximation). Then we have only boundary conditions (46-47) which can be rewritten as:

$$-\frac{1}{2}\vec{\nabla}(\bar{q}(x)\Gamma q(x)) = \frac{\vec{r}}{r}B, \tag{69}$$

$$-i\vec{\gamma}\frac{\vec{r}}{r}q(x) = \Gamma q(x). \tag{70}$$

The quarks inside the bag are described by the Dirac equation, which has in spherical coordinates the following solution:

$$q_\alpha(x) = \sum_{nkjm} a_\alpha(n,k,j,m)\Psi_{nkjm}(x,t), \tag{71}$$

where j, m are angular momentum and z-component, k-parity ($k = \pm(j+1/2)$). The boundary condition (69) can be fulfilled if the function $\bar{q}(x)\Gamma q(x)$ does not depend on angle and time. Angular independence is satisfied if the angular momentum is equal to $J = 1/2$. Then:

$$\Psi(x,t) = \frac{N}{\sqrt{4\pi}}\begin{pmatrix} ij_0(\omega_{n,-1}; r/R_0) \\ -j_1(\omega_{n,-1}; r/R_0)(\vec{\sigma}\vec{r})/r \end{pmatrix}\chi\, e^{-i\omega_{n,-1}t/R_c} \quad for\ \ k = -1,$$

$$\Psi(x,t) = \frac{N}{\sqrt{4\pi}}\begin{pmatrix} ij_1(\omega_{n,1}; r/R_0) \\ j_0(\omega_{n,1}; r/R_0)(\vec{\sigma}\vec{r})/r \end{pmatrix}\chi\, e^{-i\omega_{n,1}t/R_c} \quad for\ \ k = -1. \tag{72}$$

Here j_i are the spherical Bessel functions and the normalization constant N is equal to:

$$N = \left(\frac{\omega_{nk}^3}{2R_0^3(\omega_{nk}+k)sin^2\omega_{n,k}}\right)^{1/2}. \tag{73}$$

The boundary condition (70) gives:

$$j_0(\omega_{n,k}) = -kj_1(\omega_{n\,k}). \tag{74}$$

The frequencies $\omega_{n,k}$ can be found from the transendental equation:

$$tan\ \omega_{n,k} = \frac{\omega_{n,k}}{\omega_{n,k} + k}. \tag{75}$$

Let us turn now to the problem of the time independence implied by the first boundary condition. There are some cases when time independence is satisfied automatically:

$$\sum_\alpha a^*_\alpha(n,k)\ a_\alpha(n',k') = 0 \tag{76}$$

if $n = n', k = k', n = -n', k = -k'$ or when we have a state with negative energy. In all other cases, eq. (76) must be satisfied by a special choice of the coefficients a_α. This problem can be solved easily for classical fields but in quantum mechanics it is rather a serious problem. The MIT group suggested in their calculations that only one color state in toto can be excited. The functions $a_\alpha(n,k,m)$ must be considered as creation and annihilation operators of the quarks:

$$a_\alpha(n,k,m) = \begin{array}{ll} b_\alpha(n,k,m) & n > 0 \\ d^+_\alpha(-n,-k,m) & n < 0. \end{array} \tag{77}$$

Here the operator b_α is a creation operator, and d_α is an annihilation operator. Then the number of quarks in the system is equal to:

$$N_\alpha = \int_{bag} d^3x\ q^+_\alpha(x)q_\alpha(x) = \sum_{nkm}[b^+_\alpha b_\alpha - d^+_\alpha d_\alpha]. \tag{78}$$

Then we have the following expression for the energy of the quark system:

$$E_q = \sum_{\alpha,n,k} \frac{\omega_{n,k}}{R}[N_\alpha(n,k) + \bar{N}_\alpha(n,k)]. \tag{79}$$

The total energy of the system is equal to:

$$E_{tot} = E_q + \frac{4}{3}\pi R^3 B + \frac{Z_0}{R}, \tag{80}$$

where the second term on the right-hand side is a volume energy and the last term is the parametrization of the gluon field contribution. The mass of the hadrons can be found by minimising the total system energy E_{tot} The best description of the spectra of the low lying hadrons in such a model is achieved with

$$B^{1/4} = 0.146\ GeV, \quad Z_0 = 1.84 \qquad \alpha_s = 2.2 \quad m_s = 0.280GeV. \tag{81}$$

The masses of baryons and hadrons in the model are given in Table 3.

The bag model is a rather nice attempt to calculate the hadron spectrum on the basis of the covariant approach. Unfortunately even in the simplest cases it is necessary to imply some additional assumptions which as a rule totally destroy the relativistic invariance of the model. The classical model considered in this lecture has also another problem: namely nonconservation of the axial current. To conserve this current, the classical bag model has to be modified into the chiral bag model by the introduction of the pion field interaction with quarks on the bag surface. The chiral bag model also overcomes some other difficulties which appeared in the classical model.

Table 3

Masses of the hadron states obtained in the bag model.

Mesons	M_{BAG} (GeV)	M_{Exp} (GeV)	Barions	M_{BAG} (GeV)	M_{Exp} (GeV)
π	0.280	0.139	p	0.938	0.938
K	0.497	0.495	Λ	1.105	1.116
ρ	0.783	0.768	Σ^+	1.144	1.180
ω	0.783	0.783	Ξ^0	1.283	1.321
K^*	0.928	0.892	Δ	1.233	1.236
Φ	1.068	1.019	Σ^*	1.382	1.385
			Ξ^*	1.529	1.533
			Ω^-	1.672	1.672

5 Quark-hadron duality model

In the previous sections we considered models where the confinement potential is defined by many gluon exchanges or by some phase transition. In this section, we consider a model where confinement is connected with hadron production diagrams.

The description of the decay widths of resonances is impossible without assuming large coupling between quark-antiquark and two hadron state. But in this case the diagrams shown in fig.3b should give an appreciable contribution to the mass of the hadron states, where not only a real production of hadrons but also virtual transitions must be taken into account. It is very likely that such an interaction creates the analogues of the potential wall but with a much more complicated structure.

Another question closely connected with this subject is the question about constituent quark mass. The constituent quark mass which is calculated in the frame of the potential model is one of the most unstable parameters. The naive baryon model gives the mass of the nonstrange quark about 330 MeV. With this quark mass, the mass spectrum as well as magnetic moments of baryons can be described well. The Godfrey-Isgur model gives a quark mass 220 MeV. It was shown that taking into account the relativistic effects the baryon and meson spectra can be reproduced simultaneously. However this model has a problem to describe baryon magnetic moments and meson decay widths.

The constituent quark mass must depend rather strongly on the interaction. When quarks are situated at small distances it is likely that gluons give no contribution to the quark mass but only to the interaction forces. At large distances the gluons start to contribute to the quark mass and possibly make it infinitely large. Then the hadron interactions in the colorless channels shift the effective mass down; in this case it becomes an effective description of the collective hadron contribution. This point can be better understood using the language of singularities of the scattering amplitude.

To calculate the bound states of two particles in the framework of the relativistic approach we must solve the Bethe-Salpeter equation. If p_i, k_i and p_i' are momenta of the particle i in the initial, intermediate and final states, the amplitude of the interaction

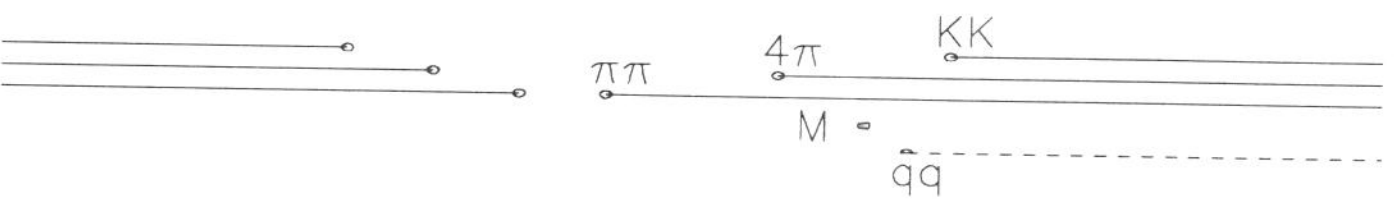

Fig.4

$A(p_1, p_2; p'_1, p'_2)$ is equal to:

$$A(p_1, p_2; p'_1, p'_2) = \int \frac{d^4k_1}{i(2\pi)^4} \frac{A(p_1, p_2; k_1, k_2) N(k_1, k_2; p'_1, p'_2)}{(m_1^2 - k_1^2)(m_2^2 - k_2^2)} + N(p_1, p_2; p'_1, p'_2). \qquad (82)$$

Here $N(p_1, p_2; p'_1, p'_2)$ is an irreducible block of the interaction which has no intermediate two particle states and m_i are quark masses. This equation can be solved by using methods like light cone variables or the dispersion relation technique. But all these methods face the same problem in the case of the quark interaction: the problem of the correct treatment of the quark-antiquark threshold singularity.

Let us consider as an example the singularities of the partial wave amplitude of the $\pi\pi$ interaction which are shown in fig.4. The amplitude has i) threshold singularities (right-hand cuts) connected with elastic scattering of the initial particles ($\pi\pi$ cut) and production of new particles (4π,$K\bar{K}$ and so on), ii) poles corresponding to bound states and iii) left-hand cuts corresponding to the interactions in the t and u channels. The presence of the threshold singularity means that particles can propagate to infinity and thus can be detected in an experiment. This singularity reflects the appearance of the imaginary part of the amplitude at energies above threshold; this can be calculated from the Bethe-Salpeter equation by changing particle propagators into δ functions.

The amplitude of eq.(82) definitely has the quark-antiquark threshold but there are no free quarks in nature. One way to solve this problem is to find a very special form of the interaction $N(p_1, p_2; p'_1, p'_2)$ which will eliminate the amplitude on the real axis, giving nonzero contribution only at the poles. But such an approach will not take into account another very important side of confinement: the amplitude of the quark-antiquark interaction has a huge number of threshold singularities connected with the hadron production processes. For example, when one describes the quark-antiquark interaction in the $^3P_0$ channel, all singularities shown in fig.4 must be reproduced.

A possible way to take this important property into account is to use the Lehman-Källen representation for the quark propagator:

$$\mathcal{G}(k) = \int_{\mu^2}^{\infty} dm^2\, \rho(m) \frac{1}{m - \hat{k}}, \qquad (83)$$

where $\rho(m)$ is a quark mass distribution function. Let us note that this representation should not be taken literally as the propagator of a single quark but reflects properties of the quark in the quark-antiquark interaction.

Experience with the quark model shows that for the calculation of some particle properties the function $\rho(m)$ may be taken independent of the type of reaction and in a form strongly localized around a certain mass. Let us consider the following parametrization of this function:

$$\rho(m) = \frac{C\sqrt{m^2-\mu^2}}{(m^2-m_0^2)^2+\Delta^4}, \tag{84}$$

where μ corresponds to the mass of the lowest hadron state and is taken equal to the pion mass for nonstrange quarks and to the kaon mass for strange quarks. For a narrow distribution $\Delta \ll m_0$, the value of m_0 effectively plays the role of the quark constituent mass. The normalization constant C is determined by the condition that the expression for the quark loop diagram reproduces the result of perturbative QCD at large energies. Indeed, when the virtuality is large, one can substitute in the quark propagator $(m - \hat{k}) \to (-\hat{k})$ and we obtain the normal QCD expression for the polarization operator:

$$\Pi_{q\bar{q}}^{\nu\nu'}(p) = -N\int \frac{d^4k}{i(2\pi)^4} Tr\{\gamma_\nu \frac{1}{-\hat{k}_1}\gamma_{\nu'}\frac{1}{-\hat{k}_2}\} \tag{85}$$

provided that the normalization condition

$$\int\limits_{\mu^2}^{\infty} dm^2 \rho(m) = 1 \tag{86}$$

is fulfilled.

Let us consider e^+e^- annihilation into hadrons. The quark loop diagram for the process $\gamma^* \to q\bar{q} \to \gamma^*$ is given by

$$\Pi_{q\bar{q}}^{\nu\nu'} = -N\int\limits_{\mu^2}^{\infty} dm_1^2\,\rho(m_1)\int\limits_{\mu^2}^{\infty} dm_2^2\,\rho(m_2)\int\frac{d^4k}{i(2\pi)^4}$$

$$Tr\{\gamma_\nu\frac{1}{m_1-\hat{k}_1}\gamma_{\nu'}\frac{1}{m_2-\hat{k}_2}\}, \tag{87}$$

where ν, ν' are photon indices and N is the number of colors (we consider the loop diagram for a fixed flavor). The imaginary part of this loop diagram then reads

$$Im\,\Pi_{q\bar{q}}^{\nu\nu'}(p) = N \int\limits_{4\mu^2}^{(m_1+m_2)^2\le s} dm_1^2\,dm_2^2\,\rho(m_1)\,\rho(m_2)\Phi(s,m_1,m_2)$$

$$\int\frac{d\Omega_{\vec{k}}}{4\pi}Tr\{\gamma_\nu(m_1+\hat{k})\gamma_{\nu'}(m_2+\hat{k}-\hat{p})\}, \tag{88}$$

where p is the photon momentum, $s = p^2$, and $\Phi(s, m_1, m_2)$ is the two particle phase space volume:

$$\Phi(s,m_1,m_2) = \frac{1}{16\pi s}\sqrt{[s-(m_1+m_2)^2][s-(m_1-m_2)^2]}. \tag{89}$$

The integration $d\Omega_{\vec{k}}$ is performed over all directions of the vector $\vec{k}$. The magnitudes of the four-vectors k and $p-k$ are fixed: $k^2 = m_1^2$ and $(p-k)^2 = m_2^2$. The integration over masses m_1 and m_2 is limited by $2\mu \le m_1 + m_2 \le \sqrt{s}$, so

$$Im\,\Pi_{q\bar{q}}^{\nu\nu'} = 0 \qquad at \qquad \sqrt{s} \le 2\mu. \tag{90}$$

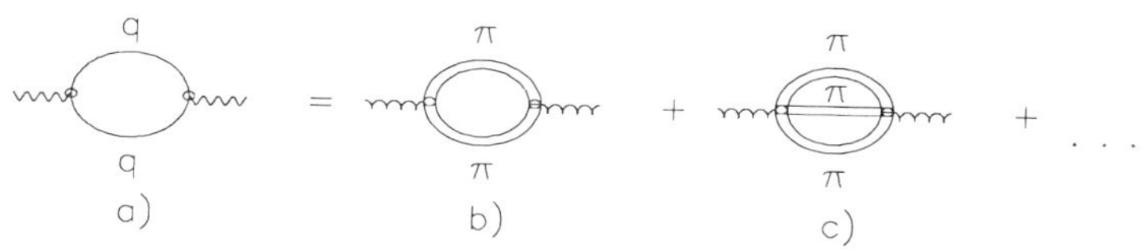

Fig.5

The total cross section of the annihilation $e^+e^- \to hadrons$ is proportional to this imaginary part:

$$\Pi_{q\bar{q}}^{\nu\nu'}(s) = (-p^2\delta_{\mu\nu} + p_\mu p_\nu)\Pi(s),$$

$$\sigma(e^+e^- \to hadrons) \sim \sum_{flavors} Im\ \Pi(s). \tag{91}$$

Observable processes are $e^+e^- \to 2\pi, e^+e^- \to 3\pi$ and so on: they are defined by the imaginary parts of diagrams Fig. 5b,5c . . . This leads to the identification:

$$\sum_{flavor} Im\ \Pi_{q\bar{q}}^{\nu\nu'}(s) = Im\ \Pi_{2\pi}^{\nu\nu'}(s) + Im\ \Pi_{3\pi}^{\nu\nu'}(s) + \ldots \tag{92}$$

where the sum on the r.h.s. includes all available hadronic self-energy parts. Then eq. (92) can be considered as the duality condition for the quark loop diagram if we neglect gluon emission. The quark-antiquark threshold shown in fig.4 plays the role of the effective language to describe the collective contribution of the hadron singularities. This idea is one of the central points in the model being discussed.

Let us first consider the short-range part of the quark interaction and the calculation of the spectrum of low-lying mesons. A standard version of the potential model was discussed in section 2 and the quark potential can be divided into two regions: a short-range part for distances about $0.3 - 0.5\ fm$ and a long-range part at distances more than $1\ fm$. The short-range part is dominated by an effective quark interaction such as gluon exchanges and/or instanton-induced forces. A long-range part is usually parametrized as a linearly rising potential $V(r) \sim br + c$ and this part of the potential is generally related to the confinement forces.

An interesting feature of the potential models is that all implementations of the residual effective quark forces are strongly attractive for the low-lying mesons (with $J^P = 0^-, 1^-$) and thus dominate the structure of the low-lying states. For these states the confinement forces can almost be ignored. The two kinds of short-range forces reproduce well the spectrum of the low-lying mesons. The exchange of the effective gluon which can be taken in the form:

$$g_V^{(qq)}(\bar{q}\vec{\lambda}\gamma_\mu q)^2 + 2g_V^{(qs)}(\bar{q}\vec{\lambda}\gamma_\mu I_f q)(\bar{s}\vec{\lambda}\gamma_\mu s) + g_V^{(ss)}(\bar{s}\vec{\lambda}\gamma_\mu s)^2. \tag{93}$$

Here q is a column of the nonstrange quark bispinors, λ are color matrices and I_f is the isospin unit matrix. SU(3) flavor symmetry breaking can be parametrized as:

$$g_V = \frac{m_q^2}{\pi^2}g_V^{(qq)} = \frac{(m_q + m_s)^2}{4\pi^2}g_V^{(qs)} = \frac{m_s^2}{\pi^2}g_V^{(ss)}. \tag{94}$$

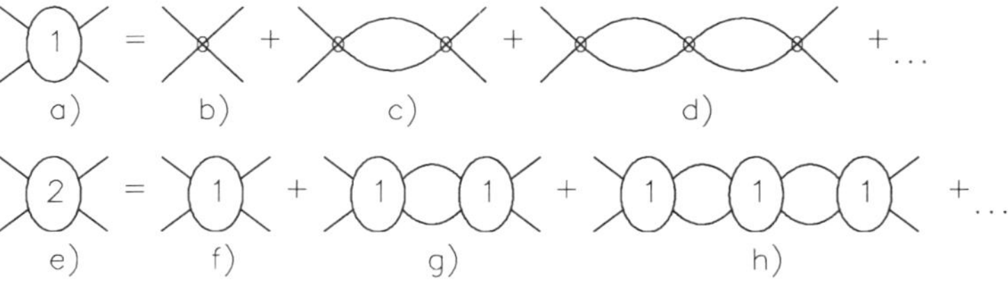

Fig.6

The interaction (93) describes rather well masses of the $J^P = 1^-$ states but fails to reproduce the pion mass and splitting of the $\eta - \eta'$ mesons. This problem can be solved by taking into account t'Hooft [5] forces which have the following form:

$$\frac{1}{2}\tilde{g}_I(\bar{q}(1+\gamma_5)I_f q)^2 - \frac{1}{2}\tilde{g}_I(\bar{q}(1+\gamma_5)\vec{\tau}_f q)^2 -$$
$$2\tilde{g}_S(\bar{s}(1+\gamma_5)q)(\bar{q}(1+\gamma_5)s) + \tilde{g}_S(\bar{s}(1+\gamma_5)s)(\bar{q}(1+\gamma_5)q), \tag{95}$$

where:

$$g_I = \frac{m_q^2}{\pi^2}\tilde{g}_I, \qquad g_S = \frac{m_q m_s}{\pi^2}\tilde{g}_S,$$

and $\vec{\tau}$ are isospin Pauli matrices.

With the interactions (93) and (95), the amplitudes have been calculated in all crossing channels corresponding to the summation of the diagrams shown in fig. 6 b,c,d. Then a bootstrap iteration procedure has been performed which means that the calculated amplitudes have been taken as an input for the next iteration (see fig. 6f,g,h). After that, the parameters of the model have been redefined with the help of the same particle masses. The iteration procedure converges rapidly (two iterations are in general quite enough; up to five iterations have been done to ensure convergence). The four parameters of this model g_V, g_I, g_S and cut-off parameter of the quark loop diagram Λ have been fixed by the empirical masses of the π, K, η' and ρ mesons.

In this approach the spectra of the low-lying mesons ($J^P = 0^-, 1^-$) can be calculated rather well, giving correct masses and mixing angle for the $\eta - \eta'$ mesons with quark masses $m_q = 385$ MeV and m_s=510 MeV. Another interesting result concerns the appearance of the constituent gluon in the model. Remember that the bootstrap procedure started from the interaction (93)-(95), where the gluon-type interaction was point-like; it corresponds to an effective gluon G_{eff} with infinite mass. However, after performing iterations taking into account the $G_{eff} \to q\bar{q} \to G_{eff}$ transitions, we obtain a finite mass for the effective gluon $M_G \sim 700 MeV$.

As we discussed above, the interaction of constituent quarks effectively takes into account the contribution of all hadron states allowed in this channel. Let us consider what happens with $q\bar{q}$ interaction if some of these meson-meson states are taken explicitly into account in addition to $q\bar{q}$ scattering channel. For the calculation of the meson spectrum and meson-meson scattering amplitude in this case, let us start from the Bethe-Salpeter equation which is shown graphically in fig.7. Here we use a single line

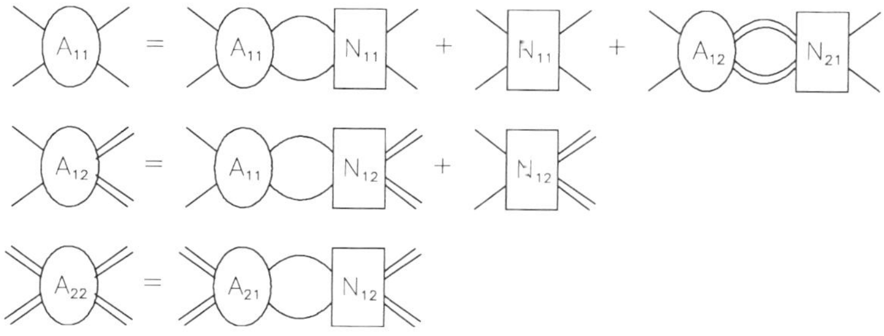

Fig.7

for quarks and a double line for a meson state. In the equation shown in fig.7, only the short-range interaction between quarks N_{11} and the $q\bar{q} \to MM$ transition ($N_{12} = N_{21}$) are taken into account. Here and below the index "1" is used for the quark-antiquark state and index "2" for a meson-meson state.

Let us consider how the transition $q\bar{q} \to MM$ can give a contribution to the long-range part of the quark-quark interaction. In order to see this we substitute the equation for the transition amplitude A_{12} into the equation for the quark amplitude A_{11}. The equation in diagrammatic form for A_{11} in this case is given in fig. 8.

One can see that the effective $q\bar{q} \to q\bar{q}$ amplitude in the model also has two parts: the direct quark-antiquark interaction part and the interaction via a box diagram with two mesons in the intermediate state. The first part was discussed above. The second part is interpreted to correspond to the confinement forces but, of course, has a more complicated structure than the potential wall in the non-relativistic quark model. The interaction which is due to the box-diagram has an imaginary part which is determined by the transition into two meson states. Let us note that, although the $q\bar{q} \to MM$ transition can be parametrized in a rather simple form, the effective $q\bar{q} \to q\bar{q}$ interaction (which is equal to the sum of the box diagrams with various mesons in the intermediate state) has a rather complicated structure.

When meson-meson scattering channels are taken into account explicitly, not only must the parameters of the quark-antiquark interaction be readjusted (as compared to the bootstrap calculations) but also the characteristics of the quark propagator. When we include the scattering channels of the lowest mesons (for example, the S-wave $q\bar{q}$ bound states 0^- and 1^-), we can expect that the average mass m_0 in the $\rho(m)$ distribution is shifted to higher values, and the distribution over quark mass becomes broader. Indeed, when all lowest hadron thresholds are taken into account explicitly, the constituent quark-antiquark threshold has to reflect the contribution of other higher hadron singularities and therefore the constituent mass becomes larger.

With this renormalized quark propagator we must then reconstruct the spectrum of these lowest mesons (self-consistency condition), and simultaneously obtain higher lying bound states (such as the P-wave $q\bar{q}$ states). This procedure can be repeated by

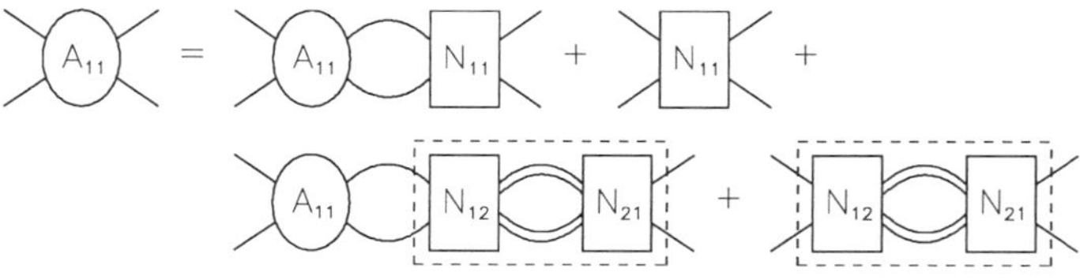

Fig.8

accounting subsequently for the scattering channels with P-wave mesons (implying a further redefinition of the quark propagator) and the calculation of still higher excited meson bound states (for example, radial excitation of the S-wave mesons). Thus, step by step, taking more and more meson degrees of freedom into account we will be able to reproduce an effective confinement interaction.

The kernel of the transition amplitude $q\bar{q} \to MM$ (N_{12}) is described by the one quark exchange diagram where the vertex for the transition of a meson into a quark-antiquark pair has the following form:

$$(\bar{q}\frac{\vec{\tau}}{2}[\gamma_5\vec{\pi}g_\pi + \gamma_\mu\vec{\rho}_\mu g_\rho]q) + (\bar{q}\frac{I}{2}[\gamma_5(\eta g_\eta \cos\alpha + \eta' g_{\eta'} \sin\alpha) + \gamma_\mu\omega_\mu g_\omega]q) + (\bar{s}[\gamma_5(-\eta g_\eta \sin\alpha + \eta' g_{\eta'} \cos\alpha + \gamma_\mu\Phi_\mu g_\Phi]s) + (\bar{q}_i[\gamma_5 K_i g_K + \gamma_\mu K^*_{i\mu} g_{K^*}]s) + (\bar{s}[\gamma_5\bar{K}_i g_K + \gamma_\mu\bar{K}^*_{i\mu} g_{K^*}]q_i); \qquad (96)$$

here α is the mixing angle for pseudoscalar mesons, while the vector mesons ω and Φ are taken as pure nonstrange and strange quark states, respectively. For simplicity we suppose SU(3)-symmetrical couplings:

$$g_\pi = g_K = g_\eta = g_{\eta'} \equiv g_1 \qquad g_\rho = g_\omega = g_\Phi = g_{K^*} \equiv g_2. \qquad (97)$$

The properties of the exchanged quark can be quite different from the properties of the constituent quark. Therefore the simplest way is to parametrize the quark exchange as a sum of partial vertices:

$$\mathcal{G}(t) = \sum_j (2l+1)P_l(cos\Theta)C_j G_{q\bar{q}}(s)G_{\mu\mu}(s), \qquad (98)$$

where C_j are parameters which depend only on the quantum number of the channel $j = J^P$ considered, l is the orbital momentum and Θ is the angle between quark and meson. The functions $G(s)$ are taken in a form which avoids divergences in the meson loop diagrams:

$$G_{q\bar{q}} = \lambda_q\sqrt{\frac{(m_1+m_2)^2 - s_q}{s - s_q}},$$

$$G_{\mu\mu} = \sqrt{\frac{(\mu_1+\mu_2)^2 - s'_q}{s - s'_q}}\left(1 + \frac{s - (\mu_1+\mu_2)^2}{s}\alpha\right) \qquad (99)$$

where

$$s_q = (m_1 + m_2)^2 - a_q$$
$$s'_q = (\mu_1 + \mu_2)^2 - a_q. \tag{100}$$

The coefficients λ_q distinguish the exchange of nonstrange and strange quarks. They are taken equal to unity for nonstrange quark exchange ($\lambda_u = \lambda_d = 1$) and λ_s is taken as a parameter for exchange of the strange quark. In a similar fashion we take $a_q = 1$ GeV^2 for the nonstrange quark exchange and $a_s = a_q\ m_s^2/m_q^2$. The parameter α is chosen to be the same for all channels.

Let us now consider the method of solution of the Bethe-Salpeter equation for the partial wave amplitude. In this case the kernel N_{12} corresponds to the production of the vertices (99) multiplied by the coefficients from (96) which characterize the produced mesons. The kernel N_{11} can be calculated by the projection of the interaction (93)-(95) into partial quark-antiquark states. After this, due to its simple form, this interaction also can be rewritten as a product of the two vertices. The expression for the one loop diagram depends on three indices which characterize the state of the system (quark-antiquark or meson-meson state) (i) before the one loop diagram, (ii) in a loop diagram and (iii) after it:

$$B_{lpk}(s, m_1, m_2) = \int\limits_{(m_1+m_2)^2}^{\infty} \frac{ds'}{\pi} G_{lp}(s', m_1, m_2) \frac{\Phi_p(s', m_1, m_2)}{s' - s} G_{pk}(s', m_1, m_2) \tag{101}$$

where $\Phi(s)$ is the two particle phase volume. In the case of the meson loop, the vertices G_{lp} are equal to $G_{\mu\mu}(s)$ from eq.(99). In the case of the quark loop, the situation is more complicated: the vertex for the $q\bar{q} \to M + M$ transition corresponds to G_{qq} while the vertex for the $q\bar{q} \to q\bar{q}$ transition must be taken from (93)-(95). To regularize the quark loop integral the cut-off function:

$$F(\Lambda, s', m_1, m_2) = \exp\left[-\Lambda\left(\frac{s'}{(m_1 + m_2)^2} - 1\right)\right] \tag{102}$$

was introduced under the integral in eq.(101).

In the case of the quark loop diagram the integration over quark masses must be performed:

$$B_{lpk}(s) = \int\limits_{\mu^2}^{\infty} dm_1^2\, \rho(m_1) \int\limits_{\mu^2}^{\infty} dm_2^2\, \rho(m_2) B_{lpk}(s, m_1, m_2) \Theta(m_1 + m_2 - M_{thr}). \tag{103}$$

Here the Θ-function represents a cut-off which should be made in the quark loop diagrams when the threshold for decay into quarks is lower then M_{thr}. This mass is connected with the threshold of the lowest hadron states produced in the channel which are not taken into account explicitly. When two-meson states have been explicitly taken into account we put M_{thr} equal to the lowest three-meson threshold in the channel considered. It means that the quark-antiquark partial width which appears in the calculations of the meson spectrum characterises the decay of this bound state into three (or more) mesons.

The amplitude can be written as:

$$A_{ik}(s) = \sum_{mp} G_{ip}(s) a_{ipk}(s) \tag{104}$$

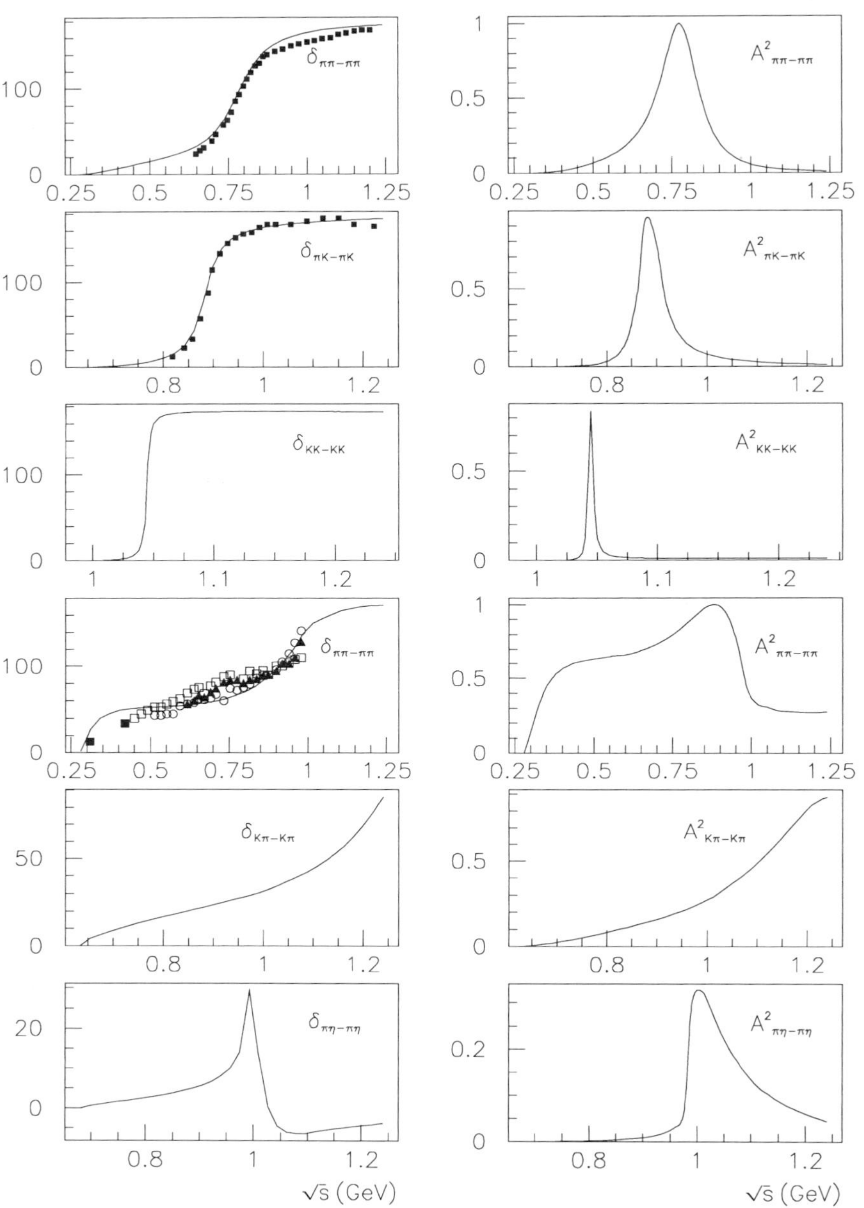

Fig.9 The phase shifts and squared amplitudes in the $^3S_1$ channel (first six diagrams) and in the $^3P_0$ channel (last six diagrams).

Table 4

Masses and widths of the meson states obtained in the framework of the quark-hadron duality model. The experimental values are given in parentheses. All units are in MeV.

$N^{2S+1}L_J$	$u\bar{d}$, $\frac{u\bar{u}-d\bar{d}}{\sqrt{2}}$, $d\bar{u}$	$u\bar{s}$, $d\bar{s}$	$\frac{u\bar{u}+d\bar{d}}{\sqrt{2}}$, $s\bar{s}$	$\frac{u\bar{u}+d\bar{d}}{\sqrt{2}}$, $s\bar{s}$
$1S_0$ $\Theta_p(grad)$	140 (140)	495 (495)	564 (548) -10 (-15 ± 5)	960 (958)
$3S_1$ Γ_{tot}	775 (768) 150 (149)	885 (892) 48 (50)	766 (782)	1046 (1020) 4 (4)
$3P_0$ Γ_{tot}	998 (985) 76 (57) Γ_{KK}=70% $\Gamma_{\pi\eta}$=30%	1242 (1430) 316 (287) $\Gamma_{K\pi}$=84% (93%) $\Gamma_{K\eta}$=16%	877 (880)	1718 (1710) 190 (150) $\Gamma_{s\bar{s}}$=50% Γ_{KK}=21%
$1P_1$ Γ_{tot}	1245 (1233) 44 (150) $\Gamma_{\omega\pi}$=100%	1235 (1270) 100 (90) $\Gamma_{K^*\pi}$=90%	1190 (1170) 150 (360) $\Gamma_{\rho\pi}$=100%	1610 (-) 160 (-)
$3P_1$ Γ_{tot}	1274 (1260) 100 (350) $\Gamma_{\rho\pi}$=60%	1310 (1400) 120 (175) $\Gamma_{K^*\pi}$=85%	1286 (1282) 38 (24) $\Gamma_{q\bar{q}}$=100%	- (1420)
$3P_2$ Γ_{tot}	1282 (1318) 44 (110) $\Gamma_{q\bar{q}}$=70% Γ_{KK}=21%	1445 (1425) 60 (98) $\Gamma_{K\pi}$=56% $\Gamma_{q\bar{s}}$=37%	1263 (1275) 194 (185) $\Gamma_{\pi\pi}$=70% Γ_{KK}=10%	- (1525)

Table 5

Model parameters. The masses of quarks and the widths of the quark mass distribution are given in MeV.

$m_q = 625$ $m_s = 800$ $\Delta = 100$	g_V=13.24 g_I=-2.31 g_S=-1.24	Λ=4.68 $C(0^-) = C(1^-) = 8.0$ $C(0^+) = 14.5$	$C(1^+) = 30.0$ $C(2^+) = 45.00$ α=0.75	λ_s=1.3 g_1/g_2=2

where the block $a_{ipk}(s)$ obeys the following equation:

$$a_{ipk} = \sum_{l} a_{ilp}B_{lpk} + G_{ik}\delta_{ip}. \tag{105}$$

The last equation can easily be rewritten in matrix form and solved accordingly.

The results obtained in the model are shown in Table 4 and examples of the meson-meson scattering amplitudes are shown in fig.9. The mass position is defined at the point where the phase shift of the scattering amplitude crosses 90°. The parameters of the model are given in Table 5. There is a rather large number of parameters in the model but as a result we obtain a huge amount of information: not only masses of the bound states but also partial widths and the meson-meson scattering amplitudes. Let us note that the present calculation reflects only the first step of the model and can be trusted only in the energy region below 1.30 GeV.

6 Conclusion

In the present lectures we considered three models of confinement which reconstruct to varying degrees the different features of this phenomenon. In the first two, we tried to go along with the classical examples which reflect most clearly the distinctive features of the models. Today the development of all models move them much closer to one another. For example, there are many efforts to rewrite the potential model in the language of the relativistic Bethe-Salpeter equation and include the hadron production processes. The development of the bag model into the chiral bag model assumes the close connection of the bag properties with the sigma-pion interaction at the surface of the bag. Let me note that the main problem for all models is a description of the scalar sector and it will be in the near future the crucial point for understanding the reliability of the model and the confinement mechanism.

References

[1] S.Sakata, Prog.Theor.Phys, **16**, 686 (1956).

[2] M.Gell-Mann, Phys.Lett, **8**, 214 (1964).

[3] G.Zweig, CERN, Rept. 8182/TH401, (1964).

[4] S.Godfrey and N.Isgur, Phys.Rev. D **32**, 189 (1985).

[5] G.t'Hooft, Nucl.Phys., B**72**, 461 (1974).

[6] A.Chodos, R.L.Jaffe, K. Johnson and C.B.Thorn, Phys.ReV., D**10**, 2599 (1974). T.De.Grand et al. D**12**, 2060 (1975).

CONCLUDING REMARKS

D.V. Bugg

Queen Mary and Westfield College, Mile End Rd., London E1 4NS, UK

Here I wish to take up a few points which I regard as important for further progress or points which concern very recent data.

1 The Effect of Inelastic Thresholds on Resonances

In these proceedings, Toby Burnett has outlined thoroughly the work of Törnqvist [1] concerning the $K\bar{K}$ and $\eta\eta$ thresholds in the C^+ sector. He has shown how these inelastic thresholds can influence both the shape of a resonance and move its mass around by substantial amounts. This is a general issue. There are many thresholds in the mass range 1 to 2 GeV, and one needs to be alert to their effects while analysing data.

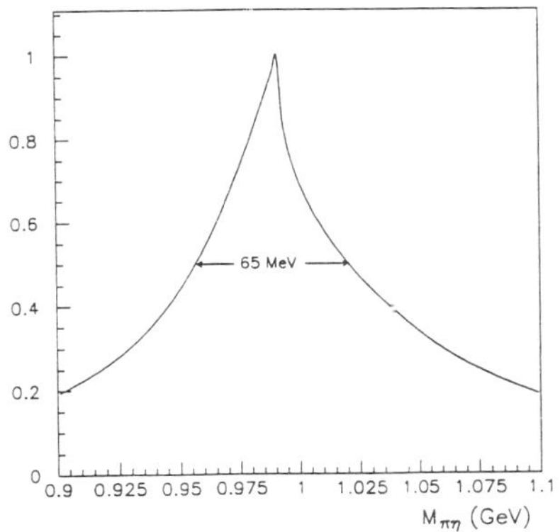

Figure 1: The shape of $a_0(980)$ v. $M_{\pi\eta}$ using the Flatté formula, fitted to Crystal Barrel data on $\bar{p}p \to \eta\pi^0\pi^0$ at rest.

Let me illustrate this with two examples. Firstly, Achasov and Shestakov [2] have remarked in a recent paper that the rapidly increasing phase space for the 4π channel could have a large effect on the masses of resonances fitted to 4π data. One should really be using a Breit-Wigner amplitude of the Flatté form [3]:

$$f = \frac{1}{M^2 - s - iM(\Gamma_{2\pi} + \Gamma_{KK} + \Gamma_{4\pi})}. \tag{1}$$

Each of the widths Γ is proportional to the available phase space for the channel in question. However, it is also multiplied by a form factor roughly $\exp(-3p^2)$, where p is the momentum of final state particles in GeV in the rest frame of the resonance. In practice, when one is well above threshold, the increasing phase space roughly cancels the form factor, and it is a good approximation to take Γ as a constant. Very close to the sharp $K\bar{K}$ threshold, Γ_{KK} rises as $\sqrt{s - 4M_K^2}$. The rapidly opening $K\bar{K}$ channel creates a rapidly increasing denominator in equn. (1) and generates a sharp cusp at the $K\bar{K}$ threshold. This is illustrated in Fig. 1 for $a_0(980)$.

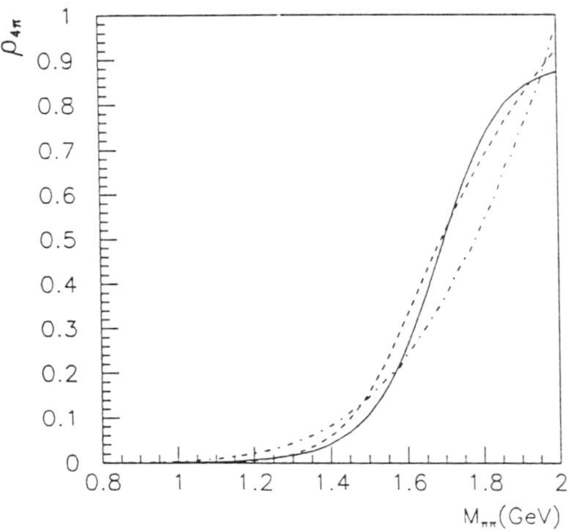

Figure 2: 4π phase space for $\rho\rho$ (dashed), $\sigma\sigma$ (dotted) and including the form factor for $\rho\rho$ (full curve).

In the mass range 1 to 2 GeV, the phase space for 4π channels is increaing rapidly. This is illustrated in Fig. 2, where the dashed curve shows $\rho\rho$ phase space and the dotted curve $\sigma\sigma$ phase space; here σ stands for the $\pi\pi$ S-wave. The full curve shows the effect of the form factor applied to $\rho\rho$ phase space. This summer, Andrei Sarantsev, Bing Song Zou and I spent some time checking what effect this has on $f_0(1335)$. The results may be of general interest. We fitted data on $\bar{p}p \to 3\pi^0$, $\eta\eta\pi^0$ and $\eta\pi^0\pi^0$ with a wide variety of assumptions about the precise form and strength of $\Gamma_{4\pi}$. With these variations, the fitted value of M in equn. (1) moved around by over 300 MeV. However, the *pole position* of the resonance was very stable between 1289 - i107 MeV and 1307 - i146 MeV. This pole position also agrees closely with the value obtained using a simple Breit-Wigner amplitude of constant width. So the message is that experimental groups should quote the pole position rather than M. The opening of the threshold affects the detailed shape of the resonance, but not the pole position to a good approximation.

Secondly I come to some very interesting new results concerning $E/\iota(1410)$. This has been studied by the Crystal Barrel group in two channels: $\bar{p}p \to \sigma\iota$, $\iota \to \eta\pi\pi$ [4] and $\bar{p}p \to \pi\iota$, $\iota \to \eta\pi\pi$ [5]. Fig. 3 shows the $\eta\pi\pi$ mass spectrum in the latter case. There is a sharp peak at 1410 $\pm$5 MeV with $\Gamma = 86 \pm 10$ MeV. Even more interesting results come from Obelix [6], who have studied $\bar{p}p \to \sigma\iota$ at rest, followed by $\iota \to K\bar{K}\pi$. They also observe a resonance at 1416 MeV, decaying to $a_0(980)\pi$ and K_S^*K, where K_S^* stands for the πK S-wave. But in addition they see a second peak at 1460 $\pm$ 10 MeV, with $\Gamma = 105 \pm 15$ MeV, decaying purely to $K^*(890)K$ with $L = 1$. A third element in the story concerns $J/\Psi \to \gamma(4\pi)$. A recent re-analysis [7] observes a strong signal for $0^- \to \rho\rho$ with $M = 1420$ MeV, $\Gamma = 160$ MeV. Presumably this is $\iota(1410)$, but with a much increased width compared to other data.

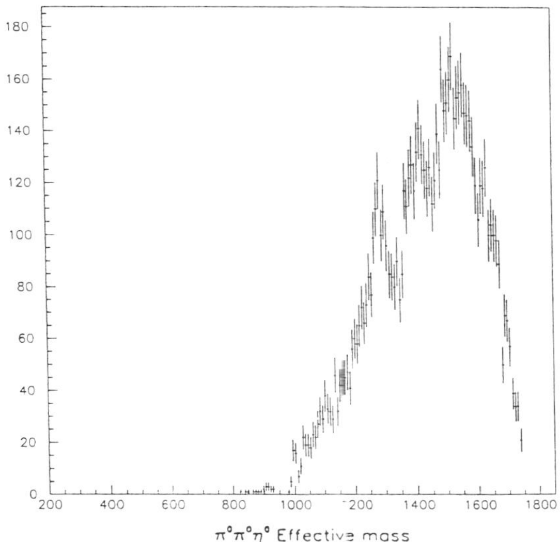

Figure 3: The $\eta\pi^0\pi^0$ mass spectrum from data at rest on $\bar{p}p \to \eta\pi^0\pi^0\pi^0$.

It is a vital question whether the second Obelix peak is distinct from the one at 1410 MeV. They claim a rapid phase variation of the $K^*(890)K$ signal relative to $a_0(980)\pi$. This is clear evidence for two separate resonances. Then the 1460 MeV resonance is a natural candidate for the $\bar{s}s$ 0^{-+} radial excitation, partner to $\eta(1295)$ and $\pi(1300)$. This makes the 1410 MeV resonance an extra state, and an obvious glueball candidate, as has been suspected for many years on the basis of its strong production in J/Ψ radiative decays.

My reason for spelling out this story in detail is that we cannot be too careful at this point, in view of the important consequences. Many fits to the ι have used Breit-Wigner amplitudes of constant Γ. I believe the time is now ripe for a coupled channel analysis of all data on 0^{-+} into $\eta\pi\pi$, $K\bar{K}\pi$ and $\rho\rho$ using the full form of equn (1). I hope that such an analysis would confirm that there are indeed two resonances at 1410 and 1460 MeV. The alternative possibility still lurks unstudied that the $K^*(890)K$ signal may be moved up from 1410 MeV by the p^3 factor for P-wave phase space in this channel.

Similar care is required in fitting 0^+ (and 2^+) resonances which have been suggested [8 – 11] in the 4π channel in the mass range 1350 to 1450 MeV mass range.

2 The 0^+ Sector

This is an area where there has been much progress recently. Having been involved, I may be biased , but I would like to make a few points.

The $f_0(975)$ and $a_0(980)$ have been well established for many years. I refer to Ted Barnes' contribution for discussion of the continuing arguments whether they are "molecules" or $\bar{q}q$ states. Recently Crystal Barrel has produced overwhelming evidence for $f_0(1505)$ in $\bar{p}p \to \eta\eta\pi^0$ [12], $3\pi^0$ [13] and $(\eta\eta')\pi^0$ [14] at rest. (There is an error in this paper; the Breit-Wigner amplitude which was fitted to the data contained only the width for the $\eta\eta'$ channel in the denominator; when the full width is used instead, the mass of the fitted resonance goes down from 1545 MeV to approximately 1505 MeV). They also claim an $I = 0$ resonance at 1335–1370 MeV and an $I = 1$ resonance at 1450 MeV [15]. Together with $K^*(1430)$, they make a natural $^3P_0$ $\bar{q}q$ nonet.

I would like to draw attention to several points which seem to have been overlooked or to have caused confusion recently. An important point is that there is good prior evidence for the f_0 resonance in the 1300-1400 MeV mass range. It is in a surprising place, namely $\pi^+\pi^- \to K^0_S K^0_S$, where three experimental groups agree on the existence of a definite peak with a width of roughly 200–250 MeV [16 –18]. These data are collected into Fig. 4. The Crystal Barrel data from ref. [12] are shown in Fig. 5. There, an arrow points to the mass of 1400 MeV, but by eye one sees that the enhancement is clearly at lower mass, and current fits give a mass in the range 1300–1330 MeV.

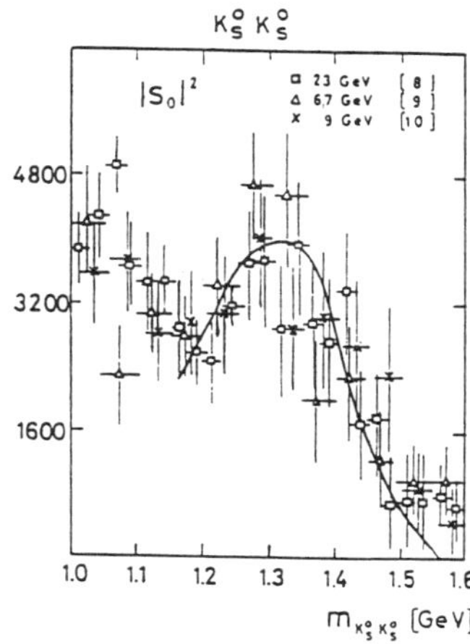

Figure 4: Intensities for $\pi^+\pi^- \to K^0_S K^0_S$ from references 16–18.

It is interesting that there is significant coupling of this resonance to $K\bar{K}$. This agrees with calculations of Barnes et al. in the flux tube model. It requires that the 0^+ nonet departs from ideal mixing, unless this resonance is the $s\bar{s}$ partner to $f_0(975)$ and $a_0(980)$. But in the latter case, where is the corresponding K^*? In these proceedings, Barnes remarks that $f_0(1335) \to \pi\pi$ may be artificially suppressed by a node in radial wave functions; in this case the $K\bar{K}$ width may not be abnormal.

A second point I would like to clarify is the remark in the Particle Data Tables [19] that $f_0(1300)$ is largely elastic. This is definitely wrong and is based on a misunderstanding. Fig. 6 shows the intensity for the $\pi\pi \to \pi\pi$ S-wave versus mass from a recent re-analysis [20] of Cern-Münich data. There are dips at $f_0(975)$ and $f_0(1505)$ and a broad peak at 1300 MeV. This broad peak has been mistaken for a resonance. What is really happening has been exposed in a paper by Zou and Bugg [21]. If one calculates the contributions to the S-wave from u-channel ρ and $f_2(1270)$ poles, one finds, Fig. 7, that the slowly rising part of the amplitude is described absolutely (with no free parameters) within 10–15% up to 1.2 GeV (full curve on the figure). The $f_0(975)$ and $f_0(1505)$ appear in Fig. 6 due to interference with this slowly varying "background". What lends further confidence to this "background" calculation is that it also reproduces the repulsive $\pi\pi$ $I = 2$ S-wave and D-wave. Barnes and Swanson [22] have obtained similar results from quark-model calculations.

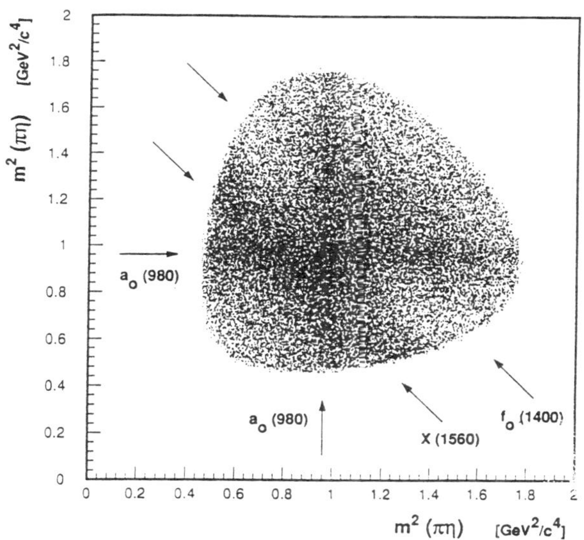

Figure 5: Data from the Crystal Barrel on $\bar{p}p \to \eta\eta\pi^0$ at rest.

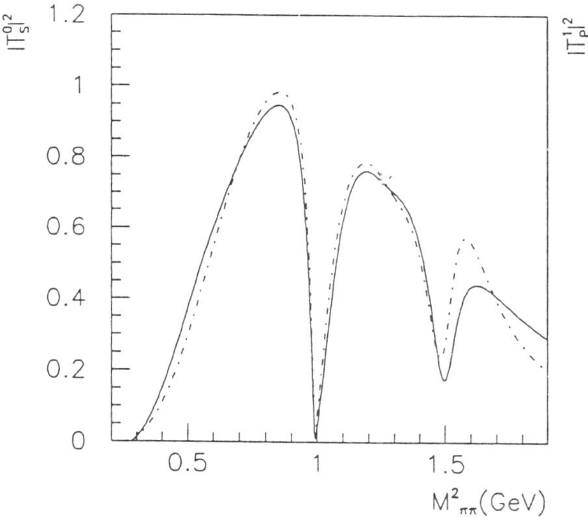

Figure 6: Two alternative fits to the $\pi^+\pi^- \to \pi^+\pi^-$ S-wave intensity v. M^2.

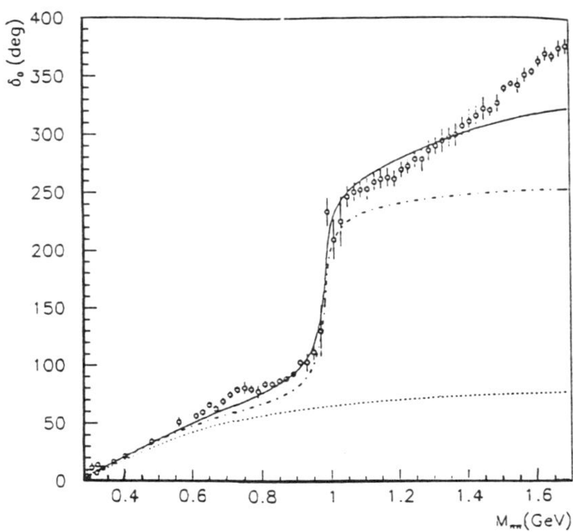

Figure 7: The $I = 0$ S-wave phase shift for $\pi\pi \to \pi\pi$. The dashed curve is the contribution from only ρ exchange. The dot-dashed curve includes also $f_0(975)$. The solid curve includes $f_2(1275)$ exchange.

The point of these remarks is that the broad peak at 1300 MeV is NOT an elastic resonance. The $f_0(1335)$ which I have described above is quite distinct. It is a highly **inelastic** resonance, coupling dominantly to 4π (and $K\bar{K}$), and is actually hard to discern in $\pi\pi$ elastic scattering. In Fig. 6, it gives a tiny dip at 1335 MeV; it is more clearly visible as a cusp on the upper left of Fig. 8.

There is another important piece of experimental data I wish to show. GAMS have measured $\pi^+\pi^- \to \pi^0\pi^0$ over an extended t range. Fig. 9 shows their data. The $f_0(975)$ appears as a dip at small t, but as a peak at large t. What is happening is that the broad "background" due to exchange processes falls rapidly with t, whereas the $f_0(975)$ varies much more slowly with t. The interpretation is that the "background" is due to long-range exchanges, while $f_0(975)$ involves at least some component coming from short-range forces, probably $\bar{q}q$.

I would like to relate this to a recent result in charmonium spectroscopy. The E760 group has located the $^1P_1$ $c\bar{c}$ state, and shown that its mass agrees within 1 MeV with the centre of gravity of the ^3P states. Barnes points out that this is strong evidence that the confining potential is scalar in nature. I suggest that this confining potential is related quantitatively to the "background" amplitude in the $\pi\pi$ S-wave. This relation is worth further close examination.

3 Glueball Candidates

Let us suppose that $f_0(1335)$, $a_0(1450)$ and $K^*(1430)$ make the $^3P_0$ nonet. (Not everyone agrees with this). It leaves $f_0(1505)$ as odd man out. It is close to the mass where Lattice QCD calculations predict a glueball. Toby Burnett has described a recent re-analysis of data on $J/\Psi \to \gamma(4\pi)$ [7] which identifies the presence of $f_0(1505)$ in these data; so it passes the test of appearing in J/Ψ radiative decays. It appears at about the rate expected for a 0^+ glueball in J/Ψ radiative decays. It is accompanied by further 0^+ peaks at 1750 and 2104 MeV. The E760 experiment [23] sees strong $\eta\eta$ peaks at just the same masses in $\bar{p}p \to (\eta\eta)\pi^0$, see Fig. 24 of Burnett's contribution to these

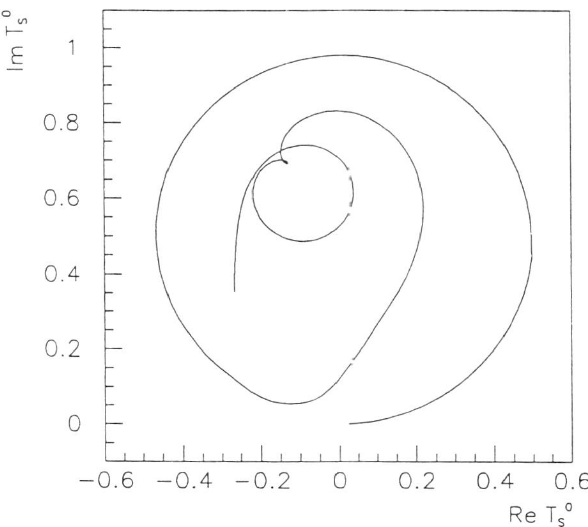

Figure 8: The Argand diagram for the $I = 0$ $\pi\pi \to \pi\pi$ S-wave amplitude from ref. 20.

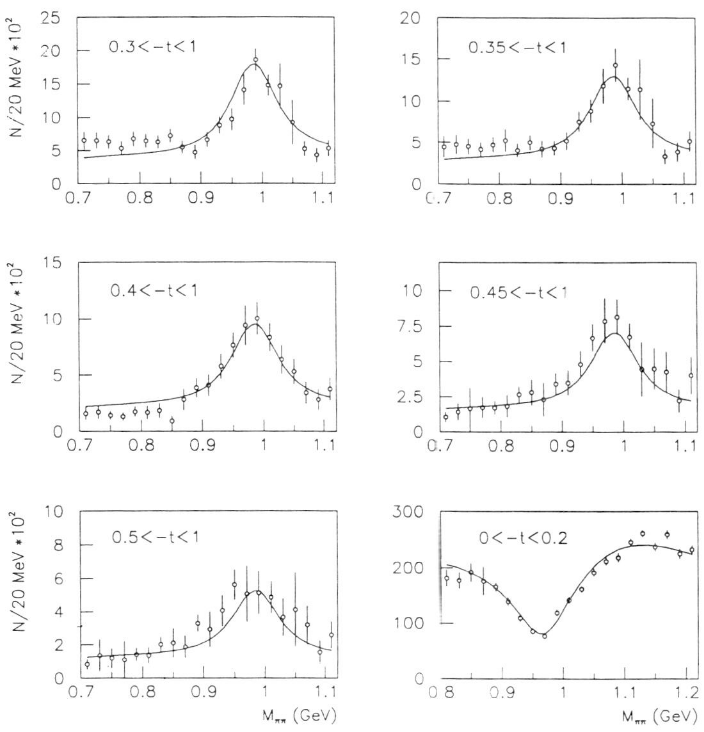

Figure 9: GAMS data on $\pi^+\pi^- \to \pi^0\pi^0$ for various t ranges.

proceedings. They are at roughly the right masses to be radial excitations of $f_0(1335)$. But I find it incredible that radial excitations should appear so strongly in E760 data. Normally one has to search desperately hard to find radial excitations, since nodes in their radial wave functions tend to lead to small matrix elements both for production and for decay to $\pi\pi$. Although there is no proof as yet, I suggest a working hypothesis is as follows. The cavity model of Jaffe and Johnson [24] suggests that the lowest two glueball states should be cavity states $[TE1]^2$ and $[TM1]^2$, with a mass ratio of 1.6. I am told by the experts that the cavity model is now frowned upon, but Lattice QCD calcuations also predict a second 0^+ glueball at about 2 GeV. I would like theorists to examine the wave functions of these two glueballs, and see if there is any relation to the cavity model. In any case, $f_0(2104)$ is a natural candidate for the second 0^+ glueball.

Amsler and Close [25] have shown that $f_0(1505)$ is unlikely to be a pure glueball, but probably mixes with nearby $q\bar{q}$ states. Such mixing with a radial excitation at 1750 MeV could explain the strong production of $f_0(1750)$. There is experimental evidence that $f_0(1750)$ is different from $f_0(1505)$ and $f_0(2104)$. In Crystal Barrel data (as yet unpublished) at 1940 MeV/c on $\bar{p}p \to \eta\eta\pi^0$, there are clear $\eta\eta$ signals at 1505 and 2110 MeV, but nothing significant at 1750 MeV.

An important experimental observation is that $f_0(1505)$, $f_0(1750)$ and $f_0(2104)$ all decay through $\sigma\sigma$, but not $\rho\rho$. This suggests that the σ acts as a filter for locating confined gluons. Frank Close has given arguments based on Lattice QCD for this result.

A *crucial* outstanding experimental issue is to locate the $s\bar{s}$ member of the 0^+ nonet. Is it $\theta(1710)$? Or is it $G(1590)$?

I have probably over-emphasised the 0^+ sector, because I know it well. But there are other clear signs of new spectroscopy in this mass range. The VES group has convincing evidence for a strong 0^- resonance at 1800 MeV, shown in Fig. 21 of Palano's contribution. It has unusual decays to $\pi f_0(1505)$ and $f_0(980)\pi$. Again I find it extraordinary that a second radial excitation of the pion should appear with such a large cross section. Many people regard it as a strong candidate for a hybrid, $\bar{q}qg$. Philip Page gives a quantitative assessment of the situation in these Proceedings. A crucial experimental issue is to search for an $I = 0$ partner.

Another point is that my research student, Andy Cooper has located two $I = 0$ $J^{PC} = 2^{-+}$ states at 1645 and 1850 MeV [26] in Crystal Barrel data at 1200 and 1940 MeV/c on $\bar{p}p \to \eta\pi^0\pi^0\pi^0$. The lower resonance decays to $a_2(1320)\pi$ and is a natural candidate for the $I = 0$ partner of $\pi_2(1670)$, i.e. $\bar{q}q\ ^1D_2$. The second state is observed in $f_2(1270)\eta$. (There is earlier evidence from CELLO [27] and Crystal Ball [28] in $\gamma\gamma$). This second state seems unlikely to be the $s\bar{s}$ member of the nonet, so if it is confirmed it requires explanation as a hybrid or glueball state. There is a peak in $J/\Psi \to \gamma(\eta\pi\pi)$ at just this mass, shown in Fig. 11(a) of Palano's contibution. The J^P analysis of this peak is a vital outstanding issue.

There are further claims for exotic states in this mass range. GAMS [29] see a strikingly narrow peak at 1910 MeV in $\eta\eta'$. Expt E818 at Brookhaven [30] suggest a broad exotic 1^{-+} resonance in $f_1(1285)\pi$.

The moral of this story is that the action lies in the mass range 1400–2200 MeV. It seems that a confined gluon has an effective mass of around 750 MeV. Lattice QCD predicts a 2^+ glueball at 2200 MeV. Crystal Barrel and Obelix are beginning to get to grips with this mass range. I am sure that 5 years of dedicated running is needed. It is a blunder of the first order of magnitude for CERN to close LEAR at the end of

1996, just when rapid progress is in sight. CERN's primary function is to build and run accelerators, and I believe that everything else should be scrutinised before the axe falls on LEAR.

4 Some Puzzles

1) Why is $\eta(1295)$ so narrow while its $I = 1$ partner $A_1(1260)$ is so broad?

2) Why do all $\bar{q}q$ states seem to have a common RMS radius of 0.8–0.9 fm? The glueball spectrum shown by Chris Michael in Fig. 2 of his contribution bears no relation to familiar spectra. This suggests that Lattice predictions do not conform to a constant RMS radius. Why? Is this spectrum trustworthy?

5 Baryons

There is a remarkable feature of baryon resonances which is well known. Masses cluster around 1690 and 1920 MeV. This is clearly visible in spectra shown by Capstick. This clustering cannot be a simple question of small $L.S$ and tensor splitting, though that will help. At 1920 MeV, all J^P are observed from $7/2^+$ down to $1/2^-$. This requires degeneracy of $[56, 2^+]$ and $[70, 1^-]$ multiplets. Why? What accounts for the D_{35} resonance from the $[70, 3^-]$ multiplet at the same mass?

It is possible that this clustering is an artefact. Manley [31] has recently completed a much needed analysis of πN inelastic scattering. Many resonances couple to ρN, ωN, $N^*\pi$ and $\Delta\pi$. It is conceivable that thresholds due to these processes produce cusps which have not been properly taken into account in analyses of πN elastic scattering. A fresh pass through this analysis is needed using Manley's results as input. It will be intriguing to see if the clustering persists. If it does, an explanation is needed.

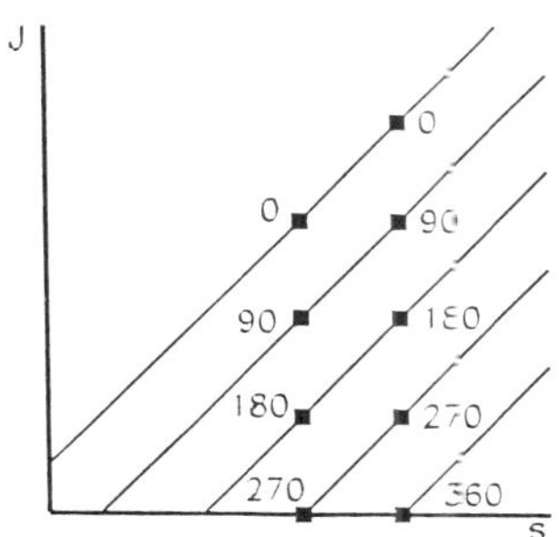

Figure 10: The phase pattern of resonances on Regge trajectories. Angles are in degrees, and data are fitted to $\bar{p}p \to \pi^+\pi^-$.

In this connection, let me draw attention to another oddity, which has not been mentioned elsewhere at this meeting. Hasan and I [32] have made an amplitude analysis of data from LEAR on $\bar{p}p \to \pi^-\pi^+$. We find a remarkable result. The amplitudes require towers of resonances, Fig. 10, where successive members of a tower differ in phase by exactly 90° within the errors. This bears some resemblance to the Veneziano model, but the 90° phase relation is new and disturbing. How do resonances know to be phase coherent, when J^P states are supposedly orthogonal? An understanding of this subtlety might well have some bearing on the strong spin-dependent effects observed in scattering in the few GeV region, effects much stronger than are explained by conventional wisdom.

6 Charm Physics

This Institute has concentrated attention on light meson spectroscopy, because this is the centre of recent experimental activity. However, theory has problems in dealing with relativistic effects in light quark spectroscopy. Calculations are much easier for heavy flavours. Spectroscopy of bottom states is tough experimentally and likely to remain so. Ted Barnes has made confident predictions for charmed hybrids in the 4.0–4.3 GeV mass range. Experimentally, this mass range should be straightforward, particularly if states are narrow. What we need is a tau-charm factory for a fresh attack on charmonium physics. Machines at SLAC, Orsay and Fràscati were closed prematurely, as is clear from the fact that new experimental results are still appearing from old data from these machines. Data on χ decays are almost non-existent. Europe could collectively afford a tau-charm factory, but it needs to be sited where there is existing expertise. How about Daresbury, where there is already an electron synchrotron of similar beam energy and intensity?

7 The Burning Issues

1) Is $\iota(1410)$ distinct fom the nearby 1460 MeV resonance? If it is a glueball, how does $\bar{q}q$ mixing move it down in mass from the 2000 MeV of Lattice QCD?

2) What is the J^P of $\theta(1710)$? What are its decay modes?

3) In order to set $I = 0$ states in context in the quark model, it is important to locate $I = 1$ $\bar{q}q$ radial 0^+ and 2^+ excitations.

4) Can the 2^+ glueball be located?

5) Are there 3g states? The cavity model would predict 0^{+-}, 1^{++}, 2^{+-} and 3^{++} in the mass range 2250–2600 MeV. The obvious place to look is in $p\bar{p}$ annihilations. A filter for the 1^{++} state would be decays through $\eta f_0(1505)$ with $L = 1$.

8 Really Concluding Remarks

It has been fun mixing with many eager new students (and old hands) at this Institute. We are all extremely grateful to NATO for financial support. I have been impressed with the quality of results which students are presenting, a quality which compares favourably with other areas of physics and high energy physics. I hope that future meetings will keep alive the present level of activity. It is particularly welcome to see theorists taking an active role in interpretation of experiments. Although experimental groups compete to produce first results, I hope that a spirit of cooperation will allow them to collaborate in joint analyses of related data from different experiments. In my experience, one learns much more from two combined sets of experimental data than from one alone.

REFERENCES

[1] N.A. Törnqvist, "Understanding the Scalar Meson $\bar{q}q$ Nonet", Helsinki preprint HU-SEFT R 1995-05, submitted to Zeit. f. Phys.

[2] N.N. Achasov and G.N. Shestakov, submitted to Phys. Rev. D.

[3] W. Flatté, Phys. Lett. 63B (1976) 224.

[4] C. Amsler et al., submitted to Phys. Lett. B

[5] C. Pinder, invited talk, Hadron'95.

[6] A. Bertin et al., Phys. Lett B (to be published)

[7] D.V. Bugg et al., Phs. Lett B353 (1995) 378.

[8] M. Gaspero, Nucl. Phys. A562 (1993) 409.

[9] A. Lanaro et al., Nucl. Phys. A558 (1993) 13c.
[10] C. Amsler et al., Phys. Lett. B322 (1994) 431.
[11] S. Abatzis et al., Phys. Lett. B324 (1994) 509.
[12] C. Amsler et al., Phys. Lett. B291 (1992) 347.
[13] V. V. Anisovich et al., Phys. Lett. B323 (1994) 233.
[14] C. Amsler et al., Phys. Lett B340 (1995) 259.
[15] C. Amsler et al., Phys. Lett. B333 (1994) 277.
[16] A.J. Pawlicki et al., Phys. Rev. D15 (1977) 3191.
[17] A. Etkin et al., Phys. Rev. D25 (1982) 2446.
[18] C. Daum et al., Zeit. Phys. C23 (1984) 339.
[19] Particle Data Group, Phys. Rev. D50 (1994) 1173.
[20] D.V. Bugg, B.S. Zou and A. Sarantsev, "New Results on $\pi\pi$ Phase-Shifts between 600 and 1900 MeV", submitted to Nucl. Phys. B.
[21] B.S. Zou and D.V. Bugg, Phys. Rev. D50 (1994) 591.
[22] T. Barnes and E.S. Swanson, Phys. Rev. D46 (1992) 131.
[23] T.A. Armstrong et al., Phys. Lett. B307 (1993) 394.
[24] R.L. Jaffe and K. Johnson, Phys. Lett. B60 (1976) 201.
[25] C. Amsler and F.E. Close, Phys. Lett. B353 (1995) 385.
[26] C. Amsler et al, submitted to Phys. Lett B; A.R. Cooper, Ph.D. Thesis, Univ. of London, 1994.
[27] M. Feindt, Proc. Int. Conf. on HEP, Singapore, 1990, eds. K.K. Phua and Y. Yamaguchi (S.E. Asia Theor. Phys. Assn. and Phys. Soc. of Japan, 1991), p591.
[28] K. Karch et al., Zeit. Phys. C54 (1992) 1328.
[29] D. Alde et al., Sov. J. Nucl. Phys. 54 (1991) 455.
[30] J.H. Lee et al., Phys. Lett. B323 (1994) 227.
[31] D.M. Manley and E.M. Saleski, Phys. Rev. D45 (1992) 4002.
[32] A.Hasan and D.V. Bugg, Phys. Lett B334 (1994) 215.

INDEX